ASCE STANDARD ASCE/EWRI 60-12

American Society of Civil Engineers

Guideline for Development of Effective Water Sharing Agreements

This document uses both the International System of Units (SI) and customary units.

AMERICAN SOCIETY OF CIVIL ENGINEERS

ENVIRONMENTAL & WATER RESOURCES INSTITUTE

Published by the American Society of Civil Engineers

Library of Congress Cataloging-in-Publication Data

Guidelines for development of effective water sharing agreements / American Society of Civil Engineers.
p. cm. — (ASCE standard)
"ASCE/EWRI 60-12."
"This document uses both the International System of Units (SI) and customary units."
Includes bibliographical references and index.
ISBN 978-0-7844-1230-5 (pbk. : alk. paper) — ISBN 978-0-7844-7691-8 (ebook)
1. Water—Law and legislation. 2. Water rights (International law) 3. Water resources development—Law and legislation. I. American Society of Civil Engineers.
K3496.G85 2012
346.04′691—dc23

2012013748

Published by American Society of Civil Engineers
1801 Alexander Bell Drive
Reston, Virginia 20191
www.asce.org/pubs

This standard was developed by a consensus standards development process which has been accredited by the American National Standards Institute (ANSI). Accreditation by ANSI, a voluntary accreditation body representing public and private sector standards development organizations in the United States and abroad, signifies that the standards development process used by ASCE has met the ANSI requirements for openness, balance, consensus, and due process.

While ASCE's process is designed to promote standards that reflect a fair and reasoned consensus among all interested participants, while preserving the public health, safety, and welfare that is paramount to its mission, it has not made an independent assessment of and does not warrant the accuracy, completeness, suitability, or utility of any information, apparatus, product, or process discussed herein. ASCE does not intend, nor should anyone interpret, ASCE's standards to replace the sound judgment of a competent professional, having knowledge and experience in the appropriate field(s) of practice, nor to substitute for the standard of care required of such professionals in interpreting and applying the contents of this standard.

ASCE has no authority to enforce compliance with its standards and does not undertake to certify products for compliance or to render any professional services to any person or entity.

ASCE disclaims any and all liability for any personal injury, property damage, financial loss, or other damages of any nature whatsoever, including without limitation any direct, indirect, special, exemplary, or consequential damages, resulting from any person's use of, or reliance on, this standard. Any individual who relies on this standard assumes full responsibility for such use.

ISBN 978-0-7844-1230-5 (paper)
ISBN 978-0-7844-7691-8 (e-book)
Manufactured in the United States of America.

18 17 16 15 14 13 1 2 3 4 5

STANDARDS

In 2006, the Board of Direction approved the revision to the ASCE Rules for Standards Committees to govern the writing and maintenance of standards developed by the Society. All such standards are developed by a consensus standards process managed by the Society's Codes and Standards Committee (CSC). The consensus process includes balloting by a balanced standards committee made up of Society members and nonmembers, balloting by the membership of the Society as a whole, and balloting by the public. All standards are updated or reaffirmed by the same process at intervals not exceeding five years.

The following standards have been issued:

ANSI/ASCE 1-82 N-725 Guideline for Design and Analysis of Nuclear Safety Related Earth Structures

ASCE/EWRI 2-06 Measurement of Oxygen Transfer in Clean Water

ANSI/ASCE 3-91 Standard for the Structural Design of Composite Slabs and ANSI/ASCE 9-91 Standard Practice for the Construction and Inspection of Composite Slabs

ASCE 4-98 Seismic Analysis of Safety-Related Nuclear Structures

Building Code Requirements for Masonry Structures (ACI 530-02/ASCE 5-02/TMS 402-02) and Specifications for Masonry Structures (ACI 530.1-02/ASCE 6-02/TMS 602-02)

ASCE/SEI 7-10 Minimum Design Loads for Buildings and Other Structures

SEI/ASCE 8-02 Standard Specification for the Design of Cold-Formed Stainless Steel Structural Members

ANSI/ASCE 9-91 listed with ASCE 3-91

ASCE 10-97 Design of Latticed Steel Transmission Structures

SEI/ASCE 11-99 Guideline for Structural Condition Assessment of Existing Buildings

ASCE/EWRI 12-05 Guideline for the Design of Urban Subsurface Drainage

ASCE/EWRI 13-05 Standard Guidelines for Installation of Urban Subsurface Drainage

ASCE/EWRI 14-05 Standard Guidelines for Operation and Maintenance of Urban Subsurface Drainage

ASCE 15-98 Standard Practice for Direct Design of Buried Precast Concrete Pipe Using Standard Installations (SIDD)

ASCE 16-95 Standard for Load Resistance Factor Design (LRFD) of Engineered Wood Construction

ASCE 17-96 Air-Supported Structures

ASCE 18-96 Standard Guidelines for In-Process Oxygen Transfer Testing

ASCE 19-10 Structural Applications of Steel Cables for Buildings

ASCE 20-96 Standard Guidelines for the Design and Installation of Pile Foundations

ANSI/ASCE/T&DI 21-05 Automated People Mover Standards—Part 1

ANSI/ASCE/T&DI 21.2-08 Automated People Mover Standards—Part 2

ANSI/ASCE/T&DI 21.3-08 Automated People Mover Standards—Part 3

ANSI/ASCE/T&DI 21.4-08 Automated People Mover Standards—Part 4

SEI/ASCE 23-97 Specification for Structural Steel Beams with Web Openings

ASCE/SEI 24-05 Flood Resistant Design and Construction

ASCE/SEI 25-06 Earthquake-Actuated Automatic Gas Shutoff Devices

ASCE 26-97 Standard Practice for Design of Buried Precast Concrete Box Sections

ASCE 27-00 Standard Practice for Direct Design of Precast Concrete Pipe for Jacking in Trenchless Construction

ASCE 28-00 Standard Practice for Direct Design of Precast Concrete Box Sections for Jacking in Trenchless Construction

ASCE/SEI/SFPE 29-05 Standard Calculation Methods for Structural Fire Protection

SEI/ASCE 30-00 Guideline for Condition Assessment of the Building Envelope

SEI/ASCE 31-03 Seismic Evaluation of Existing Buildings

SEI/ASCE 32-01 Design and Construction of Frost-Protected Shallow Foundations

EWRI/ASCE 33-09 Comprehensive Transboundary International Water Quality Management Agreement

EWRI/ASCE 34-01 Standard Guidelines for Artificial Recharge of Ground Water

EWRI/ASCE 35-01 Guidelines for Quality Assurance of Installed Fine-Pore Aeration Equipment

CI/ASCE 36-01 Standard Construction Guidelines for Microtunneling

SEI/ASCE 37-02 Design Loads on Structures during Construction

CI/ASCE 38-02 Standard Guideline for the Collection and Depiction of Existing Subsurface Utility Data

EWRI/ASCE 39-03 Standard Practice for the Design and Operation of Hail Suppression Projects

ASCE/EWRI 40-03 Regulated Riparian Model Water Code

ASCE/SEI 41-06 Seismic Rehabilitation of Existing Buildings

ASCE/EWRI 42-04 Standard Practice for the Design and Operation of Precipitation Enhancement Projects

ASCE/SEI 43-05 Seismic Design Criteria for Structures, Systems, and Components in Nuclear Facilities

ASCE/EWRI 44-05 Standard Practice for the Design and Operation of Supercooled Fog Dispersal Projects

ASCE/EWRI 45-05 Standard Guidelines for the Design of Urban Stormwater Systems

ASCE/EWRI 46-05 Standard Guidelines for the Installation of Urban Stormwater Systems

ASCE/EWRI 47-05 Standard Guidelines for the Operation and Maintenance of Urban Stormwater Systems

ASCE/SEI 48-11 Design of Steel Transmission Pole Structures

ASCE/SEI 49-07 Wind Tunnel Testing for Buildings and Other Structures

ASCE/EWRI 50-08 Standard Guideline for Fitting Saturated Hydraulic Conductivity Using Probability Density Functions

ASCE/EWRI 51-08 Standard Guideline for Calculating the Effective Saturated Hydraulic Conductivity

ASCE/SEI 52-10 Design of Fiberglass-Reinforced Plastic (FRP) Stacks

ASCE/G-I 53-10 Compaction Grouting Consensus Guide

ASCE/EWRI 54-10 Standard Guideline for Geostatistical Estimation and Block-Averaging of Homogeneous and Isotropic Saturated Hydraulic Conductivity

ASCE/SEI 55-10 Tensile Membrane Structures
ANSI/ASCE/EWRI 56-10 Guidelines for the Physical Security of Water Utilities
ANSI/ASCE/EWRI 57-10 Guidelines for the Physical Security of Wastewater/Stormwater Utilities
ASCE/T&DI/ICPI 58-10 Structural Design of Interlocking Concrete Pavement for Municipal Streets and Roadways
ASCE/SEI 59-11 Blast Protection of Buildings
ASCE/EWRI 60-12 Guidelines for Development of Effective Water Sharing Agreements

CONTENTS

FOREWORD

The Board of Direction approved revisions to the ASCE Rules for Standards Committees to govern the writing and maintenance of standards developed by ASCE. All such standards are developed by a consensus standards process managed by the ASCE Codes and Standards Committee (CSC). The consensus process includes balloting by a balanced standards committee, and reviewing during a public comment period. All standards are updated or reaffirmed by the same process at intervals between five and ten years. Requests for formal interpretations shall be processed in accordance with Section 7 of ASCE Rules for Standards Committees, which are available at www.asce.org. Errata, addenda, supplements, and interpretations, if any, for this standard can also be found at www.asce.org.

The form of this Standard Guideline reflects the goals of the Water Regulatory Standards Committee (WRSC) of the Environmental and Water Resources Institute (EWRI) of ASCE. The provisions of this document are written in permissive language and, as such, offer to the user a series of options or instructions but do not prescribe a specific course of action. Significant judgment is left to the user of this document.

Three model water sharing agreements are included in this Standard Guideline in Appendixes A, B, and C. Although the general format the appendixes follows *The Chicago Manual of Style*, imbedded within these model agreements is the format commonly used today in the drafting of proposed uniform state laws. This format, developed under the auspices of the National Conference of Commissioners of Uniform Laws, has been accepted as the format for the ASCE Model Water Codes documents (ASCE 2004a; Dellapenna 2007a). The format of the sections and subsections of each model agreement consists of statutory language in boldface that a legislature could enact with or without change. This statutory language is followed by a commentary section that describes the purpose and scope of the statutory provisions. Cross-references to other provisions in the specific agreement follow. The section or subsection is closed by a paragraph listing other interstate and international water sharing agreements that contain similar provisions.

Each section of each agreement is optional. Authorities may, however, enact the bulk of the agreement yet delete or change any particular section. Nonetheless, the WRSC made an effort to create a complete, comprehensive, and well-integrated contract between the parties capable of effectively managing shared water. The WRSC has concluded that almost every section of each agreement is necessary to achieve that goal. Some sections apply, however, only to water sharing compacts within the United States, others only to an international agreement. Others are specifically denominated optional in any jurisdiction, which indicates that the drafters consider that these sections might not be necessary or appropriate to the needs of the specific situation. A coherent and workable agreement would still result were all of the optional sections omitted. This Standard Guideline refers to current ASCE Policy Statements and to certain common references. ASCE Policy Statements normally are updated every three years and should be consulted for changes that may have occurred (www.asce.org/pressroom/news/policy.cfm).

This standard has been prepared in accordance with recognized engineering principles and should not be used without the user's competent knowledge for a given application. The publication of this standard by ASCE is not intended to warrant that the information contained herein is suitable for any general or specific use, and ASCE takes no position respecting the validity of patent rights. The user is advised that the determination of patent rights or risk of infringement is entirely his or her own responsibility.

PREFACE

The Model Water Code Project of the American Society of Civil Engineers (ASCE) was initiated in 1990 under the leadership of Ray Jay Davis of the Brigham Young University School of Law. The purpose of the project was to develop model statutory provisions intended for adoption by state governments for allocating water rights among competing interests and for resolving quantitative conflicts over water (Dellapenna 1997b).

After Davis retired from Brigham Young University in 1995, the project continued under the leadership of Joseph W. Dellapenna of the Villanova University School of Law. In 1997, ASCE published *The Regulated Riparian Model Water Code* (Dellapenna 1997b). In 2001, the code was accepted as EWRI/ASCE Standard 40-03.

Early in the project formulation and development process, the ASCE Water Laws Committee recognized that effective water allocation and management required planning and regulation by water basins. Because most basins are shared by two or more political entities and their independent legal jurisdictions (e.g., states or nations), the committee recognized the need for a companion agreement or compact to manage the waters flowing across or along the boundary between separate legal jurisdictions.

When two or more independent governments share a common water resource, the timing and magnitude of the respective individual uses can be continual sources of conflict. Water scarcity is evident throughout much of the western United States, and the use of shared water resources is often a major source of legal and political conflict. The interstate and international conflicts over the allocation of the waters of the Colorado River began early in the twentieth century and have still not been totally resolved. However, the problem is not limited to the western states. Even when water is relatively plentiful, the increasing demand for water from shared resources is growing as the population expands, dramatically increasing the needs for public water supply. This problem has been graphically shown by the recent dispute among Florida, Alabama, and Georgia over allocation of the waters of the Apalachicola–Chattahoochee–Flint River Basin. The problem is pervasive, since few river basins in the continental United States are contained within a single state's boundaries (Dellapenna 2002b; Draper 2002a, 2006).

The problem is magnified in the international arena. There are more than 250 major rivers shared between and among two or more nations. These international river basins cover almost one half of the total land surface of the globe. More than 50 rivers are shared by three or more nations, and the Danube is shared by 13 riparian countries. International river basins sustain more than 40% of the world's population. Almost 25% of the world's population lives in the Earth's semiarid and arid zones, where scarcity of water is often acute. Therefore, the potential for conflict is enormous (Draper 2002a, 2006; *The Transboundary Freshwater Dispute Database* 2012).

Among others, protracted conflict over shared waters exists among Turkey, Iraq, and Syria in the Tigris–Euphrates basin; between Jordan and Israel regarding the opposite bank sharing of the Jordan River; and among nations in the Nile River Basin. The Ganges River is a source of dispute between India and Bangladesh. Armed conflict has occurred between Ecuador and Peru over the Cenepa River. The breakup of the Soviet Union has caused conflict between former members, especially in the arid regions east of the Caspian Sea. Seven active transboundary water disputes currently exist in Africa; six in Europe and Asia; and at least five in the Americas (Dellapenna 2001a; Draper 2002a, 2006).

The need for effective cooperation among riparian countries has greatly expanded because of the growing demand for water in various international basins and the increasingly harmful effects of activities in upstream countries. Although some form of interstate compact covers most of the shared river basins in the United States, many were drafted in the first half of the twentieth century. These agreements were often one-dimensional and limited in scope, oriented to specific problems rather than holistic management of the basin's water. It can be argued that many of these interstate water compacts are inadequate to resolution of the more complex water sharing issues that will develop in the twenty-first century (Draper 2002a, 2006).

Water resource experts now recognize that the shared use of water resources is most effective when management is on the river basin level and when management of the shared resource is comprehensive and multidimensional. Internationally, the problems are more acute. More than a third of the 200 international river basins are not covered by any international agreement, and only some 30 have truly cooperative institutional arrangements. Therefore, a significant need exists for guidance and procedures that can facilitate the development of agreements that can provide a basis for effective and efficient water sharing between autonomous political entities (Draper 2002a, 2006; *The Transboundary Freshwater Dispute Database* 2012).

The creation of this Standard Guideline proceeded in two stages. First, three model water sharing agreements were published (Draper 2002a). The model agreements were then followed by a second report that provided a narrative description of suggested guidelines to be used in choosing the model agreement that might best serve as the basis of the specific water sharing agreement to be created by the parties (Draper 2006).

Although water scarcity and the increasing competition for water suggest that comprehensive management of a shared river basin is appropriate, a significant challenge to overcome is the prevailing tendency for governments at all levels to resist outside control over and interference with their internal affairs and those decisions that affect economic growth or quality of life. Each government or legally competent authority subject to the agreement, whether it is a nation-state, a state within a federal system, or a tribal entity within a national federal system, normally wants to maintain authority over the people, places, assets, and natural resources within its political boundaries. Local control is the operative word.

Strong incentives are required for a government to relinquish control of resources within its jurisdiction. Early in the process, the committee recognized that the degree to which the parties were willing to relinquish such control over water resources in or adjacent to their jurisdiction depended on the hydrologic, geographic, and political situation involved in the shared water dispute. In some situations, the parties might be willing to relinquish considerable autonomy in the search for efficient water management. In other situations, the parties may insist on maintaining control of the waters within their boundaries and may be content to coordinate water management activities in some manner (Draper 2002a). Therefore, the committee chose to develop three distinct model transboundary agreements to serve as a framework for individual agreements, each of which is based on how much control the parties might be willing to cede.

In 2002, model water sharing agreements were published as *Model Water Sharing Agreements for the 21st Century* (Draper 2002b). This publication provided three separate model agreements that may be used to provide a framework around which an agreement could be formed. It was recognized, however, that the model agreements alone would not provide sufficient guidelines to the drafters of a water sharing agreement. Because an effective agreement to share water involves a complex amalgamation of data and information, a narrative discussion was required that detailed for the drafters of a real-world water sharing agreement the research and procedures leading to the decision to use one of the model agreements as a basis for their specific agreement.

In 2006, a narrative analysis of the development of model agreements was published as the EWRI/ASCE *Sharing Water in Times of Scarcity* (Draper 2006). This publication provided guidelines to enable integration of the multiple aspects of the water resource through an analysis that synthesizes the disciplines of science, engineering, technology, economics, and law. The document sought to ensure that all pertinent factors are considered in the development of an agreement so that the agreement accommodates the physical realities of the shared resource along with the different political systems, cultures, and/or water use customs of the particular water basin.

This Standard Guideline has merged and refined these two Shared Use of Transboundary Water Resources publications (Draper 2002a, 2006). The scope of this Standard Guideline includes application for international agreements, interstate compacts, and state–tribal agreements for regulating water resources along or across political boundaries. This Standard Guideline is used to apply to any sharing of waters between independent political governments. This Standard Guideline considers a variety of issues that influence the development of interstate and/or international water sharing agreements. This publication provides guidance to enable integration of the multiple aspects of the water resource through an analysis that synthesizes the disciplines of science, engineering, technology, economics, and law. The document seeks to ensure that all pertinent factors are considered in the development of an agreement so that the agreement accommodates the physical realities of the shared resource along with the different political systems, cultures, and/or water use customs of the particular water basin.

This Standard Guideline forms a bridge between the theory and the practice of effective shared water management. It provides a process that all states and/or nations can use when creating or modifying a transboundary water sharing agreement. This process includes an assessment of the various factors influenced by the shared water use, to include correlating the geographic and political issues surrounding utilization of the water resource, inventorying the sources and uses of the water resource, analyzing the ecological impact of the transboundary use, and examining its effect on economic growth and quality of life of the various constituents. This Standard Guideline presents the various alternatives available for allocation of water among the parties, with special emphasis on extreme events (e.g., droughts or floods). The alternatives include allocation methods for surface water, underground water, and atmospheric waters. Finally, different choices for the administrative apparatus that is to supervise implementation are provided.

ACKNOWLEDGMENTS

The American Society of Civil Engineers acknowledges the work of the Water Regulatory Standards Committee (WRSC) of the Standards Development Council (SDC) of the Environmental and Water Resources Institute (EWRI). The WRSC consists of individuals from many backgrounds, including consulting engineering, law, research, the construction industry, education, government, design, and private practice. Work on this Standard Guideline material began in 1995 with the formation of the Shared Use of Transboundary Water Resources (SUTWR) Task Committee and incorporates information developed by the EWRI Laws & Institutions Committees of the American Society of Civil Engineers. Two SUTWR committee reports provide the foundation for this Standard Guideline: *Model Water Sharing Agreements for the 21st Century* (Draper 2002b) and *Sharing Water in Times of Scarcity* (Draper 2006). This Standard Guideline was prepared through the consensus standards process by balloting in compliance with rules as administered by the Codes and Standard Committee of the American Society of Civil Engineers. Those individuals who served on the Water Regulatory Standards Committee were

Robert H. Abrams, J.D., Aff.M.ASCE
Fadi Georges Comair, Ph.D., M.ASCE
Richard H. Cox
William E. Cox, Ph.D., M.ASCE
Ron Cummings, Ph.D.
Matthew D. Davis, S.M.ASCE
Joseph W. Dellapenna, J.D., LL.M., Aff.M.ASCE (WRSC Chair)
Henry Reirden Derr, P.E., M.ASCE
Stephen E. Draper, Ph.D., P.E., J.D., M.ASCE (Principal Author; Corresponding Editor)
Christopher Estes
Shou Shan Fan, Ph.D., M.ASCE
Harald D. Frederiksen, P.E., D.WRE, F.ASCE
Marshall Goulding
Ken C. Hall, P.E.
Steven C. Harris, P.E., M.ASCE
Kris G. Kauffman, P.E., M.ASCE
Conrad G. Keyes, Jr., Ph.D., P.E., L.S., D.WRE, Hon.M.ASCE
Olen Paul Matthews, Ph.D., J.D.
Nathaniel Dehass McClure, P.E., D.WRE, F.ASCE
Zachary L. McCormick
Donald M. Phelps, P.E., D.WRE, M.ASCE
Phyllis Saarinen
Gerald Sehlke, P.E., D.WRE, Aff.M.ASCE
Hal D. Simpson, P.E., D.WRE
Kenneth Stern, P.E.
C. Wesley Strickland, J.D., Aff.M.ASCE
Gregory K. Sullivan, P.E., M.ASCE
R. Peter Terrebonnne, Ph.D.
R. Timothy Weston, J.D.
Kenneth R. Wright, P.E., L.S., Dist.M.ASCE

CHAPTER 1
THE FOUNDATION FOR EFFECTIVE WATER SHARING

Economic growth and prosperity and improved quality of life require adequate supplies of high-quality water on a regular and sustained basis. This requirement means that the utilization of shared waters must be carefully and systematically coordinated among or between the parties sharing the waters. In 2007, 268 transboundary river basins were identified worldwide. These basins cover almost two-thirds of the global landmass. Forty percent (40%) of the world's population depends on these shared river basins for the water they need (International Network of Basin Organizations 2002; Sea River 2002; Draper 2007b). Of the 71 rivers in the continental United States (excluding Alaska and Hawaii) that are more than 565 km (350 mi) long, only six are not shared by one or more states and/or countries. Of more than 90,000 river km (55,800 river mi), less than 7% are not shared (Pearson Education 2002; Draper 2002b, 2007b). More than 90% of the population in the continental United States depends on waters shared with other states (Draper 2002b, 2007b).

1.1 PURPOSE AND SCOPE OF THE STANDARD GUIDELINE

An effective agreement can facilitate adequate planning, conservation, utilization, development, management, and control of water resources on a water basin basis, in a manner that is reasonable and equitable under the circumstances and that causes no significant harm to most other parties. A key challenge for the parties is to make more efficient and productive use of water and to reshape the water policies of the individual parties to better respond to periods of water shortages (Postel 1996; Draper 2002b, 2007b).

This document provides guidance to be used in the formulation of water sharing agreements between and among governments so that they may meet the challenge of devising an effective and equitable water sharing agreement. The Standard Guideline seeks to limit potential conflict while providing an appropriate balance among efficient use of the water resource for existing economic purposes, preserving the common water resource for future needs, and promoting the protection of the environment. The procedures focus on the process of creating or modifying a transboundary water sharing agreement to ensure that the parties include all pertinent factors in their negotiations.

Although this Standard Guideline has been developed predominately using the U.S. experience with transboundary water sharing, a number of international agreements were analyzed to gain an appreciation for other experiences. Consequently, this Standard Guideline has broad application for use in international agreements, interstate compacts, or state–tribal agreements for regulatory purposes along or across intergovernmental boundaries. The Standard Guideline can apply to any sharing of waters between autonomous political entities. As presented, this Standard Guideline can be applied to a variety of circumstances involving shared water. Only those sections appropriate to the particular situation and conditions need be applied (Draper 2006). The need for careful analysis of the situation and conditions that influence those sections that may apply is especially relevant when an international agreement is being developed because the hydropolitical conditions may be greatly aggravated by significant sociocultural issues. This complication may be further aggravated by the lack of an appropriate legal forum and institutions necessary to adequately ensure compliance.

The following ASCE/EWRI Standards are incorporated in this Standard Guideline and provide additional guidance when and where required:

- ASCE/EWRI Standard 33-09, *Comprehensive Transboundary Water Quality Management Agreement* (2009c);
- ASCE/EWRI Standard 34-01, *Standard Guidelines for Artificial Recharge of Ground Water* (2001);
- ASCE/EWRI Standard 42-04, *Standard Practice for the Design and Operation of Precipitation Enhancement Projects* (2004b); and
- ASCE/EWRI Standard 40-03, *Regulated Riparian Model Water Code* (2004a).

1.2 DISTINCTIVE CHARACTERISTICS OF WATER

In setting the framework for sharing water resources, it is appropriate to briefly discuss several special characteristics of the natural resource water. Water is central to the survival of life itself, and without it plant and animal life would be impossible. Water is a central component of the Earth's systems, providing important controls on the world's weather and climate. Water is also central to our economic well-being because it supports rain-fed and irrigated agriculture, forestry, navigation, waste processing, and hydroelectricity (Vörösmarty 2002). Recreation and tourism are other primary uses supported by water, especially in developed countries.

Water, unlike oil or gold, is a shared, mobile, and public resource that is used and reused for different purposes as it moves through the hydrologic cycle. Before its capture by withdrawal or diversion, a claim of exclusive ownership is difficult to sustain. Water is different from other natural resources because different users use the same water repeatedly as the water travels downstream, as in the case of surface water, or downgradient, as in the case of underground water aquifers. A case in point is Atlanta, Georgia, which withdraws most of its reliable water supply from the Chattahoochee River, consumes a portion of the water, and returns the rest of the water to the river for use by downstream users such as Lagrange or Columbus, Georgia, and even cities in Alabama and Florida (Draper 2002a, 2004, 2007b). This use and reuse is the essence of riparian law, the fundamental law theory upon which shared water use historically has been based (Draper 2006, 2007b).

It is often said that water is renewable, but the term can be misleading. Surface water and some underground water moving through aquifers like the Floridan in the southeastern United States are renewable in the sense that most water moves through the hydrologic cycle as one form or another of water (Draper 2007b). Other water, especially underground water in so-called fossil aquifers like the Ogallala in the midwest United States, is not quickly renewable (Opie 2000).

What is important from a sustainability perspective is how much of the water used by the various demands is actually renewable and available for reuse. Some analysts have concluded that globally, water availability has declined by half during the past 25 years because of population growth and drier trends in the climate (Shiklomanov 1998; Zalewski 2002). River runoff in Africa has reportedly declined by about one-third in the past 10 years, and runoff in South America has shown a significant increase (Georgeyewsky 1998). A significant decrease in precipitation has been observed since the mid-20th century in North Africa, Southeast Asia, Indonesia, and Australia. When looked at from a global perspective of the hydrologic cycle, the liquid form of H_2O is renewable. When considering water available to meet demands at a certain time and place, it is not necessarily renewable (Draper 2006, 2007b).

1.3 GUIDING PRINCIPLES FOR EFFECTIVE WATER SHARING

The uniqueness of each basin and its riparian states indicates that a universal set of principles should be fairly general (Wolf 1999). However, efficient and effective transboundary water sharing should be based on four guiding principles: coordination and cooperation, interdisciplinary analysis, watershed and river basin planning, and adaptive management (Draper 2006).

1.3.1 Coordination and Cooperation. A nation normally enters into any international agreement from a position of self-interest. In the negotiations, each party seeks the rights and authority critical to certain political, economic, or social objectives while ceding less critical rights and authority to the other nations. While accepting this fact, all parties have a duty to cooperate and negotiate in good faith. This principle is the foundation of international law, and it applies in all relations between governments or legally competent authorities subject to the agreement (Draper 1997). The parties should agree to cooperate and consult with the other parties to the agreement in their development and utilization of the water and related resources of the waters shared by the parties. Coordination and cooperation are essential to ensure effective and efficient use of those waters while minimizing harm to other parties. Additionally, the parties should agree to cooperate on the basis of equality and territorial integrity in the utilization and protection of the shared water resources. They should conduct themselves with an absence of malice and deceit, and with no intention to seek unconscionable advantage. Coordination and cooperation include the need for good faith implementation (Draper 2002b).

1.3.2 Interdisciplinary Analysis. The development of effective and efficient transboundary water sharing is a multifaceted challenge (Kenney and Lord 1994, Draper 2001). Although necessarily based on water science, i.e., knowledge of the quantity, quality, and timing of the available water supplies, the development of effective water sharing agreements is predicated on an adequate interdisciplinary analysis of water science, engineering, technology, law, and economics (Draper 2001; Jackson et al. 2001; Dellapenna and Draper 2002b).

A social component exists that should not be ignored (Fitch 2003; Levy et al. 2003). This social component relates to the cultural and noneconomic needs and wants that transcend the economic factors of society. These needs and wants would include, for example, guaranteeing that each human being, regardless of circumstances or economic efficiency, has at least 70 L of clean water daily. It might include protection of source water that is considered sacred to native populations. Many agreements initially neglect the social aspects involved in water allocation and discover that this neglect can delay or even prevent effective implementation and/or enforcement of the agreement (Draper 2007b).

In addition, this development should include an analysis of the water available for allocation to the various existing and known future water demands, to include economic growth, public health, navigation, water-based recreation, and environmental and ecological protection. Finally, the effective and efficient utilization of water requires integration of surface and underground water sharing, as well as consideration of water quantity and water quality. The potential effect of climate change should be included (Draper 2006, 2007b).

1.3.3 Watershed and River Basin Planning. Planning and management on a watershed or river basin basis, rather than artificial political boundaries, should be the foundation principle for water planning (Barlow and Clarke 2002). Watersheds are the natural spatial unit for predicting future water supply and water demands, at least as they relate to the use and reuse of water as it flows downstream (Jackson et al. 2001). The reality of political boundaries and the desire for authority and control should be respected wherever possible, but water planning and management must be based on the watershed or river basin. Watershed planning and management should include three essential components. The first component is maintaining hydrologic integrity within a regional framework. The second component is developing a multijurisdictional management structure that coordinates water planning and land use planning. The third component is the use of regional–national–state–local partnerships, as well as private–public partnerships, in the planning and management of water resources (Draper 2001, 2002a).

1.3.4 Adaptive Management. Water resource management should be based on the best information available, but it should be flexible enough to accommodate change and complexity since natural systems are inherently variable, patchy, and complex (Sophocleous 1998). In broad terms, adaptive management is "a systematic process for continually improving management policies and practices by learning from the outcomes of operational programs (Bennet and Lawrence 2002). Adaptive management requires deliberate evaluation of resource management activities and active response to what is learned (Sando 2002). In essence, adaptive management can be described as "learning by doing" (Lee and Lawrence 1986). An adaptive agreement is designed from the outset to recognize that clearly formulated hypotheses about the behavior of hydrologic systems may be changed by human use (Lee 1993).

For an agreement on water sharing to be effective, it should be able to adapt to potential changes in site-specific circumstances that may develop within the time period the agreement is in effect (Bernauer 2001). The provisions in the agreement should be sufficiently adaptable to allow them to evolve over time and to respond to changes in the climatic, hydrologic, economic, social, and even political conditions (Eaux partagées 2002; Fitch 2003). The agreement should be able to accommodate unexpected changes in water source availability that may result from extreme events, such as prolonged droughts and climate change. Likewise, the agreement should be able to accommodate changes in the demand for water that may arise from changes in the economic foundation of one or more of the parties or from sociological or political changes in the value assigned to particular water demands. As with any water management policy, the provisions and rules within a water sharing agreement can best be developed and implemented by retaining and encouraging innovation and diversity (U.S Advisory Commission on Intergovernmental Relations 1991).

1.4 GOALS AND OBJECTIVES OF WATER SHARING AGREEMENTS

The goal of water sharing agreements is to create a legal framework between the parties to establish the guiding principles of effecting water sharing, which is essential to sustaining economic growth and prosperity while enhancing the quality of life for the citizens of the parties. Quality of life requires a strong economy, good public health and safety, good recreational opportunities, and well-managed natural resources. All of these factors depend on adequate supplies of clean water (Draper 2001; McCain 2005).

Any water sharing agreement should clearly state its objectives and should consider meeting the needs of both existing and future demands and uses. The procedures to reach the agreement should include emphasis on integrated management and recognition of the economic value of water. A means of stakeholder participation in decision making is essential, as are means of providing access to water services for the poorest users. Institutionalizing an ecosystem approach and integrating private sector contribution should be developed (World Bank 1998; Eaux partagées 2002).

1.4.1 Existing Demands and Usage. The primary short-term objective of a water sharing agreement should highlight meeting existing demands and uses while maintaining reasonable environmental and ecological protection. Essentially the agreement should provide for water use during "normal" years but also provide risk management for extreme conditions, droughts or floods, and allocate water among existing uses in times of water shortages (McCormick 1994b; Draper 1997, 2001). An essential objective for the agreement, for both the short term and the long term, is equity among users (Eheart 2002).

1.4.2 Future Demands and Usage. The primary long-term objective should highlight meeting the needs of future demands and uses while meeting the long-term needs of the ecosystem. The agreement should be structured to ensure the availability of sufficient quantities of quality water for future growth as planned or predicted by all the parties involved in the agreement. The parties should avoid provisions that may favor existing uses at the expense of new, higher value uses (Draper 1997, 2001).

1.4.3 Quality of Water Allocated. The quality of the water that is allocated is as important as the quantity of water allocated. Poor quality water imposes risks that the parties should consider. First, there is the health risk to the population that uses the water for domestic purposes. Second, if the available water will not meet the standards for certain beneficial uses, there is the risk that economic growth will be impaired. Finally, there is the risk that quality degradation will have a severe impact on the ecology of the basin, resulting in the loss of long-term sustainability. Integration of water quality and quantity is essential (Draper 2002a). Riparian and aquatic ecosystems provide critical benefits for human, plant, and animal life, including improving water quality, reducing erosion and sediment losses, providing habitat (more than 75% of the animal species in arid regions require riparian habitat), creating recreation benefits and other amenities for growing populations, and providing flood control benefits (Hirji and Ibrekk 2001). AGENDA 21 from the *United Nations Conference on Environment and Development* held in Rio de Janeiro, Brazil, 1992, obligated all signatories to develop a water and sustainable development program (Ahlander 1994). The UN *Convention on the Law of the Non-Navigational Uses of International Watercourses*, United Nations Document A/51/869 (1997), establishes the criterion that "(w)atercourse States shall, individually and, where appropriate, jointly, protect and preserve the ecosystems of international watercourses."

Surface and underground water may be degraded by a variety of factors. Major problems affecting the quality of these water resources arise, for instance, from inadequate domestic sewage treatment, inadequate controls on the discharge of industrial waste and effluent, the diversion of waters resulting in insufficient water to assimilate waste, the loss and destruction of catchment areas, the improper siting of industrial plants, deforestation, and poor agricultural practices that cause leaching or runoff of nutrients and pesticides. Excessive interbasin transfer may affect the quantity of water available for waste assimilation downstream of the transfer (Draper 2002b). Transboundary water sharing should include effective plans and programs that reduce or eliminate, or at least minimize, the possible sources of water quality degradation. (ASCE 2009c).

1.4.4 Sustainable Management. Water resources should be managed in a sustainable manner so that current and future generations will have access to adequate supplies of high-quality water that supports both human needs and natural systems (Fort 1998; Eheart 2002; Georgia Joint Comprehensive Water Plan Study Committee 2002). This sustainability criterion expands the scope of water sharing beyond economic efficiency to encompass fairness to future generations as well (Tietenberg 1998). However, sustainability is a difficult concept to define, and no consensus on a definition has developed (Fort 1998). Strict interpretation of the philosophy of sustainability may deny the needs and abilities of humans to make rational choices for their immediate and long-term future.

In this light, this document defines water resource sustainability as the reasonable and beneficial use of water that avoids wasting resources and ensures reasonable protection of the environment while meeting the needs of future generations.

Sustainable outcomes for water use include the requirement that minimum water supply is guaranteed to all humans to maintain human health. Although the human minimum water needs vary according to the aridity of the region under analysis, an absolute minimum has been determined to be 50 L per day (Gleick 1996). "In November 2002, the United Nations Committee on Economic, Social and Cultural Rights affirmed that access to adequate amounts of clean water for personal and domestic uses is a fundamental human right of all people" (United Nations 2002, 2010).

The agreement should provide sufficient water to restore and maintain the health, services, and functions of ecosystems. Specific amounts vary, depending on climatic and other conditions, and flexible, adaptive management is essential. The agreement's provisions should be structured to facilitate human actions in the shared water resources that do not impair the long-term renewable freshwater stocks and flows (Hirji and Ibrekk 2001; ASCE 2009c).

1.5 SUMMARY

This Standard Guideline is to be used in the formulation of efficient and effective water sharing agreements between and among governments. The guideline is intended for use in international agreements, interstate compacts, or state–tribal agreements for regulatory purposes along or across intergovernmental boundaries. The guideline may apply to any sharing of waters between autonomous political entities.

Four guiding principles underlie the formulation and development of the final agreement (Draper 2006):

(1) Negotiations should be conducted with a commitment to coordination and cooperation,
(2) An interdisciplinary approach to water allocation among the parties should be used to overcome the inherent obstacles facing effective water sharing,
(3) The agreement should provide for management on the basis of watershed and river basins, and
(4) Adaptive management and flexible provisions should be included in the agreement.

The goal of water sharing among the parties should be to sustain economic growth and prosperity while enhancing the quality of life for the citizens of the parties. The water sharing agreement should create a legal framework that clearly states its objectives, and writers of the agreement should consider meeting the needs of both existing and future demands and uses. The procedures to reach the agreement include the following (Draper 2006):

- Emphasize integrated management,
- Recognize water's economic value,
- Enable stakeholder participation in decision making,
- Provide access to water services for the poorest users,
- Provide a systematic ecosystem approach,
- Integrate the private sector contribution, and
- Consider the needs of future generations.

CHAPTER 2
WATER SHARING ACROSS POLITICAL BOUNDARIES

Two factors are essential to develop a viable water sharing agreement across political boundaries. First, at the beginning of the formulation process, the parties should understand and analyze major obstacles facing such an agreement on effective water sharing. Second, the parties should approach the negotiation of obstacles leading to the agreement with an understanding and commitment to certain norms of conduct critical to fair and honest negotiation between parties to the agreement. Finally, an effective agreement for water sharing depends on an understanding of national and international water law, and the relationship of one with the other.

2.1 OBSTACLES TO EFFECTIVE WATER SHARING

The nature of water resource sharing by different political entities results in a number of significant obstacles to effective water sharing. The reluctance to give up dominion and control over the natural resources within the political boundaries of the parties to another party is a significant obstacle. Other obstacles include conflicts in the internal water laws and policies of the parties that make the shared goals of the agreement difficult to reach. An incomplete knowledge of the water resource availability and demand within the shared basin, as well as conflicts between the internal economic policies of the parties, may hamper an agreement between or among the parties (Clarke 1993, Draper 2001).

2.1.1 Agreement across Political Boundaries. Whether a party to the negotiations to share water is a nation-state or a state or tribal entity within a federal system, the sovereign party will normally claim the exclusive rights to all natural resources within its boundaries, including ownership of the water resources. When a government enters into an interstate or international agreement, it does so primarily from a position of self-interest (Draper 1997; Albert 2000). Water sharing requires each party to relinquish a part of its control and dominion over water that may be under its control and dominium to some degree. In negotiations, each party will seek to retain control over those rights and authorities that are critical to their political, economic, or social objectives while ceding to the other sovereign parties rights and authorities that may be less important to their primary objectives. This inherent friction underlies all negotiations. How much authority is relinquished determines the scope and possibly the effectiveness of the agreement (Eaux partagées 2002).

The political boundaries that exist between the governments present significant obstacles since water basin boundaries seldom coincide with political boundaries (Dellapenna 1994; Libiszewski 1995; World Bank 1998; Wolf 2001; Draper 2002b; Puri and Arnold 2002; Vörösmarty 2002; Matthews and St. Germain 2007). Rights to use water within a particular shared watershed or aquifer, with few exceptions, differ according to the water laws of the particular party in which the withdrawal occurs. It is a question of jurisdiction, "who has the responsibility and authority to regulate water withdrawal and use?" (Kenney and Lord 1994; McCormick 1994b).

With surface water, the conflict over shared use may arise because the upstream party's water policies and consumptive use may adversely affect the water available for use by the downstream party (Kenney and Lord 1994). For example, Nile River water use by Sudan has significant adverse effects on water use by Egypt (Dellapenna 1997b; ASCE 2004a). Because surface water often marks the boundary between political entities, the conflict may arise from the different water uses on opposite banks by adjacent states. Differing water uses by Israel and Jordan are illustrative (See *Treaty of Peace between the State of Israel and the Hashemite Kingdom of Jordan* 1994). Differing cultures and their concepts of water and its use can be an obstacle to the spirit of cooperation and trust so necessary to an effective agreement. Although the Israeli–Palestinian conflict has many causes, the inherent difference between the two culture's views of water and its allocation provides another obstacle to peace (Dellapenna 1994, 1997a; Hassoun 1998). Water is now acknowledged as a major limiting factor in the socioeconomic development of a world with a rapidly expanding population (Sibanda 2002).

Within the United States, this same conflict exists, although arguably to a lesser degree. Three elements of control and dominion clash under the constitutional powers of the federal, state, and tribal governments. Under federalism, the states maintain control over inland waters, with some important exceptions (*Kansas v. Colorado* 1902, 1907; Kenney and Lord 1994; Gelt 1997). These important exceptions include limitations imposed directly or by implication in specific legislation, such as the Endangered Species Act, 16 U.S.C. §§ 1531 et seq. (ESA) and the Clean Water Act, 33 U.S.C. §§ 1251 et seq. (CWA), limitations or obligations imposed by interstate compacts, interstate allocations established pursuant to U.S. Supreme Court decisions, and certain other doctrines established by the U.S. Supreme Court. Such doctrines include federal reserved water rights that adhere to federal lands in the west, federal control over navigability, and federal supremacy regarding federal waterpower projects.

Indian reservations also may assert dominion and control over certain waters as a result (*Arizona v. California* 1963). Thus the elements of dominion and control that obstruct effective and efficient water sharing on the international scene often obstruct effective and efficient water sharing between the states, between a state and a tribe, or even between two or more states and the federal government.

2.1.2 Conflicting Internal Water Rights. An essential requirement for economic progress is the availability of sufficient quantities of high-quality water (Draper 2001, 2002a, 2002b). Adequate supplies of water for stakeholders within the political jurisdiction are normally controlled through a legal system of water rights that allocates water. The laws usually are structured to provide a balance between providing water rights to uses that are efficient and beneficial, or at least not wasteful, while still providing legal certainty or predictability and security of tenure to those holding the water rights. Often, however, the water laws of the parties have developed differently and, consequently, conflict (Sanchez 1997; U.S. House of Representatives 1998; Grant 2001). The problem is exacerbated for laws that allocate underground water. Different laws for the allocation of surface and underground water present an issue that exists for both intra- and interjurisdictional water management (Kenney and Lord 1994).

Vested water rights within the territorial borders of the various political parties may be a source of conflict. This complication may inhibit both the development and execution of a water sharing agreement. The internal water laws of one of the parties may vest or otherwise provide to private parties legal rights to water, including private property rights. Such legal rights to water may prevent or otherwise impede the party's ability to fulfill the terms of the agreement. Within the United States, the property right associated with water varies considerably from state to state (Beck 2001; Dellapenna and Draper 2002b, c).

Consequently, different rights granted to the same source of water may result in inefficient and ineffective allocation of the water, to the detriment of all parties. The right granted by the different parties to withdraw from the same aquifer may differ, as may the rights granted to withdraw from the same surface water source. The parties may have different requirements for return flows and different standards for minimum instream flow requirements (Evans and England 1995). To prevent ineffectiveness of water allocation, all parties that have an interest in a water source should either agree on certain water-related issues or at least understand the existence of these issues to accommodate them within the framework of the agreement. These issues include determining how the parties to the agreement grant the right to use water from a specific source, the amount of water each party allocates under normal conditions from that source, and the amount of water allocated from that source during droughts (Draper 2006).

2.1.3 Conflicting Internal Water Policies. An obstacle to effective water sharing agreements may develop from differences in the water policies of the individual governments. A government's water policy should be distinguished from its water rights or water laws. Water policy is founded on the interrelationship of water law, science, engineering, technology, and economics (Draper 2001, 2002a, b, d, e). Water-related policies and programs of individual governments usually have specific objectives, often oriented to economic efficiency of water use in relation to economic plans and needs. Often the result is single-purpose laws and programs that fail to account for the interrelationship among surface, underground, and atmospheric water; quality and quantity of water allocated; and other water rights needs, such as instream flow rights (Vlachos 1997). The political influence of specific stakeholder groups may be a source of severe distortion in laws and regulations (Zusman 1997). A significant obstacle to effective water sharing arises when one party institutes a policy of private water markets while the other does not. Also, drought management and reservoir release policies of the upstream party may conflict with downstream water availability. Finally, one party's policy on interbasin diversions can have serious effects on the water available to the other party (Sherk 2000).

Separate and distinct political units within a government often make water policy independently of other policy-making entities within that same government, hindering effective water sharing agreements (Kenney and Lord 1994). In the United States, differences between and among the laws, policies, and procedures of federal and state water quality and water allocation agencies exist that undermine effective water laws and policies (Rogers 1993). More than 20 subcommittees of Congress and more than 20 federal agencies have overlapping mandates and conflicting policies, each supported by separate constituencies. The laws promulgated to implement these policies are themselves fragmented and distinct. Any agreement should acknowledge and provide resolution mechanisms for these conflicts (Postel 1996).

Scientists, economists, and legal scholars have long advocated the integrated and conjunctive management of all water resources within a watershed and river basin. After all, the hydrologic cycle is an integrated whole, and courts or administrative agencies should not segregate the parts of a mobile resource that nature integrates. Water-based activities physically affect each other and should be managed comprehensively if they are to be managed properly. On the other hand, activities related to various parts of the hydrologic cycle do have varying characteristics, including varying degrees of effects upon other uses related to other parts of the hydrologic cycle (Dellapenna 2001b).

Minimum flow policies and policies concerning interbasin transfer can have substantial consequences on the effectiveness of an agreement (Draper 2002a; Eheart 2002). There may be conflict because one party wishes to reserve from allocation and consumptive use the waters necessary for the preservation of the protected biological, chemical, or physical integrities of water sources and may insist on a mechanism to further protect instream flows. The other party may have a policy that the instream flow be maintained at a lower level only for purposes of wastewater assimilation (Draper 2002a). In other cases, a party may insist that the instream flow mimic the natural cycle of spring floods and summer low flows to sustain the biological integrity of the stream (Christensen 1996; Glennon 2002). Transferring water for use outside its basin of origin with little or no return flow to the basin of origin might pose similar problems to the basin of origin as are likely to arise in interstate transfers even when both the basin of origin and the basin of use are within a single state (Draper 2002d). Similarly, significant consumptive uses (such as the incorporation of water in a product or evaporation through cooling systems) may impose substantial downstream outcomes similar to out-of-basin transfers (Draper 2006).

Until recently, water sharing agreements have primarily been concerned with surface water sources. Now, however, with newer water withdrawal technologies and with the exponential growth in the demand for water of the past several decades, underground water has emerged as a critical regional, national, and transnational resource. The sharing of aquifers and the hydrologic interconnections between surface water and groundwater sources is increasingly becoming the focus of conflict (Dellapenna 2001a).

A significant challenge to reaching an effective agreement may arise from the different standards the parties apply to consumptive use, reuse, and conservation. Increased consumptive use can have a dramatic effect on the water available to downstream users, as can reallocation by a party from older "lower value" uses that may be less consumptive to newer "higher value" uses that have greater consumption of the water withdrawn (Draper 2002e). Water reuse policies can assist or hinder downstream users. If the reuse is structured to ensure that water available to downstream users remains unchanged or even increased, reuse policies can greatly assist the parties in reaching agreement. On the other hand, if the reuse policies can result in less water available downstream, they can greatly impede the parties from reaching an agreement. Finally, as discussed earlier, the introduction of water markets by one party can result in increased water scarcity for the other party, especially when interbasin transfers occur or the new water user engages in significantly greater consumptive use and the other party is downstream (Postel 1993).

2.1.4 Incomplete Knowledge of Water Resource Availability. Incomplete knowledge of the waters subject to the agreement is a significant obstacle to effective water sharing (Patino-Gomez et al. 2007). Effective agreements are predicated on a detailed knowledge of the quantity of water available for use and the quality of that water available for specific uses (Fort 1998). Incomplete knowledge, however, may arise from uncertainties regarding geologic matters (such as a relationship between underground water aquifer areas and surface water basins), data that are limited by the time frames for which information was collected, and uncertainties regarding prediction of future hydrologic conditions and technological developments (Patino-Gomez et al. 2007). Although incomplete knowledge hinders the agreement's effective allocation of water to the respective parties, a more significant effect is that an agreement does not adequately cover all water needs, especially during extreme events (Draper 2006).

A water sharing agreement usually focuses on allocation of water under average hydrologic conditions, as well as that allocated under drought conditions. Developing an effective agreement requires accurate and precise knowledge of these conditions as well as flood conditions. Otherwise, the result may be significant disputes over the terms of an agreement, significantly impairing the agreement's usefulness for the parties (Berman and Wihbey 1999; ASCE 2004a, 2009c; Dellapenna 2007a).

Lack of knowledge about underground water availability is especially prevalent (Hillel 1994; Tsur 1995). The hydrologic characteristics of some aquifers may be unknown, as is the case in many areas of the developing world. Substantial data gaps exist on the condition of different underground water basins, extraction amounts, current pumping practices, and recharge rates (Gleick 1998).

The question of the reliability of past hydrologic records may be a significant problem. Even what appear to be long-duration records (say the past 50 or 100 years) may not be representative of the cycles of hydrologic variation. Human-made changes in flow conditions (from storage, diversions, and changes to impervious surfaces) may skew the reliability of historic data (Draper 2006; Draper and Kundell 2007).

The uncertainty of climate change presents an equal uncertainty when the historical record may become unreliable as global warming changes precipitation patterns, volumes of snowpack, and the timing of spring thaws (Draper and Kundell 2007). ASCE has recognized the possible effect of climate change, publishing *ASCE Policy Statement 360* (ASCE 2010a):

> Climate change could pose a potentially serious impact on world wide water resources, energy production and use, agriculture, forestry, coastal development and resources, flood control, and public infrastructure. Examples include:
> *Alterations to the hydrologic patterns for multi-purpose water resource projects, of particular concern to civil engineers working in the hydroelectric industry and water supply utilities, where reservoir storage capacity may need to be increased;
> *Climate extremes, such as floods and droughts, and other significant variations in hydrologic patterns that may necessitate changes or additions to flood control infrastructure to provide adequate public safety and performance;
> *Changes in frequency and strength of tropical storms that will require changes in coastal protection systems;
> *Changes in ocean levels that will require adaptation of coastal infrastructure, including ports; and
> *Changes in permafrost conditions that require retrofitting existing foundations and alterations to foundation design.
> Such impacts could require modified agricultural practices and measures to deal with rising sea levels, water supply and quality, threats to critical infrastructure facilities, and the potential for the outbreak of disease.
>
> Rationale
> Civil engineers are responsible for design and maintenance of infrastructure projects that facilitate economic development and protect human health, welfare, and the environment. Climate change may result in significant impacts to this infrastructure. Civil engineers and government policy makers must work together to anticipate and plan for these impacts.

2.1.5 Incomplete Knowledge of the Water Resource Demands. An effective agreement requires a comprehensive understanding of the existing requirement for water, as well as projected requirements. Sound water policy planning requires a comprehensive, coordinated, and reliable source for water-related data and current reliable information (Georgia Joint Comprehensive Water Plan Study Committee 2002). The information needed includes data on water requirements that may be gathered by local, state, and national government, as well as academic institutions, the private sector, and other organizations (Draper 2001). The agreement's purpose should be to allocate the available water between and among the political parties themselves. The parties can effectively accomplish this objective only when the water requirements within the individual party's internal boundaries are known. These water requirements are often described as needs or demands (Draper 2006). Within each jurisdiction, the available water use should be divided among various, and often conflicting, uses, such as agricultural and municipal demand, waste assimilation and recreational uses, environmental protection and industrial demand, and navigation demand and hydropower. Each jurisdiction should balance its requirements with the needs of the other jurisdictions and with the available quantity of sufficient quality water for each use. Finally, an imperfect knowledge of the withdrawal amounts affecting the sustainability of the water resource may cause a non-replenished loss of water that reduces water availability, thus effectively nullifying the agreement at some future date (Nebraska Department of Natural Resources 1983; Gleick 2000b; Sherk 2000). A significant problem of inadequate knowledge of the requirements for water is especially prevalent in the agriculture sector for a variety of reasons. Even in the United States, precise knowledge of agricultural use is lacking because metering and annual reporting are not required of many irrigation users (Cummings et al. 2001b). Such knowledge is critical because agricultural use accounts for approximately two-thirds of global water use (Postel 1993). Domestic use is normally consolidated in the category of municipal use. Information on industrial and commercial water use is incomplete at best (Gleick 2002b).

2.1.6 Conflicting Internal Economic Policies. Autonomous economic policies among governments may hamper effective water sharing agreements. Different governments have different plans and priorities when balancing economic growth and the sustainability of water resources. Economic policy may be fragmented as to the means of promoting water policy effectiveness, with one government allocating according to the common law (Georgia Chamber of Commerce 2002), another allocating on the basis of codified water law and regulation (Dellapenna 2000), and a third using a free-market approach (Cummings et al. 2002) to economic planning and water use prioritization (Draper 2004). Policies regarding environmental and ecological protection and their relationship with economic growth are also a source of conflict that a water sharing agreement should accommodate (Espland 1998; Draper 2001, 2004).

2.2 NEGOTIATING OBSTACLES LEADING TO AN AGREEMENT

Certain principles of negotiations and management are essential for the development and implementation of a viable water sharing agreement. These principles are equally necessary for the agreement's ongoing effectiveness. A number of significant principles should form the framework of the agreement. They include an obligation to cooperate and negotiate in good faith, an obligation to prevent unreasonable harm to other parties, a commitment to the equitable utilization of the shared water, a commitment to the values of water resource sustainability, and an obligation to exchange all available data with the other parties (Caponera 1995; Hey 1995; Draper 1997; Albert 2000; Dellapenna 2001a).

2.2.1 Obligation to Cooperate and Negotiate in Good Faith. Normally, a government negotiates an interstate or international agreement on the basis of self-interest (Draper 1997; Albert 2000). In the negotiations, each party seeks the rights and authorities critical to certain political, economic, or social objectives while ceding less critical rights and authorities to the other governments or to an international or interstate institution. This inherent friction underlies all negotiations. While accepting this fact, however, all governments have a duty to cooperate and negotiate in good faith (United Nations 1997). This principle is the foundation of international law, and it applies in all relations between parties to the agreement. All parties are expected to conduct themselves with an absence of malice and with no intention to seek unconscionable advantage or otherwise be deceitful. This obligation is a critical foundation for water sharing agreements (Kliot et al. 1998). A commitment by all parties is essential (German Foundation for International Development 1998).

Reaching an effective and efficient agreement to share a water resource means that the parties should begin the negotiations with some forethought. Water sharing requires each party to relinquish a part of its control and dominion. How much authority and control is relinquished determines the scope and possibly the effectiveness of the agreement (Eaux partagées 2002). Self-interest may be a guiding factor in many negotiations, but such a position is not conducive to reaching an effective and efficient agreement (Draper 2006).

In most water sharing cases, the purpose should be to find a way of accommodating the interests of all parties who share in the common resource—to find rules, water development investments, and other sharing arrangements that allow the legitimate needs and desires of the respective parties to be achieved. Good agreements focus on the parties' interests rather than on a particular position. Defining a problem in terms of positions means that one party will lose the dispute. When a problem is defined in term of the parties' underlying interests, it may be possible to negotiate a solution that satisfies both parties' interests (Fisher and Ury 1983).

2.2.2 Obligation to Prevent Unreasonable Harm to Other Parties. The duty to prevent unreasonable harm requires that water allocation and use by one riparian party cannot unreasonably harm the allocation and use by other riparian parties (Kroes 1997). The duty may be best explained by drawing an analogy to the riparian doctrine of reasonable use. Under U.S. common law, this doctrine restricts water withdrawal by a riparian owner to those circumstances in which the withdrawal was for reasonable use. The withdrawal could not unreasonably harm another riparian owner. The balancing of certain factors determines reasonable use. Such factors may include the purpose of the use; the suitability of the use to the water resource; the economic value of the use; the social value of the use; the extent and amount of harm the water use causes; the practicability of avoiding the harm; the protection of existing uses; the availability of alternative sources; and the equity, or justice, of requiring the user causing the harm to bear the cost of remediation measures (Dellapenna 2001b).

Degradation of the quality of water resulting from inadequate treatment of domestic and industrial wastewater, from urban storm water runoff, or from combined sewer outflows may cause unreasonable harm. Other examples of causation of unreasonable harm include the overuse of pesticides and fertilizers on suburban lawns and/or agricultural lands. Harm may also be caused by water use by one party that reduces the amount of water available for use by other parties, or by water use by one party that causes the flooding of other parties. The water sharing agreement should describe the general standards for reasonable, beneficial use and specifically exclude certain uses or certain laws and regulations that may result in unreasonable harm to others (Draper 2006).

2.2.3 Commitment to the Equitable Utilization of the Shared Water. A fundamental aspect of transboundary water use is the mutual recognition of each party's rights to equitably share in the beneficial use of the transboundary water. At the international level, beneficial use of transboundary waters is determined according to the doctrine of Equitable Utilization (United Nations 1997; ILA 2004). In addition, many decisions from the U.S. Supreme Court in the exercise of its original jurisdiction over suits between states have decreed equitable apportionment of their shared waters. This doctrine requires governments to use transboundary water resources in a manner that is reasonable and ensures equity for all riparian parties. Certain factors are relevant to reasonable and equitable use. The weight to be given to each factor is to be determined by its importance in comparison with that of other factors. Such factors may include the geographic, hydrographic, hydrologic, climatic, environmental, ecologic, and other factors of a natural character and the social and economic needs of the governments concerned. The effects of the use or uses of the water resources by one riparian party on other riparian parties and existing and potential uses of the water resources may be considered. Other factors in the analysis could include conservation, protection, development, economy of use of the water resources, and the cost of measures taken to that effect. The availability of alternatives to a particular planned or existing use could also be relevant (Dellapenna 2001b).

2.2.4 Commitment to the Values of Water Resource Sustainability. Sustainability is a complex concept with as many definitions as there are settings and advocates (Hirji and Ibrekk 2001). It has been suggested that development is sustainable if "it meets the needs of the present without compromising the ability of future generations to meet their own needs" (Brundtland 1987). In its strictest sense, water resource sustainability requires that withdrawal and consumption of specific waters be limited in quantity to their long-term natural replenishment and that return flows be of a quality that does not adversely affect other uses of the same resource. Such a philosophy would ensure that sufficient quantities of high-quality water are available for human consumption and economic purposes far into the future (Fort 1998). However, this definition may be considered impracticable for a number of reasons. It has been suggested that an accurate measurement and quantification of sustainability is generally not possible because of limited knowledge about long-term response of natural systems. Additionally, such a philosophy does not take into account technological advances, such as economically viable desalination, that may reclaim water for additional use. The philosophy suffers from vagueness in that the geographical scope of water resource sustainability is normally not defined. It has been suggested that the philosophy of sustainability in its strictest sense denies the needs and abilities of people to make the rational choices for their immediate and long-term future (Beck 2001). These definitions also ignore ecological and social needs.

ASCE has developed the following *Policy Statement 418* (ASCE 2010b) to describe "sustainable development":

> . . . sustainable development is the challenge of meeting human needs for natural resources, industrial products, energy, food, transportation, shelter, and effective waste management while conserving and protecting environmental quality and the natural resource base essential for future development.

2.2.5 Obligation to Exchange Adequate Data with the Other Parties. Sound management of transboundary waters requires compilation of a comprehensive database (Eheart 2002). A successful transboundary water sharing program requires the combined expertise and analytical ability of experts in hydrology, hydrogeology, water resources engineering, ecology, economics, and other scientific and technological disciplines, as well as water law and policy (Draper 2002a). Effective scientific and technological analysis requires, in turn, accurate and precise information about the climate and hydrology. Water project operations should be documented. An understanding of wastewater discharge and water quality monitoring is important. Knowledge of water development plans is critical, as are water demands for consumptive uses. Water policies and programs that affect water quantity and/or quality should be well documented (Draper 2006).

To develop a useful agreement and to make it work, each party should provide the necessary data and information in an accurate and understandable form. Normally, management on a watershed basis is preferable (Draper 2002b). To be adequate, data provided by each party should provide a comprehensive view of the water resource under consideration. Data concerning water use and activities affecting both the quality and quantity of water are also essential to understanding existing resource conditions. However, data collected independently by several parties generally possess varying degrees of incompatibility that may limit full integration of the data. The willingness of all parties to adopt a common approach to minimize such problems is a basic need to facilitate transboundary management. The parties should also consider providing mechanisms within the agreement for modification, in the event that the underlying data prove inaccurate or conditions change significantly (Draper 2006).

2.3 THE INTERNATIONAL LAW OF TRANSBOUNDARY WATER SHARING

Most of the principles cited earlier are recognized as the obligations of governments when water sharing is an issue. It is important for the parties to reach an agreement to recognize the legal duties imposed upon them by international law and to conduct themselves accordingly in any negotiations involving sharing of water that crosses international boundaries (Draper 2006).

2.3.1 Substantiating International Legal Norms. International law demands that nations, in their relationships with other nations, conduct themselves according to specific standards of conduct, or legal norms. Nation-states rarely openly or directly defy generally accepted principles of international law but rather rely on those principles in their diplomatic exchanges (McCaffrey 1993). A nation not following these legal norms may be ostracized and sanctions of one sort or another may be imposed. However, the specific international legal norms are not as easy to ascertain as are the legal duties and obligations imposed by national laws (Draper 2006).

International law differs fundamentally from national (domestic) law. The domestic laws of any government derive from a variety of sources. These sources can include a written constitution, civil codes or statutes, common law (judge-made law), religious law, tradition, and administrative regulations. The citizens are obliged to follow these laws. These sources of law do not exist in the international arena. The basic characteristic of international law is that it is consensual. A nation usually consents to a rule only if it is in the country's best interests in terms of the long-term benefits outweighing the long-term costs. An example is the 1921 *Convention and Statute on the Regime of Navigable Waterways of International Concern* (the Barcelona Convention). That convention obliged "each riparian State . . . to refrain from all measures to prejudice the navigability of the (international) waterway, or reduce the facilities for navigation." This obligation was clearly in the trading interest of the different parties to the convention, and a large number of nations signed the convention (Draper 2006).

Article 38(1) of the *Statute of the International Court of Justice* (1989) describes how that court determines international legal obligations:

> The Court, whose function is to decide in accordance with international law such disputes as are submitted to it, shall apply:
> a. International conventions, whether general or particular, establishing rules expressly recognized by the contesting states;
> b. International custom, as evidence of a general practice accepted as law;
> c. The general principles of law recognized by civilized nations;
> d. Judicial decisions and the teachings of the most highly qualified publicists of the various nations, as subsidiary means for the determination of rules of law.

2.3.2 Legal Norms of Water Sharing. Some obligations may readily be described as international norms, as defined in the *Statute of the International Court of Justice* (1989). To cooperate and negotiate in good faith is an international legal obligation expected of all nations. A number of different international treaties, conventions, and resolutions substantiate this premise. This obligation is applicable to a wide range of areas and is not limited to transboundary water sharing. The other duties cannot be satisfied without the willingness of the parties to cooperate and act in good faith efforts. The duty to prevent unreasonable harm is also an international norm that can be substantiated by a number of international documents (Draper 2006).

Other principles, however, rely on international custom and the writings of highly qualified publicists. The United Nations' International Law Commission is an eminent example of one means of determining customary law. The commission and other legal scholars have declared the international law of water sharing to include two other obligations: (1) the duty of equitable utilization of the shared water and (2) the obligation to exchange all available data with the other parties. Enforcement of these legal norms can, of course, be problematic (Overseas Development Institute 2001; Draper 2006).

2.3.3 Legal Evolution of Shared Water Agreements. Water sharing agreements between or among governments create a legal structure for the legally competent authority in which they are drafted. Internationally the agreement itself creates the legal structure, while in the context of the United States, the U.S.

Constitution provides the necessary legal structure. A review of both U.S. and international law is appropriate for an understanding of the applicability of the model codes and the extent to which their modification may be necessary (Draper 2006).

Within the United States, the federal–state partnership established by the U.S. Constitution provides a stable legal relationship between the states upon which interstate conflict resolution may be based. Three levels of authority arise under the constitutional powers of the federal, state, and tribal governments. Under federalism, the states maintain authority over inland waters unless Congress has preempted them through the exercise of its powers or a direct command of the Constitution. Examples include legislation, such as the Clean Water Act (CWA 1987) and the Endangered Species Act (ESA 1982), and certain legal doctrines established by the courts. Such legislation and doctrines include federal reserved water rights that attach to federal lands in the west, federal control over navigability, and federal supremacy for federal waterpower projects. Independently of the states, Indian tribes also may assert authority over certain waters as the result of the treaties the various tribes entered into with the United States. Thus, under some conditions, the state controls use of waters within its boundaries. Under other conditions, an Indian tribe may control certain waters within a state. Finally, under still other conditions, the federal government may control certain waters within a state. Added to this "brew," the use of the waters within one state may significantly affect the quality or quantity of waters within another state. Inevitably, conflicts over use of the water develop. Significantly, the Constitution provides a mechanism for conflict resolution, and a forum for final conflict resolution exists in the U.S. Supreme Court (Draper 2002a, d, 2006).

There are three significant methods of resolving interstate water allocation disputes: (1) interstate water apportionment compacts, (2) acts of Congress, and (3) U.S. Supreme Court original jurisdiction suits between the states. The first method is an agreement between the states to allocate or manage the shared waters. The three model water sharing agreements presented in (Draper 2002b) relate primarily to the first method, interstate water apportionment compacts. The second method is an allocation of interstate waters by congressional action. There has been but one instance of this method of transboundary water sharing, the *Boulder Canyon Project Act* (1928). In this act, Congress authorized construction of the massive Hoover Dam on the lower main stem of the Colorado River and allocation of the waters of the lower basin, predicated on ratification of the *Colorado River Compact* (1928) (McDonald 1997). In the *Truckee–Carson–Pyramid Lake Water Rights Settlement Act* (1990), Congress "authorized the federal purse" to help resolve water rights allocation disputes (Grant 2001). Other methods for Congress to affect interstate water allocation exist. Federally authorized water projects, and the resulting contracts for delivery of water to users, can result in a de facto apportionment; an example is the apportioning of water stored in federal dams on the North Platte River (Grant 2001).

The third method of resolving interstate water allocation disputes is adjudication before the U.S. Supreme Court. It is important that the drafters of an interstate compact understand these original jurisdiction suits between the states because any interstate dispute that is not resolved by compact or act of Congress would be resolved by the court, likely based on the principles of equitable apportionment. This situation creates a baseline against which each party measures what it is asked to give up under the compact (Draper 2006).

The basis of the equitable apportionment doctrine is that interstate waters should be apportioned by balancing equities. Equity (or "fairness," as opposed to "equality") is achieved by balancing the stake (or value) each party may have in utilizing the shared resource (Grant 2001). The factors to be analyzed to determine a "fair" allocation of water between or among the parties include (Draper 2006):

1. Physical and climatic conditions;
2. The consumptive use of water;
3. The character and rate of return flows;
4. The extent of established uses and economies built on them;
5. The availability of storage water;
6. The practical effect of wasteful uses on the downstream area; and
7. Damage to upstream areas compared to the benefits to downstream areas if limitations were to be imposed on the former.

However, there is no requirement that an interstate compact be based on the principles of equitable apportionment, and the states can apportion waters in any manner or through any procedure the party states choose. Once approved by Congress, the compact becomes federal law, according to the terms of the compact. Should a dispute arise over interpretation of the compact and the matter be referred to the Supreme Court for resolution, it is doubtful the principles of equitable apportionment would be applied unless the compact explicitly stated that such principles formed the basis of the compact. The Supreme Court has stated that "courts have no power to substitute their own notions of an 'equitable apportionment' for the apportionment chosen by Congress" (*Arizona v. California* 1963). Because "congressional consent transforms an interstate compact . . . into a law of the United States, . . . no court may order relief inconsistent with its expressed terms" (*Texas v. New Mexico* 1983). Additionally, it should be noted that, to date, no state has sued for apportionment of a water source to which that state is not riparian (Grant 2001).

Within the international arena, the problems with effective conflict resolution are magnified for two reasons. First, no forum for final conflict resolution of water disputes exists in relation to water sharing conflicts. The United Nations, the World Court, and the World Trade Organization (WTO) provide forums for conflict resolution in some subject matters and under certain conditions, but no forum for conclusive judgment for conflicts involving water sharing exists. Secondly, no clear standards exist in customary international law that relate to transboundary water sharing (Draper 2006).

Under long-standing customary international law, a state had the exclusive rights to natural resources within its boundaries. This customary rule of international law does not apply to water within political boundaries. Although no definitive international treaty or agreement has been ratified on a worldwide basis to change the rule of absolute sovereignty, customary international law clearly embraces the rules of equitable utilization and no significant harm. The *Helsinki Rules on the Uses of the Waters of International Rivers* (1966) crystallized the consensus that each co-riparian state is entitled to "a reasonable and equitable share in the beneficial use of the waters of an international drainage basin." Since 1966, a variety of other sources of international law have elaborated on this view. The 1972 Stockholm Declaration articulated the view that states have "the responsibility to ensure that activities within their jurisdiction or control do not cause damage to the environment of other States or of areas beyond the limits of national jurisdiction." However, a companion declaration in this 1972 declaration also articulated the need to protect "the sovereign right [of states] to exploit their own resources pursuant to their own environmental policies" without articulating how to coordinate these rules.

The *Treaty for Amazonian Cooperation* (1978) proclaims "that the exclusive use and utilization of natural resources inside their own territory is a right inherent to the states' sovereignty and that its exercise will not be subject to restrictions other than those imposed by international law." The standards of international law were not described, however. The 1992 United Nations *Convention on the Protection and Use of Transboundary Watercourses and International Lakes* declared that the parties "shall take all appropriate measures to prevent, control and reduce any transboundary impact . . . (and) ensure that transboundary waters are used in a reasonable and equitable way, taking into particular account their transboundary character, in the case of activities which cause or are likely to cause transboundary impact." Oriented to European transboundary water quality, this convention calls for "sustainable water-resources management" but does not discuss allocation of water (Draper 2006).

In 1992, the United Nations Commission on Sustainable Development reiterated the right to exploit internal resources but also emphasized the obligation not to cause transboundary harm (*Report of the United Nations Conference on Environment and Development* 1992). This report, published as Agenda 21, stated to the Secretary-General that

> States have, in accordance with the Charter of the United Nations and the principles of international law, the sovereign right to exploit their own resources pursuant to their own environmental policies and have the responsibility to ensure that activities within their jurisdiction or control do not cause damage to the environment of other States or areas beyond the limits of national jurisdiction.

Clearly, without a definitive international treaty, signed and ratified by a significant number of the world's powers, current customary law concerning the nonnavigational use of international watercourses is still ambiguous, providing two governing principles (equitable utilization and no significant harm) that often in practice contradict each other. The *Convention on the Law of the Non-Navigational Uses of International Watercourses* (United Nations 1997) may end this confusion about customary law. The UN General Assembly approved this convention in May 1998 by a vote of 104 to 3. Although it will not enter into force until it is ratified by 35 nations, the convention provides clearer standards for agreements concerning the shared use of international watercourses. The convention states:

> Article 5
> Equitable and reasonable utilization and participation
> 1. Watercourse States shall in their respective territories utilize an international watercourse in an equitable and reasonable manner. In particular, an international watercourse shall be used and developed by watercourse States with a view to attaining optimal and sustainable utilization thereof and benefits therefrom, taking into account the interests of the watercourse States concerned, consistent with adequate protection of the watercourse.
> 2. Watercourse States shall participate in the use, development and protection of an international watercourse in an equitable and reasonable manner. Such participation includes both the right to utilize the watercourse and the duty to cooperate in the protection and development thereof, as provided in the present Convention.

The Convention's doctrine of "equitable and reasonable utilization" is similar to but more expansive than the U.S. Supreme Court's doctrine of "equitable apportionment." This is Article 6 of the Convention (1998):

> Article 6
> Factors relevant to equitable and reasonable utilization
> 1. Utilization of an international watercourse in an equitable and reasonable manner within the meaning of article 5 requires taking into account all relevant factors and circumstances, including:
>   (a) Geographic, hydrographic, hydrological, climatic, ecological and other factors of a natural character;
>   (b) The social and economic needs of the watercourse States concerned;
>   (c) The population dependent on the watercourse in each watercourse State;
>   (d) The effects of the use or uses of the watercourses in one watercourse State on other watercourse States;
>   (e) Existing and potential uses of the watercourse;
>   (f) Conservation, protection, development and economy of use of the water resources of the watercourse and the costs of measures taken to that effect;
>   (g) The availability of alternatives, of comparable value, to a particular planned or existing use.
> 2. In the application of article 5 or paragraph 1 of this article, watercourse States concerned shall, when the need arises, enter into consultations in a spirit of cooperation.
> 3. The weight to be given to each factor is to be determined by its importance in comparison with that of other relevant factors. In determining what is a reasonable and equitable use, all relevant factors are to be considered together and a conclusion reached on the basis of the whole.

Article 7 establishes the requirement for all parties to "prevent the causing of significant harm to other watercourse States."

> Article 7
> Obligation not to cause significant harm
> 1. Watercourse States shall, in utilizing an international watercourse in their territories, take all appropriate measures to prevent the causing of significant harm to other watercourse States.
> 2. Where significant harm nevertheless is caused to another watercourse State, the States whose use causes such harm shall, in the absence of agreement to such use, take all appropriate measures, having due regard for the provisions of articles 5 and 6, in consultation with the affected State, to eliminate or mitigate such harm and, where appropriate, to discuss the question of compensation.

The convention further declares that all watercourse parties should "cooperate on the basis of sovereign equality, territorial integrity, mutual benefit and good faith in order to attain optimal utilization and adequate protection of an international watercourse" (Article 8).

One word of caution is appropriate. It has been noted that the "equitable and reasonable utilization" standards are vague and there is an absence of a neutral enforcement mechanism in the convention (Dellapenna 1997a). Both points demonstrate that the convention itself cannot stand alone in resolving conflicts over shared water resources. Rather, the convention should provide the standards and structure of a water sharing agreement developed for a specific watershed or river basin (Draper 2006).

In 2004, the International Law Association proposed changes to existing international laws concerning water resources (ILA 2004). In addition to recognizing the need for conjunctive management and integrated management, the rules added to the list compiled for Article 6 of the UN Convention (1998). In the ILA's view, relevant factors and circumstances to be evaluated to define "utilization of an international watercourse in an equitable and reasonable manner" would include (ILA 2004)

> (h) The sustainability of proposed or existing uses; and
> (i) The minimization of environmental harm.

Notably, the *Convention on the Law of the Non-Navigational Uses of International Watercourses* (United Nations 1997) applies only to coriparians on international watercourses. Attention is directed to the terminology used in this Convention, especially regarding the term "equitable and reasonable utilization." Nowhere in the convention does the term "beneficial use" appear. This term of art is used extensively in those U.S. states adhering, in whole or in part, to the prior appropriation water allocation doctrine. Because the term "beneficial use" raises significant policy issues, especially with regard to the classification of instream uses as valid and legally protected water uses, and because current international law uses the term "equitable and reasonable utilization" to denote those water uses that are permissive, the latter phrase is used in this model agreement (Draper 2006).

The significant difference between international law and the laws of the United States is that, for the United States, the U.S. Supreme Court provides the institutional legal framework with a "court of last resort" to resolve interstate water disputes. No such forum exists for international water disputes; reliance should be placed on internationally accepted legal norms, such as a ratified the *Convention on the Law of the Non-Navigational Uses of International Watercourses* (United Nations 1997).

2.4 SUMMARY

A number of obstacles exist in developing an effective agreement to share water resources between governments. These obstacles include conflicts in the internal water laws and policies of the parties, an incomplete knowledge of the water resource availability and demand within the shared basin, and conflicts between the internal economic policies of the parties.

To negotiate these obstacles, efficient and effective water sharing across political boundaries requires that development of the agreement follow a certain pattern. The parties should acknowledge certain standards of conduct for negotiation. The parties should assume the obligation to cooperate and negotiate in good faith and the obligation to prevent unreasonable harm to other parties. They should commit to the equitable utilization of the shared water and to the values of water resource sustainability. The parties should assume the obligation to exchange adequate data with the other parties and make a commitment to do so. The obligation to cooperate and negotiate in good faith and the obligation to prevent unreasonable harm are standards of conduct clearly mandated by international law. Because the validity of the other standards of conduct should rely on international custom and the writings of highly qualified publicists, the parties to the agreement should explicitly address those obligations in the agreement process. Negotiating the obstacles presented by conflicting economic objectives and priorities is especially difficult to overcome, in no small part because of the conflicting philosophical concepts of economic efficiency (ASCE 2004a, 2009a; Dellapenna 2007a).

Once these obligations are delineated, along with a mechanism to ensure that the parties embrace these obligations, the specifics of the agreement can be formulated. Significant sections of the agreement should include a mechanism to allocate the shared waters between the parties according to some objective function that includes the economic, social, environmental, and ecological objectives of the parties. An institution should then be developed that can effectively implement the agreement at a reasonable cost without unduly infringing on individual states and their separate authority over internal waters. Equally important, the source water needed to fulfill those objectives and the water available in the shared resource should be compared and the objectives modified as necessary to reflect the reality of the yield of the shared resource. This step requires significant scientific and engineering studies to estimate the water available under normal conditions, as well as under conditions of drought and flood. Finally, some dispute resolution mechanism should be constituted to minimize the potential for destabilization of the agreement (ASCE 2004a, 2009a; Dellapenna 2007a).

Necessarily, an effective agreement may only be achieved with an understanding of, and adherence to, the established laws of water sharing. Within the sense of federal law, an understanding of the application of federal law to water sharing between states is essential. Likewise, within the international context, an understanding of the norms of international law is essential. As supported by the American Society of Civil Engineers (ASCE 2009a), the techniques of alternative dispute resolution can often be used to avoid costly time delays that may disproportionately affect the economic, social, environmental, and ecological programs of affected states and/or nations.

> The American Society of Civil Engineers (ASCE) supports dispute avoidance and alternative dispute resolution techniques such as those contained in the Engineers Joint Contract Documents Committee's (EJCDC) documents to bring disputes related to engineering and construction to a fair, timely, and cost-effective conclusion without litigation.

Chapter 3
RESOURCE ASSESSMENTS

The foundation of a water sharing agreement requires accurate and precise information about the availability of water, the quality of that water, and the demands that should be accommodated. A number of assessments should be conducted at the beginning of the formulation of the agreement. These assessments include definition of the geographical and political boundaries of the water sources to be shared and analysis of the water policies, laws, and regulations of the individual parties that may conflict. An assessment of the yield of the water sources for normal conditions should be made, as well as for extreme conditions of droughts and floods. An accurate accounting for existing demands and one for predicting future water demands should be made, as well as an assessment of the water quality and quantity necessary for ecological integrity. An economic impact assessment should then be made that correlates the other assessments. From a procedural standpoint, strategies for the development of a water sharing agreement should initially proceed with the exchange of data and information between technical experts, with the objective of reaching a technical consensus on both the actual and potential problems known or envisioned. Only then can alternative strategies to respond to the problems be developed (Eaux partagées 2002).

3.1 GEOGRAPHICAL AND POLITICAL BOUNDARIES

One criterion of the *Convention on the Law of the Non-Navigtional Uses of International Watercourses* (United Nations 1997) makes the definition of the waters to which an agreement applies a mandatory provision in the agreement. The parties should identify and document the type and geographical extent of the water resources to be subject to the agreement. This description allows all appropriate and necessary parties within the geographic–political region to be included in the agreement and negotiation process. Only then can an investigation of any conflicts in water law and policy proceed (Dellapenna 1994, 1997a, 1997b; ASCE 2004a).

To formulate an effective agreement, the parties should analyze the factors that influence the water resource in question, including the climatology, physiology, geology, and the interaction between underground and surface water resources. The analysis should identify pollution sources and the resulting effect on water quality. The geographic scope of the water resources to be covered by the agreement should be sufficiently expansive to fully address all the water sharing issues involved. It is important to acknowledge that the agreement reflects the particular circumstances and compromises reached in its formulation and that it applies only to the waters shared between the parties (Draper 2002b, 2006). The agreement should include the total surface area of drainage throughout the basin, including the aquifers underlying the surface drainage. Some tributaries could be connected to the underlying aquifers, and some of the aquifers could be connected to more than one of the surface water basins (Draper 2002b).

The purposes and scope of an agreement weigh heavily on this assessment, delineating the parties to be included. All principal stakeholders, such as persons, institutions, or governments directly affected by the shared use should be identified and included in the negotiations and the agreement, including any government that has direct access to the sources of water. Both governmental and private stakeholders should have a voice in formulation of the agreement. The *Treaty for Amazonian Cooperation* (1978), involving a balance of development and environmental protection, is a good example. Without the combined signatures of Bolivia, Brazil, Colombia, Ecuador, Guyana, Peru, Surinam, and Venezuela, the agreement could not be effective.

In terms of atmospheric water, the analysis of meteorological and climatological factors should identify those states or nations that will be affected. The applications of new techniques for weather modification or of atmospheric water management might have effects beyond any one state's borders (Gochis 2001; ASCE 2004b; Keyes 2006). For instance, significant extraction of atmospheric water in western Canada may affect other provinces in Canada, as well as the United States. However, because the effect of extraction of atmospheric water may extend far distances in ever-decreasing or increasing significance, the parties should adopt some measure of "reasonable impact" (Draper 2006).

Although private water rights holders within the various jurisdictions are not parties to the agreement, the various persons and organizations associated with the various water demands should have a voice in formulating the agreement. Although the extent of consultation with private groups depends largely on the political nature of the governments themselves, some recognition of existing rights is required. In most situations, a conflict exists between demands for water for economic purposes and the needs of environmental and ecological protection. Water users frequently need external incentives to accept the reservation of some water for environmental and ecological protection. Unless the governmental entities involved with formulating the agreement actively seek to include the various interest groups, the effectiveness of the agreement may be compromised (Draper 2001, 2006; Keyes 2006).

3.2 ASSESSMENT OF WATER LAWS AND POLICIES

Different water issues may exist for the individual parties, both within the state and across the state's boundaries. Each party may have developed, therefore, certain laws and policies to deal with these issues, which may be at variance with the laws and policies of other parties. Comparing and contrasting the different laws to expose conflicts that may affect water sharing is essential (Dellapenna and Draper 2002b). Within the United States, the extent of conflicts varies not only between the states involved but also with an array of federal and tribal water rights and issues. It has been observed that current United States water-use policy is in such disarray that many critical interstate water basins are left "dangling under Solomon's sword" (Tarlock 2001). This situation does not make water sharing between political entities any easier (Draper 2006).

3.2.1 Water Rights. Differences in law should be identified and accounted for in the agreement. By comparing and contrasting the laws of the individual parties, drafters of the agreement can identify potential conflicts and determine the areas requiring meticulous coordination. Each jurisdiction should have mechanisms to implement and enforce the laws. Knowledge of the administration and management institutions of each is necessary to administer the water sharing agreement and to ensure that its provisions are enforced. Participants should review and analyze each jurisdiction's institutional mechanisms that affect any form of water management. For instance, definition of legal terms within the individual jurisdictions is a prerequisite for understanding the conflicts and similarities in their laws. Of specific concern are the definitions of "reasonable use" and/or "beneficial use" and the term "equitable" (Draper 2006).

Whereas most jurisdictions maintain governmental control over the allocation of water and its use, the use of private water markets has been contemplated as a means to make the allocation process more efficient. The existence of such a private water market policy may result in significant differences in the way individual water rights are transferred. How water rights are lost is a concern in each jurisdiction. Each of these matters affects the final form of the agreement (Draper 1997).

The analysis to compare and contrast the water laws of the different participants can be efficiently accomplished within the context of the legal theories of water law and property rights. Legal theory establishes three basic approaches to water law and property rights: common property, private property, and public property (Freyfogle 1999; Dellapenna 2000; Gleick 2000a; Dellapenna and Draper 2002b).

Although the disciplines of science and hydrology have long recognized that surface water rivers and underground water aquifers are interrelated waters, the law has not (Eckstein 1998; Dellapenna and Draper 2002b; Glennon 2002). Traditionally, the framework of water law for surface water and underground water has been separate and distinct. Rights to use surface water (rivers and streams) are generally provided in one of three water right doctrines: riparian rights, appropriative rights, and regulated riparianism (Dellapenna 2001b).

The riparian rights doctrine is a surface water rights doctrine, typified by common law, that treats water as a common resource. This surface water rights doctrine, also referred to as the doctrine of "reasonable use" or "eastern water law," permits a landowner adjacent to a watercourse to withdraw and use the water as long as the use is "reasonable." The doctrine leaves the courts to resolve conflicting claims of right to use the common resource. Shortfalls in water availability are shared by all riparians, with courts deciding each user's share according to its reasonableness at the time of the decision (Draper 2006).

The appropriative rights doctrine is a second surface water rights doctrine that treats water essentially as private property. Water is appropriated according to "first in time, first in right." The senior appropriator is the one who first puts a specific amount of water to beneficial use. All others who put water to beneficial use later are junior appropriators. In cases of water scarcity, there is no sharing of the shortfall in water; the junior appropriator forfeits his or her water use until the senior appropriator has fully satisfied his or her appropriation. This doctrine is also referred to as "western water law."

The regulated riparianism doctrine is a third surface water rights doctrine that has begun to supplant common law riparian rights. This supplanting has occurred because states have recognized the need for active public management of water as the needs of large water users (e.g., commercial, municipal, industrial, and agricultural) increase and begin to compete for the available water supplies. Rather than treating water as a common resource (as under common law riparian rights) or as a private resource (as under appropriative rights), the regulated riparian rights doctrine treats water as a public resource. The central requirement of regulated riparianism is that water users should obtain state permission (time-limited permits) to withdraw and use large amounts of water for a specific purpose. Water withdrawals above a certain threshold are required to have a permit; withdrawals below the threshold are exempted. The threshold varies from state to state (Draper 2006).

The western doctrine of prior appropriation is premised on shortages allocated by priority schedules that provide a clear and complete risk allocation scheme before the shortages occur (Tarlock 2000). The water user need not be adjacent to the water source. In contrast, riparian rights are premised on shortages being shared in "reasonable use" by all riparians. The water user normally should be adjacent to the watercourse. International water law is similar to riparian rights and normally requires the equitable sharing of transboundary water resources (Draper 2006).

Five legal doctrines apply to underground water (Dellapenna and Draper 2002b; Draper 2002a). The absolute dominion doctrine allows an overlying landowner to withdraw an unlimited amount of water, unconstrained (in its purest form) by any concern about injury to another landowner. This doctrine is actually a form of the "water as private property" model, in which each common owner has the unlimited right to capture the underground water. The reasonable use doctrine for underground water requires the landowner to make "reasonable use" of water; any unreasonable use of underground water (uses that unreasonably injure adjacent landowners as determined in court) is actionable. However, the landowner may withdraw and consume the underground water for reasonable uses without regard to the effect on others. There is no sharing of the aquifer among landowners, as incorporated in surface water riparianism (Tarlock 1989), except as a court may determine that continuing a use is unreasonable. Implicit in the doctrine is that the "reasonable use" should take place on the land itself. Under the correlative rights doctrine, landowners may withdraw only their share of the underground water for "reasonable use." If there is not sufficient water available, each landowner shares in the shortfall in proportion to the amounts allocated to that landowner's use. This doctrine treats water as a common resource and resembles the common law riparian rights model described earlier (Trelease 1979; Tarlock 1989). Defining and enforcing the shares is a complex and often unsatisfactory process. Two other legal doctrines of underground water have evolved. Under the appropriative rights doctrine, underground water is allocated according to the "first in time, first in right," as with appropriative rights doctrine for surface water. It treats water as private property. Finally, a growing number of states apply their regulated riparian system, created for surface water, or a similar system specifically to underground water. Under a regulated riparian approach, underground water is allocated by a statutory system of a comprehensive water withdrawal permit system (Draper 2006).

There can be significant differences in water laws even between parties that have similar water law systems. The difference may be especially important in how certain water rights are vested by the individual internal water laws. Initial identification of vested water rights and knowledge of laws that may vest such rights in the future are necessary to prevent extremely complex water problems, such as those experienced by Los Angeles as Arizona and Nevada exercise their legal rights to the waters of the Colorado River (Dellapenna 2001b). These differences in the definition of the type, quantity, and extent of water subject to water laws can include the extent and duration of water rights; the permissible quantity of use; and the timing, pattern, and location of use.

Without careful consideration when the parties to the agreement have differences in their systems of water law and property, the understanding of specific provisions in the agreement by one party may be very different than the understanding of other parties. This difference could quickly lead to an unenforceable agreement.

3.2.2 Water Policies. An assessment is also necessary of water policies that are an extension of water law. However, an assessment is broader in scope and less tied to the formalities of legal provisions. These policies range from specific water policies, such as requirements for instream flow and safe yield, to economic policies that may indirectly affect water demand. It has been suggested that eastern and western water policies have begun to converge (Kundell and Tetens 1997), but for the foreseeable future, dramatic differences in water policy within the United States exist.

Minimum Flow and Safe Yield. A significant policy conflict may arise with regard to the individual party's requirements for minimum flows for surface water and safe yield for underground water (and reservoirs). Instream flow requirements can mean many things. It can be the amount of water remaining in a stream, without diversions, that is required to satisfy a particular aquatic environment (Nevada Division of Water Planning 1999) or the flows and levels reserved from allocation to protect the appropriate biological, chemical, and physical integrity of water sources (Georgia Water Coalition 2002). Required minimum flow may be defined as the minimum flow necessary to assimilate projected wastewater discharges. It can also mean the stream flows required for swimming, floating a canoe, supporting fisheries, sustaining waterfowl habitat, providing downstream water supplies, assimilating wastes, or generating hydroelectric power. These uses require different quantities of water at different times and at different levels of water quality (Sherk 2002). Other definitions set variable minimum flows that mimic natural flows (Poff et al. 1997). If the individual parties to the agreement do not agree to a common definition or requirement, or at least ensure that the agreement accommodates the difference, significant differences may arise (Draper 2006).

Safe yield or sustainable yield (Helweg 2000), as applied to underground water, is usually more limited in scope. Policies of maintaining safe yield usually refer to a management concept that allows water users to pump only the amount of groundwater that is replenished naturally through precipitation and surface-water seepage (Sophocleous 1998). As with differing policies regarding instream flow, differing policies of whether to restrict withdrawals can seriously damage the effectiveness of an agreement unless the difference is addressed in the agreement (Draper 2006).

Interbasin Transfer of Water. Interbasin transfer of water is the transfer of water rights and/or a diversion of water (either underground water or surface water) from one drainage or hydrographic basin (the basin of origin) to another (Nevada Division of Water Planning 1999; Draper 2004). As competition for limited water resources has increased, conflicts concerning interbasin water transfers have increased (Kundell and Tetens 1997; Draper 2004). One party to the agreement may have a policy of meeting water supply needs in one region by transporting abundant supply from another (Hirji and Ibrekk 2001; Draper 2004). This policy may result in negative effects on both the environment and the economy of the basin of origin that may include alteration of stream flow affecting instream uses of the water and preclusion of potential future instream and offstream uses of water that support economic activity (Georgia Joint Comprehensive Water Plan Study Committee 2002; Draper 2004). For purposes of water sharing, unless the receiving basin is in a common river basin whose waters are shared by the parties, interbasin transfer can reduce the available water to be shared. Clearly, policies of interbasin transfer by one party could have a significant effect on the water to be shared. To reach an effective agreement, the interbasin policies of the respective parties should be closely analyzed. It may be necessary for the agreement to detail how and when interbasin transfers of shared waters can occur (Draper 2006).

Various governments have imposed different criteria for interbasin transfer. Commonly, such transfers are allowed only after a showing that, among other criteria, there is no other reasonable alternative and that programs of conservation and efficiency have been implemented (Georgia Joint Comprehensive Water Plan Study Committee 2002; Glennon 2002; Draper 2004).

Flood Control. Flood control policies and works may also have a dramatic effect on the timing and elevation of water levels and thus may become a contentious issue between the parties. The issue should be addressed as an individual area of coordination. This provision recognizes the right of each party to make efforts to safeguard its people and economic forces from flood damages, but it also establishes an avenue for the sharing of data on flood control efforts as well as an independent analysis of the effects of those efforts on other parties (Draper 2000a).

Priority of Use. The *Convention on the Law of the Non-Navigational Uses of International Watercourses* (United Nations 1997) advises that "(i)n the absence of agreement or custom to the contrary, no use of an international watercourse enjoys inherent priority over other uses." If sufficient water is available to meet all equitable and reasonable uses, all demands should be met and no need for prioritization exists. Even if sufficient water is available during normal conditions, restrictions on water rights during times of water shortages or dry periods should be analyzed (Kundell and Tetens 1997). If sufficient water is not available, it has been recommended that a priority of allocation be established. A number of prioritization schemes have been proposed. One suggestion includes a priority list of human needs, vested water rights, and minimum instream flow for environmental and ecological protection and water quality purposes, followed by allocations that balance demands that provide maximum economic benefit with demands that have a direct influence on quality of life issues but cannot meet the economic threshold (Sterner 1994). The Delaware River Basin Commission (DRBC 2003) and the Susquehanna River Basin Commission (SRBC 1970) provide priority for (1) domestic uses (drinking and bathing), (2) public safety, and (3) the minimum amounts necessary for preservation of livestock and property; thereafter they attempt to balance reductions in water use to maximize productivity and preservation of employment and preserve essential instream flows for water quality and biological conditions. The state of Pennsylvania prioritizes water use under drought conditions with domestic use (including drinking, bathing, cooking, laundry, livestock watering, and other uses necessary for life and health), which is given priority with no regard for the amount of water left in the stream or lake after these uses are satisfied. The next priority is the public's right to navigate, followed by all nondomestic water uses. Upstream nondomestic uses, including irrigation, manufacturing, and power production, can be slowed if there is insufficient water to meet downstream domestic or navigational needs (Abdalla et al. 2007).

It can be argued, however, that placing domestic uses (drinking and bathing) or human needs as top priority does not match the reality of a rapidly urbanizing society. Most water for domestic uses is provided through a municipal system, and it is difficult to segregate what portion is for domestic use and what portion is for other uses, such as commercial. Once the law fixes domestic use as the top priority, authorities must then publish detailed requirements of what is and what is not allowed. Municipalities managing water use during drought emergencies may limit or ban lawn watering, car washing, and golf course irrigation, but enforcement is difficult at best. Moreover, it is difficult for political and factual reasons to curtail commercial or retail use. Separating water that is used within a shopping mall among the grocery store (food as a human need), the drug store (medicines as a human need), and the clothing store, for instance, is a challenge (Draper 2006).

3.3 ASSESSMENT OF WATER RESOURCES

Establishing equitable principles and standards for managing the shared use of water between governments requires detailed knowledge of the quantity and quality of the shared water resources and a comprehensive understanding of how the use of the water resources by individual parties will affect others (Draper 1997; Kaya 1998). Even more importantly, a mechanism to predict accurately future sources and demands is essential (GJCWPSC 2002). Such detailed knowledge is rarely available without extensive research and funding, even in the United States or other developed countries. If an effective water sharing agreement is really intended, the parties need to devote the human and financial resources necessary to develop a broad understanding of the extent and limits of the water sources available to be shared (U.S. Army Corps of Engineers 1995; Kundell and Tetens 1997).

The ultimate goal of the assessment of water sources is to determine whether existing or "reasonably anticipated" water sources are sufficient to meet the expected future demands (Ryan et al. 1998). Both natural factors and human effects may change the amount of water available. Factors that affect water availability include "climatic variability and change, population growth that reduces per capita water availability, contamination that reduces usable water supplies, physical overuse of a stock, such as underground water overdraft, and technical factors" (Dellapenna 1997a; Gleick 1998). The negative effect of human activities on water sources has been noted (Shiklomanov 1993):

> Despite the ability of stream flow to renew and self-purify, in recent decades the intensive development of industry and agriculture throughout the world, population growth, the opening of new territories, the sharp increase in water withdrawals on all continents (except most recently in the United States and parts of Europe), and the transformation of the earth's natural cover have begun to exert a significant impact on the natural fluctuations of the stream flow and the state of fresh water resources.

3.3.1 Data Collection and Uncertainty. Accurate and reliable data and information are key requirements in effective water planning and management (Dellapenna 1997b; Draper 1997; GJCWPSC 2002; Vörösmarty 2002; ASCE 2004a). Given the complexities of water resource decisions, collection efforts should focus on collecting the data and information needed to support those decisions by providing input for computer models and geographic information systems used in making those decisions. Unfortunately, in many countries collection of hydrographic information has deteriorated (Vörösmarty 2002).

Accessibility to the data and information is equally important. Numerous national, state, and local agencies collect water-related data for various purposes and manage those data using incompatible formats and systems. Many records are maintained in paper files; on microform, microfiche, or microfilm; or on individual desktop computers. Demand for public access to data has skyrocketed as use of the Internet has become more commonplace. Although access to water-related data and information has improved significantly, limitations still remain. Successful development and implementation of a comprehensive, integrated data and information management system for the shared water resource depend on the strong commitment of participating agencies and strong executive and legislative leadership (Georgia JSC 2002).

Predicting the yield of shared water sources based on the historical record can result, however, in significant future problems. Future overallocation caused by an inaccurate historical record is illustrated by the 1922 Colorado River water allocation that was based on the 30 years of hydrological record available at that time. Later research, based on an analysis of tree rings, determined that the amounts of water used in the allocation were abnormally high, resulting in overallocation of the water source (Stockton and Jacoby 1976; Murakami 1995).

The issue of climate change has surfaced as a potential impediment to effective long-range policies and management of water resources (Nicholls 2000; Draper and Kundell 2007).

The challenge for the water manager is how a change in climate might dramatically alter how water as a scarce resource is allocated. Nowhere is this challenge greater than the allocation of shared waters between or among politically independent governments (Draper and Kundell 2007).

Sustainable water yields may or may not be reduced in the long-term average, but they are almost certainly less reliable in the short term (Sophocleous 1998; Draper and Kundell 2007). A rapidly growing body of evidence suggests that we are entering a somewhat warmer and definitely more variable world. The *ASCE Policy Statement* (ASCE 2010a), in Section 2.1.4, recognizes the problems associated with water resource planning and climate change.

Accurate and reliable data and information are key requirements in effective water planning and management (Dellapenna 1997b; Draper 1997, 2006; Vörösmarty 2002; ASCE 2004a; Draper and Kundell 2007). The effects of climate change extend across political boundaries. Therefore, the parties to a water sharing agreement should recognize that the water sharing agreement may need to be adapted, even significantly adapted or replaced, as more accurate and precise predictability becomes available on those changes (Frederick and Schwarz 1999). Climate change affects water sharing in many of the same ways it affects the management of water within a specific legal jurisdiction (Draper and Kundell 2007).

3.3.2 Water Supply. Surface and underground water resources are often managed as separate and distinct sources (Eckstein 1998; Draper 2002d). However, such individualized management does not provide a comprehensive understanding of either the water sources or their use. In reality, however, the three components of water—surface, underground, and atmospheric—are part of one hydraulic system (Glennon 2002). Before implementing transboundary water use allocation or transfer, participants should conduct a comprehensive inventory of the shared water resource. Such an assessment requires a sound historical record of the natural water cycle, to the extent it is available, as altered by human intervention (Vörösmarty 2002).

In most situations, each component of water, whether it be surface, underground, or atmospheric water, has historically been assessed and managed as separate and distinct sources from the others. However, such individualized assessments do not provide an integrated understanding of either the water sources or their use (German Foundation for International Development 1998; Illinois State Water Survey 2001). The flows of surface waters and underground water are often closely linked and need to be studied and managed conjunctively (Winter et al. 1998; Illinois State Water Survey 2001; Glennon 2002). This interrelationship is significant because the interface between surface water and underground water is often direct, with streams sometimes recharging the aquifer and sometimes gaining water from the aquifer. It is estimated that the contribution of underground water to small- and medium-sized streams is between 40% and 50% (Fort 1998). Clearly, then, the available waters should be managed conjunctively (Hayton and Utton 1989).

Effective water sharing agreements need to have their provisions based on accurate and reliable data and information. The goal of the parties should be to collect and make available data and information that enable them to make timely, accurate, and complete decisions in the development of the water sharing agreement. Successful data development depends on a strong commitment of participating parties (Water Plan Advisory Committee 2002).

A comprehensive assessment requires that three conditions be met. First, the parties should define the critical hydrologic conditions that frame the water sharing agreement. Second, adequate and reliable data concerning the water sources and demands should be available. Third, a shared water resource database should be developed and maintained (GJCWPSC 2002). Certain parameters define the framework of the agreement. The parties need to specify the extreme hydrologic events that the agreement will cover. This requirement suggests that the parties should clearly establish the quantitative measures of "normal" or "average" hydrologic conditions and establish the levels when special management for drought (dry) or flood (wet) conditions arise. The parties should also define the levels of water quality degradation they are willing to accept as a result of meeting the demands for the water.

A number of factors influence the availability of water within the shared resource. Water in its different forms moves between the atmosphere, through precipitation and condensation, to rivers, streams, lakes, and reservoirs, to vegetation and soils, to aquifers or oceans, returning to the atmosphere through evaporation. Any efficient management of one form cannot be attained without linkages to the other, preferably dealing with both underground water and surface waters as a unitary whole (Haddadin 2000). The timing and amount of available water to be shared should be based on a careful analysis of this linkage (Illinois State Water Survey 2001). The timing and amount of water available for shared use is also linked to human intervention, both by the addition of supply augmentation facilities and programs, such as reservoirs and aquifer storage and recovery systems, and by the loss of reusable water consumed by individual uses and unavailable to others (Draper 2002e).

The goal of collecting data and information on precipitation amounts, recharge rates, and other data should be to determine the reliable available quantities of water to meet projected needs (Illinois State Water Survey 2001). The word *reliable* is emphasized because, in the absence of storage devices, the demand for water by the individual water user sectors should be reliable and consistent for extended periods of time. It is not sufficient that the statistical average yield of the water source equal the projected demand. The yield of the shared water source should be able to meet the demand under most climatological, meteorological, and hydrological conditions.

Water utilization by one party can have serious effects on the quantity and/or quality of the water available for use by other parties. A significant source of controversy develops when one party undertakes the construction and operation of water supply reservoirs and hydropower facilities that may dramatically lower the flow or decrease the quality in the shared water resource. Although this reduction may be limited to the initial start-up period and may be limited to a period of several years, severe economic and social effects may occur to other parties. Return flows should be coordinated to ensure that downstream users are not significantly affected in an adverse manner. In the case of water supply reservoirs, this coordination may become critical during periods of drought. In the case of hydropower dams, especially those that provide power principally during peak electrical demand, usually in the morning and early evening, the timing of release may be critical. Other parties may also be affected by significant changes resulting from interbasin transfers or increased water consumption (Draper 2002b, 2004). Water supply reservoirs and hydropower facilities also often have significant biological effects that may or may not be related to total water supply or water quality.

3.3.3 Underground and Atmospheric Water Supply. Conjunctive analysis, considering surface and underground water supplies together, coordinates use and storage to increase the total water yield over time, thereby increasing the reliability of the water supply. Integrated analysis, considering water supplies and water quality together, coordinates the adequacy of the water supplies for those water uses that demand a certain level of quality (U.S. Advisory Commission on Intergovernmental Relations 1991).

Failure to deal specifically with the issue of underground water withdrawal has led to litigation in the context of existing interstate compacts in the United States. It is therefore recommended that underground and atmospheric water be included within the scope of the agreement. As technology advances, use of and control of atmospheric water may also become more commonplace (ASCE 2004b; Bomar 2006), and consideration should be given to dealing with the potential for such use and control before it becomes established and results in unexpected interference with the provisions made for underground and surface water (Draper 2002b). Shared aquifers require careful evaluation because there is a general lack of detailed data (ASCE 2001). The effects of the use of transboundary aquifers are subtle and widely spread geographically and often delayed in time (Puri and Arnold 2002).

Withdrawing water from shallow aquifers for public and domestic water supply, irrigation, and industrial uses is widespread. Withdrawing water from shallow aquifers can diminish the available surface water supply by capturing some of the underground water flow that otherwise would have discharged to surface water or by inducing flow from the surface water into the surrounding aquifer (Winter et al. 1998). Conversely, underground water resources can be used as an effective tool to mitigate the effects of drought. Because underground water aquifers are less affected by short-term variability in climate conditions than are surface water rivers and streams, they can be artificially recharged during high water periods and utilized during low water periods (ASCE 2001; Keyes 2006). However, increased use of underground water during droughts may have unanticipated effects. If an increase in underground water withdrawal is continued after the drought, a permanent and unanticipated change in the level of underground water may develop (Alley et al. 1999).

Provisions in water sharing agreements of transboundary underground water aquifers should distinguish between two types of aquifers. The first type includes aquifers that are hydrologically connected to surface waters, in which a conjunctive approach may be used directly. For this type, typified by the Floridian aquifer shared by many southeastern U.S. states where rates of aquifer recharge may be as high as 20 in. per year (Miller 1990), the transboundary concern is twofold. The first concern is the transboundary effect of excessive water withdrawal by the upgradient party and its effect on the water available to the downgradient party. Second is the possible reduction in the recharge of the aquifer by developmental activities within the recharge areas by one of the parties, usually the upgradient party. Clearly, knowledge of the aquifer recharge areas is extremely important (Dellapenna 1997b; ASCE 2009c).

The second type of aquifer includes those with no direct surface water connection (so-called "ancient" or "fossil" aquifers). In this case, typified by the Ogallala aquifer shared by many high plains U.S. states, recharge is negligible because recharge of these aquifers is normally quite slow, ranging in the tens of thousands of years (Cummings et al. 2001a). An estimate of the magnitude and extent of the water available for use in this case cannot be made without considering the risk to future users of exhaustion of the water source, an issue of sustainability (Donkers 1997). A problem that is even more significant in the sharing of such an aquifer involves the controlled depletion of storage. Increased use by one party may result in, or exacerbate, existing "mining," where annual pumping exceeds natural recharge, effectively depleting the source (Cummings et al. 1996). In some areas of the Ogallala, more than 50% of the predevelopment saturated zone has been dewatered (Alley et al. 1999). The parties, however, can mitigate this issue of increased underground water "mining" by including in the water sharing agreement a requirement for more water-efficient irrigation systems, incentives for the introduction of crops that use less water, and the irrigation of fewer acres by underground water in some areas because of the declining water levels (Solley 1997).

3.3.4 Quality of Water Supply. The issue of water quality is also pertinent to an analysis of water resources. In many cases, providing adequate supplies of water for a particular demand is not sufficient. The need may well be adequate supplies of water at an acceptable level of quality (Draper 2001). The parties should recognize a level of water quality within the shared waters that meets a number of needs. At a minimum, the shared water should maintain ecosystem integrity and preserve aquatic ecosystems while sustaining economic growth and prosperity (ASCE 2009c). Integration of water quality and quantity programs is one aspect of, and is inseparable from, comprehensive, coordinated, multipurpose, basinwide water and related resource planning and management (Goldfarb 1993).

Surface and underground water may be degraded by a variety of factors. Major problems affecting the quality of these water resources arise, for instance, from inadequate domestic sewage treatment, inadequate controls on the discharge of industrial waste and effluent, the diversion of waters resulting in insufficient water to assimilate waste, the loss and destruction of catchment areas, the improper locating of industrial plants, deforestation, and poor agricultural practices that cause leaching of nutrients and pesticides (Draper 2002b). In coastal regions, saltwater intrusion can nullify the underground water supply as a viable water source to meet the needs of agricultural irrigation and municipal water supply (State of Georgia 1997).

Poor-quality water imposes risks that the parties should recognize as a common threat. There is, for instance, the potential health risk to the population that uses the water for domestic purposes. If the available water does not meet the minimum standards for certain industrial purposes, there is the risk that economic growth may be impaired. Also, there is the risk that quality degradation may have a severe effect on long-term sustainability. Therefore, integration of water quality and quantity is essential (Ahlander 1994). The *Convention on the Law of the Non-Navigational Uses of International Watercourses* (United Nations 1997) establishes the criterion that "watercourse States shall, individually and, where appropriate, jointly, protect and preserve the ecosystems of international watercourses."

Within the United States, the Clean Water Act (1987) provides a baseline for surface water quality programs, but the same cannot be said of underground water. Treaties between Canada and the United States and between Mexico and the United States have established water quality norms in North America (Killgore and Eaton 1995). Similar norms exist for the European Union and the Amazon Basin (Draper 2002b). However, in many other regions of the world, water quality criteria are not prescribed (Draper 2006).

3.3.5 Water Supply Augmentation. Water augmentation projects and programs are techniques to change the timing and/or quantity of water available for use. Examples of effective water augmentation techniques include surface water reservoirs, aquifer storage and recovery, precipitation enhancement, interbasin transfer, evaporation suppression, desalination, and wastewater reclamation and reuse (Keyes 1977).

Reservoirs can significantly augment water supply in areas or periods of water scarcity. Therefore, the parties may need to assess both existing dams and potential construction of new dams. With regard to existing reservoirs, almost half the world's large reservoirs have been built exclusively or primarily for irrigation. An estimated 30% to 40% of the 268 million hectares (662 million acres) of irrigated lands worldwide rely on dams. Dams are estimated to contribute from 12% to 16% of world food production (World Commission on Dams 2000). However, in addition to growing demand for municipal water supply, a combination of social pressures is emerging, with the potential to shift the operational priorities of reservoirs significantly in the near future between, among others, hydropower and water supply (McMahon and Farmer 2002). However, reallocation of reservoir storage by one party may cause unintended consequences on the quantity, timing, and quality of water available to the other party. One of the specific risks associated with reservoirs is degraded water quality, resulting from increased levels of mercury pollution early in the postconstruction life of the reservoir. Another potential problem that may have a detrimental effect on long-term water supply availability is the tendency of reservoirs to emit greenhouse gases, carbon dioxide, and methane, adding to the projected increase in climate change (Pielou 1998).

Recreation has also developed into a significant competing purpose with other traditional uses, at least for reservoirs in the United States. Both lake and river recreation has been noted as a significant national resource and public benefit, making an important contribution to local, state, and national economies. However, recreation at federal lakes has not heretofore been treated as a high priority, or as an equal with other reservoir uses, even if it is designated as an authorized purpose (National Recreational Lakes Study Commission 1999).

As water scarcity continues to grow and needs change, reallocation of the purposes of reservoirs mounts. Although reallocation often considers changes in purpose from hydropower to municipal use in the United States (McMahon and Farmer 2002), in other areas reallocation from irrigation to industrial use might be appropriate.

Construction of new reservoirs by one party can also significantly affect the quantity, timing, and quality of water available to other parties, especially downstream parties. A large number of aquatic and ecosystem effects exist that should be considered, including the effects of reservoirs and flow modifications on terrestrial ecosystems and biodiversity, the emission of greenhouse gases associated with reservoirs, and the accumulation of pollutants. The effects of altering the natural flood cycle on downstream floodplains and the effects of reservoirs on fisheries in the upstream, reservoir, and downstream areas can be significant. The cumulative effects of a series of reservoirs on a river system should be carefully considered (World Commission on Dams 2000).

Aquifer storage and recovery (ASR) is the storage of water in aquifers during times when water is plentiful and recovery of the water during times when it is needed. Proponents claim that this technique provides a cost-effective solution to many of the world's water management needs, storing water during times of flood or when water is plentiful and quality is good, and recovering it later during emergencies or times of water shortage, or when water quality from the surface water source may be poor. Large water volumes can be stored deep underground, reducing or eliminating the need to construct large and expensive surface reservoirs. In many cases, the storage zones are aquifers that have experienced long-term declines in water levels because of heavy pumping to meet increasing urban and agricultural water needs. Underground water levels can then be restored if adequate water is recharged (Pyne 2008). The primary benefits include significant cost savings from not having to construct expensive surface reservoirs. Potential disadvantages include lowering water quality because of mixing higher and lower quality water sources. From the shared water perspective, a good understanding of the regional hydrogeologic framework is essential to assess the regional costs and benefits of any proposed regional ASR systems (Water Science and Technology Board 2001). The identification and protection of the sources of aquifer recharge is essential (ASCE 2001).

Artificially induced precipitation is another water supply augmentation technique that, if harmfully practiced by one party, may affect the water available to the other party. The moisture suspended in the atmosphere is a natural resource. It has been estimated that only about 2% to 3% of the moisture in the atmosphere falls as natural precipitation (*Colorado Weather Modification Permit Program* 2002). Techniques to increase precipitation artificially have become increasingly more viable (ASCE 2006; Keyes 2006).

For effective water sharing, an understanding of the potential effects of such an inducement both within and outside of the shared resource in question is necessary (ASCE 2004b). A number of economic, environmental and ecological, social, and legal factors determine the acceptability of a cloud seeding program (Keyes 2006). A considerable amount of research has been conducted in the western United States on the potential environmental and ecological effects of cloud seeding projects (Reinking et al. 1995; Keyes 2006). Some of the more common concerns that could be addressed include toxicity of seeding materials (i.e., silver iodide), extension of the snowmelt in higher elevations, and increases in soil erosion. Published results from these projects indicate no significant effect from winter or summer projects. The parties to any agreement should recognize that the effects of artificially induced precipitation might extend beyond the borders of the party boundaries, depending on the scale of the program to augment precipitation. Any party that augments precipitation within the basin might be entitled to full and exclusive use of additional water supplies resulting from such augmentation, if those supplies can be effectively measured, notwithstanding any other standard of allocation set forth in the agreement, as long as the downwind party is satisfied that the augmentation does not affect precipitation elsewhere in the shared water resource (Draper 2002b, 2006).

Interbasin transfer of water, long a fixture in the western United States, is an increasingly significant water augmentation technique in eastern states. In the international arena, Canada has passed legislation widely prohibiting interbasin transfers. Interbasin transfer with the Great Lakes as the basin of origin has been increasingly controversial as the Great Lakes states and provinces seek to define what might or might not be considered acceptable interbasin transfers (Draper 2004).

Interbasin transfer of water involves the withdrawal, diversion, or pumping of surface water from one river basin or the withdrawal of underground water from a point located within or beneath one river basin and release of all or any part of the water into a river basin different from the basin of origin (Draper 2004). The transfer of water to outside the boundaries of a river basin has effects on the water and other resources in the basin of origin and on the receiving basin. Such effects differ from those caused by uses of water within the same basin because the nonconsumed water may not be returned to the stream from which it is taken for further use in that river basin (Georgia JSC 2002). In a favorably configured basin that is well managed, most water supplies can be utilized a number of times as the water is withdrawn, used, and the nonconsumed water returned after treatment to the original water source. Water transferred by interbasin transfer is essentially fully consumed and unavailable for further use by those downstream in the basin of origin.

The parties should consider three aspects of interbasin transfer. The first is interbasin transfer to another basin outside the boundaries of the shared water resource. Such a transfer by one party diminishes the water available to some downstream parties by the elimination of returned flows into the shared basin of origin. The second aspect is transfer of water from a watershed within the shared water basin to another watershed within the shared water basin. Such a transfer definitely affects the water available to the downstream users in the watershed of origin. The third aspect is a transfer of water into the shared water basin. In this case, an importation of water from outside the basin may be excluded from the provisions set forth elsewhere in the agreement, and the party importing such water should have the right to full and complete use and consumption of such imported water, as determined by the party's internal laws (Draper 2002b, 2004). However, water quality issues may be significant. It has been claimed that interbasin transfers are not a "supply augmentation" but rather a reallocation that can be characterized as changes in use (Hall and Meral 2002).

Other water augmentation techniques include water reuse and desalination of seawater and brackish underground water (El-Naser 2001; Pankratz 2004). Water reuse involves water reclamation and the treatment of wastewater to make it reusable for other applications, such as irrigation, industrial cooling, washing, and specific environmental and ecological purposes (Basin Advisory Council 2002). In some cases, direct water reuse, for cultural or perceived public health reasons, cannot be used for municipal supply. However, by using reclaimed water for irrigated agriculture or industry, it can reduce the demands on present and future water supplies (El-Naser 2001). The primary concern for the parties to a water sharing agreement is the scale and scope of water reuse by one of the parties, especially the upstream party. Because any reuse results in some percentage of the water being consumed and unavailable for further use downstream, the potential exists that this water augmentation technique, while augmenting the upstream party's available water, may reduce the water available to downstream users if such reuse is at a large enough scale. Desalination is an expensive augmentation technique, at least under current technology. In Toronto, the cost of traditional surface water withdrawal is approximately $0.87/m^3$, whereas desalination costs are approximately $8.00/m^3$ (Zimmerman 2002). A primary reason for this cost differential has been reported as the concentrate disposal techniques used (Hightower 2004; Keyes 2007). This difference suggests that significant engineering advances must be made before the technique reaches its full potential. As desalination technology advances (Wiseburgh 2002), the technique will become a viable option, especially in those areas that experience increasing costs for water (Pankratz 2004).

3.4 ASSESSMENT OF WATER DEMANDS

The need for water reaches into all aspects of human life. The existence of life requires adequate supplies of clean water. Estimates of the human need for water to sustain a minimum acceptable quality of life range from 50 L/day to 100 L/day (Falkenmark and Lindh 1974; Gleick 1996). Food production for the world's expanding population depends largely on irrigation. Agriculture demand is estimated to account for 60–80% of water consumption worldwide (Postel 1993; Rothfeder 2001; Barlow and Clarke 2002). It requires approximately 3.0 liters (0.79 gallons) of water a day to grow the food for one person's nutritious but low-meat diet (de Villiers 2001). Beyond survival, adequate supplies of clean water are required for economic prosperity (Draper 2001). Water needs in the economic sector can be enormous. It requires almost 543,000 L (143,000 gal.) to produce one ton of paper (Gleick 1993) and more than 150,000 L (40,000 gal.) to produce an automobile (Barlow and Clarke 2002; Gleick 1993). Global water withdrawal and use by industry has been estimated at 22% of withdrawals (World Bank 2001). Water is important for other aspects of the quality of life for the human species, as well. The demands for good public health, recreation, and a sustainable environment depend largely on adequate amounts of instream water being available (Draper 2001). Forty percent of the world's population and more than 90% of the people in the United States depend on waters shared with other states or countries to fulfill their water needs and demands (International Network of Basin Organizations 2002; Sea River 2002; Draper 2003).

3.5 SUMMARY

Effective water sharing is based on the ability to allocate reliable, or consistent, supplies of source water to fulfill the existing and future demands of the parties. Therefore, the parties should identify and document the type and geographical extent of the water resources to be subject to the agreement. This description should allow all appropriate and necessary parties within the geographic and/or political region to be included in the agreement and negotiation process. It also allows an analysis of the different water laws and policies of the parties to expose conflicts that may affect water sharing.

Having defined the hydrogeographic boundaries of the shared water resources, the parties may proceed to develop an accurate estimation of water supplies of sufficient quality that may be reliably available to meet the geographic and temporal demands within the shared water resources. Identification of existing demands for the shared waters is essential to enable estimates of future water demands. As with the assessment of water sources, it is important that the parties devote the human and financial resources to determine existing as well as future demands within the context of the agreement (Draper 2006).

Chapter 4
ADMINISTRATION OF AGREEMENTS

The importance of the administration and institutional provisions of the agreement cannot be overemphasized. The effectiveness of any agreement or law depends on how effectively it is administered and how rigorously its provisions are enforced. These institutional provisions involve far more than just hydrologic or engineering management of the water resources in question. They also involve establishing procedures to manage complicated coordination among a number of private and public institutions, addressing critical social issues, and responding to complex environmental and sustainability needs. The institutional provisions should provide for effective mechanisms for cooperation, coordination, and communication. These provisions should also establish a structure that resolves conflicts in a timely and cost-effective manner. Although no single legal model or approach to cooperation is appropriate for all or even most situations (*Berlin Recommendations for the International Roundtable on Transboundary Water Management* 1998), certain standards are appropriate for any water sharing institution.

4.1 THE CHALLENGE

Proper administration and management of shared water resources requires reconciling a variety of issues. Political issues are an important consideration. State and local government water management agencies may disagree among themselves on specific provisions because the agencies respond to different constituencies and interest groups that often diverge with respect to water needs. Not only may political issues divide agencies, constituencies, and interest groups horizontally (e.g., state agencies for agriculture, environmental and ecological protection, and industry and trade may disagree), but vertically as well (i.e., national, state, and local agencies may disagree). The administrative and institutional provisions should also address geographical issues that refer to the scale of the agreement: river basin, national subdivisions, national, regional, or global. There is little disagreement among experts that the appropriate administrative entity for international water sharing is the river basin (Gooch et al. 2003). Environmental and ecological issues should address integrated analysis of water, as well as biological and wildlife issues. Functional issues focus on consideration of diverse water use applications, such as urban water supply, wastewater management, navigation, and irrigation (Kliot and Shmueli 2001).

Because basins throughout the world exhibit a wide range of existing water resource issues and management institutions, no single type of coordinating institution is appropriate for all (Eaux partagées 2002). Without question, the specific water resource problem at hand should drive the institutional design. Water resource problems are customarily defined through the political process, but once the water resource problem has been defined, certain principles and practices guide institutional design to ensure that the solution to the problem is effective and permanent. A critical need is the active support and long-term commitment on the part of top-level political leaders and representatives involved in the establishment of such an institution (Kliot and Shmueli 2001).

4.2 THE HISTORICAL RECORD OF INTERNATIONAL WATER SHARING INSTITUTIONS

Typically in the international context, water sharing agreements empower a joint agency that consists of national sections, each of which obtains funding from within its own government to cover the internal office, staff, printing, travel, and other administrative expenses (Hayton and Utton 1989; Dellapenna 1994; Killgore and Eaton 1995). Two different legal model institutions are highlighted.

4.2.1 Coordination and Cooperation Agreements. The *Treaty between the United States and Great Britain Relating to Boundary Waters* (1909), between Canada and the United States, has been offered as a good model for an international water sharing agreement (Nile Management Roundtable 2001). This model provides excellent "cooperation and coordination" capabilities that provide an umbrella under which agreements for specific projects may be created. The treaty manages transboundary water sharing issues across and along the longest undefended boundary between two countries in the world (Draper 2002b). The treaty established the International Joint Commission (IJC), whose mission is to prevent or resolve disputes by resolving issues between the two countries over boundary and transboundary waters and making recommendations about their management. Its authority is limited to boundary waters themselves, not to tributaries or waters lying wholly within one of the nations. However, the treaty provides a forum for subsidiary agreements that relate to a specific geographical area or specific problem. It sets forth a process and an administering body for future decision making with regard to issues of the defined transboundary water resources. It has been used over the past nine decades to successfully resolve more than 95% of the issues addressed under its jurisdiction. Key organizational features include six commissioners (three from each country) who make decisions by consensus, supported by small staffs in both countries and by boards and task forces composed of equal members from each country serving in their personal and professional capacity rather than as representatives of their countries or organizations and operating based on joint fact finding and consensus (Killgore and Eaton 1995).

4.2.2 Comprehensive Agreements. At the other extreme is the *Agreement on the Cooperation for the Sustainable Development of the Mekong River Basin* (1995), which evolved from the 1957 adoption of the *Statute of the Committee for the Coordination of Investigations of the Lower Mekong Basin* by the four lower riparian states of Cambodia, Laos, South Vietnam, and Thailand (Radosevich and Olson 1999). The agreement's original purpose was cooperation and coordination with the intent of enlarging its mandate to encompass the development of tributaries and possibly mainstream projects. Because of the political instabilities over the ensuing decades, it was not until 1995 that the *Agreement on the Cooperation for the Sustainable Development of the Mekong River Basin* was signed. The agreement created the Mekong River Commission (MRC), composed of three permanent bodies: the council, the joint committee, and the secretariat (Radosevich and Olson 1999). The council membership includes one member from each riparian state at the ministerial and cabinet level who is authorized to make policy decisions on behalf of his or her government. The joint committee membership includes one member from each riparian state who performs technical decision making for the commission. It oversees implementation of the policies and decisions of the council and supervises the secretariat. Subcommittees of the joint committee focus on three critical areas: the basin development plan, water allocation, and water quality. The secretariat provides technical and administrative services to the council and joint committee. A nonriparian CEO recommended by the joint committee and approved by the council directs the secretariat. Multilateral and bilateral donor agencies, primarily the United Nations Development Programme and the World Bank, provide grant funding for almost all of the MRC program and project costs; administrative costs are apportioned among member states (Radosevich and Olson 1999; Draper 2002b, 2007a).

4.2.3 Analysis. International water sharing agreements display large differences in administration because of different purposes, political influences, hydrology, and economic conditions, among other factors. However, the successful agreement between Israel and Jordan to share the Jordan River is due, in large part, to the establishment of a joint agency, the Joint Water Committee, which provides a forum for ongoing communication between the two parties (Dellapenna 1994, 1997a; McCaffrey 1997). That being said, though, simply providing a forum for ongoing communication by itself does not result in optimum water sharing (Albert 2000). A survey of the water sharing agreements in force in Africa suggests that three other principles, in addition to providing for a joint agency, are important for effective agreement to share international waters. First, agreements should be multipurpose rather than single purpose. Second, there should be shared financial responsibility and adequate funding for administration and enforcement. Finally, those agreements with joint research initiatives within the agreement, with the implied financial responsibility, seem to be more successful. This same study concluded that the principal problems to be faced when developing a joint agency for water sharing included

(1) an overpoliticization of the institutions;
(2) the tendency for proliferation of institutions that lead to duplications and increased financial requirements;
(3) the distortion of economic needs with perceived political needs;
(4) overcentralization of institutions and the stifling of public initiatives; and
(5) financial mismanagement (Okidi 1997).

A different analysis of nine international water sharing agreements has noted, however, that a formal multipurpose joint development agency by itself does not ensure the effectiveness of a water sharing agreement. Other organizational principles should be applied to the formation of the administrative mechanism developed in the agreement. The Ganges has been noted as exemplifying the shortcomings of any unwillingness to cooperate in the whole basin (Kliot et al. 1998).

A recent study in Europe has reported similar criteria for effective water sharing agreements. This study listed five minimum requirements for the administration of a water sharing agreement:

(1) The agreement should provide for an assembly (similar to a commission or council), composed of members appointed by the states and associated authorities, which elects its president, who may alternate among the members of each state.
(2) A board of trustees to oversee the administration and enforcement of the agreement on a regular basis should be appointed.
(3) A dispute resolution mechanism should be included.
(4) The institution should include the formation of various working groups directed at the principal purposes of the agreement, and the groups should meet periodically in places set out in the agreement.
(5) A permanent secretariat (commission or council) is essential. The agreement should set forth the role of each component of the structure as well as the rules for decision making, such as unanimity, majority, or qualified majority (Eaux partagées 2002).

These criteria find support in a previous study of the Mekong River Agreement (Radosevich and Olson 1999).

4.3 THE HISTORICAL RECORD OF U.S. WATER SHARING INSTITUTIONS

Within the United States, four institutional models have been used for interstate water sharing. The predominant administrative mechanism for action or coordination of interstate rivers and streams has been the interstate compact, a water sharing agreement between states without active federal participation, or federal participation at the technical level only (Grant 2001). A second mechanism is the federal–interstate compact, an agreement between states that includes active federal participation. A third form, federal administration of shared waters without specific state involvement, has been used in two instances (U.S. Army Corps of Engineers 1999). Finally, a fourth mechanism, the Title II river basin planning commission, was introduced beginning in 1965, but federal funding ended in 1981 (Wade et al. 1994). The model is presently in disfavor because it failed to achieve their objectives in the past, primarily because of reluctance by the states and federal agencies to provide sufficient authority to the river basin commissions to accomplish their missions (Eisel and Aiken 1997).

4.3.1 Interstate Compacts. Interstate compacts administer the shared waters according to agreement among the riparian states. These water sharing mechanisms are contracts among the states themselves. Unless the agreement is so inconsequential that it does not encroach upon federal authority, the compact should have the consent of the U.S. Congress (*Virginia v. Tennessee* 1893; Sundeen and Runyon 1998). The federal government, though not a signatory, may assist the states in the development and operation of the compact. The decision making is normally made with a commission whose members are usually the state governors or their representatives; each state has one vote. The rules of decision typically require unanimity. Although the federal government does not have voting powers in the operational decisions of the compact, it has veto power where federal interests exist. An interstate compact is a contract enforceable in the U.S. Supreme Court by specific performance and by damage awards (DuMars 1990; McCormick 1994a, b).

Although an interstate compact may serve a "wide range of functions (U.S. Army Corps of Engineers 1999), usually the purpose of the compact is specific because the formation of the compact occurs only when the need exists for an urgent solution to a specific problem. Normally, interstate compacts do not create mechanisms for the actual management of water sources or for allocation of water among particular users. Rather, the agreement allocates the shared water between and among the states that then allocate water to specific uses and users. The effectiveness of interstate compacts, at least with regard to drought, is "predicated on having well-publicized plans and rules for planning purposes" (Walker et al. 1991). Although there are exceptions, the determination of whether an interstate compact formed under water scarcity conditions is successful seems to turn on whether the compact provides an active mechanism for compact administration and whether a dispute resolution mechanism exists (Draper 2006, 2007a).

Among interstate compacts that have no active administration, the 1943 *Republican River Compact* is a good example. The compact originally provided only that "it shall be the duty of the three states to administer the compact." Such an absence of active administration proved unworkable. The states subsequently found it necessary to create a Republican River Compact Administration to manage compact issues (McCormick 1994b). Because no dispute resolution mechanism exists within the compact, it would appear that absent an informal agreement among the parties, the only recourse is the U.S. Supreme Court. The most significant issue relates to underground water pumping and its effect on river flows, but the issue was not explicitly addressed in the compact (McCormick 1994b). A similar situation of no active compact administration exists with the ongoing dispute among Alabama, Florida, and Georgia. A compact has been signed and ratified by each state pending an agreement on an allocation formula for the Apalachicola–Chattahoochee–Flint rivers. Although the compact creates a commission of the three governors, neither an active administrative mechanism nor a dispute resolution mechanism exists. The 1923 *South Platte River Compact* likewise established no administrative or dispute resolution mechanism and has the potential for protracted litigation since irrigation demand exceeds supply (Draper 2007a).

The 1955 *Bear River Compact* is similar, having no active administration or dispute resolution mechanism, although the fact that unanimity is not the rule in decision making minimizes disputes. Also, this compact explicitly contained a provision requiring review and possible revision at no less than 20-year intervals. The success of this compact can be traced to these compact characteristics and the fact that the compact is limited to division of rights to flowing water and storage among users (Draper 2006, 2007a).

4.3.2 Federal–Interstate Compacts. Federal–interstate compacts are similar to interstate compacts with the exception that the water sharing agreement is between the riparian states themselves and the federal government, in equal partnership. The exercise of federal powers is subject to the terms and conditions of the compact and the authority of any compact-created commission or agency (Wade et al. 1994; Dellapenna 2001b). Two instances of the federal–interstate compact exist: the *Delaware River Basin Compact* (1961) and the *Susquehanna River Basin Compact* (1970). These compacts provide for multijurisdictional commissions to manage water sources with authority to allocate water among particular users during times of shortage (Dellapenna 2001b). In granting its consent to these compacts, Congress included reservations to prevent impairment of future exercise of federal power and avoid future limitations on congressional power to pass laws inconsistent with the provisions of either compact (U.S. Army Corps of Engineers 1999).

The *Delaware River Basin Compact* (1961) involves the states of Delaware, New Jersey, New York, and Pennsylvania, and the *Susquehanna River Basin Compact* (1970) involves New York, Maryland, and Pennsylvania. Each compact creates a regional entity composed of the governor of each of the involved states and one member appointed by the president of the United States. Each member is entitled to one vote, and no action may be taken unless a majority is present and votes in favor of the proposal. The jurisdiction of the commissions is limited to the geographical boundaries of the basins. The compacts provide the commissions with regulatory authority and responsibilities relating to water supply, pollution control, flood protection, watershed management, recreation, hydroelectric power, regulation of withdrawals and diversions, intergovernmental relations, capital financing, and planning and budgeting. The commission can regulate individual water users within the basins if necessary to accomplish their responsibilities. The compacts provide the commissions with power to allocate the waters of the basins among the signatory states and to impose conditions, obligations, and release requirements (Wade et al. 1994; Dellapenna 2001a). A unique feature is the compacts' power to allocate the waters of their respective basins among the signatory states in accordance with the U.S. Supreme Court's doctrine of equitable apportionment (Muys 2001).

Analysis of the federal–interstate legal model suggests that it can be effective in maximizing the benefits of the waters of the basin when all parties are in agreement. According to some commentators, however, several issues have limited the full potential of the Delaware River Basin Commission (DRBC). These issues relate primarily to the lack of a sufficiently strong commitment by the federal representative and the federal agencies to the compact objectives, as well as noncompliance by the federal government regarding its financial commitment (Muys 2001). Because no DRBC provision provides the commission with alternate means to generate sufficient operating funds, this lack of federal funding has a significant impact on the effectiveness of the compact (Bush 2004).

This model does require the individual parties to cede a significant amount of their authority and control over internal waters to decision makers who are not directly accountable to the state governments, possibly making the development of individual economic and environmental and ecological plans and programs difficult.

4.3.3 Federal Administration of Shared Waters. Federal administration of river basins provides management of shared waters without state participation. Two instances of this model have been applied in the United States. These instances are the extensive federal management of the Tennessee River and of the Colorado River. The Tennessee Valley Authority (TVA) was created as a federal regional agency to operate a basinwide system of dams and reservoirs to (1) control flooding along the Tennessee River and its major tributaries, (2) maintain navigation depths on rivers, and (3) generate power (TVA 1933). Its mandate has since been broadened to include pollution control and provision of recreational opportunities (Freeman and Lesesne 1980). The TVA is governed by a board of directors, composed of three members appointed to overlapping nine-year terms by the president of the United States, with confirmation by the Senate. The board of directors appoints a general manager, who operates as the administrative head of the corporate agency, with general policy direction supplied by the board (TVA 1933). The TVA is structured hierarchically, with the board of directors over the office of the general manager, which in turn oversees seven offices: general counsel, agricultural and chemical development, power, engineering design and construction, natural resources, community development, and management services. The powers given to the TVA are considerable, in keeping with their mandate for planning and operations of navigation, flood control, and hydropower facilities, as well as responsibilities for water quality and multipurpose reservoir operations (Freeman and Lesesne 1980). Essentially, the TVA provides comprehensive water management of the entire river basin without state involvement (other than political) (TVA 1933).

The second form of exclusively federal management of interstate water resources is the single federal administrator, used for allocation of the waters of the Colorado River Basin. The U.S. Secretary of the Interior on a yearly basis allocates the waters of the Colorado River to the riparian states, as stipulated in the *Colorado River Basin Project Act* of 1956 and the *Colorado Project Act* of 1968 as well as the *Colorado River Compact* (1928; Anderson 2002). The U.S. Bureau of Reclamation (USBR) manages the allocation program (Carter and Morehouse 2001). The USBR operates eight dams on the Lower Colorado River and bases the allocation on a reservoir routing analysis of the discharge from two federal projects, the Glen Canyon Dam and the Hoover Dam. The allocation to the states is based on (1) the amount of water already in storage; (2) the anticipated releases for various uses, such as hydroelectric power generation; and (3) the need to equalize storage behind the two dams (Carter and Morehouse 2001). The seven Colorado River Basin states and other stakeholders are consulted (Anderson 2002). The Lower Colorado Basin (Arizona, California, and Nevada) is managed according to strict adherence to inflexible rules, primarily because of continuing tensions among the three Lower Basin States and because portions of the river below Lake Mead are relatively heavily populated (Pagano et al. 1999). Within the Upper Colorado Basin (Colorado, New Mexico, Utah, and Wyoming), however, the USBR operates the reservoirs under an adaptive management approach that emphasizes monitoring the effects of allocations on the environment, followed by adjustments to reflect the findings of such monitoring.

Analysis of this single federal administrator model suggests that it does not lead to efficient and equitable resource allocation because it does not provide for appropriate representation and participation of all major affected interests. It does not facilitate communication and bargaining among states to adopt and implement operating rules that resolve conflict (Lord et al. 1995).

4.3.4 Title II River Basin Planning Commissions. Title II river basin planning commissions provide passive management of shared waters under authority of Congress. River basin planning commissions were established under the authority of Title II of the *Water Resources Planning Act of 1965* (subsequently repealed in 1981). Although no longer authorized, lessons may be learned from a consideration of the strengths and weaknesses of the Title II model.

The 1965 act established the Water Resources Council, composed of the U.S. Secretary of the Interior; the U.S. Secretary of Agriculture; the U.S. Secretary of the Army; the U.S. Secretary of Health, Education, and Welfare; and the chair of the Federal Power Commission. It also established planning commissions composed of federal and state representatives with equal voting power for the Missouri, New England, Great Lakes, Ohio, Columbia, and Upper Mississippi regions or river basins. The council, among other things, reviewed plans from comprehensive regional or river basin commissions and established the principles, standards, and procedures for federal participants in those plans. The river basin planning commissions were responsible for coordinating federal, state, interstate, local, and nongovernmental plans for development of a comprehensive plan for all federal, state, interstate, local, and nongovernmental development of water and related land resources. The planning commissions developed recommendations for long-range schedules of priorities for data collection and analysis, planned for specific projects, and undertook studies of water necessary in the preparation of the comprehensive plan (Draper 2006).

The chair for each river basin commission was appointed by the president of the United States to serve as the representative of the federal government and coordinator for federal members of the commission. Commissioners included one member appointed by each state, one member appointed by any interstate agency created by an interstate compact with jurisdiction in the area, a representative for Native American tribes within the region or basin, and an international representative, when appropriate. Commission actions required the consensus of all members. Significantly, however, its decisions were not binding to the participants. This aspect necessarily hindered the effectiveness of the river basin commissions.

The planning commission's authority was limited primarily to passive powers: planning, internal administrative, and operational functions. The planning functions of the Title II commissions varied, although most included preparing and updating a comprehensive plan for water and related land resources development within the basin (U.S. Army Corps of Engineers 1999). For example, the New England Commission focused on power plant siting, coastal zone coordination, and nonstructural flood control, whereas the Upper Mississippi Commission was specifically oriented toward navigation improvements, among other elements (Mitchell 1990). Commission expenses were shared among federal and state members. Each commission recommended what portion of expenses should be borne by the federal government, subject to approval by the Water Resources Council. The commissions were responsible for apportioning the remainder of the commission's expenses among the various members (Mitchell 1990). The *Water Resources Planning Act* (1995) also authorized appropriations for the council and for financial assistance to the states in the form of grants to assist states to participate in the development of comprehensive water and related land resource plans. The basis for allotments to the states included evaluation of several factors, including population, land area, the need for comprehensive water and related land resource planning programs, and the financial needs of the respective states (Mitchell 1990). As noted, this legal model was abandoned in 1981 by U.S. presidential executive order.

4.4 ORGANIZATIONAL DESIGN PRINCIPLES

The design of the institution should consider each of the following characteristics: interdisciplinary analysis, legal basis, political validity and equitable participation, financial viability, implementation and enforcement competence, stakeholder involvement, and adaptability.

4.4.1 Interdisciplinary Analysis. Regardless of the institutional framework, it is clear that no single discipline—neither law, nor economics, nor engineering—provides all of the answers for resolving water disputes (Wolf 1998; Draper 2001).

4.4.2 Legal Basis. The institution should possess adequate legal authority to cope with the problem (Dellapenna 1994; Eheart 2002). In the United States, the legal basis of interstate compacts, including ratification by state legislators and consent by the U.S. Congress, provides legitimacy for the institution as it fulfills its responsibilities and authorities. In the international arena, legitimacy may be established by the acknowledgment of and consent to the agreement by the national governments and by any regional government involved (Draper 2006, 2007a).

4.4.3 Political Validity and Equitable Participation. The institution should be validated by a political consensus among the parties involved, as well as by the nongovernment organizations representing the various private stakeholders (Kakebeeke et al. 2000). "The issue with many water institutions is their failure to adequately address issues of equity" (Gleick 1998). The institutional provisions should be constructed through a bottom-up planning process that includes the balancing of the interests of all possible stakeholders. The process should be able to ensure equitable treatment of all interests involved with or affected by the agreement. Mechanisms should exist for equitable participation, including participation by politically weak stakeholders and consideration of ecosystems and future generations (Draper 2006, 2007a).

4.4.4 Financial Viability. The agreement, or the cooperating parties themselves, should provide financial capability to perform the authorities and responsibilities outlined in the agreement (ARCADUS 2001; *Berlin Recommendations for the International Roundtable on Transboundary Water Management* 1998). The role of the private sector, or public–private partnerships, should be considered when appropriate (Draper 2001). This consideration is especially critical when a disparity in the financial capabilities of the individual parties exists. Unequal financial resources available for data collection, hydrologic modeling, human resources, and other items may destroy the effectiveness of the agreement. It may be necessary for one party to subsidize another, or the institution itself could be charged with data collection, financed as appropriately befits the needs and benefits accruing to the individual parties (Draper 2006, 2007a).

4.4.5 Implementation and Enforcement Competence. The institution should be structured to ensure that implementation of the agreement is feasible, efficient, and effective. The parties should provide the means for resolving intra- and interagency, intra- and interstate, and international conflicts. At the political level, policy decision-making mechanisms should be included in the agreement that satisfy, or at least acknowledge, all political viewpoints associated with the potential problems. At the environmental and functional levels, the institution should possess management capacity and institutional culture consistent with problem solving (Dellapenna 1994).

4.4.6 Stakeholder Involvement. A top-down administrative mechanism is not conducive to effective water allocation, and all relevant stakeholders should be involved in the decision-making process (Olem and Duda 1995; Fort 1998; Bandaragoda 2000; Kakebeeke et al. 2000; Lubell 2000; Planning and Management Consultants 2000; World Commission on Dams 2000; Gooch et al. 2003). These stakeholders should include governmental and nongovernmental organizations, as well as private stakeholders (Fort 1998; German Foundation for International Development 1998; Barlow and Clarke 2002). Public education programs should be developed to create awareness, shape public attitudes, and drive behavior change (Bandaragoda 2000; Draper 2001).

4.4.7 Adaptability. Just as the agreement should be adaptable to conform to changing economic, hydrologic, environmental, and ecological conditions, so should the institutional framework to administer the agreement be adaptable to changing institutional needs (Draper 1997, 2002a, 2004, 2007a; Bandaragoda 2000).

4.5 CHOICES IN INSTITUTIONAL DESIGN

The choice of institutional framework requires detailed consideration. The long-term success of commissions should be based on the careful design of management structures that provide for effective planning and management, allow administrators and technical staff to operate efficiently, and are cost-effective (Dellapenna 1994; German Foundation for International Development 1998).

4.5.1 Purpose and Objectives. The purpose and objectives of the institution itself should be clearly and effectively defined and described since they will determine the powers assigned to and authorized for the institution. They may involve either passive or "soft" administration of the agreement, or they may involve active or "hard" management (Derthick 1974; Teclaff 1987).

Passive administrative objectives are those that require little or no substantive decision making by the institution. Such activities include but are not limited to collecting, correlating, and reporting data and information; maintaining records; advising decision makers; providing independent research and/or analysis; coordinating and cooperating in studies; and investigating alleged violations (Draper 2007a).

In addition to these passive management objectives, active administrative objectives include those activities involving substantive decision making. Such activities include but are not limited to planning, developing projects, operating and maintaining facilities, formulating and implementing regulations, and taking remedial action for violations (Draper 2006, 2007a).

4.5.2 Rules of Decision. The rules of decision include the membership of the decision-making body, the levels of decision making, and, when appropriate, the voting rules. The agreement should carefully define membership in the substantive decision-making bodies (Dellapenna 1994). Normally the agreement designates specific representatives of the signatory parties who may modify and interpret the agreement itself. The nature of the membership and/or representatives changes depending on the purpose and scope of the agreement. The model water sharing agreements provide typical examples (Draper 2007a).

The agreement should also describe when unanimity, consensus, or majority voting is applicable (Dellapenna 1994; McCormick 1994a). The agreement may delegate other substantive decisions to the institution designed to manage the agreement. The rules of decision should provide for adequate consideration of the interests of all stakeholders (Roll and Lopman 2001). Several options exist for the voting process for decision makers. Many existing water sharing agreements provide for making decisions informally through consensus. Discussions during the consensus-building process often prevent disputes that might arise from lack of information or misunderstandings. Rules that require unanimity effectively give one party a veto in any decisions, but unanimous voting may be the only politically feasible option for certain decisions. Decisions requiring nonunanimous votes can be extremely effective for decisions that are not politically sensitive or those involving technical matters. The rules of decision can call for a simple majority or some higher percentage (Dellapenna 1994).

4.5.3 Powers and Duties. The agreement should describe in detail the powers authorized for and the duties assigned to the institution. In addition to the concern for whether the institution should have a passive or active management style, the agreement should clearly specify the "affirmative" or "restrictive" nature of the powers that should be considered (Dellapenna 1994; Eaux partagées 2002). Affirmative powers are those that are proactive in their effect on water resource users, such as arbitrating disputes, responding to emergencies, modifying outdated policies, implementing new agreements, streamlining permitting processes, and innovating conservation measures. Restrictive powers are those that have negative impacts on users, such as taxation and user fees. Technical support by agencies within the governmental structure of the individual parties may be used as appropriate (Dellapenna 1994; German Foundation for International Development 1998).

4.5.4 Accountability and Responsiveness. The agreement should provide for accountability and responsiveness, including accountability to decision makers, water resource users, and the general public. Except in rare cases, internal meetings should be open to the public, and the agreement should establish full disclosure provisions and require publishing all institutional records. The institution should respond to the concerns of all interest groups (Draper 2006, 2007a).

4.5.5 Compliance. The water sharing agreement should have "the clout to enforce its mandates" (Dellapenna 1997a). At least in the international arena, the respective national governments normally perform this function (Hayton and Utton 1989), and most water sharing agreements do not even require the monitoring of compliance (Kakebeeke et al. 2000; van Edig et al. 2001). Consequently, the agreement should provide for a compliance review procedure to ensure effective compliance with the agreement on transboundary waters in a manner that avoids complexity or confrontation, that is transparent, and that provides the parties with the right to make decisions relating to the compliance verification and control. Such a compliance review may be informal, or it may be a formal compliance review committee established in the agreement (Kakebeeke et al. 2000).

4.5.6 Dispute Resolution. Disputes inevitably arise as an agreement is implemented and enforced (Dellapenna 1994; McCormick 1994b; Bandaragoda 2000). They may involve differences in interpretation of the agreement's provisions or noncompliance with the agreement itself. The disputes may also arise because of changing conditions that alter the effectiveness of the agreement for one or more of the parties. Therefore, the institutional provisions should provide for a process to resolve disputes quickly and effectively (Hayton and Utton 1989; ASCE 2009c). The mechanism should emphasize a streamlined process of dispute resolution that minimizes costly, time-consuming litigation.

The parties should consider nonjudicial conflict resolution because use of the courts in most instances holds significant disadvantages. Judges are generalists who are, in most cases, dependent on the testimony and evidence presented to them. The judicial process, with its manifold procedural safeguards, is normally too slow for effective natural resource management. Judicial decrees are often retrospective, geographically limited, and quite specific. Unlike administrative agencies, courts, by their nature, do not provide prospective, uniform regulations of general applicability. Courts often lack the ability to consistently monitor and evaluate solutions that they have devised (Goldfarb 1993).

Alternative dispute resolution processes are nonjudicial dispute resolution techniques that are designed to resolve disputes as quickly as possible and at the lowest cost to the parties involved (Fort 1998). The process consists of a successive series of techniques that become increasingly time-consuming and expensive (Hayton and Utton 1989). These processes include four techniques: negotiation, mediation, arbitration, and litigation. With each successive step, the parties spend more time and more money for a result over which the parties have less and less control (Draper 2002b, 2006).

Negotiation is a process in which the conflicting parties engage in face-to-face discussions to develop a mutually satisfactory agreement on the issues or problems at hand. No outside, independent party or individual is involved. If negotiations between the parties themselves are not effective, the process evolves into mediation (ASCE 2009c; Draper 2002b, 2006).

Mediation is the intervention of a third party in a dispute between two other parties in an attempt to reconcile their differences, usually upon the request of the parties. The qualifications of the mediator should include knowledge, experience, and background in the water resource issues themselves, as well as an understanding and background in the legal issues involved. Mediation typically moves through three stages. The mediator identifies and develops a factual discussion of the disputed and undisputed issues with the parties, both individually and collectively. The purpose is to ensure that all parties understand the strengths and weaknesses of their cases and the perceived weaknesses and strengths of the opposing parties. The mediator then explores with the parties their goals, objectives, and interests, attempting to create alternative solutions to their perceived concerns. Often, this portion of the process also involves discussions with the parties individually and collectively. Finally, after the mediator intervenes, the parties themselves may then reassume a negotiating posture and possibly agree on a mutually acceptable alternative solution, or they may proceed to arbitration (ASCE 2009c; Draper 2002b, 2006).

Arbitration is an informal trial, with the parties choosing the judge and using a process that has a less formal evidentiary process than a formal trial. Arbitration differs from mediation only in the rules of decision as to the solution chosen to resolve the dispute. Whereas the mediator seeks to persuade the parties to agree on a mutually acceptable solution, the parties agree to allow the arbitrator to make decisions that are binding on the parties (ASCE 2009c; Draper 2002b, 2006).

In the case where a dispute cannot be resolved within a prescribed time period, either through good faith negotiations or independent dispute resolution, the agreement should designate a specific legal forum to resolve the dispute through litigation (World Commission on Dams 2000). In the United States, the federal court system provides the forum (Draper 1997). In the international context, the forum might be a fact-finding commission, as detailed in Article 33 of the *Convention on the Law of the Non-Navigational Uses of International Watercourses* (United Nations 1997), or the International Court of Justice (World Commission on Dams 2000).

In creating an effective institution to administer a water sharing agreement, the choices depend on several questions. It depends on whether the agreement is single-purpose or multipurpose. The agreement determines whether the administration of the agreement is active or passive. Finally, effectiveness of the institution depends on whether the parties are willing to surrender a degree of authority in the interest of efficient use of the water resource (Draper 2006).

4.6 SUMMARY

The importance of the administration and institutional provisions of the agreement cannot be overemphasized. Without effective administration, implementation of the agreement falters and enforcement of its provisions is limited. Without an effective institutional framework, the parties spend much of their time and resources in dispute resolution rather than effective water management.

The effectiveness of any agreement or law depends on how effectively it is administered and how rigorously its provisions are enforced. These institutional provisions involve far more than just engineering management of the water resources in question. They also involve establishing procedures to manage complicated coordination among a number of private and public institutions, addressing critical social issues, and responding to complex environmental and ecological sustainability needs. The institutional provisions should provide for effective mechanisms for cooperation, coordination, and communication (Draper 2006).

Chapter 5
WATER ALLOCATION STRATEGIES

The choice of how the shared waters are allocated between the parties determines the success of the agreement. Although the variation among allocation strategies for effective water sharing may be large in theory, historical experience has demonstrated that the number of successful strategies is limited. The historical record shows that distinct allocation strategies have been applied to allocating surface, underground, and atmospheric water. Few examples exist of allocation strategies that combine surface, underground, and atmospheric waters (Draper 2006).

5.1 STRATEGIES FOR WATER ALLOCATION

The objective of water allocation should be to provide adequate supplies of high-quality water for the various instream and off-stream water needs of the political units sharing the basin while preserving the hydrological, biological, and chemical elements of the ecological functioning. The allocation should evaluate and accommodate the social, environmental, and cultural, as well as the economic, issues that may affect the agreement (Fort 1998). It should apportion the water justly and fairly and should minimize the potential for conflict. Ideally, allocation should consolidate the analysis of water availability and demand at the subbasin level, where quantity, quality, and sustainability can be integrated (ASCE 2009c). Consideration of the consumptive uses of the shared resource, not solely the water withdrawals, is important (Kliot and Shmueli 2001), and some people have advocated allocation of consumptive use as the one appropriate method of allocation (Draper 2002e). Although economic benefits have not been explicitly used in allocating transboundary water, economic principles have helped guide transboundary water sharing agreements (Hamner and Wolf 1998). Proposals have been made to introduce the sharing of economic, environmental, ecological, and social benefits as the primary basis for cost sharing, thus creating incentives for cost efficiency (*Berlin Recommendations for the International Roundtable on Transboundary Water Management* 1998; ASCE 2009c).

Conceptually, the purpose of water allocation is to allocate the risk of water scarcity among the parties. One risk to be allocated is the risk of a water shortage during drought conditions. Another risk to be allocated is the risk of a shortage in water that might be needed for future economic growth and intergenerational quality of life. In allocating water, the parties are actually determining to what extent each party bears these risks. Allocation strategies should also take into account the possibility of a "political shortfall," that is, a shortfall created not by a physical shortage of water but by some social or political factor, such as new regulations, that reduces the quantity actually available for use (McCormick 1994b).

5.1.1 Surface Water. Although the specific manner of allocating water can vary according to a variety of influences, most can be classified as either flow or storage allocation. Choice of the method depends on what the parties want to accomplish and how they want to divide the risk of shortage (McCormick 1994b; Kenney 1995).

Priority of a Particular Demand. This allocation strategy sets priorities by type of use rather than by user location within the water basin and provides certain quantitative limitations on those priorities. This method can be considered more economically efficient since it allocates water on the basis of value rather than location of use. Care should be taken that new uses are not foreclosed because old uses have consumed the resource. This type of allocation requires accurate forecasting of the future values of differing uses of water. The allocation of the risk of shortage is apportioned among types of users rather than among hydrogeographic areas. Lower value uses, with lower priorities, bear the burden in a period of shortage. If the parties are willing to view the basin as an integral whole, this allocation strategy provides a mechanism to integrate the mechanics and economic planning of water use. The approach allows potential optimization of the economic value of water allocation, although the issues of environmental, ecological, social, and cultural demands are still problematic. Compacts adopting this strategy in whole or in part include *Belle Fourche River Compact* (1944), *Costilla Creek Compact* (1946), *Klamath River Basin Compact* (1957), and *Kansas–Nebraska Big Blue River Compact* (1972).

Storage Limitations. A storage limitation strategy limits the amount of water that an upstream entity may impound annually, seasonally, or on some other time period basis. It may be combined with some other method of allocation. A storage limitation strategy has several advantages. It is relatively easy to administer, and it allows conservation of upstream water and storage for use at different seasons. Its disadvantages arise from its simplicity. The method does not by itself specify limits on consumption or depletion, and withdrawals downstream from reservoirs are not necessarily controlled. With regard to risk allocation, the downstream party assumes the majority of the risk because stream flow may not be sufficient to do more than fill upstream reservoirs. Adding users upstream may further diminish flow available downstream. Finally, the downstream party assumes the risk of shortage during prolonged periods of drought or during flood control operations. Compacts adopting this strategy in whole or in part include *Rio Grande Compact of 1938* (1938); *Arkansas River Compact*, 63 Stat. 145 (1949); *Canadian River Compact* (1952); *Bear River Compact* (1955); *Arkansas River Basin Compact of 1965* (1966); *Upper Niobrara River Compact* (1969); and *Kansas–Nebraska Big Blue River Compact* (1972).

Guaranteed Quantity at a Point. Using this strategy, a guaranteed quantity is to be delivered at certain points. The upstream party guarantees that a fixed amount of water will pass a certain point every year or other time period. This method's advantage is that it is relatively simple to administer. However, it requires accurate knowledge of flow history at the various reference points. The risk of water shortage falls upon the upstream parties, which guarantee the minimum flow, although the upstream parties may also obtain benefit of extra water in periods when stream flow exceeds the base amount. To modify the division of risk, the parties may agree to divide any surplus over a specified minimum flow so that both upstream and downstream parties share part of the surplus. Alternatively, the risk assumed by the upstream party may be lessened by limiting what it should do to ensure the minimum flow. Compacts adopting this strategy in whole or in part include *South Platte River Compact* (1923), *Colorado River Compact* (1928), *Rio Grande Compact of 1938* (1938), *Bear River Compact* (1955), *Arkansas River Basin Compact of 1965* (1966), and *Sabine River Compact* (1992).

Percentage of Flow. With this strategy, the parties allocate water by either a fixed percentage or a formula based on different flow levels. Each participant is entitled to take its specified percentage of the flow. This strategy's significant advantage is that the parties share any surplus or deficit, and it is easy to administer once data are available. However, certain disadvantages also exist. The method requires more complex administration. Gauging stations should be established at agreed-upon points to determine the amounts being withdrawn and the starting amounts available. If existing allocations are not grandfathered, those existing rights may be impaired if the allocated percentage results in insufficient water for those rights. In some jurisdictions, this arrangement could result in legal "takings" claims. This strategy should include provisions for instream flow maintenance, which may result in less usable water than the strict percentages might indicate. The method is, however, relatively benign with regard to risk allocation because each party shares in surplus or deficit in proportion to its allocated percentage. Compacts adopting this strategy in whole or in part include *La Plata River Compact* (1925), *Belle Fourche River Compact* (1944), *Snake River Compact* (1949), *Upper Colorado River Basin Compact* (1949), *Yellowstone River Compact* (1950), *Narmada Water Disputes* (1979), and *Red River Compact* (1980).

Hydrologic Models. As a complement in choosing the appropriate allocation strategy, the parties may find it helpful to use hydrologic modeling of the shared water resource. In this allocation technique, the parties develop a hydrologic model to explain the flow of the river and allocate water according to differences in flow under different precipitation regimes. Parties share water according to schedules derived from the model. For rivers with great flow variability, the hydrologic model can provide a system that takes variation of precipitation into account and can apportion the flow based on conditions in particular areas. The method may also appear to be more scientifically valid than others. However, the uncertainty of hydrologic modeling poses a significant risk that the model is inaccurate, in which case the allocation does not accomplish what the parties intended. By using the method, the parties may divide risk of shortage any way they see fit, but the likely result is a division designed to reflect local, rather than basinwide, variations in natural conditions. Compacts adopting this strategy in whole or in part include *Pecos River Compact* (1948) and *Republican River Compact* (1943).

Comprehensive Basin Management. In this strategy, an independent commission, under supervision and policy control of the states involved, provides comprehensive basinwide watershed management by protecting water quality, resolving interstate water disputes without costly litigation, allocating and conserving water, managing river flow, and providing numerous other services to the signatory parties. The independent commission usually manages surface-water quality, including both point and nonpoint sources of pollution; groundwater and surface-water quantity, including water demands, water withdrawals, water allocations, water conservation, and protected areas; drought management; and instream flow management (Delaware River Basin Commission 2003). This strategy applies equally as a underground water allocation strategy or as a strategy to integrate surface and underground water. Underground water management has also been included in this strategy in some agreements. Compacts adopting this strategy include the *Delaware River Basin Compact* (1961) and the *Susquehanna River Basin Compact* (1970).

5.1.2 Underground Water. In most cases, underground water allocation should be included in any water allocation strategy because of the hydrologic connection between underground and surface water (Hayton and Utton 1989). Until recent years, however, little attention was given to underground water and its conjunctive use with surface water, although it is very much an active issue in the development of water resources agreements in the Middle East (Dellapenna 1994a, 1997a, 1997b; ASCE 2004a). International law on transboundary underground water sharing is weak and poorly codified (FAO/UNDP/World Bank 1995). "International competence over aquifers divided by the [U.S.–Mexico] frontier is largely undefined; it is fair to say that the legal and institutional situation is chaotic" (Utton and Atkinson 1979).

The unintended consequences of underground water withdrawals by one party on the other party are a special consideration. Excessive or unregulated underground water withdrawals by one party may deplete a jointly used aquifer and result in aquifer depletion that affects the other party. Dewatering of wetlands within the shared river basin is a threat, as is reduction of flows in streams (Cherry and Badr 1998), especially during drought periods. Land subsidence and degradation of water quality may result. In coastal areas, saltwater intrusion may disrupt water supplies for industrial and municipal uses.

In those cases when both significant underground and surface water use exist, the agreement should include some requirement for the conjunctive management of surface and underground water resources by the parties. By considering both underground and surface water, it may be possible to optimize the water available for use by both parties while minimizing costs (Basagaolu et al. 1999; Barlow et. al. 2003; Sehlke 2004; Wilson 2006).

Maximum Withdrawal Rate. This type of allocation strategy does not divide the water itself but limits the rates of extraction by the parties. The significant advantage of this strategy is that it controls the rate of mining of underground water and may prevent one party from depleting the aquifer before the other can make use of it. However, because no division of the water itself occurs, this method encourages each party to withdraw water before the other does. Consequently, the maximum rate becomes the actual withdrawal rate. This disadvantage does not arise if the withdrawal rates are less than or equal to the recharge rate. With regard to risk allocation, each party risks losing the opportunity to use the water unless it pumps at the maximum rate (McCormick 1994b). Maximum underground water withdrawal limits were developed in the Treaty Addendum (1973).

Planned Depletion. Planned depletion is an economic strategy that has been proposed in which water is withdrawn from the aquifer at whatever rate is necessary to support economic efficiency as long as it is cost-efficient to do so, without regard to the recharge rate of the aquifer. Under the strategy, once it is no longer cost-efficient to withdraw the water, other, unspecified, alternatives should be found (Cummings et al. 1996). In a case where recharge is limited or nonexistent, the parties could agree that the underground water source is, in effect, a nonrenewable resource and divide the aquifer itself in terms of total withdrawal over a period of time. This method's significant advantage is that the parties need not compete against each other to be first to withdraw, at least to the extent noted in the previous paragraph. However, the parties may have an incentive to withdraw water as rapidly as possible to guard against the possibility that the estimate of the recoverable quantity is incorrect and to avoid having to pump the deeper water after the other parties have removed the shallower resources. By using this method, the parties each assume the risk that the estimate of resources is wrong, and the last one to pump may either lose water or face increased cost (McCormick 1994a, b; Hayton and Utton 1989). Because this strategy is the antithesis of sustainability and would appear to embody the problems associated with the "tragedy of the commons," such a strategy over the long term is ill advised (Draper 2006, 2007a).

Hydrologic Models. Conceptually identical to hydrologic models for surface water allocation among the parties, this strategy involves modeling the inflow–outflow hydrogeologic characteristics of the aquifer and allocating water on an annual, seasonal, or other time-based period. Parties share water according to schedules derived from the model. This technique is appropriate, however, only for those aquifers, such as the Floridan aquifer in the southeastern United States, that have relatively quick recharge. The method may be considered to be based on sound science, although modeling poses a significant risk of being inaccurate. By using the method, the parties may divide risk of shortage any way they see fit, but the likely result is a division designed to reflect local, rather than basinwide, variations in natural conditions (Draper 2006, 2007a).

Comprehensive Basin Management. This strategy uses the same principles and procedures as the similar model used for surface water.

5.1.3 Atmospheric Water. As a starting point for atmospheric water strategies, *Policy Statement 275* of the American Society of Civil Engineers (2009b) is instructive:

> The American Society of Civil Engineers (ASCE) supports and encourages the protection and prudent development of atmospheric water (also known as "weather modification" or "cloud seeding") for beneficial uses. Sustained support for atmospheric water data collection, research and operational programs, and the careful evaluations of such efforts including the assessment of extra-area and long-term environmental effects, is essential for prudent development. ASCE recommends that the results and findings of all atmospheric water-management programs and projects be freely disseminated to the professional community, appropriate water managers and to the public.

ASCE developed materials providing guidance in the use of atmospheric water-management technology with weather modification organizations and now recognizes that continued development of atmospheric water-management technology is essential (ASCE 2004b). However, the essential difficulty in drafting a water sharing agreement with respect to atmospheric moisture is in determining whether efforts to induce precipitation by one party are actually responsible for any particular quantity of precipitation falling within the river basin (Draper 2006, 2007a; Keyes 2006).

Payer Benefits. In this strategy, the party that induces the precipitation may receive full credit for all such induced water supplies, up to a certain maximum amount. Thus, the party paying for the inducement efforts would receive the full benefit of that investment, encouraging further efforts to enhance precipitation. This method's disadvantage is that estimation of the additional increment of available water could be inaccurate. Risk is allocated solely to the party seeking to augment water supplies, as are the benefits. Any party that attempts to augment precipitation within the basin obtains the full benefit of those efforts but also bears the entire cost (Bomar 2006; Draper 2006).

Undifferentiated Rainfall. This strategy treats all induced precipitation the same as naturally occurring precipitation. The method then allocates water according to existing separate surface or underground water rights. Therefore, no dispute can occur over whether the precipitation was induced or not. Usually collective or governmental corporations as beneficiaries share the costs and benefits of inducing precipitation in proportion to their expected benefits. However, this method reduces the incentive to augment precipitation because the benefits of such investments are not allocated directly or proportionally to those who assumed the costs and risks. Rather, the benefits are spread among all parties and may require further allocation with another surface water allocation strategy (ASCE 2004b; Draper 2006; Keyes 2006).

5.1.4 Integrating Surface, Underground, and Atmospheric Water Allocation. The realities of the hydrologic cycle suggest that any agreement should acknowledge the link among underground, surface, and atmospheric water (Tsur 1997; Hirji and Ibrekk 2001; ASCE 2009c).

When the parties know and understand the connections among atmospheric, underground, and surface water, they can make informed decisions on the allocation of risk. In these circumstances, the parties can treat hydrologically connected surface and underground water as if they were a single source. Problems arise, however, when the connections are unknown or when the assumed connection is wrong. Agreements should consider which party has the risk of shortage in these circumstances. Some of the surface allocation methods avoid this problem, but these are not necessarily the best solutions in all circumstances. For example, underground water use may be unimportant when the agreement guarantees delivery of a given quantity at a specified point or when the quantity is guaranteed by the priority of a particular demand. Incorporating atmospheric water may cause additional uncertainty because there is likely a problem in proving that weather inducement efforts caused a particular increase in water quantity and then in showing the location of that quantity within the hydrologic cycle of only one basin.

5.1.5 Allocating Water Resulting from Supply Augmentation. The construction of instream and offstream reservoirs, underground water storage, wastewater reuse, and diversions from outside the basin can augment local supplies (Keyes 1977; Hayton and Utton 1989; Cherry and Badr 1998; Asheesh 2000). Water sharing agreements should contain provisions that determine how additional supplies are to be allocated. In addition, the parties may consider the equity of existing conservation programs and past expenditures on supply augmentation in the initial allocation negotiation.

Conservation measures can help provide additional supplies of water locally and modify availability elsewhere in a river basin (Draper 2002b, 2006). At the time of negotiating an agreement, the parties should consider whether their existing conservation programs are equal in scope. This equality may or may not be a major factor in the negotiation, depending on how the parties treat increased supplies from new conservation measures. In general, the party paying the cost of implementing a conservation practice should benefit from the increased supply. This principle would allow one party to pay for another party's conservation measures and receive additional allocation as a result (Draper 2002b, 2006).

5.2 ADAPTIVE MANAGEMENT

Water sharing agreements should be formulated with built-in flexibility, and they need to allow for modifications in the role of these institutions over time to meet changing conditions and to address emerging issues (*Berlin Recommendations for the International Roundtable on Transboundary Water Management* 1998). Such changing conditions may be the result of natural events, such as diminishing rainfall over a long period of time, or of social factors, such as increased development in one part of a basin with demand greater than what was originally anticipated (McCormick 1994b). Adaptive management strategies might include changing water allocation rules; increasing water use efficiency; and long-term aspects, such as crop breeding and selection of plants to exploit possible differences in rainfall, temperature, and carbon dioxide concentrations (Fort 1998; Hassall et al. 1998). At the least, provisions should be included in the agreement for periodic review (Muys 1971).

Adaptive management emerged in the mid-1960s as a response to the need for not just looking at the effect of decisions but also actually using science to help guide decision making (Light 2002). Adaptive management of water sharing is essentially the application of the scientific method to the inherently complex and divergent goals and objectives of the parties. The scientific method is a process by which hypotheses on the nature of a physical event are developed from historical observations. The hypotheses are then tested against new observations of the event and altered as necessary to accommodate the new observations. By using this flexible model, the parties may establish specific objectives and develop hypotheses on the expected outcomes of the water sharing agreement. Monitoring programs may then be used to test whether existing management approaches are achieving those outcomes, and then the agreement can be altered, depending on the monitoring results. The key features of carrying out adaptive management in the natural resource context are

1. establishing the mutually agreed-upon objectives and outcomes in the water sharing agreement,
2. proposing a management regime designed to achieve the desired objectives;
3. developing hypotheses and experiments to test whether the proposed management regime is in fact achieving the objectives;
4. setting up monitoring and testing programs to carry out the experiments and test the hypotheses "on the ground"; and
5. adjusting the management regime in response to the information received from the monitoring and testing, if the outcomes turn out not to be as desired (Neuman 2001).

An adaptive approach requires flexible, adaptive policies (not rigid and locked-in policies); integrated approaches (rather than piecemeal approaches); planning and management for learning (not just for economic gains or social products); and experimentation and monitoring to test policies and identify necessary changes and responses (Holling 1978; Coleman 1998).

Adaptive management techniques are increasingly being used, especially for environmental programs such as the *Cooperative Agreement for Platte River Research and Other Efforts Relating to Endangered Species Habitats Along the Central Platte River, Nebraska* (1997) among Nebraska, Wyoming, and the U.S. Department of the Interior (Strickland et al. 2002). The 2000 National Marine Fisheries Service Federal Columbia River Power System biological opinion (Bonneville Power Administration, U.S. Bureau of Reclamation, U.S. Army Corps of Engineers 1999) uses adaptive management to specify a combination of improvements at federal dams, coupled with an offsite mitigation program to increase survival in other life stages (Toole 2002). Adaptive management is used to achieve the objectives of the strategic plan of the California Department of Fish and Game's CALFED Ecosystem Restoration Program: (1) achieve recovery and increase populations of native species, (2) rehabilitate natural processes to favor native species with minimum ongoing intervention, (3) maintain and enhance populations of selected species for harvest, (4) protect or restore functional habitat types throughout the watershed, (5) prevent the establishment of additional non-native invasive species and reduce the negative ecological and economic impacts of established non-native species in the Bay-Delta Estuary and its watershed, and (6) improve and/or maintain water and sediment quality conditions that fully support healthy and diverse aquatic ecosystems in the Bay-Delta Estuary and watershed, and (7) eliminate, to the extent possible, toxic impacts to aquatic organisms, wildlife, and people (Pawley 2000).

The technique may be expanded to consider more than environmental and ecological objectives. An adaptive management project for the Klamath River Basin has been undertaken by the U.S. Department of Agriculture. The objectives of the project include (1) decreasing the amount of water needed for agriculture, (2) increasing water storage, (3) improving water quality, and (4) developing fish and wildlife habitat (Natural Resources Conservation Service 2003). The number of such programs is growing and includes major federally managed processes in the Everglades in Florida, the operation of the Colorado River Storage Project (especially Glen Canyon Dam) in Arizona, and the multiple-state cooperative programs on the Rio Grande and Platte River Basins (Dwyer 2001; U.S. Army Corps of Engineers 2007).

5.3 SUMMARY

A large number of water sharing strategies exists in theory. However, as the historical record reveals, only a few general methodologies have actually been used. These methodologies are generally separated according to whether they apply to surface, underground, or atmospheric water. For surface water, six strategies have been applied in water sharing agreements:

1. priorities of use are set according to specific water demands, such as agricultural or municipal;
2. limitations are placed on water storage by upstream parties;
3. delivery of a specific quantity of water by the upstream party is mandated at a particular location on the shared resource;
4. the shared resource is divided among the parties according to a certain percentage of the flow;
5. hydrologic modeling of the outflows of the shared water is used to explain the flow of the river and to allocate water according to differences in flow under different precipitation regimes; and
6. comprehensive basin management in which an independent commission, under supervision and policy control of the states involved, allocates water according to a predetermined objective function (Draper 2006).

Four strategies have been used that relate to underground water:

1. maximum withdrawal rates that limit the rates of extraction by the parties;
2. a strategy of planned depletion, an economic strategy in which water is withdrawn from the aquifer at whatever rate is necessary to support economic efficiency as long as it is

cost-efficient to do so, without regard to the recharge rate of the aquifer;
3. hydrologic modeling, as done for surface water; and
4. comprehensive basinwide management.

Strategies for atmospheric water sharing include (1) allocation of additional flows resulting from induced precipitation in which the benefits accrue to the party that funds the precipitation inducement and (2) a strategy that treats all induced precipitation the same as naturally occurring precipitation. In most respects, however, the realities of the hydrologic cycle suggest that any agreement should acknowledge the link among underground, surface, and atmospheric water. All agreements should adopt the principles of adaptive management (Draper 2006).

Chapter 6
MODEL WATER SHARING AGREEMENTS

The sequence in developing water sharing agreements is straightforward. First the assessments are conducted. Second a value is placed on the various uses to which the water may be used. Third, the goals to be achieved are determined and strategies to reach those goals developed. Fourth, the general framework for administration is developed. Once these four steps are completed, the parties are prepared to develop a written agreement that codifies how the water shall be allocated. Depending on the degree to which the parties are willing to submit to joint management of the shared resource and the level of efficient use of the resource they are willing to accept, they may choose the appropriate allocation strategy. Commensurate with the chosen strategy, the parties should establish an appropriate institution to manage or administer the allocation, including effective dispute resolution mechanisms (Draper 2002b, 2006).

Once the agreement has been signed and ratified, the parties should develop an institutional mechanism to assess the adequacy of the data used, the values and goals established by the agreement, and progress toward achieving goals, followed by modification of the agreement, as appropriate (Draper 2002b, 2006).

6.1 PURPOSE AND SCOPE OF MODELS

Although water scarcity and the increasing competition for water suggest that comprehensive management of a shared river basin is appropriate, a significant challenge to overcome is the prevailing tendency for governments at all levels to resist outside control over and interference with their internal affairs and those decisions that affect economic growth or quality of life. Each party, whether it is a nation-state or state within a federal system or a tribal entity within a national federal system, wants to maintain authority over the people, places, assets, and natural resources within its political boundaries (Draper 2002b).

The substance of a water sharing agreement largely depends on the willingness of the parties to relinquish absolute authority and control over that portion of the shared resource that lies within their borders. In some cases, the parties are willing only to keep each other informed of their plans to develop that portion of the resource within their borders. In other cases, the parties may be willing to take part in joint ventures to develop the resource. Finally, the parties may be willing to cede their separate authorities and jointly manage the resource. Therefore, a single legal model of a water sharing agreement is inappropriate (Draper 2002b, 2006).

Strong incentives are required for a government to relinquish control to others. The willingness of the parties to relinquish authority over water resources depends on the hydrologic, geographic, and political situation. In some situations, the parties may relinquish considerable autonomy in the search for efficient water management. In other situations, the parties insist on maintaining control of the waters within their boundaries, and they are content to coordinate water management activities in some manner. Therefore, three distinct model transboundary agreements are provided in this chapter to serve as a framework for individual agreements (Draper 2002b, 2006). These model agreements are presented in detail below.

The *Model Water Sharing Agreement A (Coordination and Cooperation)* in Appendix A provides the framework for agreements in situations in which each party prefers to retain complete management authority over that portion of the shared water resource that lies within its boundaries. However, the parties accepting Model A should acknowledge that effective management of those resources requires significant cooperation and coordination. They recognize that a mechanism is needed for exchanging water-related information concerning shared waters, to include hydrologic data and proposed projects that may affect extraterritorial waters. Model A makes clear provision for the parties to adequately protect their individual interests while coming to the table to resolve, where possible, transboundary water issues. The model can be described as the "state's rights" model. This model ensures that each party is aware of the quantity and quality of water that is available for its use. It provides for the parties to share information about the activities and plans that may affect that availability. This "coordination and cooperation" model makes clear provision for parties to protect their individual interests while resolving transboundary water issues peacefully (Draper 2002b, 2006).

The *Model Water Sharing Agreement B (Limited Purpose)* of Appendix B is concerned with water management needs that are limited in scope. It is a model wherein the parties enter into an agreement to achieve certain limited purposes. The willingness to concede authority over management of internal water resources is limited to what is necessary to achieve a specific goal. The limited purpose goals may vary from simple allocation of water released from a single reservoir, to prioritization of needs during droughts, to other matters, such as water quality control (Draper 2002b, 2006).

An interesting water management strategy may be developed using Model A as the framework for identifying individual water resources projects, which may be defined by separate individual agreements that are based on Model B. The 1909 *Treaty between the United States and Great Britain Relating to Boundary Waters* is an example of such a water management strategy. The treaty itself provides the mechanism for coordination and cooperation between the two parties, while specific side agreements have been developed, such as the *Agreement between the United States and Canada on Great Lakes water quality (*1978); the Reference Letter of the International Souris–Red River Engineering Board, forming the basis of an agreement to allocate the waters of those rivers (International Souris River Board 1948); and the 1941 agreement on the International Columbia River Board to monitor the effects of the operation of the Grand Coulee Dam and Roosevelt Lake on the levels and flows of the Columbia River at the international boundary (International Joint Commission 1941; Draper 2002b, 2006).

The Model Water Sharing Agreement C (Comprehensive) in Appendix C provides a model for comprehensive planning and management of shared water resources. It is the model that best conforms to efficient water management, according to science-based management, and is the model most of the experts on the working group recommend. In this model, the parties share authority over the water resource at issue. This model is appropriate for those situations in which the parties wish to achieve optimal use of the shared water resources by multidimensional management on a river basin basis. The agreement is extensive and considers all aspects of management of the water resources. The goal of this comprehensive, integrated agreement is to achieve allocation that is equitable and fair to all prospective users. It requires that each party restrict practices to the reasonable use of water and provide sufficient data to the other parties to verify that its use is reasonable. Prime examples of Model A are the 1961 *Delaware River Basin Compact* and the 1970 *Susquehanna River Basin Compact* (Draper 2002b, 2006).

The Utton Transboundary Resources Center at the University of New Mexico School of Law has published a model interstate compact similar in purpose but different in scope to the ASCE Model Water Sharing Agreement C (Muys et al. 2006). The purpose of the model compact is to "provide a mechanism by which interstate water conflicts may be resolved in an amicable, efficient, equitable and effective manner" (Utton Transboundary Resources Center 2007). It provides provisions on establishment and structure of the commission, water apportionment, water quality protection and water resources management programs, dispute resolution, interagency coordination, and budgeting and funding. As is the case with the three ASCE model compacts, a commentary section follows each article to explain the rationale for each provision. The question of state authority is not addressed (Hanson 2006; Utton Transboundary Resources Center 2007). Those institutions developing transboundary water sharing agreements should review and analyze the Utton model as well as the ASCE models presented below (Draper 2002b, 2006).

Regardless of the model chosen, however, a final word of caution is extended to the parties drafting any agreement. On many occasions, each source of water, whether it be surface, underground, or atmospheric water, is assessed and managed as a separate and distinct source from the others. However, such individualized assessments do not provide an integrated understanding of either the water sources or their uses. Before implementing a transboundary water use apportionment or allocation, participants should conduct a comprehensive water resources assessment. The assessment should evaluate the following:

- The sources of water;
- The volume, flow, and distribution of water within the basin;
- The water quality; and
- Water demand, including environmental, ecological, and anthropogenic need (Draper 2006).

The goal should be to determine the hydrologic mass balance for the basin and to establish a detailed conceptual model of the basin that defines the water resources available for use, carefully defining the time period and area covered and the present and future water demands (Draper 2006).

Appendix A
MODEL WATER SHARING AGREEMENT A (COORDINATION AND COOPERATION)

The *Model Water Sharing Agreement A (Coordination and Cooperation)* provides the framework for agreements in situations where each party prefers to retain complete management authority over that portion of the shared water resource that lies within its boundaries.

A.1 ARTICLE 1A: DECLARATION OF POLICIES AND PURPOSES

§1A-1-01 General Policies

(a) The water resources of the _______ River Basin have local, regional, and national significance and their management, development, and control by the individual Parties and under appropriate arrangements for intergovernmental cooperation are public purposes for the respective signatory Parties.

(b) The major purposes of this Agreement are to provide the coordination and cooperation necessary for the management and development of the ________ River and its tributaries, the exchange of data and other information pertinent to independent water planning and development by the respective Parties of the _____ River and its tributaries, to promote interstate comity, and to remove causes of present and future controversy.

(c) The general policies of this Agreement include facilitating the equitable and reasonable use of the water resources shared between the Parties, the exchange of data and other information pertinent to water utilization by the respective Parties, and the cooperation and consultation necessary to achieve the purposes of this Agreement.

(d) The bases for this Agreement are the geophysical, climatic, meteorological and other conditions peculiar to the _____ River Basin, and application of its provisions is limited to those waters.

Commentary: Economic growth and prosperity require adequate supplies of high-quality water on a regular and sustained basis. This need requires that utilization of shared waters be well coordinated among or between the parties sharing the waters. An effective agreement can facilitate adequate planning, conservation, utilization, development, management, and control of water resources on a water basin basis, in a manner that is reasonable and equitable under the circumstances and that causes no significant harm to other parties. A key challenge for the parties is to make more efficient and productive use of water and to reshape the water policies of the individual parties to better respond to periods of water shortages (Postel 1996).

One criterion of the *Convention on the Law of the Non-Navigational Uses of International Watercourses,* United Nations Document A/51/869 (United Nations 1997) makes the definition of the waters to which an agreement applies a mandatory provision in the agreement. The parties should carefully frame the extent of the specific water resources involved in the agreement. They should identify the type and geographical extent of the water resources to be subject to the agreement. To formulate an effective agreement, the parties should analyze the factors that influence the water resource in question, including the climatology, physiology, geology, and the interaction between underground and surface-water resources. The analysis should identify pollution sources and the resulting effect on water quality. The geographic scope of the water resources to be covered by the agreement should be sufficiently expansive to fully address all the water sharing issues involved. The use of the term "_________ River Basin," if objectionable for any reason by one or more parties, may be changed to "_______ River and its tributaries," "boundary waters," "shared waters," "frontier waters," or any phrase that accurately describes and encompasses the entirety of the water resources subject to the agreement (Draper 2002b).

It is important to acknowledge that the agreement reflects the particular circumstances and compromises reached in its formulation and that it applies only to the waters shared between the parties. The inclusion of §1A-1-01(b) may avoid later claims that other rivers and streams, or other bodies of water, are subject to the agreement (Draper 2002b).

Cross References: §1A-1-02 (coordination and cooperation); §1A-1-03 (good faith implementation); §1A-1-04 (preservation of federal rights); §2A-1-01 (effective date); §2A-1-02 (consent to jurisdiction); §2A-1-03 (duration of agreement); §2A-1-04 (powers of signatory parties; withdrawal); §2A-1-05 (amendments and supplements); §2A-1-06 (existing agencies); §2A-1-07 (severability); §2A-1-08 (annexes); §2A-2-01 (____ Basin); §2A-2-05 (party or parties); §3A-1-01 (alternative 1, use of party officials); §3A-1-02 (alternative 1, substitution of officials); §3A-1-03 (alternative 1, implementation and verification of agreement); §3A-1-04 (alternative 1, funding); §3A-1-01 (alternative 2, commission created); §3A-1-02 (alternative 2, commission jurisdiction); §3A-1-03 (alternative 2, commissioners); §3A-1-04 (alternative 2, status, immunities, and privileges); §3A-1-05 (alternative 2, commission organization and staffing); §3A-1-06 (alternative 2, advisory boards); §3A-1-07 (alternative 2, rules of procedure); §3A-2-01 (alternative 2, general powers and duties); §3A-2-02 (alternative 2, special powers and duties); §3A-3-01 (alternative 2, meetings, hearings and records); §3A-3-02 (alternative 2, funding and expenses of the commission); §4A-1-01 (exclusive jurisdiction and control); §4A-1-02 (data exchange); Article 5A (dispute resolution).

Similar Agreements: *Pecos River Compact*, 63 Stat. 159 (1948); *Delaware River Basin Compact*, Pub. L. 87-328, 75 Stat. 688-716 (1961); *Susquehanna River Basin Compact*, Pub. L. 91-575, 84 Stat. 1509 (1970); *Alabama–Coosa–Tallapoosa River Basin Compact*, O.C.G.A. 12-10-110 (1997); *Apalachicola–Chattahoochee–Flint River Basin Compact*, O.C.G.A. 12-10-100 (1997); *Agreement Between the People's Republic of Bulgaria and the Republic of Turkey Concerning Co-operation* *in the Use of the Waters of Rivers Flowing through the Territory of Both Countries*, UNTS, Vol. 807, 117 (1968); *Convention on the Law of the Non-Navigational Uses of International Watercourses*, United Nations Document A/51/869 (United Nations 1997).

§1A-1-02 Coordination and Cooperation
The Parties agree

(a) **To cooperate and consult with the other Parties to this Agreement in their development and utilization of the water and related resources of the waters shared by the Parties in order to ensure use of those waters while minimizing harm to other Parties.**
(b) **To cooperate on the basis of autonomous equality and territorial integrity in the utilization and protection of the shared water resources.**
(c) **To conduct themselves with an absence of malice and deceit, and with no intention to seek unconscionable advantage.**

Commentary: This provision provides a framework for the parties in the development of their individual water policy planning. It recognizes that each party should follow in its rational management of water resources certain fundamental principles. These general objectives and principles improve the likelihood of accomplishing the purposes of the agreement (Draper 2002b).

A nation normally enters into any international agreement with a position of self-interest. In the negotiations, each party seeks the rights and authority critical to certain political, economic, or social objectives while ceding less critical rights and authority to the other nations. While accepting this fact, all parties have a duty to cooperate and negotiate in good faith. This principle is the foundation of international law, and it applies in all relations between autonomous states (Draper 2002b).

Cross References: §1A-1-01 (general policies); §2A-1-06 (existing agencies); §3A-1-01 (alternative 1, use of party officials); §3A-1-02 (alternative 1, substitution of officials); §3A-1-03 (alternative 1, implementation and verification of agreement); §3A-1-04 (alternative 1, funding); §3A-1-01 (alternative 2, commission created); §3A-1-02 (alternative 2, commission jurisdiction); §3A-1-03 (alternative 2, commissioners); §3A-1-04 (alternative 2, status, immunities, and privileges); §3A-1-05 (alternative 2, commission organization and staffing); §3A-1-06 (alternative 2, advisory boards); §3A-1-07 (alternative 2, rules of procedures); §3A-2-01 (alternative 2, general powers and duties); §3A-2-02 (alternative 2, special powers and duties); §3A-3-01 (alternative 2, meetings, hearings and records); §3A-3-02 (alternative 2, funding and expenses of the commission); §4A-1-01 (exclusive jurisdiction and control); §4A-1-02 (data exchange); Article 5A (dispute resolution).

Similar Agreements: *Agreement Between the People's Republic of Bulgaria and the Republic of Turkey Concerning Co-operation in the Use of the Waters of Rivers Flowing through the Territory of Both Countries*, UNTS, Vol. 807, 117 (1968); *Convention between Switzerland and Italy Concerning the Protection of Italo-Swiss Waters Against Pollution*, UNTS, Vol. 957, 277 (1972); *Stockholm Declaration of the United Nations Conference on the Human Environment*, 11 ILM 1416 (United Nations 1972); *Treaty for Amazonian Cooperation*, 17 ILM 1046 (1978); *Convention Between the Federal Republic of Germany and the Czech and Slovak Federal Republic and the European Economic Community on the International Commission for the Protection of the Elbe*, International Environmental Law, Multilateral Agreements, 976:90/1 (1990); *Convention on the Protection and Use of Transboundary Watercourses and International Lakes*, 31 ILM 1312 (1992); *Canada–Mexico–United States: North American Agreement on Environmental Cooperation*, 32 ILM 1480 (1993); *Treaty of Peace Between the State of Israel and the Hashemite Kingdom of Jordan*, 34 ILM 43 (1994); *Agreement on the Cooperation for the Sustainable Development of the Mekong River Basin*, 34 ILM 864 (1995); *Convention on the Law of the Non-Navigational Uses of International Watercourses*, United Nations Document A/51/869 (United Nations 1997).

§1A-1-03 Good Faith Implementation
The Parties agree to implement immediately all provisions of this Agreement, and each Party covenants that its officers and agencies will not hinder, impair, or prevent any other Party from carrying out any provision of this Agreement.

Commentary: This provision complements the provision concerning the duty to cooperate and negotiate in good faith. It should be noted, however, that good faith misinterpretation of compact obligations does not excuse a party from damage liability (*Texas v. New Mexico*, 482 U.S. 124, 1987). In that case, the U.S. Supreme Court reasoned that a compact is a contract, and standard contract law does not allow a defense based on misinterpretation of contract obligations (Grant 2001, §45.07(c), §46.05(d)).

Cross References: §1A-1-01 (general policies); §1A-1-02 (coordination and cooperation); §2A-1-06 (existing agencies); §2A-2-05 (party or parties); §3A-1-01 (alternative 1, use of party officials); §3A-1-02 (alternative 1, substitution of officials); §3A-1-03 (alternative 1, implementation and verification of agreement); §3A-1-04 (alternative 1, funding); §3A-1-01 (alternative 2, commission created); §3A-1-02 (alternative 2, commission jurisdiction); §3A-1-03 (alternative 2, commissioners); §3A-1-04 (alternative 2, status, immunities, and privileges); §3A-1-05 (alternative 2, commission organization and staffing); §3A-1-06 (alternative 2, advisory boards); §3A-1-07 (alternative 2, rules of procedures); §3A-2-01 (alternative 2, general powers and duties); §3A-2-02 (alternative 2, special powers and duties); §3A-3-01 (alternative 2, meetings, hearings and records); §3A-3-02 (alternative 2, funding and expenses of the commission); §4A-1-01 (exclusive jurisdiction and control); §4A-1-02 (data exchange); Article 5A (dispute resolution).

Similar Agreements: *Helsinki Rules on the Uses of the Waters of International Rivers*, 52 I.L.A. 484 (1966); *Stockholm Declaration of the United Nations Conference on the Human Environment*, 11 ILM 1416 (United Nations 1972); *Convention on the Law of the Non-Navigational Uses of International Watercourses*, United Nations Document A/51/869 (United Nations 1997).

§1A-1-04 Preservation of Federal Rights (Optional, for U.S. use)
Nothing in this Agreement shall be deemed:

(a) **To impair or affect any rights or powers of the United States, its agencies or instrumentalities, in and to the use of the waters of the _____ River Basin nor its capacity to acquire rights in and to the use of said waters;**
(b) **To subject any property of the United States, its agencies, or instrumentalities to taxation by any party, nor to create an obligation on the part of the United States, its agencies, or instrumentalities, by reason of the acquisition, construction, or operation of any property or works of whatsoever kind, to make any payments to any State or political subdivision thereof, State agency, municipality, or entity whatsoever in reimbursement for the loss of taxes;**

(c) **To subject any property of the United States, its agencies, or instrumentalities to the laws of any State to an extent other than the extent to which these laws would apply in the absence of this Agreement.**

Commentary: These sections may be included in agreements (between states of the United States. They are probably unnecessary to preserve federal rights, but inasmuch as the U.S. Congress should approve the agreement, the inclusion of these provisions may make it easier to obtain that approval (Draper 2002b).

Cross References: §1A-1-01 (general policies); §2A-1-04 (powers of signatory parties; withdrawal); §2A-1-06 (existing agencies); §2A-2-01 (____ Basin); §3A-1-02 (commission jurisdiction).

Similar Agreements: *Rio Grande Compact of 1938*, 53 Stat. 785, 938 (1938); *Republican River Compact*, 57 Stat. 86 (1943); *Belle Fourche River Compact*, 58 Stat. 94 (1944); *Pecos River Compact*, 63 Stat. 159 (1948); *Snake River Compact*, 64 Stat. 29 (1949); *Upper Colorado River Basin Compact*, 63 Stat. 31 (1949); *Yellowstone River Compact,* 65 Stat. 663 (1950); *Canadian River Compact*, 66 Stat. 74 (1952); *Klamath River Basin Compact*, 71 Stat. 497 (1957); *Bear River Compact*, 72 Stat. 38 (1955), amended 94 Stat. 4, Art. XIII(2) (1980).

§1A-1-05 National Security (Optional, for international use)

(a) **Nothing in this Agreement shall be construed to require any Party to make available or provide access to information the disclosure of which it determines to be contrary to its essential security interests.**

(b) **Nothing in this Agreement shall be construed to prevent any Party from taking any actions that it considers necessary for the protection of its essential security interests relating to a formal declaration of war.**

Commentary: National security concerns necessarily take precedence over any program of water management and the exchange of data among the parties. The exchange of data related to each party's essential security interests would not be provided to the other parties (Draper 2002b).

Cross References: §1A-1-01 (general policies); §2A-1-04 (powers of signatory parties; withdrawal); §2A-1-06 (existing agencies); §3A-1-02 (alternative 2, commission jurisdiction).

Similar Agreements: *Canada–Mexico–United States: North American Agreement on Environmental Cooperation*, 32 ILM 1480 (1993).

A.2 ARTICLE 2A: GENERAL

Part 1 General Obligations

§2A-1-01 Effective Date

This Agreement shall become operative when approved by the appropriate governing authorities of all Parties. The Agreement will go into full force and effect at 12:01 a.m. [time zone] on the day immediately following the final act necessary for approval of the Agreement, as defined by the domestic law of each Party.

Commentary: Any agreement of this nature should specify the date or conditions upon which it will take effect. In the case of agreements between states of the United States, the conditions with respect to U.S. Congress are designed to provide some measure of protection against subsequent federal action that might disturb the allocation system agreed upon by the contracting parties. Despite the requirement of federal approval of interstate compacts, the federal government is not normally a party to those agreements and may not be bound by the provisions of those agreements unless there is specific legislation committing the federal government to be so bound. The provisions of Article 2-1-01, modeled after the *Republican River Compact*, 57 Stat. 86 (1943), and the *Bell Fourche River Compact*, 58 Stat. 94 (1944), condition the effectiveness of the agreement on passage of such legislation by Congress and also establish a basis for compensation for takings under the Fifth Amendment should a subsequent Congress decide to take action contrary to that commitment. For instance, a later Congress has the power to set aside the actions of an earlier Congress, but the question of takings and just compensation then arises. If these conditions are not incorporated, the states making the agreement may later find that federal actions render their agreement ineffective (Draper 2002b).

Cross References: §2A-1-03 (duration of agreement); §2A-1-04 (amendments and supplements); §2A-2-05 (party or parties).

Similar Agreements: *Republican River Compact*, 57 Stat. 86 (1943); *Belle Fourche River Compact*, 58 Stat. 94 (1944); *Delaware River Basin Compact*, Pub. L. 87-328, 75 Stat. 688-716 (1961); *Susquehanna River Basin Compact*, Pub. L. 91-575, 84 Stat. 1509-1541 (1970); *Convention on the Protection and Use of Transboundary Watercourses and International Lakes*, 31 ILM 1312 (1992); *Convention on the Law of the Non-Navigational Uses of International Watercourses*, United Nations Document A/51/869 (United Nations 1997).

§2A-1-02 Consent to Jurisdiction (Optional, for U.S. use)

This Agreement shall be effective once the U.S. Congress gives its consent for the United States to be named and joined as a Party defendant or otherwise in any case or controversy involving the construction or application of this Agreement in which one or more of the States is a plaintiff, without regard to any requirement as to the sum or value in controversy or diversity of citizenship of Parties to the case or controversy.

Commentary: The predominance of federal interests in water resources makes it likely that any litigation concerning the agreement between states will involve federal interests. The doctrine of sovereign immunity could prevent joinder of the federal interests as parties to the suit absent a waiver of sovereign immunity. The discretionary decision not to join federal parties led to dismissal of a suit filed by Texas against New Mexico in 1951 to enforce certain provisions of the *Rio Grande Compact of 1938*, 53 Stat. 785, 938 (1938). The U.S. Supreme Court dismissed the case because the federal government was not joined as a party but had important interests that would be affected by any such suit (*Texas v. New Mexico*, 352 U.S. 991, 1957). In 1952, Congress enacted the *McCarran Amendment,* 43 U.S.C. 66, which waived federal sovereign immunity to be joined in general stream adjudications. However, in most cases, the agreement covers management and development issues that reach beyond general stream adjudication. The parties should consider including such a waiver of sovereign immunity as a condition to effectiveness of the agreement. They may also wish to add a provision granting jurisdiction over any such cases to the district courts, which may be preferable to the U.S. Supreme Court as the initial forum for resolving certain types of disputes. The *Red River Compact*, 94 Stat. 3305 (1980) takes this approach (Draper 2002b).

Cross References: §1A-1-04 (preservation of federal right); §1A-1-05 (national security); §2A-2-05 (party or parties).

Similar Agreements: *Kansas–Nebraska Big Blue River Compact*, 86 Stat. 193 (1972); *Red River Compact*, 94 Stat. 3305 (1980).

§2A-1-03 Duration of Agreement (Optional)

(a) **The Parties intend that the duration of this Agreement shall be for an initial period of (__) years from its effective date. Notification of the withdrawal should be made (__) months in advance of the prospective withdrawal. If none of the signatory Parties notifies the Commission of intention to terminate the Agreement at the end of the then current (__) year period, the Agreement shall be continued for an additional period of (__) years.**

(b) **In the event that this Agreement should be terminated by operation of paragraph (a) above, the management structure for the Agreement shall be dissolved, its assets and liabilities transferred equitably to the Parties, and its corporate affairs completed in such manner as may be provided by agreement of the signatory Parties.**

Commentary: The parties may prefer to establish no duration and rely on later provisions to modify or terminate the agreement. However, two significant principles are established by this provision. First, setting the duration for an extended period of time allows for predictability in terms of water resources development; it also allows sufficient time to recover capital costs in the financing of projects. Second, establishing the duration might ensure that the parties reconsider the agreement only after a sufficient hydrologic record is established (Draper 2002b). It should be noted, however, that this provision greatly affects the exercise of authority of the parties involved.

Cross References: §2A-1-01 (effective date); §2A-1-05 (amendments and supplements); §2A-1-04 (powers of signatory parties; withdrawals).

Similar Agreements: *Delaware River Basin Compact*, Pub. L. No. 87-328, 75 Stat. 688-716 (1961); *Susquehanna River Basin Compact*, Pub. L. 91-575, 84 Stat. 1509-1541 (1970).

§2A-1-04 Powers of Signatory Parties; Withdrawal (Optional, for international use)

Nothing in this Agreement shall be construed to relinquish the functions, powers, or duties of the government of any signatory Party with respect to the control of any waters solely within its territory, nor shall any provision hereof be construed in derogation of any of the powers of the Parties to regulate commerce within their borders.

Commentary: This provision acknowledges the inherent authority of the individual parties for internal water sources and recognizes that any relinquishment of authority is limited solely to the purposes of this agreement (Draper 2002b).

Cross References: §1A-1-02 (coordination and cooperation); §1A-1-03 (good faith implementation); §2A-2-06 (waters of the basin); §4A-1-01 (exclusive jurisdiction and control).

Similar Agreements: *Delaware River Basin Compact*, Pub. L. 87-328, 75 Stat. 688-716 (1961); *Susquehanna River Basin Compact*, Pub. L. 91-575, 84 Stat. 1509-1541 (1970); *Apalachicola–Chattahoochee–Flint River Basin Compact*, O.C.G.A. 12-10-100 (1997); *Alabama–Coosa–Tallapoosa River Basin Compact*, O.C.G.A. 12-10-110 (1997); *Convention on the Protection and Use of Transboundary Watercourses and International Lakes*, 31 ILM 1312 (1992); *Canada–Mexico–United States: North American Agreement on Environmental Cooperation*, 32 ILM 1480 (1993); *Agreement on the Cooperation for the Sustainable Development of the Mekong River Basin*, 34 ILM 864 (1995).

§2A-1-05 Amendments and Supplements (Optional)

The provisions of this Agreement shall remain in full force and effect until amended by action of the governing bodies of each of the Parties and consented to and approved by any other necessary authority in the same manner as this Agreement is initially ratified.

Commentary: Agreements may, over time, fail to operate as well as initially intended. Therefore, some amendment process should be specified. In some cases, the approval of another institution may be required. If, for example, the agreement is between states of the United States, the U.S. Constitution arguably requires congressional approval of any amendment as well as approval of the original agreement, unless the agreement provides for a different method of amendment. In this latter case, the congressional approval of the initial agreement would implicitly grant consent to modify the agreement in accordance with the terms of the agreement. If the agreement is between independent nations, the references to other necessary authority may be omitted, but the particular circumstances of each case should be considered (Draper 2002b).

Cross References: §2A-1-07 (limited applicability).

Similar Agreements: *Delaware River Basin Compact*, Pub. L. 87-328, 75 Stat. 688-716 (1961); *Susquehanna River Basin Compact*, Pub. L. 91-575, 84 Stat. 1509-1541 (1970); *Convention on the Protection and Use of Transboundary Watercourses and International Lakes*, 31 ILM 1312 (1992); *Agreement on the Cooperation for the Sustainable Development of the Mekong River Basin*, 34 ILM 864 (1995).

§2A-1-06 Existing Agencies (Optional)

It is the purpose of the signatory Parties to preserve and utilize the functions, powers, and duties of existing offices and agencies of the individual governments to the extent not inconsistent with the Agreement, and the institution established to enforce this Agreement is authorized and directed to utilize and employ such offices and agencies for the purpose of this Agreement to the fullest extent it finds feasible and advantageous.

Commentary: The use of existing offices and agencies prevents duplication, and consequently reduces the costs, of data collection and management of the water resource subject to the agreement (Draper 2002b).

Cross References: §3A-1-01 (alternative 1, use of party officials); §3A-1-05 (alternative 2, commission organization and staffing); §3A-1-06 (alternative 2, advisory boards); §3A-1-07 (alternative 2, rules of procedures); §3A-2-01 (alternative 2, general powers and duties).

Similar Agreements: *Delaware River Basin Compact*, Pub. L. 87-328, 75 Stat. 688-716 (1961); *Susquehanna River Basin Compact*, Pub. L. 91-575, 84 Stat. 1509-1541 (1970); *Convention on the Protection and Use of Transboundary Watercourses and International Lakes*, 31 ILM 1312 (1992).

§2A-1-07 Severability

Should a tribunal of competent jurisdiction hold any part of this Agreement to be void or unenforceable, it shall be considered severable from those portions of the Agreement capable of continued implementation in the absence of the voided provisions. All other severable provision capable of continued implementation shall continue in full force and effect.

Commentary: The drafters of the agreement should consider whether they wish this clause to be included. The advantage of such a clause is that it avoids the possibility of having the entire agreement become null and void if any part is found to be void or unenforceable (Draper 2002b).

Cross References: §2A-1-04 (amendments and supplements).

Similar Agreements: *Yellowstone River Compact*, 65 Stat. 663 (1950); *Sabine River Compact*, 68 Stat. 690 (1953), amended 76 Stat. 34 (1962), 91 Stat. 281 (1977), 106 Stat. 4661 (1992); *Klamath River Basin Compact*, 71 Stat. 497 (1957); *Delaware River Basin Compact*, Pub. L. 87-328, 75 Stat. 688-716 (1961); *Susquehanna River Basin Compact*, Pub. L. 91-575, 84 Stat. 1509-1541 (1970).

Part 2 Definitions

§2A-2-01 Atmospheric Water

"Atmospheric water" means all available moisture above the surface of the earth, and water surface and all forms of precipitation but not including water projecting from irrigation systems.

Commentary: This definition is consistent with the definitions usually used in state and federal laws on weather modification. See generally Beck (2001); Davis (1987); Gochis (2001); Keyes (2006).

§2A-2-02 __________ Basins

"__________ Basin" means the area of drainage into the _______ River and its tributaries, [and] aquifers underlying the drainage, or only the aquifers themselves.

Commentary: The agreement could include the total surface area of drainage throughout the basin and the aquifers underlying the surface drainage. Some tributaries can be connected to the underlying aquifers holding the underground water. Some of the aquifers could be connected to more than one of the surface water basins. The geographic scope of the agreement should be defined to ensure that there are no future disagreements about what lands are or are not covered by the agreement. A map may be incorporated, but care should be taken that the map is cartographically accurate. Because the map is likely to be at a scale too small for precise delineation of boundaries, it should be made clear that it is for general reference only. In the event of a dispute over land or within the defined river and its tributaries, the actual limits of the watershed as determined on the ground should be controlling (Draper 2002b).

Cross References: §1A-1-01 (general policies); §1A-1-04 (preservation of federal rights).

§2A-2-03 Drought

"Drought" conditions means conditions of abnormal water scarcity in a specific area, resulting from natural conditions.

Commentary: Management action arises from a drought, or lack of mean annual rainfall, but could arise from other causes as well, such as the collapse of a dam with the resulting draining of a reservoir on which the commission users depend. Mather (1974) indicates that the number of quite different definitions is large. The definition should be determined, in large measure, by the use intended. Then a "drought management strategy" would be a specific course of conduct planned by the commission as a necessary or appropriate response to the lack of precipitation (Draper 2002b; ILA 2004).

Cross References: §3A-2-01 (general powers and duties).

§2A-2-04 Flood

"Flood" means a rising of water to levels that have detrimental effects on or in one or more basin states with a frequency agreeable to the Parties.

Commentary: The flood condition is almost the opposite of a drought. A large amount of water is to be controlled by facilities of the commission. The parties are to agree as to the frequency of the flow of high waters in the basin. Most of the time, these flows are during periods that exceed the amount of flow that occurs during the years of mean annual precipitation (Draper 2002b).

Cross References: §3A-2-01 (general powers and duties); §4A-1-02 (data exchange).

§2A-2-05 Interbasin Transfer

An "interbasin transfer" is any bulk transfer of water from one water basin to another.

Commentary: This definition merely makes clear that the term "interbasin transfer" is not limited in any fashion but refers to all transfers from one water basin to another. The provisions regarding interbasin transfers allow regulations to exempt certain small transfers. In many states within the United States, 100,000 gal. per day (approximately 380,000 L per day) are exempt from regulation (Draper 2002b; ILA 2004).

Cross References: §4A-1-02 (data exchange).

§2A-2-06 Party or Parties

"Party or parties" means, unless the text otherwise indicates, those governmental units signatory to this Agreement.

Commentary: Defining the terms in this way avoids the need to include similar language at numerous points throughout the agreement. As a matter of law, it may be unnecessary to state this principle (Draper 2002b).

Cross References: §1A-1-02 (coordination and cooperation); §1A-1-03 (good faith implementation); §1A-1-05 (national security); §2A-1-01 (effective date); §2A-1-02 (consent to jurisdiction); §2A-1-04 (powers of signatory parties); §3A-1-01 (alternative 1, use of party officials); §3A-1-02 (status, immunities, and privileges); §3A-1-05 (commission organization and staffing); §3A-1-06 (advisory boards); §3A-2-01 (general powers and duties); §3A-2-02 (special powers and duties); §4A-1-01 (exclusive jurisdiction and control); §4A-1-02 (data exchange); Article 5A (dispute resolution) (Draper 2002b).

§2A-2-07 Waters of the Basin

"Waters of the Basin" shall include all water found within the Basin, whether surface, underground, or atmospheric water.

Commentary: This definition should be included to make it clear that underground water and atmospheric water are included within the scope of the agreement, if that is the intent of the parties. The technological questions relating to atmospheric water may result in uncertainty regarding its allocation, but to the extent the parties wish to reach a complete agreement, the matter should be addressed, or recognition should be given to the fact that the parties have chosen to reserve that issue for later resolution. The parties should also decide whether water imported from other basins should be included within the scope of the agreement. If it is not to be so included, that exclusion should be noted in this article (Draper 2002b; ILA 2004).

Cross References: §2A-1-04 (powers of signatory parties; withdrawal).

A.3 ARTICLE 3A: ADMINISTRATION

Part 1 Administrative Authority

Alternative 1 (administration by party officials)

§3A-1-01 Use of Water Management Officials of the Parties

It shall be the duty of the Parties to administer this Agreement through the official of each Party who is now or may hereafter be charged with the duty of administering the public water resources management and to collect and correlate through such officials the data necessary for the proper administration of the provisions of this agreement. Such officials may, by unanimous action, adopt rules and regulations consistent with the provisions of this agreement.

Commentary: This alternative provides an austere means of administering the agreement. It conceives of no additional administrative apparatus beyond the administrative machinery that exists with the parties.

Cross References: §1A-1-01 (general policies); §1A-1-02 (coordination and cooperation); §1A-1-03 (good faith implementation); §2A-2-05 (party or parties); §3A-1-02 (alternative 1, substitution of officials); §3A-1-03 (alternative 1, implementation and verification of agreement); §3A-1-04 (alternative 1, funding).

Similar Agreements: *La Plata River Compact*, 43 Stat. 796 (1925); *Republican River Compact*, 57 Stat. 86 (1943); *Snake River Compact*, 64 Stat. 29 (1949); *Costilla Creek Compact*, 60 Stat. 246 (1946), amended 77 Stat. 350 (1963); *Upper Niobrara River Compact*, 83 Stat. 86, Art. V (1969).

§3A-1-02 Substitution of Officials

Whenever any official of any Party is designated to perform any duty under this Agreement, such designation shall be interpreted to include the Party's official or officials upon whom the duties now performed by such official may hereafter devolve.

Commentary: This article is included to guard against confusion in the event there is a subsequent reorganization of a party's government (Draper 2002b).

Cross References: §1A-1-01 (general policies); §1A-1-02 (coordination and cooperation); §1A-1-03 (good faith implementation); §2A-2-05 (party or parties); §3A-1-01 (alternative 1, use of party officials); §3A-1-03 (alternative 1, implementation and verification of agreement); §3A-1-04 (alternative 1, funding).

Similar Agreements: *La Plata River Compact*, 43 Stat. 796 (1925); *South Platte River Compact*, 44 Stat. 195 (1923).

§3A-1-03 Implementation and Verification of Agreement

(a) Each Party shall identify or maintain the administrative machinery necessary to implement the provisions of this Agreement, and, where several governmental institutions are involved, create the necessary coordinating mechanism for the authorities dealing with designated aspects of the Agreement.

(b) Each Party shall have the duty to establish, maintain, and operate such suitable water gauging stations and facilities for measuring water quantity and quality as it finds necessary to administer and effect verification of this Agreement. Reliance on any existing facility operated by a regional authority, national government, or international organization shall not relieve the Party from ensuring that data of sufficient quality and quantity is available to administer and verify this agreement.

Commentary: Implementation and verification of the agreement require administrative and technical support that should be provided by the parties. This article obligates the parties to provide that support (Draper 2002b).

Cross References: §1A-1-01 (general policies); §1A-1-02 (coordination and cooperation); §1A-1-03 (good faith implementation); §3A-1-01 (alternative 1, use of party officials); §3A-1-04 (alternative 1, funding).

Similar Agreements: *La Plata River Compact*, 43 Stat. 796 (1925).

§3A-1-04 Funding

Each Party shall allocate sufficient qualified personnel with adequate enforcement powers and sufficient funds to accomplish the tasks necessary for the implementation of this Agreement.

Commentary: In the case of simple allocation agreements in which no commission is established, funding provisions are not normally included. However, to ensure that no misunderstanding exists concerning the responsibilities of each party, an explicit provision may be preferable. Article 3B-1-04 is designed to avoid disputes over financing by requiring that each party operate the necessary facilities within its borders (Draper 2002b).

Cross References: §1A-1-01 (general policies); §1A-1-02 (coordination and cooperation); §1A-1-03 (good faith implementation); §3A-1-01 (alternative 1, use of party officials); §3A-1-02 (alternative 1, substitution of officials); §3A-1-03 (alternative 1, implementation and verification of agreement).

Similar Agreements: *Agreement on the Conservation of Nature and Natural Resources* (ASEAN 1985).

Alternative 2 (administration by commission)

§3A-1-01 Commission Created

(a) The ____________ Commission (hereinafter the Commission) is hereby created as a body politic and corporate, with succession for the duration of this Agreement, as an agency and instrumentality of the governments of the respective signatory Parties.

(b) The Commission shall develop and effectuate policies for cooperation, coordination, and provision of information, concerning all planning and management activities and water projects affecting their common water resources, in order to assist the Parties in sharing their shared water resources in an equitable and reasonable manner.

Commentary: The name of the commission should reflect the geographical setting of the particular water resources, usually a river designating the border between the parties or a river running across their common border, to which the agreement refers (Draper 2002b).

To be effective in coordinating water utilization and precluding conflict, the institutional structure should necessarily stimulate cooperation between or among governmental institutions of the signatory parties. It should also be able to (1) determine the facts of water use in the territory of each party, (2) resolve disputes across the boundaries between or among the parties, (3) guide responses to unusual temporary water emergencies, (4) regulate or design long-term solutions, and (5) enforce its decision (e.g., United Nations 1975; Le Marquand 1978).

The organizational structure of the commission should be constituted according to the specifics of the water resource itself and the political structures of the parties involved. What works for wealthy nations may not work for developing countries, and, in some shared water situations, cultural differences (e.g., the Jordan River) may require different management structures than those used in situations involving similar cultures (e.g., the Rhine River) (Williams 1993). Sterner (1994) states that "there should be an appropriate legal structure and set of institutions that define property rights and establish the framework within which an environmental authority can function."

Cross References: §1A-1-01 (general policies); §3A-1-02 (alternative 2, commission jurisdiction); §3A-1-03 (alternative 2, commissioners); §3A-1-04 (alternative 2, status, immunities, and privileges,); §3A-1-05 (alternative 2, commission organization and staffing); §3A-1-06 (alternative 2, advisory boards); §3A-1-07 (alternative 2, rules of procedures); §3A-2-01 (alternative 2, general powers and duties); §3A-2-02 (alternative 2, special powers and duties); §3A-3-01 (alternative 2, meetings, hearings and records); §3A-3-02 (alternative 2, funding and expenses of the commission); §4A-1-01 (exclusive jurisdiction and control); §4A-1-02 (data exchange); Article 5A (dispute resolution).

Similar Agreements: *Klamath River Basin Compact*, 71 Stat. 497 (1957); *Delaware River Basin Compact*, Pub. L. 87-328, 75 Stat. 688-716 (1961); *Susquehanna River Basin Compact*, Pub. L. 91-575, 84 Stat. 1509-1541 (1970); *Apalachicola–Chattahoochee–Flint River Basin Compact*, O.C.G.A. 12-10-100 (1997); *Alabama–Coosa–Tallapoosa River Basin Compact*, O.C.G.A. 12-10-110 (1997); *Treaty between the United States and Great Britain relating to Boundary Waters*, 36 Stat. 2451 (1909); *Treaty between the United States and Mexico. Utilization of Waters of the Colorado and Tijuana Rivers and of the Rio Grande*, 59 Stat. 1219 (1944); *The Indus Waters Treaty 1960*, 419 UNTS 126 (1960); *Canada–Mexico–United States: North American Agreement on Environmental Cooperation*, 32 ILM 1480 (1993); *Agreement on the Cooperation for the Sustainable Development of the Mekong River Basin*, 34 ILM 864 (1995).

§3A-1-02 Commission Authorities and Responsibilities

The Commission shall coordinate all programs of data collection and dissemination within the _________ River Basin and shall facilitate the resolution of any controversy or conflict involving the use, obstruction, or diversion of those waters within the Basin that affects water availability and quality.

Commentary: This provision describes the geographic and hydrologic authorities and responsibilities of the commission. Similar provisions appear in all agreements to clarify and define the limits of authority ceded by the parties to an independent administrative unit (Hayton and Utton 1989).

Cross References: §1A-1-01 (general policies); §2A-1-02 (consent to jurisdiction); §3A-1-01 (alternative 2, commission created); §3A-1-03 (alternative 2, commissioners); §3A-1-04 (alternative 2, status, immunities, and privileges); §3A-1-05 (alternative 2, commission organization and staffing); §3A-1-06 (alternative 2, advisory boards); §3A-1-07 (alternative 2, rules of procedures); §3A-2-01 (alternative 2, general powers and duties); §3A-2-02 (alternative 2, special powers and duties); §3A-3-01 (alternative 2, meetings, hearings and records); §3A-3-02 (alternative 2, funding and expenses of the commission); §4A-1-01 (exclusive jurisdiction and control); §4A-1-02 (data exchange); Article 5A (dispute resolution).

Similar Agreements: *Delaware River Basin Compact*, Pub. L. 87-328, 75 Stat. 688-716 (1961); *Susquehanna River Basin Compact*, Pub. L. 91-575, 84 Stat. 1509-1541 (1970); *Treaty between the United States and Great Britain Relating to Boundary Waters*, 36 Stat. 2451 (1909); *Treaty between the United States and Mexico. Utilization of Waters of the Colorado and Tijuana Rivers and of the Rio Grande*, 59 Stat. 1219 (1944); *Agreement on the Cooperation for the Sustainable Development of the Mekong River Basin*, 34 ILM 864 (1995).

§3A-1-03 Commissioners

Alternative 1

(a) The Commission shall be governed by a Board of Commissioners, consisting of the Governor or other Chief Executive Officer of each of the signatory Parties plus a federal representative appointed by the President of the United States.

(b) Each Commissioner may appoint an alternate to act in his place and stead, with authority to attend all meetings of the Commission, and with power to vote in the absence of the Commissioner. Unless otherwise provided by law of the signatory Party for which he is appointed, each alternate shall serve during the term of the Commissioner appointing him, subject to removal at the pleasure of the Commissioner. In the event of a vacancy in the office of alternate, it shall be filled in the same manner as an original appointment for the unexpired term only.

(c) All matters requiring a policy decision affecting the substance of the Agreement shall be decided upon by the Commissioners and unanimity shall be required. For all other matters, the rule of decision shall be simple majority. Each member shall be entitled to one vote on all matters that may come before the Commission. No action by the Commissioners shall be taken at any meeting unless a majority of the membership shall vote in favor thereof. The federal representative shall have no vote.

Alternative 2

(a) The signatory Parties agree to establish and maintain the _____________ Commission composed of (___) commissioners, (___) on the part of each signatory party, appointed by the (President or Prime Minister or Head of State) thereof.

(b) The Chair of the Commission shall be for a term of (___) years and rotate according to the alphabetical listing of the signatory parties.

Commentary: Alternative 1 recognizes that the responsibility (and authority) for all policy decisions affecting the substance of the agreement remains with the principal executive office of the respective parties. The membership of the typical commission in the United States includes one or more members from each party plus a federal representative. However, in most instances, the federal representative has no vote (Grant 2001, §46.03). The *Delaware River Basin Compact* and the *Susquehanna River Basin Compact* are notable exceptions (Draper 2002b).

Alternative 2 is presented under the assumption that two parties are involved. When more than two parties are involved, appropriate changes to the composition of the commission should be made. Alternative 2 is less specific and consequently may be more appropriate for the international context.

Cross References: §1A-1-01 (general policies); §1A-1-04 (preservation of federal rights); §1A-1-05 (national security); §2A-1-02 (consent to jurisdiction); §2A-1-04 (powers of signatory parties; withdrawal); §2A-1-05 (amendments and supplements); §2A-1-06 (existing agencies); §2A-1-07 (limited applicability); §2A-2-05 (party or parties); §3A-1-01 (alternative 2, commission created); §3A-1-02 (alternative 2, commission jurisdiction); §4A-1-01 (exclusive jurisdiction and control).

Similar Agreements: *Klamath River Basin Compact*, 71 Stat. 497 (1957); *Delaware River Basin Compact*, Pub. L. 87-328, 75 Stat. 688-716 (1961); *Susquehanna River Basin Compact*, Pub. L. 91-575, 84 Stat. 1509-1541 (1970); *Treaty between the United States and Great Britain Relating to Boundary Waters*, 36 Stat. 2451 (1909); *Treaty between the United States and Mexico. Utilization of Waters of the Colorado and Tijuana Rivers and of the Rio Grande*, 59 Stat. 1219 (1944); *Convention between the Federal Republic of Germany and the Czech and Slovak Federal Republic and the European Economic Community on the International Commission for the Protection of the Elbe*, International Environmental Law, Multilateral Agreements, 976:90/1 (1990); *Convention on the Protection and Use of*

Transboundary Watercourses and International Lakes, 31 ILM 1312 (1992); *Canada–Mexico–United States: North American Agreement on Environmental Cooperation*, 32 ILM 1480 (1993); *Agreement on the Cooperation for the Sustainable Development of the Mekong River Basin*, 34 ILM 864 (1995).

§3A-1-04 Status, Immunities, and Privileges (Optional)
To enable the Commission to fulfill its purpose and the functions with which it is entrusted, the status, immunities, and privileges set forth in this Article shall be accorded to the Commission in the territories of each Party.

(a) The Commission, its property and its assets, wherever located, and by whomsoever held, shall enjoy the same immunity from suit and every form of judicial process as is enjoyed by the Parties, except to the extent that the Commission may expressly waive its immunity for the purposes of any proceedings or by the terms of any contract.
(b) Property and assets of the Commission, wheresoever located and by whomsoever held, shall be considered public property and shall be immune from search, requisition, confiscation, expropriation, or any other form of taking or foreclosure by executive or legislative action.
(c) To the extent necessary to carry out the purpose and functions of the Commission and to conduct its operations in accordance with this Agreement, all property and other assets of the Commission shall be free from restrictions, regulations, controls, and moratoria of any nature affecting the implementation of this Agreement, except as may otherwise be provided in this Agreement.
(d) The official communications of the Commission shall be accorded by each Party the same treatment that it accords to the official communications of the other Parties.
(e) (Optional, for international use only) The Commissioners and other personnel engaged directly in the affairs of the Commission shall have the following privileges and immunities:
  (1) Immunity from legal process with respect to acts performed by them in their official capacity except when the Commission expressly waives this immunity;
  (2) When not citizens of one of the signatory Parties, the same immunities from immigration restrictions, alien registration requirements, and national service obligations and the same facilities as regards exchange provisions as are accorded by each Party to the representatives, officials, and employees of comparable rank of the other Party; and
  (3) The same privileges in respect of traveling and facilities as are accorded by each Party to representatives, officials, and employees of comparable rank of the other Party.
(f) The Commission, its property, other assets, income, and the operations it carries out pursuant to this Article shall be immune from all state taxation. The Commission shall also be immune from any obligation relating to the payment, withholding or collection of any tax or customs duty. No state tax shall be levied on or in respect of salaries and benefits paid by the Commission to officers or staff of the Commission who are not local citizens.
(g) Each Party, in accordance with its juridical system, shall take such action as is necessary to make effective in its own territories the principles set forth in this Article, and shall inform the Commission of the action which it has taken on the matter.

Commentary: This provision provides the commissioners and their personnel with the same legal protections that normally exist for governmental officials of the parties (Draper 2002b).

Cross References: §1A-1-01 (general); §2A-2-05 (party or parties); §3A-1-01 (alternative 2, commission created); §3A-1-02 (alternative 2, commission jurisdiction); §3A-1-03 (alternative 2, commissioners); §3A-1-05 (alternative 2, commission organization and staffing); §3A-1-06 (alternative 2, advisory boards); §3A-1-07 (alternative 2, rules of procedures); §3A-2-01 (alternative 2, general powers and duties); §3A-2-02 (alternative 2, special powers and duties); §3A-3-01 (alternative 2, meetings, hearings and records); §3A-3-02 (alternative 2, funding and expenses of the commission); §4A-1-01 (exclusive jurisdiction and control); §4A-1-02 (data exchange).

Similar Agreements: *Agreement between the Government of the United States of America and the Government of the United Mexican States Concerning the Establishment of a Border Environment Cooperation Commission and a North American Development Bank*, 19 U.S.C. §473 (1993).

§3A-1-05 Commission Organization and Staffing

(a) The Commission shall meet and organize as promptly after the members thereof are appointed, and when organized the Commission may fix such times and places for its meetings as may be necessary, subject at all times to special call or direction by agreement of the Parties. Each Commissioner, upon the first joint meeting of the Commission after his appointment, shall, before proceeding with the work of the Commission, make and subscribe a solemn declaration in writing that he will faithfully and impartially perform the duties imposed upon him under this Agreement, and such declaration shall be entered on the records of the proceedings of the Commission.
(b) The respective Commissioners may each appoint a secretary, and these shall act as joint secretaries of the Commission at its joint sessions, and the Commissioners may utilize professional and administrative personnel from existing agencies of the respective Parties from time to time, as it may deem advisable. Their respective Governments shall pay the salaries and personal expenses of the Commissioners and their supporting staff.

Commentary: This provision provides the authority and instructions for the organization and initiation of commission undertakings (Draper 2002b).

Cross References: §1A-1-01 (general policies); §1A-1-02 (purposes of agreement); §1A-1-03 (objectives of agreement); §2A-2-05 (party or parties); §3A-1-01 (alternative 2, commission created); §3A-1-02 (alternative 2, commission jurisdiction); §3A-1-03 (alternative 2, commissioners); §3A-1-04 (alternative 2, status, immunities, and privileges); §3A-1-06 (alternative 2, advisory boards); §3A-1-07 (alternative 2, rules of procedures); §3A-2-01 (alternative 2, general powers and duties); §3A-2-02 (alternative 2, special powers and duties); §3A-3-01 (alternative 2, meetings, hearings and records).

Similar Agreements: *Treaty between the United States and Great Britain Relating to Boundary Waters*, 36 Stat. 2451 (1909);

Convention between the Federal Republic of Germany and the Czech and Slovak Federal Republic and the European Economic Community on the International Commission for the Protection of the Elbe, International Environmental Law, Multilateral Agreements, 976:90/1 (1990).

§3A-1-06 Advisory Boards (Optional)

- **(a) The Commission may appoint an advisory board or boards, composed of qualified persons to conduct on its behalf investigations and studies that may be necessary or desirable and to report to the Commission regarding any questions or matters involved in the subject matter of the reference.**
- **(b) Such boards ordinarily will have an equal number of members from each Party.**
- **(c) The Commission will make copies of the main or final report of such board or a digest thereof available for examination by all Parties.**

Commentary: This provision acknowledges that the resolution of some disputes involving water may require specialists and experts with particular knowledge or talent not available within the commission structure (Draper 2002b).

Cross References: §1A-1-01 (general policies); §2A-2-05 (party or parties); §3A-1-01 (alternative 2, commission created); §3A-1-02 (alternative 2, commission jurisdiction); §3A-1-03 (alternative 2, commissioners); §3A-1-04 (alternative 2, status, immunities, and privileges); §3A-1-05 (alternative 2, commission organization and staffing); §3A-1-07 (alternative 2, rules of procedures); §3A-2-01 (alternative 2, general powers and duties); §3A-2-02 (alternative 2, special powers and duties); §3A-3-01 (alternative 2, meetings, hearings and records); §4A-1-01 (exclusive jurisdiction and control); §4A-1-02 (data exchange).

Similar Agreements: *Treaty between the United States and Great Britain Relating to Boundary Waters*, 36 Stat. 2451 (1909).

§3A-1-07 Rules of Procedure

The Commission shall adopt its own Rules of Procedure, and may seek technical advisory services, as it deems necessary.

Commentary: Some compacts contain detailed rules of procedure. Such detail can, however, cause unnecessary complications to effective, efficient, and timely response to emergencies and extreme hydrological conditions, such as floods or droughts. Except when necessary for policy reasons, the agreement should not bind the commission to specific procedural requirements (Draper 2002b).

Cross References: §3A-1-01 (alternative 2, commission created); §3A-1-02 (alternative 2, commission jurisdiction); §3A-1-03 (alternative 2, commissioners); §3A-1-04 (alternative 2, status, immunities, and privileges); §3A-1-05 (alternative 2, commission organization and staffing); §3A-1-06 (alternative 2, advisory boards); §3A-2-01 (alternative 2, general powers and duties); §3A-2-02 (alternative 2, special powers and duties); §3A-3-01 (alternative 2, meetings, hearings and records).

Similar Agreements: *Klamath River Basin Compact*, 71 Stat. 497 (1957); *Delaware River Basin Compact*, Pub. L. 87-328, 75 Stat. 688-716 (1961); *Susquehanna River Basin Compact*, Pub. L. 91-575, 84 Stat. 1509-1541 (1970); *Canada–Mexico–United States: North American Agreement on Environmental Cooperation*, 32 ILM 1480 (1993); *Agreement on the Cooperation for the Sustainable Development of the Mekong River Basin*, 34 ILM 864 (1995); *Convention on the Law of the Non-Navigational Uses of International Watercourses*, United Nations Document A/51/869 (United Nations 1997).

Part 2 Powers and Duties

§3A-2-01 General Powers and Duties

The Commission shall have the power and responsibility

- **(a) To adopt bylaws and procedures governing its conduct;**
- **(b) To sue and be sued in any court of competent jurisdiction;**
- **(c) To retain and discharge professional, technical, clerical, and other staff and such consultants as are necessary to accomplish the purposes of this Agreement;**
- **(d) To receive funds from any lawful source and expend funds for any lawful purpose;**
- **(e) To enter into agreements or contracts, where appropriate, in order to accomplish the purposes of this Agreement;**
- **(f) To create committees and delegate responsibilities;**
- **(g) To plan, coordinate, monitor, and make recommendations for the use of the water resources shared by the Parties for the purposes of, but not limited to, coordination and cooperation for the purposes of minimizing adverse impacts of floods and droughts, and facilitating the utilization of the waters as may be deemed appropriate by the Commission;**
- **(h) To participate with other governmental and non-governmental entities in carrying out the purposes of this Agreement;**
- **(i) To perform all functions required of it by this Agreement and to do all things necessary, proper, or convenient in the performance of its duties hereunder, either independently or in cooperation with any Party.**

Commentary: Certain authorities and responsibilities pertain to any organizational structure, whether the structure is a business corporation, a governmental agency, or an institution developed to manage international relations. This list provides the commission with the minimum powers necessary to achieve the purposes assigned to it.

Cross References: §1A-1-04 (preservation of federal rights); §1A-1-05 (national security); §2A-1-02 (consent to jurisdiction); §2A-1-03 (duration of agreement); §2A-1-04 (powers of signatory parties; withdrawal); §2A-1-05 (amendments and supplements); §2A-1-06 (existing agencies); §2A-1-07 (limited applicability); §2A-2-02 (drought); §2A-2-03 (flood); §2A-2-05 (party or parties); §3A-1-07 (alternative 2, rules of procedures); §3A-2-02 (alternative 2, special powers and duties); §3A-3-01 (alternative 2, meetings, hearings and records); §3A-3-02 (alternative 2, funding and expenses of the commission); §4A-1-01 (exclusive jurisdiction and control); §4A-1-02 (data exchange); Article 5A (dispute resolution).

Similar Agreements: *Apalachicola–Chattahoochee–Flint River Basin Compact*, O.C.G.A. 12-10-100 (1997); *Alabama–Coosa–Tallapoosa River Basin Compact*, O.C.G.A. 12-10-110 (1997); *Delaware River Basin Compact*, Pub. L. 87-328, 75 Stat. 688-716 (1961); *Susquehanna River Basin Compact*, Pub. L. 91-575, 84 Stat. 1509-1541 (1970); *Klamath River Basin Compact*, 71 Stat. 497 (1957); *Treaty between the United States and Mexico. Utilization of Waters of the Colorado and Tijuana Rivers and of the Rio Grande*, 59 Stat. 1219 (1944); *Convention between the Federal Republic of Germany and the Czech and Slovak Federal Republic and the European Economic Community on the International Commission for the Protection of the Elbe*, International Environmental Law, Multilateral Agreements, 976:90/1 (1990); *Canada–Mexico–United States: North American Agreement on Environmental Cooperation*, 32 ILM 1480 (1993); *Agreement on the Cooperation for the Sustainable Development of the Mekong River Basin*, 34 ILM 864 (1995).

§3A-2-02 Special Powers and Duties (Optional)

(a) **The Commissioners shall develop and implement a comprehensive system to facilitate the exchange of information between the Parties. This system of information exchange shall be sufficient in scope to adequately promote the utilization of the water resources shared by the Parties.**

(b) **The Parties further agree that any other questions, conflicts or matters of difference arising between them involving the rights, obligations, or interests of either in relation to the other or to the inhabitants of the other, across or along the boundary between the Parties, shall be referred to the Commission for evaluation and report, whenever any of the Parties shall request that such questions or matters of difference be so referred.**

(c) **(Optional, for use with §3A-1-03, Alternative 2) The Commissioners shall make a joint report to the respective Governments in all cases in which all or a majority of the Commissioners agree, and in case of disagreement the minority may make a joint report to both Governments, or separate reports to their respective Governments. Such reports of the Commission shall not be regarded as decisions of the questions or matters so submitted either on the facts or the law, and shall in no way have the character of an arbitral award. In case the Commission is evenly divided upon any question or matter referred to it for report, the Commissioners on each side shall make separate reports to their own Government.**

(d) **(Optional, for use with §3A-1-03, Alternative 2) Unanimity of the Commissioners shall be required to render a decision. In case the Commission is divided upon any question or matter presented to it for decision, the Commissioners on each side shall make separate reports to their own Government. The Parties shall thereupon endeavor to agree upon an adjustment of the question or matter of difference, and if an agreement is reached between them, it shall be reduced to writing in the form of a protocol, and shall be communicated to the Commissioners, who shall take such further proceedings as may be necessary to carry out such agreement.**

Commentary: The commission is charged with certain purposes that are specific to the parties' agreement to the equitable and reasonable use of the water resources shared between the parties, the exchange of data and other information pertinent to water use by the respective parties, and the cooperation and consultation necessary to achieve the purposes of this agreement. This provision provides the necessary special powers to achieve the goals of the agreement (Draper 2002b).

Cross References: §1A-1-04 (preservation of federal rights); §1A-1-05 (national security); §2A-1-02 (consent to jurisdiction); §2A-1-03 (duration of agreement); §2A-1-04 (powers of signatory parties; withdrawal); §2A-1-05 (amendments and supplements); §2A-1-06 (existing agencies); §2A-1-07 (limited applicability); §2A-2-05 (party or parties); §3A-1-07 (alternative 2, rules of procedures); §3A-2-01 (alternative 2, general powers and duties); §3A-3-01 (alternative 2, meetings, hearings and records); §3A-3-02 (alternative 2, funding and expenses of the commission); §4A-1-01 (exclusive jurisdiction and control); §4A-1-02 (data exchange); Article 5A (dispute resolution).

Similar Agreements: *Treaty between the United States and Great Britain Relating to Boundary Waters*, 36 Stat. 2451 (1909); *Convention between the Federal Republic of Germany and the Czech and Slovak Federal Republic and the European Economic Community on the International Commission for the Protection of the Elbe*, International Environmental Law, Multilateral Agreements, 976:90/1 (1990).

Part 3 Administrative Procedures

§3A-3-01 Meetings, Hearings and Records (Optional)

(a) **The signatory Parties recognize the importance and necessity of public participation in promoting utilization of the water resource of the _____ River Basin. Consequently, all meetings of the Commission shall be open to the public except with respect to issues of personnel.**

(b) **The minutes of the Commission shall be a public record open to inspection at the respective offices of the Commissioners or their alternates during regular business hours.**

Commentary: Although optional, inclusion of this provision is essential to maximum effectiveness of coordination and cooperation. Incorporation of this article memorializes the public nature of the enterprise. Effective water sharing demands that all stakeholders have information upon which they can rely to make rational decisions about water use. Without sufficient public participation, the parties are unable to maximize the use of the water resource. The U.S.–Canada International Joint Commission has recognized the need for "engaging public support" (International Joint Commission 1984). Agenda 21 (United Nations 1992) recognizes the need for the widest cooperation between governmental and nongovernmental organizations (Draper 2002b).

Cross References: §3A-1-07 (rules of procedures); §3A-2-01 (general powers and duties); §3A-2-02 (alternative 2, special powers and duties).

Similar Agreements: *Delaware River Basin Compact*, Pub. L. 87-328, 75 Stat. 688-716 (1961); *Susquehanna River Basin Compact*, Pub. L. 91-575, 84 Stat. 1509-1541 (1970); *Apalachicola–Chattahoochee–Flint River Basin Compact*, O.C.G.A. 12-10-100 (1997); *Alabama–Coosa–Tallapoosa River Basin Compact*, O.C.G.A. 12-10-110 (1997); *Convention on the Protection and Use of Transboundary Watercourses and International Lakes*, 31 ILM 1312 (1992); *Canada–Mexico–United States: North American Agreement on Environmental Cooperation*, 32 ILM 1480 (1993).

§3A-3-02 Funding and Expenses of the Commission

Commissioners shall serve without compensation from the Commission. All general operational funding required by the Commission and agreed to by the members shall obligate each Party to pay an equal (equitable) share of such agreed upon funding. Funds remitted to the Commission by a Party in payment of such obligation shall not lapse; provided, however, that if any Party fails to remit payment within 90 days after payment is due, such obligation on the part of the other Parties shall terminate and any Party which has made payment may have such payment returned.

Commentary: Effective operation of the commission, and hence effective implementation of the agreement, requires a dependable source of funding. If a program of financing is uncertain, the agreement will fail. Although this provision suggests equal funding by the parties, other possibilities exist. "Equitable" funding may balance national economic power and the ability to pay for the commission. Alternatively, "equitable" funding may base funding on some economic measure of the benefits accruing to each of the parties as a result of the agreement. The term once decided upon should be defined in Article 2, Part 2 (Draper 2002b).

Cross References: §3A-1-01 (alternative 2, commission created); §3A-1-02 (alternative 2, commission jurisdiction); §3A-1-03 (alternative 2, commissioners); §3A-1-04 (alternative 2, status, immunities, and privileges); §3A-1-05 (alternative 2, commission organization and staffing); §3A-1-06 (alternative 2, advisory boards); §3A-3-01 (alternative 2, meetings, hearings and records).

Similar Agreements: *Convention between the Federal Republic of Germany and the Czech and Slovak Federal Republic and the European Economic Community on the International Commission for the Protection of the Elbe*, International Environmental Law, Multilateral Agreements, 976:90/1 (1990); *Canada–Mexico–United States: North American Agreement on Environmental Cooperation*, 32 ILM 1480 (1993); *Agreement on the Cooperation for the Sustainable Development of the Mekong River Basin*, 34 ILM 864 (1995); *Apalachicola–Chattahoochee–Flint River Basin Compact*, O.C.G.A. 12-10-100 (1997); *Alabama–Coosa–Tallapoosa River Basin Compact*, O.C.G.A. 12-10-110 (1997).

A.4 ARTICLE 4A: COORDINATION OF WATER ISSUES

§4A-1-01 Exclusive Jurisdiction and Control

(a) Each of the Parties reserves to itself, unless otherwise mandated by U.S. federal law or contractually agreed upon by the Parties, the exclusive control over the utilization, consumption or diversion of all waters within its jurisdiction.

(b) The Parties covenant, however, that any interference with or diversion from its natural channel of shared waters which results in injury to the other Party's utilization of the shared waters shall give rise to the injured Parties the same rights and entitlements to the same legal remedies as if such injury took place in the jurisdiction of the Party where such use, diversion or interference occurs.

(c) It is understood, however, that none of the Parties intends by the foregoing provision to surrender any right that it may have to enjoin or otherwise object to any interference with or diversion by the other Party of shared waters that has a reasonable potential to cause material injury to the utilization of shared waters within its jurisdiction.

Commentary: This provision establishes the principle of the right of each party to allocate or otherwise use and control the waters within its jurisdiction, constrained only by the requirement that such use be reasonable and equitable. Enforcement of this principle shall be based on those causes of actions and remedies available in tort within the legal system of the party causing the injury. This article does provide for prospective relief from use, interference, or diversion that may have a reasonable potential to cause material harm. A "reasonable potential" to cause material injury would be determined as a matter of law (Draper 2002b).

This provision provides legal remedies for the citizens of one party who may be injured by the activities involving the shared waters by citizens of another party. This provision provides the injured citizen with standing to sue or standing to seek administrative remedies in the jurisdiction of another party. Without such a provision, an injured individual would only be able to seek remedies by petitioning his or her own government to seek remedies from the other government. The complexities of such a political and administrative solution dramatically limit timely redress of the injury (Draper 2002b).

Cross References: §1A-1-04 (preservation of federal rights); §1A-1-05 (national security); §2A-1-02 (consent to jurisdiction); §2A-1-04 (powers of signatory parties; withdrawal); §2A-2-05 (party or parties); §4A-1-02 (data exchange).

Similar Agreements: *Treaty between the United States and Great Britain Relating to Boundary Waters*, 36 Stat. 2451 (1909); *Sabine River Compact*, 68 Stat. 690 (1953), amended 76 Stat. 34 (1962), 91 Stat. 281 (1977), 106 Stat. 4661 (1992).

§4A-1-02 Data Exchange

(a) The Parties agree to exchange complete documentation, data, and information on the status and condition of water in the various phases of the hydrologic cycle and on the development and utilization of those waters within their respective borders, if such development and utilization affect the shared water resources. The documentation, data, and information shall be sufficient in scope to allow the other Parties to determine the potential damages that may therefrom arise. Such exchange shall include, but is not limited to, the following activities that may substantially raise or lower the historical levels, or change the natural periodic fluctuations, of the shared resource:

(1) Planned remedial or protective works or any dams, reservoirs or other diversions or obstructions to waters;

(2) Significant interbasin transfers;

(3) Changes to consumptive water utilization;

(4) Existing and planned flood protection programs and works which increase the risk of flood damages to other Parties;

(5) Water development or utilization that has an adverse effect on the biological, physical, and chemical quality of the shared resource, so as to affect the public health and safety, the recreational potential, environmental sustainability or the quality of life of the other Parties;

(6) Any augmentation of water supply that may affect the other party.

(b) (Optional) Meetings to exchange complete documentation, data, and information shall take place, upon the call of the Party originating such data, within 60 days of such data being available. Exchange of data and any consultation resulting from provision of the data shall occur prior to the data and information becoming part of the public record.

Commentary: Water utilization by one party can have serious effects on the quantity and/or quality of the water available for use by the other parties. A significant source of controversy develops when one party undertakes the construction and operation of water supply reservoirs and hydropower facilities that may dramatically lower the flow in the shared water resource. Although this reduction may be limited to the initial start-up period and may be limited to a period of several years, severe economic and social effects may occur to other parties. As importantly, releases from the works should be coordinated to ensure that downstream users are not affected. In the case of water supply reservoirs, this coordination may become critical during periods of drought. In the case of hydropower dams, especially those of a "peaking power" nature, the timing of release may be critical. Other parties may also be affected by significant changes resulting from interbasin transfers or increased water consumption (Draper 2002b).

Flood control policies and works also have a dramatic effect on the timing and elevation of water levels and thus may become a major contentious issue among the parties. The issue should be addressed as an individual area of coordination (Draper 2002b). This provision recognizes the right of each signatory party to make efforts to safeguard its people and economic forces from flood damages but establishes an avenue for the sharing of data on flood control efforts, as well as an independent analysis of the effects of those efforts on other parties.

Changes and/or additions to the industrial base of one party, or increased urbanization of a party, can result in substantially degraded discharges into the shared waters. Poor-quality water imposes risks that the parties should recognize as a common threat. First there is the health risk to the population that uses the water for domestic purposes. Second, if the available water does not meet the standards for certain industrial purposes, there is the risk that economic growth may be impaired. Finally, there is the risk that quality degradation may have a severe effect on the ecology of the basin, resulting in long-term sustainability complications. Integration of water quality and quantity is essential. Agenda 21 (United Nations 1992) obligated all signatories to develop a program of water and sustainable development (Ahlander 1994). United Nations (1998) establishes the criterion that "watercourse States shall, individually and, where appropriate, jointly, protect and preserve the ecosystems of international watercourses" (Draper 2002b).

Surface and underground water may be degraded by a variety of factors. Major problems affecting the quality of these water resources arise, for instance, from inadequate domestic sewage treatment, inadequate controls on the discharge of industrial waste and effluent, the diversion of waters resulting in insufficient water to assimilate waste, the loss and destruction of catchment areas, the improper siting of industrial plants, deforestation, and poor agricultural practices that cause leaching of nutrients and pesticides. Transboundary water sharing should include effective plans and programs that eliminate, or at least minimize, the possible sources of water quality degradation (Draper 2002b).

The timing of the data exchange is important for all parties. It is recommended that the data and information be provided to other parties before the information becomes a part of the public record because changing public plans and programs is difficult at best (Draper 2002b).

Cross References: §1A-1-04 (preservation of federal rights); §1A-1-05 (national security); §2A-1-02 (consent to jurisdiction); §2A-2-03 (flood); §2A-2-04 (interbasin transfer); §2A-2-05 (party or parties); §4A-1-01 (exclusive jurisdiction and control).

Similar Agreements: *Klamath River Basin Compact*, 71 Stat. 497 (1957); *Arkansas River Basin Compact of 1965*, 80 Stat. 1409 (1966); *Kansas–Nebraska Big Blue River Basin Compact,* 86 Stat. 193 (1972); *Arkansas River Basin Compact of 1970*, 87 Stat. 569 (1973); *Delaware River Basin Compact*, Pub. L. 87-328, 75 Stat. 688-716 (1961); *Susquehanna River Basin Compact*, Pub. L. 91-575, 84 Stat. 1509-1541 (1970); *Red River Compact*, 94 Stat. 3305 (1980); *Oregon–California Goose Lake Interstate Compact*, 98 Stat. 291 (1984); *Treaty between the United States and Great Britain Relating to Boundary Waters*, 36 Stat. 2451 (1909); *Treaty between the United States and Mexico. Utilization of Waters of the Colorado and Tijuana Rivers and of the Rio Grande*, 59 Stat. 1219 (1944); *Pecos River Compact*, 63 Stat. 159 (1948); *Agreement between the United States and Canada on Great Lakes Water Quality*, 1153 UNTS 187 (1978); *Convention on the Protection and Use of Transboundary Watercourses and International Lakes*, 31 ILM 1312 (1992); *Convention on the Law of the Non-Navigational Uses of International Watercourses*, United Nations Document A/51/869 (United Nations 1997).

A.5 ARTICLE 5A: DISPUTE RESOLUTION (OPTIONAL)

Disputes inevitably arise as agreements are implemented. These disputes may involve differences in interpretation of an agreement's provisions or noncompliance with the agreement itself. The disputes may also arise because of changing conditions that alter the effectiveness of the agreement for one or more of the parties. Although a speedy and equitable process of dispute resolution serves all parties well, some governments do not wish to enter into an obligatory process. In such a case, Article 5A may be omitted. In other instances, the parties may recognize the need to institutionalize a dispute resolution process (Draper 2002b).

§5A-1-01 Resolution by Signatory Parties

Whenever any difference or dispute may arise between two or more Parties to this Agreement regarding any matters covered by this Agreement, particularly as to the interpretations of the Agreement and the legal rights of the parties, the Parties shall first make every effort to resolve the issue through negotiations and consultations based on the powers and duties herein described.

Commentary: This alternative is the least restrictive upon the autonomy of the parties yet expresses their recognition of the need for peaceful resolution of disputes. The alternative is appropriate in those cases where no commission has been established to manage the data exchange (Draper 2002b).

Cross References: §1A-1-01 (general policies); §1A-1-02 (coordination and cooperation); §1A-1-03 (good faith implementation); §1A-1-04 (preservation of federal rights); §1A-1-05 (national security); §2A-1-04 (powers of signatory parties; withdrawal); §2A-1-06 (existing agencies); §2A-1-07 (limited applicability); §2A-2-05 (party or parties); §5A-1-02 (alternative 1, right to litigate).

Similar Agreements: *Agreement on the Cooperation for the Sustainable Development of the Mekong River Basin*, 34 ILM 864 (1995).

§5A-1-02 Right to Litigate

Nothing in this Agreement shall be construed to limit or prevent either Party from instituting or maintaining any action or proceeding, legal or equitable, in any tribunal of competent jurisdiction for the protection of any right under this Agreement or the enforcement of any of its provisions.

Commentary: The existence of an appropriate tribunal may pose a problem in cases not involving an entity like the United States. It may be advisable to specify the tribunal in the agreement itself to avoid dispute over jurisdictional questions at a later date (Draper 2002b).

Cross References: §1A-1-01 (general policies); §1A-1-02 (coordination and cooperation); §1A-1-03 (good faith implementation); §1A-1-04 (preservation of federal rights); §1A-1-05 (national security); §2A-1-04 (powers of signatory parties; withdrawal); §2A-1-06 (existing agencies); §2A-1-07 (limited applicability); §2A-2-05 (party or parties); §5A-1-01 (alternative 1, resolution by signatory parties).

Similar Agreements: *Belle Fourche River Compact*, 58 Stat. 94 (1944); *Colorado River Compact*, 45 Stat. 1057 (1928); *Snake River Compact*, 64 Stat. 29 (1949).

A.6 SIGNATURES

IN WITNESS WHEREOF, and in evidence of the adoption and enactment into law of this Agreement by the signatory Parties, the representatives of the States of ____________, ____________, and ________ do hereby, in accordance with authority conferred by law, sign this Agreement in (_____) duplicate original copies, as attested by the appropriate authorities of the respective Parties, and have caused the seals of the respective States to be hereunto affixed this ____ day of ______.

Appendix B
MODEL WATER SHARING AGREEMENT B (LIMITED PURPOSE)

The *Model Water Sharing Agreement B (Limited Purpose)* should be used when water management needs are limited in scope. It is a model agreement to achieve certain specific, limited purposes.

B.1 ARTICLE 1B: DECLARATION OF POLICIES AND PURPOSES

§1B-1-01 Purposes and Scope of Agreement

(a) **The waters of the _______ River Basin have local, regional, and national significance; equitable and reasonable allocation of the shared waters of the Basin are public purposes for the respective signatory Parties.**

(b) **The purposes of this Agreement are to promote interstate [international] comity; to remove causes of present and future controversy; to make secure and protect water resource developments within the Parties; to equitably and reasonably allocate the shared waters of the Basin; and to augment the benefits of the shared waters of the Basin through joint planning and management of specific projects.**

(c) **The physical and other conditions peculiar to the Basin constitute the basis for this Agreement, and its provisions are applicable only [to the surface waters of the Basin] [to underground waters and atmospheric water augmentation as well as the surface waters affecting the Basin].**

Commentary: This article is critical to future interpretation and implementation of the agreement. The nature of this product-oriented compact presupposes the existence of specific purposes for which the agreement is made, and those should be set out here. If the agreement is to focus only on allocation, for example, references to drought or flood control strategies would be excluded (Draper 2002b). In paragraph (b), drafters may wish to use the words ". . . to promote international equity" rather than "interstate equity." In paragraph (c), the choice of parenthetical phrase to be used will depend on whether the agreement concerns only surface waters or all water sources.

It is important to acknowledge that the agreement reflects the particular circumstances and compromises reached in its formulation, as applied to the particular basin. Care should be taken to ensure that this agreement could not be applied to other basins, unless the intent of the parties is otherwise. The inclusion of §1-1-01(c) avoids later claims that other rivers and basins, or other bodies of water, should be dealt with in a similar manner. If underground water and atmospheric water are to be included within the scope of the agreement, it may be mentioned here. It is particularly important to address the atmospheric and underground water issues in this paragraph to avoid later disputes over whether or not underground water and atmospheric water are included within the scope of the agreement.

Cross References: §1B-1-02 (coordination and cooperation); §1B-1-03 (good faith implementation); §1B-1-04 (preservation of federal rights); §1B-1-05 (national security); §2B-1-01 (effective date); §2B-1-02 (duration of agreement); §2B-1-03 (consent to jurisdiction); §2B-1-04 (amendments and supplements); §2B-1-05 (limited applicability); §2B-1-06 (annexes); §2B-2-01 (atmospheric water); §2B-2-03 (conservation measures); §2B-2-04 (drought); §2B-2-06 (flood); §2B-2-07 (party or parties); §2B-2-08 (underground water); §2B-2-09 (waters of the basin); §3B-1-01 (use of party officials); §3B-1-02 (substitution of officials); §3B-1-03 (implementation and verification of agreement); §3B-1-04 (funding); §3B-2-01 (rights in territory of other party); §3B-2-02 (storage and diversion); §3B-2-03 (eminent domain); §3B-2-04 (navigation); §4B-1-01 (exclusive jurisdiction and control); §4B-1-02 (water allocation); §4B-1-03 (water levels protected); §4B-1-04 (underground water; limit on withdrawals); §4B-1-05 (flood protection works); §4B-1-06 (augmentation of supply); §4B-1-07 (water quality).

Similar Agreements: *Arkansas River Compact*, 63 Stat. 145 (1949); *Belle Fourche River Compact*, 58 Stat. 94 (1944); *Canadian River Compact*, 66 Stat. 74 (1952); *Colorado River Compact*, 45 Stat. 1057 (1928); *Costilla Creek Compact*, 60 Stat. 246 (1946), amended 77 Stat. 350, Art. I (1963); *Kansas–Nebraska Big Blue River Compact*, 86 Stat. 193 (1972); *Klamath River Basin Compact*, 71 Stat. 497 (1957); *Pecos River Compact*, 63 Stat. 159 (1948); *Red River Compact*, 94 Stat. 3305 (1980); *Snake River Compact*, 64 Stat. 29 (1949); *Upper Colorado River Basin Compact*, 63 Stat. 31 (1949); *Upper Niobrara River Compact*, 83 Stat. 86, Art. V (1969).

§1B-1-02 Coordination and Cooperation
Alternative 1

(a) **Each of the Parties pledges to support implementation of the provisions of this Agreement, and covenants that its officers and agencies will not hinder, impair, or prevent any other Party from carrying out any provision or recommendation of this Agreement.**

(b) **The Parties shall at all times endeavor to agree on the interpretation and application of this Agreement, and shall make every attempt through cooperation and consultations to arrive at a mutually satisfactory resolution of any matter that might affect its operation.**

(c) **The Parties agree that their respective governmental organizations shall provide the information necessary to assist in the equitable and reasonable utilization of those resources. Such information shall include, but not be limited to, all planning and management activities and water projects affecting their shared water resources.**

(d) **The Parties further acknowledge that all states are expected to conduct themselves with an absence of malice and deceit, with no intention to seek unconscionable advantage.**

Alternative 2
The Parties agree to the following objectives:

(a) **To cooperate and consult with the other Parties to this Agreement in their development, utilization, consumption, and conservation of the water and related resources shared by the Parties in order to ensure equitable and reasonable use of those waters while minimizing harm to other Parties.**
(b) **To cooperate on the basis of equality and territorial integrity in the utilization and protection of the shared water resources.**
(c) **To conduct themselves with an absence of malice or deceit and with no intention to seek unconscionable advantage.**

Commentary: Normally a state or nation enters into any international agreement with a position of self-interest. In the negotiations, each party seeks the rights and authorities critical to certain political, economic, or social objectives while ceding less critical rights and authorities to the other parties. While accepting this fact, all parties have a duty to cooperate and negotiate in good faith (Draper 2002b). This principle is the foundation of international law, and it applies in all relations between autonomous states.

This provision provides a framework for the parties in the development of their individual water policy planning. It recognizes that there are certain fundamental principles that each party should follow in their rational management of water resources. It would be irrational for one party to agree to "equitable and reasonable utilization" when it does not follow a similar philosophy within its own borders. These general objectives and principles improve the likelihood of accomplishing the purposes of the water sharing.

Cross References: §1B-1-01 (purposes and scope of agreement); §1B-1-03 (good faith implementation); §1B-1-04 (preservation of federal rights); §1B-1-05 (national security); §2B-1-01 (effective date); §2B-1-02 (duration of agreement); §2B-1-03 (consent to jurisdiction); §2B-1-04 (amendments and supplements); §2B-1-05 (limited applicability); §2B-1-06 (annexes); §2B-2-05 (equitable and reasonable utilization); §3B-1-01 (use of party officials); §3B-1-02 (substitution of officials); §3B-1-03 (implementation and verification of agreement); §3B-1-04 (funding); §3B-2-01 (rights in territory of other party); §3B-2-02 (storage and diversion); §3B-2-03 (eminent domain); §3B-2-04 (navigation); §4B-1-01 (exclusive jurisdiction and control); §4B-1-02 (water allocation); §4B-1-03 (water levels protected); §4B-1-04 (underground water; limit on withdrawals); §4B-1-05 (flood protection works); §4B-1-06 (augmentation of supply); §4B-1-07 (water quality); §5B-1-01 (resolution by signatory parties); §5B-1-02 (right to litigate).

Similar Agreements: *Agreement Between the People's Republic of Bulgaria and the Republic of Turkey Concerning Co-operation in the Use of the Waters of Rivers Flowing through the Territory of Both Countries*, UNTS, Vol. 807, 117 (1968); *Convention between Switzerland and Italy Concerning the Protection of Italo-Swiss Waters Against Pollution*, UNTS, Vol. 957, 277 (1972); *Stockholm Declaration of the United Nations Conference on the Human Environment*, 11 ILM 1416 (United Nations 1972); *Treaty for Amazonian Cooperation*, 17 ILM 1046 (1978); *Convention between the Federal Republic of Germany and the Czech and Slovak Federal Republic and the European Economic Community on the International Commission for the Protection of the Elbe*, International Environmental Law, Multilateral Agreements, 976:90/1 (1990); *Convention on the Protection and Use of Transboundary Watercourses and International Lakes*, 31 ILM 1312 (1992); *Canada–Mexico–United States: North American Agreement on Environmental Cooperation*, 32 ILM 1480 (1993); *Treaty of Peace between the State of Israel and the Hashemite Kingdom of Jordan*, 34 ILM 43 (1994); *Agreement on the Cooperation for the Sustainable Development of the Mekong River Basin*, 34 ILM 864 (1995); *Convention on the Law of the Non-Navigational Uses of International Watercourses*, United Nations Document A/51/869 (United Nations 1997).

§1B-1-03 Good Faith Implementation
The Parties agree to implement all provisions of this Agreement, and each covenants that its officers and agencies will not hinder, impair, or prevent any other Party carrying out any provision or recommendation of this Agreement.

Commentary: It should be noted that good faith misinterpretation of compact obligations does not excuse a party from damage liability (*Texas v. New Mexico*, 482 U.S. 124, 1987). In that case, the U.S. Supreme Court reasoned that a compact is a contract, and standard contract law does not allow a defense based on misinterpretation of contract obligations (Grant 2001, §45.07(c), §46.05(d)).

Cross References: §1B-1-01 (purposes and scope of agreement); §1B-1-02 (coordination and cooperation); §1B-1-04 (preservation of federal rights); §1B-1-05 (national security); §2B-1-01 (effective date); §2B-1-02 (duration of agreement); §2B-1-03 (consent to jurisdiction); §2B-1-04 (amendments and supplements); §2B-1-05 (limited applicability); §2B-1-06 (annexes); §3B-1-01 (use of party officials); §3B-1-02 (substitution of officials); §3B-1-03 (implementation and verification of agreement); §3B-1-04 (funding); §3B-2-01 (rights in territory of other party); §3B-2-02 (storage and diversion); §3B-2-03 (eminent domain); §3B-2-04 (navigation); §4B-1-01 (exclusive jurisdiction and control); §4B-1-02 (water allocation); §4B-1-03 (water levels protected); §4B-1-04 (underground water; limit on withdrawals); §4B-1-05 (flood protection works); §4B-1-06 (augmentation of supply); §4B-1-07 (water quality); §5B-1-01 (resolution by signatory parties); §5B-1-02 (right to litigate).

Similar Agreements: *Helsinki Rules on the Uses of the Waters of International Rivers*, 52 I.L.A. 484 (1966); *Stockholm Declaration of the United Nations Conference on the Human Environment*, 11 ILM 1416 (United Nations 1972); *Convention on the Law of the Non-Navigational Uses of International Watercourses*, United Nations Document A/51/869 (United Nations 1997).

§1B-1-04 Preservation of Federal Rights (Optional, for U.S. use)
Nothing in this Agreement shall be deemed:

(a) **To impair or affect any rights or powers of the United States, its agencies or instrumentalities, in and to the use of the waters of the _____ River Basin nor its capacity to acquire rights in and to the use of said waters;**
(b) **To subject any property of the United States, its agencies, or instrumentalities to taxation by either State or subdivision thereof, nor to create an obligation on the part of the United States, its agencies, or instrumentalities, by reason of the acquisition, construction, or operation of any property or works of whatsoever kind, to make any payments to any State or political subdivision thereof, State agency, municipality, or entity whatsoever in reimbursement for the loss of taxes;**

(c) To subject any property of the United States, its agencies, or instrumentalities, to the laws of any State to an extent other than the extent to which these laws would apply without regard to the Agreement.

Commentary: This article should be included in agreements between states of the United States. Arguably, it may be unnecessary to preserve federal rights, but inasmuch as Congress should approve the agreement, the inclusion of these provisions may facilitate obtaining that approval.

Cross References: §1B-1-01 (purposes and scope of agreement); §1B-1-05 (national security); §4B-1-01 (exclusive jurisdiction and control).

Similar Agreements: *Republican River Compact*, 57 Stat. 86 (1943); *Belle Fourche River Compact*, 58 Stat. 94 (1944); *Pecos River Compact*, 63 Stat. 159 (1948); *Snake River Compact*, 64 Stat. 29 (1949); *Upper Colorado River Basin Compact*, 63 Stat. 31 (1949); *Yellowstone River Compact*, 65 Stat. 663 (1950); *Canadian River Compact*, 66 Stat. 74 (1952); *Klamath River Basin Compact*, 71 Stat. 497 (1957); *Bear River Compact*, 72 Stat. 38 (1955), amended 94 Stat. 4, Art. XIII(2) (1980).

§1B-1-05 National Security (Optional, for international use)

(a) Nothing in this Agreement shall be construed to require any Party to make available or provide access to information the disclosure of which it determines to be contrary to its essential security interests.

(b) Nothing in this Agreement shall be construed to prevent any Party from taking any actions that it considers necessary for the protection of its essential security interests relating to a formal declaration of war.

Commentary: National security concerns necessarily take precedence over any program of water management and the exchange of data.

Cross References: §1B-1-01 (purposes and scope of agreement); §1B-1-04 (preservation of federal rights); §4B-1-01 (exclusive jurisdiction and control).

Similar Agreements: *Canada–Mexico–United States: North American Agreement on Environmental Cooperation*, 32 ILM 1480 (1993).

B.2 ARTICLE 2B: GENERAL

Part 1 General Obligations

§2B-1-01 Effective Date

Alternative 1: (For international use)

This Agreement shall become operative when approved by the appropriate governing authorities of all Parties. The Agreement will go into full force and effect at 12:01 a.m. [time zone] on the day immediately following the final act necessary for approval of the Agreement, as defined by the domestic law of each Party, by the last Party to give such approval.

Alternative 2: (For U.S. use)

This Agreement shall become operative when, subsequent to approval by the Legislature of each of the States, the Congress of the United States adopts legislation providing, among other things, that:

(a) Any equitable and reasonable uses hereafter made by the United States, or those acting by or under its authority, within a State, of the waters allocated by this Agreement, shall be within the allocations hereinabove made for use in that State and shall be taken into account in determining the extent of use within that State.

(b) The United States shall recognize, to the extent consistent with the best utilization of the waters for multiple purposes that equitable and reasonable use of the waters within the Basin is of paramount importance to development of the Basin. This shall pertain to the exercise of rights or powers arising from whatever jurisdiction the United States has in, over, and to the waters of the ___________ River and all its tributaries. The United States government shall exercise no power that may interfere with the full equitable and reasonable use of the waters unless the exercise of such power is in the interest of the best utilization of such waters for multiple purposes.

Commentary: Any agreement of this nature should specify the date or conditions upon which it will take effect. In the case of agreements between states of the United States, the conditions with respect to Congress are designed to provide some measure of protection against subsequent federal action that might disturb the allocation system agreed upon by the contracting parties. Despite the requirement of federal approval of interstate compacts, the federal government is not normally a party to those agreements and may not be bound by the provisions of those agreements unless there is specific legislation committing the federal government to be so bound. The provisions of §2-1-01, modeled after the *Republican River Compact*, 57 Stat. 86 (1943), and the *Belle Fourche River Compact*, 58 Stat. 94 (1944), condition the effectiveness of the agreement on passage of such legislation by Congress and also establish a basis for compensation for takings under the Fifth Amendment should a subsequent Congress decide to take action contrary to that commitment. A later Congress has the power to set aside the actions of an earlier Congress, but the question of takings and just compensation then arises. If these conditions are not incorporated, the states making the agreement may later find that federal actions render their agreement ineffective (Draper 2002b).

Cross References: §1B-1-01 (purposes and scope of agreement); §2B-1-02 (duration of agreement); §2B-1-03 (consent to jurisdiction); §2B-1-04 (amendments and supplements); §2B-1-05 (limited applicability); §2B-1-06 (annexes); §3B-1-01 (use of party officials); §3B-1-02 (substitution of officials); §3B-1-03 (implementation and verification of agreement).

Similar Agreements: *Republican River Compact*, 57 Stat. 86 (1943); *Belle Fourche River Compact*, 58 Stat. 94 (1944); *Delaware River Basin Compact*, Pub. L. 87-328, 75 Stat. 688-716 (1961); *Susquehanna River Basin Compact*, Pub. L. 91-575, 84 Stat. 1509-1541 (1970); *Convention on the Protection and Use of Transboundary Watercourses and International Lakes*, 31 ILM 1312 (1992); *Convention on the Law of the Non-Navigational Uses of International Watercourses*, United Nations Document A/51/869 (United Nations 1997).

§2B-1-02 Duration of Agreement (Optional)

The Parties intend that the duration of this Agreement shall be for an initial period of [___] years from its effective date.

Commentary: The parties may prefer to establish no duration and rely on later provisions to modify or terminate the agreement. However, two significant principles are established by this provision. First, setting the duration for an extended period of time allows for predictability on terms of water resource development; it also allows sufficient time to recover capital costs in the financing of projects. Second, establishing an extended duration ensures that the parties reconsider the agreement only after a sufficient hydrologic record is established. However, an extended duration does somewhat constrain the parties if significant climate change occurs and dramatically alters the hydrology of the shared water resource. Additionally, significant changes in water demands or changes in water technology could make the terms of the agreement unworkable (Draper 2002b).

Cross References: §2B-1-01 (effective date); §2B-1-04 (amendments and supplements).

Similar Agreements: *Delaware River Basin Compact*, Pub. L. 87-328, 75 Stat. 688-716 (1961); *Susquehanna River Basin Compact*, Pub. L. 91-575, 84 Stat. 1509-1541 (1970).

§2B-1-03 Consent to Jurisdiction (For U.S. use only)

This Agreement shall be effective upon the United States Congress giving its consent for the United States to be named and joined as a Party defendant or otherwise in any case or controversy involving the construction or application of this Agreement in which one or more of the States is a plaintiff, without regard to any requirement as to the sum or value in controversy or diversity of citizenship of Parties to the case or controversy.

Commentary: The predominance of federal interests in water resources makes it likely that any litigation concerning an agreement between states will involve federal interests. The doctrine of sovereign immunity could prevent joinder of the federal interests as parties to the suit absent a waiver of sovereign immunity. The inability to join federal parties led to dismissal of a suit filed by Texas against New Mexico in 1951 to enforce certain provisions of the *Rio Grande Compact of 1938*, 53 Stat. 785, 938 (1938). The U.S. Supreme Court dismissed the case because the federal government was not joined as a party but had important interests that would be affected by any such suit (*Texas v. New Mexico*, 352 U.S. 991 (1957). The parties should consider including such a waiver of sovereign immunity as a condition to effectiveness of the agreement. They may also wish to add a provision granting jurisdiction over any such cases to the district courts, which may be preferable to the U.S. Supreme Court as the initial forum for resolving certain types of disputes. The *Red River Compact*, 94 Stat. 3305 (1980) takes this approach (Draper 2002b).

Cross References: §1B-1-04 (preservation of federal right); §1B-1-05 (national security).

Similar Agreements: *Kansas–Nebraska Big Blue River Compact*, 86 Stat. 193 (1972); *Red River Compact*, 94 Stat. 3305 (1980).

§2B-1-04 Amendments and Supplements (Optional)

The provisions of this Agreement shall remain in full force and effect until amended by action of the governing bodies of the Parties and consented to and approved by any other necessary authority in the same manner as this Agreement is required to be ratified to become effective.

Commentary: Agreements may, over time, fail to operate as well as initially intended. Therefore, some amendment process should be specified. In some cases, the approval of another institution may be required. If, for example, the agreement is between states of the United States, the U.S. Constitution arguably requires congressional approval of any amendment as well as approval of the original agreement, unless the agreement provides for a different method of amendment (Draper 2002b). In this latter case, the congressional approval of the initial agreement would implicitly grant consent to modify the agreement in accordance with the terms of the agreement. If the agreement is between nation-states, the references to other "necessary authority" may be omitted, but the particular circumstances of each case should be considered.

Cross References: §1B-1-01 (purposes and scope of agreement); §2B-1-05 (limited applicability); §2B-1-06 (annexes); §3B-1-03 (implementation and verification of agreement).

Similar Agreements: *Delaware River Basin Compact*, Pub. L. 87-328, 75 Stat. 688-716 (1961); *Susquehanna River Basin Compact*, Pub. L. 91-575, 84 Stat. 1509-1541 (1970); *Convention on the Protection and Use of Transboundary Watercourses and International Lakes*, 31 ILM 1312 (1992); *Agreement on the Cooperation for the Sustainable Development of the Mekong River Basin*, 34 ILM 864 (1995).

§2B-1-05 Limited Applicability (Optional)

Should a tribunal of competent jurisdiction hold any part of this Agreement to be void or unenforceable, all other severable provisions shall continue in full force and effect.

Commentary: The drafters of the agreement should consider whether they wish this clause to be included. The advantage of such a clause is that it avoids the possibility of having the entire agreement become null and void if any part is found to be void or unenforceable. On the other hand, the agreement may be viewed as such an integrated package that the parties would choose to have the entire agreement fall if any part falls (Draper 2002b).

Cross References: §1B-1-01 (purposes and scope of agreement); §2B-1-04 (amendments and supplements); §2B-1-06 (annexes); §3B-1-01 (use of party officials); §3B-1-02 (substitution of officials); §3B-1-03 (implementation and verification of agreement); §3B-1-04 (funding); §3B-2-01 (rights in territory of other party); §3B-2-02 (storage and diversion); §3B-2-03 (eminent domain); §3B-2-04 (navigation); §4B-1-01 (exclusive jurisdiction and control); §4B-1-02 (water allocation); §4B-1-03 (water levels protected); §4B-1-04 (underground water; limit on withdrawals); §4B-1-06 (augmentation of supply); §4B-1-07 (water quality); §5B-1-01 (resolution by signatory parties); §5B-1-02 (right to litigate).

Similar Agreements: *Yellowstone River Compact*, 65 Stat. 663 (1950); *Sabine River Compact*, 68 Stat. 690 (1953), amended 76 Stat. 34 (1962), 91 Stat. 281 (1977), 106 Stat. 4661 (1992); *Klamath River Basin Compact*, 71 Stat. 497 (1957); *Delaware River Basin Compact*, Pub. L. 87-328, 75 Stat. 688-716 (1961); *Susquehanna River Basin Compact*, Pub. L. 91-575, 84 Stat. 1509-1541 (1970).

Part 2 Definitions

§2B-2-01 Atmospheric Water

The phrase "atmospheric water" means all available moisture above the surface of the earth, including underground and surface water and all forms of precipitation but not including water projecting from irrigation systems.

Commentary: This definition is consistent with the definitions usually used in state and federal laws on weather modification. See generally Beck (2001), §3.04; Davis (1987); Gochis (2001); Keyes (2006).

Cross References: §1B-1-01 (purposes and scope of agreement); §2B-2-09 (waters of the basin).

§2B-2-02 __________ Basin

"__________ Basin" means the area of drainage into the _______ River and its tributaries, [and] aquifers underlying the drainage, or only the aquifers themselves.

Commentary: The agreement could include the total surface area of drainage throughout the basin and contain aquifers underlying the surface drainage. Some tributaries can be connected to the underlying aquifers holding the underground water. Some of the aquifers could be connected to more than one of the surface water basins. The geographic scope of the agreement should be defined to ensure that there are no future disagreements about what lands are or are not covered by the agreement. A map may be incorporated, but care should be taken that the map is cartographically accurate. Because the map is likely to be at a scale too small for precise delineation of boundaries, it should be made clear that it is for general reference only. In the event of a dispute over land or within the defined _____ River and its tributaries, the actual limits of the watershed as determined on the ground should be controlling (Draper 2002b).

Cross References: §2B-2-05 (equitable and reasonable utilization); §3B-1-02 (substitution of officials); §4B-1-03 (water levels protected).

§2B-2-03 Conservation Measures

"Conservation measures" refers to any measures adopted by a water right holder, or several water right holders acting in concert pursuant to a conservation agreement reviewed and approved by the Commission as being appropriate water-saving strategies for purposes of the Comprehensive Water Management Plan, to reduce the withdrawals and/or consumptive uses, including, but not limited to:

(a) Improvements in water transmission and water use efficiency;
(b) Reduction in water use;
(c) Enhancement of return flows; and
(d) Reuse of return flows.

Commentary: Sustainable development requires steps to conserve the waters of the river basin. This definition limits the application of the term "conservation measures" to practices that have been reviewed and approved by the commission as being appropriate water-saving strategies for the purposes of the comprehensive water management plan. Specifically excluded from this definition are practices applied to native or naturally occurring waters, return flows from other water rights, or other water sources not associated with the water right holder or sought by the applicant (Draper 2002b).

Nothing in this model agreement attempts to spell out in detail what steps might actually qualify as appropriate conservation measures. Such efforts as improved efficiency in manufacturing processes, the substitution of drip irrigation for sprinklers, or the introduction by a public water supply enterprise of a requirement that customers use low-flow toilets or showerheads would all be appropriate examples. The model agreement leaves the precise details regarding the suitability of these or other possible conservation measures to be developed by the regulatory and planning processes prescribed for the state agency (Draper 2002b).

Cross Reference: §1B-1-01 (purposes and scope of agreement).

§2B-2-04 Drought

"Drought" means conditions of abnormal water scarcity in a specific area, resulting from natural conditions.

Commentary: Management action arises from a drought, or lack of mean annual rainfall, but could arise from other causes as well, such as the collapse of a dam with the resulting draining of a reservoir on which the commission users depend. The definition should be determined, in large measure, by the use intended. Then a "drought management strategy" would be a specific course of conduct planned by the commission as a necessary or appropriate response to the lack of precipitation (Draper 2002b).

Cross Reference: §1B-1-01 (purposes and scope of agreement).

§2B-2-05 Equitable and Reasonable Utilization Consumption or Diversion

Utilization, consumption or diversion of a transboundary water resource in an equitable and reasonable manner requires taking into account all relevant factors and circumstances, including:

(a) Geographic, hydrographic, hydrological, climatic, ecological and other factors of a natural character;
(b) The social and economic needs of the Parties concerned;
(c) The population dependent on the water resource in each of the Parties;
(d) The effects of the use or uses of the water resources in by one Part on other Parties;
(e) Existing and potential uses of the water resource;
(f) Conservation, protection, development and economy of use of the water resource and the costs of measures taken to that effect;
(g) The availability of alternatives, of comparable value, to a particular planned or existing use.
(h) The potential or actual material injury or harm to other Parties utilizing the shared water resource.

Commentary: This definition is based in the 1997 *Convention on the Law of the Non-Navigational Uses of International Watercourses,* which was approved by the General Assembly of the United Nations by a vote of all but three nations. Drafters may wish to revert to the use of the term "equitable apportionment." However, this phraseology and definition are recommended since it is more encompassing and more descriptive of water-related activities that will affect water availability to the Parties.

Cross-references: §1B-1-02 (coordination and cooperation); §2B-2-02 (equitable and reasonable utilization, consumption or diversion); §4B-1-01 (exclusive jurisdiction and control).

§2B-2-06 Flood

"Flood" means a rising of water to levels that have detrimental effects on or in one or more Basin States with a frequency agreeable to the Parties.

Commentary: The flood condition is almost the opposite of a drought. A large amount of water is to be controlled by facilities of the commission. The parties are to agree as to the frequency of the flow of high waters in the basin. Most of the time, these flows are during periods that exceed the amount of flow that occurs during the years of mean annual precipitation (Draper 2002b; ILA 2004). As an example, 10-year flood, meaning a flood that has a 10% probability of occurring during any one year.

Cross References: §1B-1-01 (purposes and scope of agreement); §4B-1-03 (water levels protected); §4B-1-05 (flood protection works).

§2B-2-07 Party or Parties

"Party or Parties" means, unless the text otherwise indicates, those governments signatory to this Agreement.

Commentary: Defining the terms in this way avoids the need to include similar language at numerous points throughout the agreement (Draper 2002b).

Cross References: §3B-1-01 (use of party officials); §3B-1-02 (substitution of officials); §3B-1-03 (implementation and verification of agreement); §3B-1-04 (funding); §3B-2-01 (rights in territory of other party); §3B-2-02 (storage and diversion); §3B-2-03 (eminent domain); §3B-2-04 (navigation); §4B-1-01 (exclusive jurisdiction and control); §4B-1-02 (water allocation); §4B-1-03 (water levels protected); §4B-1-04 (underground water; limit on withdrawals); §4B-1-05 (flood protection works); §4B-1-06 (augmentation of supply); §4B-1-07 (water quality); §5B-1-02 (right to litigate).

§2B-2-08 Underground Water

"Underground water" means water beneath the surface of the ground located in a saturated zone and in direct contact with the ground or soil.

Commentary: This definition of "underground water" includes all forms of water in the ground, being equivalent to terms such as "groundwater" and similar expressions. It excludes soil (capillary) moisture that might be drawn upon by plants but cannot practically be withdrawn by direct human activity. A somewhat more precise definition is found in the *Illinois Water Use Act*: "water under the ground where the fluid pressure in the pore space is equal to or greater than atmospheric pressure" (Illinois Water Use Act 1983; Dellapenna 2001 §6.04; Murphy 2001).

Cross References: §1B-1-01 (purposes and scope of agreement); §2B-2-09 (waters of the basin); 4B-1-07 (water quality).

§2B-2-09 Waters of the Basin

"Waters of the Basin" shall include all water found within the Basin, whether surface, underground, or atmospheric water other than marine waters.

Commentary: This definition would be included to make it clear that underground water and atmospheric water are included within the scope of the agreement, if that is the intent of the parties. The technological questions relating to atmospheric water and, possibly, underground water, may result in uncertainty regarding its allocation, but to the extent the parties wish to reach a complete agreement, the matter should be addressed, or recognition should be given to the fact that the parties have chosen to reserve that issue for later resolution. The parties should also decide whether water imported from other basins should be included within the scope of the agreement. If it is not to be so included, that exclusion should be noted in this article.

Cross References: §1B-1-01 (purposes and scope of agreement); §2B-2-01 (atmospheric water); §2B-2-08 (underground water); §4B-1-02 (water allocation); §4B-1-07 (water quality).

B.3 ARTICLE 3B: ADMINISTRATION

<u>Part 1 Administration Officials</u>

§3B-1-01 Use of Water Management Officials of the Parties

It shall be the duty of the Parties to administer this Agreement through the official of each Party who is now or may hereafter be charged with the duty of administering the public water supplies, and to collect and correlate through such officials the data necessary for the proper administration of the provisions of this Agreement. Such officials may, by unanimous action, adopt rules and regulations consistent with the provisions of this agreement.

Commentary: This article is one of two articles that provide a minimal means of administering the agreement. If a more structured or active administration is desired, a commission and authority may be established; *see* Model A (Coordination and Cooperation) and Model C (Comprehensive Management) for appropriate provisions and commentary (Draper 2002b).

Cross References: §1B-1-01 (purposes and scope of agreement); §1B-1-02 (coordination and cooperation); §1B-1-03 (good faith implementation); §2B-2-07 (party or parties); §3B-1-02 (substitution of officials); §3B-1-04 (funding).

Similar Agreements: *La Plata River Compact*, 43 Stat. 796 (1925); *Republican River Compact*, 57 Stat. 86 (1943); *Snake River Compact*, 64 Stat. 29 (1949); *Costilla Creek Compact*, 60 Stat. 246 (1946), amended 77 Stat. 350 (1963); *Upper Niobrara River Compact*, 83 Stat. 86, Art. V (1969).

§3B-1-02 Substitution of Officials

Whenever any official of any Party is designated to perform any duty under this Agreement, such designation shall be interpreted to include the Party's official or officials upon whom the duties now performed by such official may hereafter devolve.

Commentary: This is the second of two articles that provide a minimal means of administering the agreement. Article 3B-1-02 is included to guard against confusion in the event that there is a subsequent reorganization of a party's government (Draper 2002b).

Cross References: §1B-1-01 (purposes and scope of agreement); §1B-1-02 (coordination and cooperation); §1B-1-03 (good faith implementation); §2B-2-02 (basin); §2B-2-07 (party or parties); §3B-1-01 (use of party officials); §3B-1-04 (funding).

Similar Agreements: *La Plata River Compact*, 43 Stat. 796 (1925); *South Platte River Compact*, 44 Stat. 195 (1923).

§3B-1-03 Implementation and Verification of Agreement

(a) Each Party shall identify or maintain the administrative machinery necessary to implement the provisions of this Agreement and, where several governmental institutions are involved, create the necessary coordinating mechanism for the authorities dealing with designated aspects of the environment.

(b) Each Party shall establish, maintain, and operate such suitable water gauging stations and facilities for measuring water quantity and quality as it finds necessary to administer and effect verification of this Agreement.

Commentary: Implementation and verification of the agreement require administrative and technical support that should be provided by the parties. This article obligates the parties to provide that support (Draper 2002b).

Cross References: §1B-1-01 (purposes and scope of agreement); §1B-1-02 (coordination and cooperation); §1B-1-03 (good faith implementation); §2B-2-07 (party or parties); §3B-1-04 (funding); §3B-2-01 (rights in territory of other party); §3B-2-02 (storage and diversion); §3B-2-03 (eminent domain); §3B-2-04 (navigation); §4B-1-01 (exclusive jurisdiction and control); §4B-1-02 (water allocation); §4B-1-03 (water levels protected); §4B-1-04 (underground water; limit on withdrawals); §4B-1-05 (flood protection works); §4B-1-06 (augmentation of supply); §4B-1-07 (water quality).

Similar Agreement: *La Plata River Compact*, 43 Stat. 796 (1925).

§3B-1-04 Funding

Each Party shall allocate sufficient qualified personnel with adequate enforcement powers and sufficient funds to accomplish the tasks necessary for the implementation of this Agreement.

Commentary: In the case of simple allocation agreements in which no commission is established, funding provisions are not normally included. However, to ensure that no misunderstanding exists concerning the responsibilities of each party, an explicit provision may be preferable. Article 3B-1-04 is designed to avoid disputes over financing by requiring that each party will operate the necessary facilities within its borders (Draper 2002b).

Cross References: §1B-1-01 (purposes and scope of agreement); §1B-1-02 (coordination and cooperation); §1B-1-03 (good faith implementation); §2B-2-07 (party or parties); §3B-

1-03 (implementation and verification of agreement); §3B-2-01 (rights in territory of other party); §3B-2-02 (storage and diversion); §3B-2-03 (eminent domain); §3B-2-04 (navigation); §4B-1-01 (exclusive jurisdiction and control); §4B-1-02 (water allocation); §4B-1-03 (water levels protected); §4B-1-04 (underground water; limit on withdrawals); §4B-1-05 (flood protection works); §4B-1-06 (augmentation of supply); §4B-1-07 (water quality).

Similar Agreement: *Agreement on the Conservation of Nature and Natural Resources* (ASEAN 1985).

Part 2 Activities Within the Territory of Other Party

§3B-2-01 Rights in Territory of Other Party (Optional)

Either Party shall have the right, in accordance with the laws of the other Party, to file applications for and obtain consent to construct or participate in the construction and use of any dam, storage reservoir, or diversion works in the territory of the other Party for the purpose of conserving and regulating the apportioned water without prejudice based on extraterritorial status, provided, that such right is subject to the rights of the other Party to control, regulate, and use water apportioned to it.

Commentary: To achieve efficient use of allocated water, it may be desirable for one party or its citizens to construct reservoirs or other works within the boundaries of the other party. This is the first of two articles that strive to gain concurrence and consent to do so while preserving the rights of the party in which the works are constructed to control resource use within its territory (Draper 2002b).

Cross References: §1B-1-01 (purposes and scope of agreement); §1B-1-04 (preservation of federal rights); §1B-1-05 (national security); §2B-1-03 (consent to jurisdiction); §2B-2-07 (party or parties); §3B-1-03 (implementation and verification of agreement); §3B-2-02 (storage and diversion); §3B-2-03 (eminent domain); §3B-2-04 (navigation); §4B-1-01 (exclusive jurisdiction and control).

Similar Agreements: *Belle Fourche River Compact*, 58 Stat. 94 (1944); *Republican River Compact*, 57 Stat. 86 (1943); *Snake River Compact*, 64 Stat. 29 (1949); *Upper Colorado River Basin Compact*, 63 Stat. 31 (1949); *Yellowstone River Compact*, 65 Stat. 663 (1950).

§3B-2-02 Storage and Diversion (Optional)

Each claim hereafter initiated for storage or diversion of water in the territory of one Party for use by another Party shall be filed in the appropriate office of the Party in which the water is to be diverted, and a duplicate copy of the application including a map showing the character and location of the proposed facilities and the location(s) of the proposed uses shall be filed in the appropriate office of the Party from which the water is to be withdrawn. Any such construction or diversion by one Party within the territory of a second Party shall be subject to all appropriate laws and regulations of the second Party.

Commentary: This is the second of two articles that strive to gain concurrence and consent to do so, while preserving the rights of the party in which the works are constructed to control resource use within its territory (Draper 2002b).

Cross References: §1B-1-01 (purposes and scope of agreement); §1B-1-04 (preservation of federal rights); §1B-1-05 (national security); §2B-1-03 (consent to jurisdiction); §2B-2-07 (party or parties); §3B-1-03 (implementation and verification of agreement); §3B-2-01 (rights in territory of other party); §3B-2-03 (eminent domain); §3B-2-04 (navigation); §4B-1-01 (exclusive jurisdiction and control).

Similar Agreements: *Belle Fourche River Compact*, 58 Stat. 94 (1944); *Snake River Compact*, 64 Stat. 29 (1949); *Yellowstone River Compact*, 65 Stat. 663 (1950).

§3B-2-03 Eminent Domain (Optional)

(a) **Any Party, or person, or other entity claiming water pursuant to the allocation of water to either Party, shall have the right to acquire necessary property rights in the territory of another Party by purchase or through the exercise of the power of eminent domain for the construction, operation, and maintenance of storage reservoirs and of appurtenant works, canals, and conduits required for the enjoyment of the privileges granted by Article 4B, provided, however, that the Party, person, or entity exercising such rights shall pay to the political subdivisions of the Party in which such works are located, each and every year during which such rights are enjoyed for such purposes, a sum of money equivalent to the average annual amount of current year taxes assessed against the lands and improvements thereon during the years preceding the use of such lands in reimbursement for the loss of taxes to said political subdivision of the Party.**

(b) **Such power of condemnation shall be exercised in accordance with the provisions of any law applicable to the jurisdiction in which the property is located.**

(c) **Nothing in this Agreement authorizes the taking of any existing vested property right in the use of water except for just compensation, in accordance with the internal laws of the Party in which the property or usufructuary right exists.**

Commentary: To actually use the water allocated by the agreement, it may be necessary for one of the parties or its citizens to construct reservoirs or other works within the boundaries of the other party. This provision allows that to be done, through eminent domain if necessary, but also provides for payments in lieu of property taxes to avoid problems, which might arise if one party attempted to tax property belonging to another. If this provision is not included, the use of eminent domain presents a question of party law in the party in which the works are to be constructed. This alternative is likely to be adopted only within the United States or another federal system; issues of authority may preclude use of this alternative as between autonomous states on the international level (Draper 2002b).

This provision expressly requires "just compensation" for any taking of property rights. The "just compensation," however, depends largely on the individual internal laws of the parties themselves. In the United States, the recent rulings in regulatory affairs by the U.S. Supreme Court have held that a serious impairment of the value of land by a regulation of its use should be compensated, but lawmakers have noted that the state could diminish the value of a water right by as much as 95% without incurring liability, at least when a system of regulated riparian rights exist (Lucas 112 S. Ct. 2886, 1992; Byrne 1995; Houck 1995; Sax 1990).

Cross References: §1B-1-01 (purposes and scope of agreement); §1B-1-04 (preservation of federal rights); §1B-1-05 (national security); §2B-1-03 (consent to jurisdiction); §2B-2-07 (party or parties); §3B-1-03 (implementation and verification of agreement); §3B-2-01 (rights in territory of other party); §3B-2-02 (storage and diversion); §4B-1-01 (exclusive jurisdiction and control).

Similar Agreements: *Belle Fourche River Compact*, 58 Stat. 94 (1944); *Republican River Compact*, 57 Stat. 86 (1943); *Snake River Compact*, 64 Stat. 29 (1949); *Upper Colorado River Basin Compact*, 63 Stat. 31 (1949); *Yellowstone River Compact*, 65 Stat. 663 (1950).

§3B-2-04 Navigation (For international use)

(a) **The Parties agree that the navigation of the _______ River shall forever continue free and open for the purposes of commerce to the inhabitants and to the ships, vessels, and boats of both Parties equally, subject, however, to any laws and regulations of either Party, within its own territory, not inconsistent with such privilege of free navigation and applying equally and without discrimination to the inhabitants, ships, vessels, and boats of both Parties.**

(b) **It is further agreed that so long as this Agreement shall remain in force, this same right of navigation shall extend to all water bodies, tributaries, and canals connecting boundary waters, now existing or which may hereafter be constructed on either side of the line. Either of the Parties may adopt rules and regulations governing the use of such connecting water bodies, tributaries, and canals within its own territory and may charge tolls for the use thereof, but all such rules and regulations and all tolls charged shall apply alike to the subjects or citizens of the Parties and the ships, aid vessels, and boats of both of the Parties, and they shall be placed on terms of equality in the use thereof.**

Commentary: A clear statement of navigational interests of each party is set forth. Should some other paramount use between the parties dominate at the time of treating or negotiations, this utilization would be set forth herein (Draper 2002b).

Cross References: §1B-1-01 (purposes and scope of agreement); §1B-1-04 (preservation of federal rights); §1B-1-05 (national security); §2B-1-03 (consent to jurisdiction); §2B-2-07 (party or parties); §3B-1-03 (implementation and verification of agreement); §3B-2-01 (rights in territory of other party); §3B-2-02 (storage and diversion); §4B-1-01 (exclusive jurisdiction and control); §4B-1-03 (water levels protected).

Similar Agreement: *Treaty between the United States and Great Britain Relating to Boundary Waters*, 36 Stat. 2451 (1909).

B.4 ARTICLE 4B: EQUITABLE AND REASONABLE USE OF WATER

§4B-1-01 Exclusive Jurisdiction and Control

(a) **Each of the Parties reserves to itself, unless otherwise mandated by federal law or contractually agreed upon by the Parties, the exclusive control over the equitable and reasonable utilization, consumption, or diversion of all waters within its borders.**

(b) **The Parties agree that any use or diversion from their natural channel of shared waters which result in injury to one Party's equitable and reasonable utilization of the shared waters shall give to the injured Party, or citizens of that Party, the same legal rights and entitlements to the same legal remedies as if such injury took place within the jurisdiction of the Party where such use or diversion occurs. Requirements for legal standing of citizens of the Party incurring injury by actions within the territory of the other Party shall be identical to those established for the citizens of the other Party.**

(c) **It is understood, however, that neither Party intends by the foregoing provision to surrender any right that it may have to enjoin or otherwise object to any interference with nor diversion by the other Party of shared waters that has a reasonable potential to cause material injury to the equitable and reasonable utilization of shared waters within its jurisdiction.**

Commentary: This provision establishes the principle of the right of each party to allocate or otherwise use and control the waters within its borders, constrained only by the requirement that such use be reasonable and equitable. Enforcement of this principle shall be based on those causes of actions and remedies available in tort within the legal system of the party causing the injury. This article provides for prospective relief from use, interference, or diversion that may have a reasonable potential to cause material harm. A "reasonable potential" to cause material injury would be determined as a matter of law. The provision also resolves any legal standing issues that may arise in the case of citizens of one party requesting legal intervention within the territory of the other party (Draper 2002b).

Cross References: §1B-1-01 (purposes and scope of agreement); §1B-1-04 (preservation of federal rights); §1B-1-05 (national security); §2B-1-03 (consent to jurisdiction); §2B-2-05 (equitable and reasonable utilization); §2B-2-07 (party or parties); §3B-1-03 (implementation and verification of agreement); §3B-2-01 (rights in territory of other party); §3B-2-02 (storage and diversion); §3B-2-03 (eminent domain); §3B-2-04 (navigation); §4B-1-01 (exclusive jurisdiction and control).

Similar Agreements: *Treaty between the United States and Great Britain Relating to Boundary Waters*, 36 Stat. 2451 (1909); *Sabine River Compact*, 68 Stat. 690 (1953), amended 76 Stat. 34 (1962), 91 Stat. 281 (1977), 106 Stat. 4661 (1992).

§4B-1-02 Water Allocation (Optional)
[See Annex I for specific allocation alternatives]

Commentary: The basic allocation is a matter for negotiation. It may be based upon the relative geographic areas of the parties, the relative contribution of each party to the flow of the boundary stream, or any other method that is agreed upon as being equitable. A significant issue to be resolved in the negotiations is the means or methods used to verify compliance with the allocation provisions. A simple expedient of measuring flow at a specific point may be acceptable in cases in which the river flow is consistently stable. However, in cases in which extreme variations in flow occur or in geographic regions in which unstable meteorological and climatic changes are the norm, verification may require a sophisticated, complex scheme based on consistently measured consumption by the parties (McCormick 1994a, b; Draper 2002b).

The actual agreement may be relatively simple, as in the case of the *Sabine River Compact,* 68 Stat. 690 (1953), amended 76 Stat. 34 (1962), 91 Stat. 281 (1977), 106 Stat. 4661 (1992), between Texas and Louisiana. The Sabine originates in Texas, and then forms the border between the two states. Texas is given the right to unrestricted use of the water above the gauging station at Logansport, where the river becomes the state boundary, except for an essentially de minimis flow requirement. The water in the boundary reach is allocated equally between the two states, and any withdrawal from tributaries to that reach of the river is charged against the withdrawing state's allocation. A more complicated approach is seen in the *Red River Compact* among Texas, Oklahoma, Arkansas, and Louisiana (1980). The Red River serves not only as a state boundary but also flows across borders. In addition, both appropriative and riparian water rights are recognized, depending on which state is involved. The resulting allocation is based on a division of the river into five reaches, with separate allocations for subbasins within each reach (McCormick 1994a, b; Draper 2002b).

Annex I provides a number of alternatives for allocation formulas that have been utilized in interstate compacts in the United States (McCormick 1994a, b).

Cross References: §1B-1-01 (purposes and scope of agreement); §1B-1-04 (preservation of federal rights); §1B-1-05 (national security); §2B-1-03 (consent to jurisdiction); §2B-2-07 (party or parties); §2B-2-09 (waters of the basin); §3B-1-03 (implementation and verification of agreement); §3B-2-01 (rights in territory of other party); §3B-2-02 (storage and diversion); §3B-2-03 (eminent domain); §3B-2-04 (navigation); §4B-1-02 (water allocation); §4B-1-03 (water levels protected); §4B-1-04 (underground water; limit on withdrawals); §4B-1-05 (flood protection works); §4B-1-06 (augmentation of supply); §4B-1-07 (water quality).

Similar Agreements: See Annex I.

§4B-1-03 Water Levels Protected (Optional)

(a) The Parties agree that, except for the dams, reservoirs, obstructions, and diversions heretofore permitted or hereafter provided for by special agreement between the Parties hereto, no additional dams, reservoirs, obstructions or diversions of the shared waters of the ________ River Basin which affect the natural level or flow of boundary waters on the other side of the line shall be made except by approval of the other Party.

(b) The signatory Parties agree to furnish the other Party with complete documentation of any planned remedial or protective works or any dams, reservoirs, or other diversions or obstructions to waters flowing into or from waters of the Basin. The other Party shall analyze the documentation to determine the potential flood damages that may therefrom arise and consult with the signatory Parties concerning the findings of such analysis.

(c) The foregoing provisions are not intended to limit or interfere with the existing rights of the Parties to undertake and carry on governmental works in shared waters for water development activities for economic growth, public health, recreational activities, or environmental protection, provided that such works are wholly within its jurisdiction and do not materially affect the level or flow of the waters available to the other Party.

Commentary: A significant source of controversy develops during the construction and operation of water supply reservoirs and hydropower facilities. In both cases, significant reduction in flow may dramatically lower the flow in the shared water resource. Although this reduction may be limited to the initial start-up period and may be limited to a period of several years, severe economic and social effects may occur to other parties. As importantly, releases from the works should be coordinated to ensure that downstream users are not affected. In the case of water supply reservoirs, this requirement may become critical during periods of drought. In the case of hydropower dams, especially those of a "peaking power" nature, the timing of release may be critical. This provision requires the sharing of data concerning such works and establishes a means of analysis of potential effects by specifically allowing for "special agreements" regarding raised levels above the natural level of transboundary rivers between the parties. It also prohibits "dams or other obstructions" from raising the natural level of waters on the other side without such works having the approval of the other party (Draper 2002b).

Cross References: §1B-1-01 (purposes and scope of agreement); §2B-1-03 (consent to jurisdiction); §2B-2-02 (basin); §2B-2-06 (flood); §2B-2-07 (party or parties); §3B-1-03 (implementation and verification of agreement); §3B-2-01 (rights in territory of other party); §3B-2-02 (storage and diversion); §3B-2-04 (navigation); §4B-1-01 (exclusive jurisdiction and control); §4B-1-02 (water allocation); §4B-1-03 (water levels protected); §4B-1-04 (underground water; limit on withdrawals); §4B-1-05 (flood protection works); §4B-1-06 (augmentation of supply); §4B-1-07 (water quality).

Similar Agreements: *Treaty between the United States and Great Britain Relating to Boundary Waters*, 36 Stat. 2451 (1909); *Pecos River Compact*, 63 Stat. 159 (1948).

§4B-1-04 Underground Water; Limit on Withdrawals (Optional)

When such action is necessary [to maintain an allocation set out elsewhere], the Parties shall regulate, in the same manner that surface flow is regulated, withdrawal of water from irrigation wells located within ___ miles of the river or its tributaries.

Commentary: If underground water is subject to the allocation provisions of the agreement, it may be useful to specifically address the steps to be taken with respect to withdrawals. This provision, adapted from the *Kansas–Nebraska Big Blue River Compact* between Kansas and Nebraska, 86 Stat. 193 (1972), uses a distance limitation to determine which wells fall within the scope of the agreement. If it is possible to establish the hydrological connection among all wells and the surface flow, the mileage limitation may be replaced with references to wells with such connection. In the absence of definitive hydrologic information, the mileage limitation may make administration easier, if less accurate (Draper 2002b).

No further specific allocation systems for underground water are provided because it is assumed that if underground water is allocated by agreement, that allocation is in conjunction with allocation of related surface water sources and the allocation of underground water is incorporated as part of the overall allocation of water. If underground water is allocated independently from surface water, the parties might use the surface models as a guide with respect to types of allocations (e.g., proportional or guaranteed minimum) (Draper 2002b).

Cross References: §1B-1-01 (purposes and scope of agreement); §2B-1-03 (consent to jurisdiction); §2B-2-07 (party or parties); §3B-1-03 (implementation and verification of agreement); §3B-2-01 (rights in territory of other party); §3B-2-02 (storage and diversion); §3B-2-04 (navigation); §4B-1-01 (exclusive jurisdiction and control); §4B-1-02 (water allocation); §4B-1-03 (water levels protected); §4B-1-05 (flood protection works); §4B-1-06 (augmentation of supply); §4B-1-07 (water quality).

Similar Agreement: *Kansas–Nebraska Big Blue River Compact*, 86 Stat. 193 (1972).

§4B-1-05 Flood Protection Works (Optional)

(a) As a general concept, the use of the channels of the waters of the _____ River Basin for the discharge of flood or other excess waters shall be free and not subject to limitation by either country, and neither country shall have any claim against the other in

respect of any damage caused by such use. However, the signatory Parties declare their intention to manage flood control programs and activities in such manner, consistent with the normal operations of its hydraulic systems, as to avoid, as far as feasible, material damage in the territory of the other.

(b) Each Party agrees to furnish the other Party with complete documentation of existing and planned flood protection programs and works. The other Party shall analyze the documentation to determine the potential flood damages that may therefrom arise and enter into consultations and negotiations as necessary concerning the findings of such analysis.

Commentary: Flood control policies and works can have a dramatic effect on the timing and elevation of water levels and thus may become a major contentious issue between the parties. The issue often goes beyond "equitable and reasonable utilization" and needs to be addressed as an individual area of coordination. This provision recognizes the right of each party to make efforts to safeguard its people and economic forces from flood damages but also establishes an avenue for the sharing of data on flood control efforts, as well as an independent analysis by a third party of the effects of those efforts on other parties (Draper 2002b).

Cross References: §1B-1-01 (purposes and scope of agreement); §1B-1-04 (preservation of federal rights); §1B-1-05 (national security); §2B-1-03 (consent to jurisdiction); §2B-2-06 (flood); §2B-2-07 (party or parties); §3B-1-03 (implementation and verification of agreement); §3B-2-01 (rights in territory of other party); §3B-2-02 (storage and diversion); §3B-2-03 (eminent domain); §3B-2-04 (navigation); §4B-1-01 (exclusive jurisdiction and control); §4B-1-02 (water allocation); §4B-1-03 (water levels protected); §4B-1-04 (underground water; limit on withdrawals); §4B-1-06 (augmentation of supply); §4B-1-07 (water quality).

Similar Agreement: *Treaty between the United States and Mexico. Utilization of Waters of the Colorado and Tijuana Rivers and of the Rio Grande*, 59 Stat. 1219 (1944).

§4B-1-06 Augmentation of Supply (Optional)

(a) Any importation of water from outside the Basin shall be excluded from the provisions set forth elsewhere in this Agreement, and the Party importing such water shall have the right to full and complete use and consumption of such imported water.

(b) Any Party which augments precipitation within the Basin shall be entitled to full and exclusive use of additional water supplies resulting from such augmentation, notwithstanding any other standard of allocation set forth in this Agreement. In the event the Parties cannot agree on whether or to what extent precipitation has been augmented, the Party asserting the right to such increased supplies shall bear the burden of proving that the increase, if any, was the result of the Party's augmentation efforts and not simply the result of natural variation in precipitation amounts.

(c) Any Party implementing a conservation program with respect to water supplies shall be entitled to full and complete use and consumption of all increased supplies resulting from such conservation program. The burden of showing such increase shall rest on the Party claiming such increase.

Commentary: Article 4B-1-06 (a) makes clear that if a party arranges to increase supplies by importing water, it need not share those additional supplies. Article 4B-1-06(b) applies the same principle for precipitation augmentation. Article 4B-1-06(c) provides encouragement for conservation by rewarding the party that undertakes that effort. Caution should be exercised in incorporating this provision, however, inasmuch as the level of conservation efforts between the parties may be unequal at the time the agreement is negotiated. A party that has already made significant efforts should not be placed at a disadvantage relative to a party that, before the agreement, made little effort to conserve (Draper 2002b).

Cross References: §1B-1-01 (purposes and scope of agreement); §2B-2-07 (party or parties); §3B-2-01 (rights in territory of other party); §3B-2-02 (storage and diversion); §3B-2-03 (eminent domain); §4B-1-01 (exclusive jurisdiction and control); §4B-1-02 (water allocation); §4B-1-03 (water levels protected); §4B-1-04 (underground water; limit on withdrawals); §4B-1-05 (flood protection works); §4B-1-07 (water quality).

Similar Agreements: *Rio Grande Compact of 1938*, 53 Stat. 785, 938 (1938); *Pecos River Compact*, 63 Stat. 159 (1948).

§4B-1-07 Water Quality (Optional)

Alternative 1 (For U.S. use)

The Parties shall:

(a) Manage the waters of the _______ River Basin within their jurisdiction to maintain ecosystem integrity, preserve and protect aquatic ecosystems effectively from any form of (significant) degradation on a drainage Basin or sub-Basin basis. Natural water quality solutions, such as riparian vegetated buffers along the River and its tributaries, will be utilized to the maximum extent possible.

(b) Publish biological, health, physical, and chemical quality criteria for all water bodies (surface and underground water), according to Basin capacities and needs, with a view to an ongoing improvement of water quality.

(c) Establish standards for the discharge of effluents and for the receiving waters, no less stringent than the effluent limitations established by the U.S. Environmental Protection Agency, including standards for land use management.

(d) Establish minimum flow criteria to ensure nourishment of wetlands and riparian buffers as necessary to properly filter nitrates and phosphorus arising from nonpoint runoff.

(e) Maintain the quality of the Waters of the Basin at or above water quality standards as may be adopted, now or hereafter, by the water pollution control agencies of the respective Parties in compliance with the provisions of the Clean Water Act, 33 U.S.C. §§1251 *et seq.*, and amendments thereto.

Alternative 2 (For international use)

The Parties mutually agree to:

(a) Comply with the principle of individual Party efforts to control natural and man-made water pollution within each Party and to the continuing support of both Parties in active water pollution control programs.

(b) Cooperate, through their appropriate Party agencies, in the investigation, abatement, and control of sources of alleged interparty pollution within the Basin.

(c) Cooperate in maintaining the quality of the Waters of the Basin at or above water quality standards as may be developed and agreed to by the Parties.

Commentary: The quality of the water that is allocated is as important as the quantity of water that is allocated. Poor-quality water imposes risks that the parties should consider. First there is the health risk to the population that uses the water for domestic purposes. Second, if the available water does not meet the standards for certain industrial purposes, there is the risk that economic growth will be impaired. Finally, there is the risk that quality degradation may have a severe impact on the ecology of the basin, resulting in long-term sustainability complications. Integration of water quality and quantity is essential. *Agenda 21* (United Nations 1992) obligated all signatories to develop a program of water and sustainable development (Ahlander 1994). The *Convention on the Law of the Non-Navigational Uses of International Watercourses*, United Nations Document A/51/869 (United Nations 1997), establishes the criterion that "watercourse States shall, individually and, where appropriate, jointly, protect and preserve the ecosystems of international watercourses."

Surface and underground water may be degraded by a variety of factors. Major problems affecting the quality of these water resources arise, for instance, from inadequate domestic sewage treatment; inadequate controls on the discharge of industrial waste and effluent; the diversion of waters resulting in insufficient water to assimilate waste; the loss and destruction of catchment areas; the improper siting of industrial plants; deforestation; and poor agricultural practices, which cause leaching of nutrients and pesticides. Transboundary water sharing should include effective plans and programs that eliminate, or at least minimize, the possible sources of water quality degradation (Draper 2002b).

The complex interconnected nature of freshwater systems suggests that freshwater management should be systemically integrated, taking a catchment management approach that balances the needs of people and the environment. The parties should manage the waters of the basin, preserve aquatic ecosystems, and protect them effectively from any form of degradation on a drainage basin or subbasin basis. Scientific research has shown that the use of riparian vegetative buffers can have an extremely salutary effect on reducing pollution, especially sediments, nitrates, and phosphorus. Such natural solutions for quality control should be used to the maximum possible extent. Periodic flushing flows at periodic intervals may be an effective technique to quickly rejuvenate degraded waters (Draper 2002b).

The parties should establish biological, health, physical, and chemical quality criteria for all significant water bodies in the basin to continually improve water quality. The parties should establish minimum standards both for discharge of effluents and for receiving waters. We recommend that the parties institute standards for land use management, such as limits on agrochemical use, deforestation, and wasteful irrigation practices. Such rational land use standards should prevent land degradation, erosion, and siltation of lakes and other water bodies (Draper 2002b).

Cross References: §1B-1-01 (purposes and scope of agreement); §1B-1-04 (preservation of federal rights); §1B-1-05 (national security); §2B-1-05 (limited applicability); §2B-1-06 (annexes); §2B-2-07 (party or parties); §2B-2-08 (underground water); §2B-2-09 (waters of the basin); §3B-1-03 (implementation and verification of agreement); §3B-2-01 (rights in territory of other party); §3B-2-02 (storage and diversion); §3B-2-03 (eminent domain); §3B-2-04 (navigation); §4B-1-01 (exclusive jurisdiction and control); §4B-1-02 (water allocation); §4B-1-03 (water levels protected); §4B-1-04 (underground water; limit on withdrawals); §4B-1-05 (flood protection works); §4B-1-06 (augmentation of supply); §4B-1-07 (water quality).

Similar Agreements: *Rio Grande Compact of 1938*, 53 Stat. 785, 938 (1938); *Klamath River Basin Compact*, 71 Stat. 497 (1957); *Arkansas River Basin Compact of 1965*, 80 Stat. 1409 (1966); *Kansas–Nebraska Big Blue River Compact*, 86 Stat. 193 (1972); *Arkansas River Basin Compact of 1970*, 87 Stat. 569 (1973); *Delaware River Basin Compact*, Pub. L. 87-328, 75 Stat. 688-716 (1961); *Susquehanna River Basin Compact*, Pub. L. 91-575, 84 Stat. 1509-1541 (1970); *Red River Compact*, 94 Stat. 3305 (1980); *Oregon–California Goose Lake Interstate Compact*, 98 Stat. 291 (1984); *Treaty between the United States and Great Britain Relating to Boundary Waters*, 36 Stat. 2451 (1909); *Agreement Between the United States and Canada on Great Lakes Water Quality*, 1153 UNTS 187 (1978); *Convention on the Protection and Use of Transboundary Watercourses and International Lakes*, 31 ILM 1312 (1992); *Convention on the Law of the Non-Navigational Uses of International Watercourses*, United Nations Document A/51/869 (United Nations 1997).

B.5 ARTICLE 5B: DISPUTE RESOLUTION (OPTIONAL)

Disputes inevitably arise as an agreement is implemented. These disputes may involve differences in interpretation of the agreement's provisions or noncompliance with the agreement itself. The disputes may also arise because of changing conditions that alter the effectiveness of the agreement for one or more of the parties. Although a speedy and equitable process of dispute resolution serves all parties well, some governments do not wish to enter into an obligatory process. In such a case, Article 5A may be omitted. In other instances, the parties may recognize the need to institutionalize a dispute resolution process (Draper 2002b).

§5B-1-01 Resolution by Signatory Parties

Whenever any difference or dispute may arise between two or more Parties to this Agreement regarding any matters covered by this Agreement, particularly as to the interpretations of the Agreement and the legal rights of the Parties, the Parties shall first make every effort to resolve the issue through negotiations and consultations based on the powers and duties herein described.

Commentary: This alternative is the least restrictive upon the authority of the parties yet expresses their recognition of the need for peaceful resolution of disputes. The alternative is appropriate in those cases in which no commission has been established to manage the data exchange (Draper 2002b).

Cross References: §1B-1-01 (purposes and scope of agreement); §1B-1-02 (coordination and cooperation); §1C-1-03 (good faith implementation); §1B-1-04 (preservation of federal rights); §1B-1-05 (national security); §5B-1-02 (right to litigate).

Similar Agreement: *Agreement on the Cooperation for the Sustainable Development of the Mekong River Basin*, 34 ILM 864 (1995).

§5B-1-02 Right to Litigate

Nothing in this Agreement shall be construed to limit or prevent either Party from instituting or maintaining any action or proceeding, legal or equitable, in any tribunal of competent jurisdiction for the protection of any right under this Agreement or the enforcement of any of its provisions.

Commentary: The existence of an appropriate tribunal may pose a problem in cases not involving an entity like the United States or the European Union. It may be advisable to specify the tribunal in the agreement itself to avoid dispute over jurisdictional questions at a later date (Draper 2002b).

Cross References: §1B-1-01 (purposes and scope of agreement); §1B-1-02 (coordination and cooperation); §1C-1-03 (good faith implementation); §1B-1-04 (preservation of federal rights); §1B-1-05 (national security); §5B-1-01 (resolution by signatory parties).

Similar Agreements: *Belle Fourche River Compact*, 58 Stat. 94 (1944); *Colorado River Compact*, 45 Stat. 1057 (1928); *Snake River Compact*, 64 Stat. 29 (1949).

B.6 SIGNATURES

IN WITNESS WHEREOF, and in evidence of the adoption and enactment into law of this Agreement by the signatory Parties, the representatives of the States of ____________, ____________, ________ do hereby, in accordance with authority conferred by law, sign this Agreement in (__) duplicate original copies, as attested by the appropriate authorities of the respective *signatory* Parties, and have caused the seals of the respective States to be hereunto affixed this ____ day of ______.

B.7 ALLOCATION ALTERNATIVES

Annex I: Allocation Alternatives (McCormick 1994a, b; Draper 2002b).

§4B-1-02 WATER ALLOCATION
Alternative 1: Percentage Allocation

(a) The Parties agree that the unappropriated Waters of the _____ River Basin as of the date of this Agreement shall be allocated to each Party as follows:

__ % to [Upstream Party]

__ % to [Downstream Party]

(b) For storage of its allocated water, either Party shall have the right to purchase at cost up to ___% of the total storage capacity of any reservoir or reservoirs constructed by the other Party in its territory for irrigation of lands, or may construct reservoirs itself for the purpose of utilizing such water. Either Party may temporarily divert, or store for beneficial use, any unused part of the above percentages allotted to the other, but no continuing right shall be established thereby.

(c) Rights to the use of the waters of the Basin, whether based on direct diversion or storage, are hereby recognized as of the date of this agreement to the extent these rights are valid under the law of the Party in which the use is made, and shall remain unimpaired hereby. These rights, together with the additional allocations made under this Article, are agreed to be an equitable apportionment between the Parties of the Waters of the Basin.

(d) The waters allocated under this Article and the other rights recognized under Article 4B are hereinafter referred to collectively as the apportioned water. For the purposes of the administration of this Agreement and determining the apportioned water at any given date within a given calendar year, there shall be taken the sum of:

(1) The quantity of water in acre-feet [cubic meters] that passed the boundary lines of the Parties during the period ______________;

(2) The quantity of water in acre-feet [cubic meters] in storage on that date in all reservoirs built in [Upstream Party] in the Basin subsequent to the date of this Agreement; and

(3) The quantity of underground water withdrawn from aquifers tributary to surface water sources in the Basin during the period ______________.

Commentary: Article 4B-1-02(a) is a simple proportional allocation for use in upstream/downstream conflicts. Both parties share the risk of shortage in proportion to their percentage of allocation. Determination of the percentages is a matter of negotiation. It may be based upon amount of irrigable land, historic use, population, or some other factor, or it may simply be derived by bargaining.

In §4B-1-02(b), the right of the upstream party to use reservoirs constructed by the downstream party is similarly a matter for negotiation but may result in the construction of fewer reservoirs than would otherwise be necessary.

Article 4B-1-02(c) excludes existing uses from the allocation. Inclusion of such a grandfather clause may avoid claims that water rights or other property interests have been taken as a result of the allocations. Inclusion of such a provision is reasonable only if there remains unallocated supply within the basin. If only new uses are affected by the allocation, the actual proportion of the river's flow available to each party may be different than the allocation for future use. This situation would be the case where one party is using considerably more water than the other at the time the allocation is agreed upon.

Article 4B-1-02(d), or something similar, should be included because there should be some quantity against which the percentages can be applied. In this model, taken from the *Belle Fourche River Compact*, 58 Stat. 94 (1944), the state line is the measuring point, and a particular time period is used.

The state line may not, however, be appropriate because the upstream party can divert water before it reaches that point. It may be necessary to specify the use of gauging stations at particular points upstream of the relevant boundary, as is the case with the *Rio Grande Compact of 1938*, 53 Stat. 785, 938 (1938). The *Rio Grande Compact of 1938* uses a set of tables to calculate delivery obligations. The amount required to be delivered by the states at specified locations is determined by the flow measured at those locations. Where there are tributaries of significance, as on the upper Rio Grande, it may be necessary to specify particular locations on each tributary.

If underground water is included in the allocation, provision should be made to account for underground water withdrawals and availability. No specific mention is made of atmospheric water or water imported from other basins. In practice, this inclusion would mean that additional water from those sources would be subject to the allocation, if it were to enter into the flow of the river, because it would pass the gauging station. If such water is not to be included in the allocation, specific exclusions should be incorporated in the agreement to avoid future disputes over whether underground water was intended to be subject to the allocation system.

Cross Reference: §4B-1-02 (water allocation).

Similar Agreements: *Belle Fourche River Compact*, 58 Stat. 94 (1944); *Snake River Compact*, 64 Stat. 29 (1949); *Yellowstone River Compact*, 65 Stat. 663 (1950).

Alternative 2: Proportionate Allocation

(a) The waters of the ______ River Basin are hereby equitably apportioned between the Parties as follows:

(1) At all times between the ___ day of [month] and the ___ day of the succeeding [month], each Party shall have the unrestricted right to the use of all water which may flow within its boundaries;

(2) By reason of the usual annual rise and fall, the flow of said river between the ____ day of [month] and the ____ day of [month] of each year, shall be apportioned between the Parties in the following manner:

(A) Each Party shall have the unrestricted right to use all the waters within its boundaries on each day when the mean daily flow at the [Downstream] station is _______ cubic feet [cubic meters] per second, or more;

(B) On all other days [Upstream Party] shall deliver at the [Downstream] Gauging Station a quantity of water equivalent to ____ percent of the mean flow at the [Upstream] Gauging Station for the preceding day, but not to exceed ______ cubic feet [cubic meters] per second;

(3) [Downstream Party] shall not at any time be entitled to receive nor shall [Upstream Party] be required to deliver any water not then necessary for beneficial use in [Downstream Party];

(4) A substantial delivery of water under the terms of this Article shall be deemed a compliance with its provisions and minor and compensating irregularities in flow or delivery shall be disregarded.

(b) Verification of the allocation of waters through the use of gauging stations established by the Parties.

(1) [Upstream Party], at its own expense, shall establish and maintain two permanent stream-gauging stations upon the ________ River for the purpose of measuring and recording its flow, which shall be known as the [Upstream] Gauging Station and the [Downstream] Gauging Station, respectively.

(2) The [Upstream] Gauging Station shall be located at some convenient place near [geographic reference]. Suitable devices for ascertaining and recording the volume of all diversions from the river above Station, [_____] shall be established and maintained (without expense to [Downstream Party]), and whenever in this Agreement reference is made to the flow of the river at [Upstream] Gauging Station, it shall be construed to include the amount of the concurrent diversions and underground water withdrawals above said station.

(3) The [Downstream] Gauging Station shall be located at some convenient place within one mile of the boundary line between [Upstream Party] and [Downstream Party]. Whenever in this Agreement reference is made to the flow of the river at the [Downstream] Gauging Station, it shall be construed to include the volume of any other water that may hereafter be diverted from said river or underground water tributary to said river in [Upstream Party] for use in [Downstream Party].

(4) Each of said stations shall be equipped with suitable devices for recording the flow of water in said river at all times between the ____ day of ________ and the ____ day of ______ of each year. The designated officials of the Parties shall make provision for cooperative gauging at the two stations, for the details of the operation, exchange of records and data, and publication of the facts.

Commentary: Alternative 2, adapted from the *La Plata River Compact*, 43 Stat. 796 (1925), calls for apportionment only during certain periods of the year and only at times when the flow of the river falls below a certain level. It is directed toward upstream/downstream conflicts. The La Plata is a relatively small river, whose flow can fluctuate widely. Allocation was deemed necessary only in times of low flow. The state engineers of Colorado (the upstream state) and New Mexico (the downstream state) are given some power under the compact to coordinate the use of whatever water is available during those low-flow periods. This alternative provides for the actual allocation of the water when the triggering event has occurred. It contains an additional provision to allow for more efficient use of the water by allowing the parties to allocate the entire flow to each party in rotation. Finally, a de minimis clause is included to avoid argument over relatively minor problems. The alternative specifies how flows are to be calculated and who bears the cost of the gauging stations. It provides for adjustments of the measured amounts to take into account diversions and underground water withdrawals not recorded at the gauging stations. These diversions and underground water withdrawals are in effect charged to the party that benefits from them.

Cross Reference: §4B-1-02 (water allocation).

Similar Agreement: *La Plata River Compact*, 43 Stat. 796 (1925).

Alternative 3: Guaranteed Minimum Quantities Apportioned

(a) There is hereby apportioned from the _______ River to [Downstream Party] the exclusive reasonable use of [______] acre-feet [cubic meters] per annum, measured at [geographic location] which shall include all water necessary for the supply of any rights which may now exist.

(b) [Upstream Party] shall not cause the flow of the river at [geographic location] to fall below an aggregate of [ten times the quantity above] for any period of [ten] consecutive years determined in progressive continuing series, beginning with the ___ day of ____ next succeeding the effective date of this Agreement.

Commentary: This alternative is derived from part of the *Colorado River Compact* of 1922. It is directed to upstream/downstream conflicts. The *Colorado River Compact* involves seven states, and the water is allocated not between states, but between the upper and lower basins of the river. The agreement calls for the lower basin to receive 7,500,000 acre-ft per year, but in apparent recognition of the variable nature of the Colorado's flow, the compact also provides that the states of the upper basin will not cause the flow at Lee Ferry (the gauging station) to fall below 75,000,000 acre-ft in any 10-year period. The risk of shortage is therefore on the states of the Upper Basin, but shortages in one year can be made up in another so long as the 10-year requirement is met. In this example, the provision might read thus:

(a) There is hereby apportioned from the Colorado River to Arizona the exclusive reasonable use of 7,500,000 acre-ft [925,000 hectare-meters] per annum, measured at Lee Ferry, which shall include all water necessary for the supply of any rights which may now exist.

(b) Nevada shall not cause the flow of the river at Lee Ferry to fall below an aggregate of 75,000,000 acre-ft for any period of 10 consecutive years determined in progressive continuing series, beginning with the 1st day of January next succeeding the effective date of this Agreement.

This provision places the risk of shortfall on the upstream state, which is bound to deliver a specified amount regardless of changes in supply. A rolling average is used to determine compliance. The *Colorado River Compact* uses a 10-year period, but the actual number is a matter for negotiation. Although it is not absolutely necessary that some averaging period be chosen, the potential for variation in the flow of most rivers makes it a reasonable provision.

Cross Reference: §4B-1-02 (water allocation).

Similar Agreements: *South Platte River Compact*, 44 Stat. 195 (1923); *Colorado River Compact*, 45 Stat. 1057 (1928).

Alternative 4: Guaranteed Quantity; Vested Rights Protected

(For appropriative rights states only)

(a) Between the ___ day of [month] and the ____ day of [month] in each year, [Upstream Party] shall not permit diversions from the river to supply [Upstream Party] appropriations with a priority date subsequent to [date] to the extent that such diversions will diminish the flow of the river at the interstate gauging station below a mean flow of _____ cfs, except as limited by §4B-1-02(d).

(b) [Upstream Party] and [Downstream Party], at their joint expense, shall maintain a stream gauging station upon the _______ River near [geographic reference] for the purpose of measuring the amount of water flowing in said river from [Upstream Party] into [Downstream Party]. The location of said station may be changed from year to year as the river channels and flow conditions of said river may require.

(c) The officials responsible for administering public water supplies in each of the Parties shall make provision for the cooperative gauging at and the details of operation of said gauging station and for the exchange and publication of records and data.

(d) [Downstream Party] shall not be entitled to receive, and [Upstream Party] shall not be required to deliver, on any day any part of the flow of the river not then necessary for Beneficial Use in [Downstream Party]; provided, however, that a minimum of _____ cfs [cubic meters per second] shall be released, if available, to maintain minimum flows.

Commentary: This alternative is adapted from the *South Platte River Compact*, 44 Stat. 195 (1923), between Colorado and Nebraska. The provision is applicable only to those states within which appropriative rights law exists. The South Platte agreement provides for a minimum flow at the state boundary, but Colorado (the upstream state) did not actually guarantee that the flow would equal the minimum. Instead, Colorado agreed that it would not permit any diversions with a priority date later than June 14, 1897, if those diversions would reduce the flow at the boundary below 120 cfs [3.40 m^3/sec]. Under the compact, Nebraska would bear the risk that the river would have insufficient flow to meet the demands of Colorado appropriators with a priority date earlier than June 14, 1897, and still provide 120 cfs [3.40 m^3/sec] at the border. The result is a sharing of the risk of shortage. Nebraska bears the risk up to a certain point; thereafter, the risk of shortage falls on later Colorado appropriators.

This alternative seeks to accomplish the same result. It provides for a minimum flow at the party boundary but limits the obligation of the upstream party to provide that water by requiring only that the upstream party not permit diversions with a priority after a given date, if those diversions would reduce the flow below the minimum called for. Senior appropriators, whose rights presumably vested before the date of the compact, are still able to withdraw water even if those withdrawals result in a boundary line flow below the minimum. The downstream party thus bears the risk that the available water will be less than those vested rights plus the minimum.

Article 4B-1-02(d) is included to ensure that water that can be used in the upstream party is not required to pass the border between the parties unless it will be put to beneficial use in the downstream party. The term "beneficial use" is restricted to those jurisdictions that adhere to the law of appropriate rights and thus determine whether the downstream party is entitled to release of water for purposes such as instream use in those jurisdictions. Minimum flows may be maintained by virtue of the provision. In addition, the beneficial use is referred to as "in" the downstream party. This specification could preclude a requirement for water to be released to satisfy the demands of water transfers (water markets) if those markets resulted in transfers outside the boundaries of the downstream party.

Cross Reference: §4B-1-02 (water allocation).

Similar Agreement: *South Platte River Compact*, 44 Stat. 195 (1923).

Alternative 5: Storage Allocations

(a) No reservoir or other storage facility, whether surface or underground, built after the date of this Agreement solely to utilize the water allocated to either Party shall have a capacity in excess of ______ acre-feet [cubic meters], without the approval of both Parties.

(b) [Upstream Party] shall have free and unrestricted use of all waters originating in the Basin above [name] dam.

(c) [Downstream Party] shall have free and unrestricted use of all waters originating in the Basin below [name] dam.

Commentary: If the upstream party builds excess storage capacity, the downstream party may not receive the full benefit of its bargain. If that is a concern, a storage limitation such as this can be included. Underground storage is within the scope of this provision, to allow for the possibility of storage in other than surface reservoirs.

In some cases, the parties may determine that limitations on storage satisfy any need for allocation. Storage is relatively easy to measure, and such agreements have the virtue of simplicity. Careful attention should be paid to definitions. The latter two sections of this article are derived from the *Canadian River Compact*, 66 Stat. 74 (1952). This agreement among New Mexico, Texas, and Oklahoma eventually required interpretation by the U.S. Supreme Court in Oklahoma and *Texas v. New Mexico*, 111 S. Ct. 2281 (1991). The case required the court to decide the meaning of the term "originating"; more specifically, a key issue was whether water that entered the river above Conchas Dam but was then spilled or released from the dam "originated" above or below the dam. The court found that this water originated below the dam. Even seemingly unambiguous terms such as "originating" should be used with care.

Cross Reference: §4B-1-02 (water allocation).

Similar Agreements: *Kansas–Nebraska Big Blue River Compact*, 86 Stat. 193, §5.2(c) (1972); *Upper Niobrara River Compact*, 83 Stat. 86 (1969); *Canadian River Compact*, 66 Stat. 74 (1952).

Appendix C
MODEL WATER SHARING AGREEMENT C (COMPREHENSIVE)

Model Water Sharing Agreement C (Comprehensive Water Sharing Agreement) provides a model for comprehensive planning and management of shared water resources.

C.1 ARTICLE 1C: DECLARATION OF POLICIES AND PURPOSES

§1C-1-01 General Policies

(a) The water resources of the __________ River Basin have a local, regional, and national significance and their comprehensive management for purposes of conservation, utilization, development, and control, under appropriate arrangements for intergovernmental cooperation, are public purposes for the respective signatory Parties.

(b) The purposes of this Agreement are to promote interstate [international] comity, to remove causes of present and future controversy, and to efficiently and effectively utilize the shared water resources through integrated management and control.

(c) The general policies of this Agreement include the equitable and reasonable use of the water resources of the _________ River Basin, the need to exchange data and the intent to engage in consultation on the possible effects of planned measures on the condition of the water resource.

(d) The geophysical, climatic/meteorological, and other conditions peculiar to the _______ River Basin constitute the basis for this Agreement and its provisions are applicable only to the Basin.

Commentary: Economic growth and prosperity require adequate supplies of high-quality water on a regular and sustained basis. To achieve this goal, the planning, conservation, utilization, development, management, and control of water resources must include the entire water basin , extending beyond political boundaries. In most cases, optimum achievement of the goal means the basin states must give up some degree of political authority. Planning at the river basin scale is a recognized basis for adequate water management, especially with regard to access to potable water (Zylicz 1994). A key challenge for the parties is to make more efficient and productive use of water and to reshape the institutions managing water allocation to better adapt to the increase in water issues, whether they be concerned with water quality or quantity (Postel 1996).

One criterion of the *Convention on the Law of the Non-Navigational Uses of International Watercourses*, United Nations Document A/51/869 (United Nations 1997), makes the definition of the waters to which an agreement applies a mandatory provision in the agreement. The parties should carefully frame the extent of the specific water resources involved in the agreement. They should identify the type and geographical extent of the water resources to be subject to the agreement. To formulate an effective agreement, the parties should analyze the factors that influence the water resource in question, such as the climatology, physiology, geology, and the interaction between underground and surface water resources. The analysis should identify pollution sources and the resulting effect on water quality. The purposes and scope of the agreement should weigh heavily on this assessment because this section delineates the parties to be included. The geographic scope of the water resources to be covered by the agreement should be sufficiently expansive to fully address all the water sharing issues involved. The use of the term "river basin," if objectionable for any reason by one or more parties, may be changed to "region," "area," "system," or any word that accurately describes and encompasses the entirety of the water resources subject to the agreement (Draper 2002b).

It is important to acknowledge that the agreement reflects the particular circumstances and compromises reached in its formulation, as applied to the particular basin, so that it recognizes that the provisions of the agreement are restricted only to those water resources the parties intend. The inclusion of §1C-1-01(d) may avoid later claims that other rivers and basins, or other bodies of water, should be dealt with in a similar manner (Draper 2002b).

Cross References: §1C-1-02 (purposes of agreement); §1C-1-03 (objectives of agreement); §1C-1-04 (coordination and cooperation); §1C-1-05 (reservation of federal rights); §1C-1-06 (national security; for use in the international context); §2C-1-01 (activation of agreement); §2C-1-02 (consent to jurisdiction); §2C-1-03 (duration of agreement); §2C-1-04 (amendments and supplements); §2A-1-04 (powers of signatory parties; withdrawal); §2C-1-06 (existing agencies); §2C-1-07 (limited applicability); §2C-1-08 (annexes); §2C-2-11 (waters of the basin); §3C-1-01 (commission created); §3C-1-02 (jurisdiction of the commission); §3C-1-03 (commissioners); §3C-1-04 (status, immunities, and privileges); §3C-1-05 (commission organization and staffing); §3C-1-06 (rules of procedure); §3C-1-07 (commission administration); §3C-2-01 (general powers and duties); §3C-2-02 (powers and duties reserved to the commissioners); §3C-2-03 (obligations of the commission); §3C-2-04 (regulations; enforcement); §3C-2-05 (prohibited activities); §3C-2-06 (referral and review); §3C-2-07 (advisory committees); §3C-2-08 (reports); §3C-2-09 (condemnation proceedings); §3C-2-10 (meetings, hearings and records); §3C-2-11 (tort liability); §3C-3-01 (coordination and cooperation); §3C-3-02 (project costs and evaluation standards); §3C-3-03 (projects of the signatory parties); §3C-3-04 (cooperative services); §4C-1-01 (joint exercise of authority); §4C-1-02 (interrelationship of water resources); §4C-1-03 (comprehensive water management plan); §4C-1-04 (purpose and objectives of comprehensive water management plan); §4C-1-05 (conditions of comprehensive water management plan); §4C-1-06 (deviation from comprehensive water management plan); §4C-1-07 (allocation during flood conditions); §4C-1-08 (allocation during drought conditions); §4C-1-09 (non-impairment of comprehensive water management plan by state action); §4C-2-01 (water allocation, generally); §4C-2-02 (waters not subject to allocation); §4C-2-03 (basin water sources); §4C-2-04 (basin water demands and needs); §4C-2-05 (allocation to equitable and reasonable uses); §4C-2-06 (watershed management); §4C-3-01 (existing rights recognized); §4C-3-02 (flood protection works); §4C-3-03 (minimum flows); §4C-3-04 (withdrawals and diversion; protected areas); §4C-3-05 (water levels protected); §4C-3-06 (augmentation of supply); §4C-3-07 (water quality); §4C-3-08 (underground water; limit on withdrawals); §4C-3-09 (atmospheric water); §4C-3-10 (recreation); Article 5C (financing); §6C-1-01 (good faith implementation); §6C-1-02 (modification of agreement); §6C-1-03 (material breach); §6C-1-04 (alternative dispute resolution); §6C-1-05 (negotiations and consultations); §6C-1-06 (consultations and mediation); §6C-1-07 (arbitration); §6C-1-08 (right to litigate).

Similar Agreements: *Delaware River Basin Compact*, Pub. L. 87-328, 75 Stat. 688-716 (1961); *Susquehanna River Basin Compact*, Pub. L. 91-575, 84 Stat. 1509-1541 (1970); *Convention on the Law of the Non-Navigational Uses of International Watercourses*, United Nations Document A/51/869 (United Nations 1997).

§1C-1-02 Purposes and Scope of Agreement

By this Agreement, the signatory Parties agree to the following:

(a) to provide for the most efficient use of the Waters of the _____ River Basin through comprehensive, basin-wide management;

(b) to provide for the equitable and reasonable utilization of the water resources, including surface, underground, and atmospheric supplies;

(c) to provide for the maintenance of the quality of water within the Basin;

(d) to remove all causes, present and future, which might lead to controversies or conflicts; to promote interstate [international] comity;

(e) to minimize economic and social disruption and dislocation through implementation of drought strategies and flood control programs;

(f) to limit this Agreement to the water resources arising within the drainage basin of the _________ River; and

(g) the Parties to this Agreement are the States of [Party 1] and [Party 2].

Commentary: The purpose statement is intended to be indefinite to allow flexibility in the management of the shared water resource. Although the parties may wish to be more specific when using this model agreement, the need for such specificity may mean the agreement presented at Section 6.3, Model B, is more appropriate (Draper 2002b).

Goldfarb (1993) has noted, "integration of water quality and quantity programs is one aspect of, and is inseparable from, comprehensive, coordinated, multipurpose, basinwide water and related resource planning and management."

Many recent international agreements, most notably the *Convention on the Law of the Non-Navigational Uses of International Watercourses*, United Nations Document A/51/869 (United Nations 1997), establish as a principle purpose of all transboundary water sharing agreements the "equitable and reasonable utilization" of water (Draper 2002b).

If the agreement is to control the use of underground and atmospheric water as well as surface water, the intent to encompass underground and atmospheric water within the scope of the agreement should be made clear at this point. Failure to specifically deal with the issue of underground water withdrawal has led to litigation in the context of existing interstate compacts in the United States. It is therefore recommended that underground and atmospheric water be included within the scope of the agreement. As technology advances, use of and control of atmospheric water may also become more commonplace, and consideration should be given to dealing with the potential for such use and control before it becomes established and results in unexpected interference with the provisions made for underground and surface water (Draper 2002b).

This provision further defines the geographical description of the international watercourse in question, as required by the *Convention on the Law of the Non-Navigational Uses of International Watercourses*, United Nations Document A/51/869 (United Nations 1997).

Cross References: §1C-1-01 (general policies); §1C-1-03 (objectives of agreement); §1C-1-04 (coordination and cooperation); §1C-1-05 (reservation of federal rights); §1C-1-06 (national security; for use in the international context); §2C-1-01 (activation of agreement); §2C-1-02 (consent to jurisdiction); §2C-1-03 (duration of agreement); §2A-1-04 (powers of signatory parties); §2C-1-05 (amendments and supplements); §2C-1-06 (existing agencies); §2C-1-07 (limited applicability); §2C-1-08 (annexes); §2C-2-06 (equitable and reasonable apportionment); §2C-2-07 (flood); §2C-2-09 (party or parties); §2C-2-10 (underground water); §2C-2-11 (waters of the basin); §3C-1-01 (commission created); §3C-1-02 (jurisdiction of the commission); §3C-1-03 (commissioners); §3C-1-04 (status, immunities, and privileges); §3C-1-05 (commission organization and staffing); §3C-1-06 (rules of procedure); §3C-1-07 (commission administration); §3C-2-01 (general powers and duties); §3C-2-02 (powers and duties reserved to the commissioners); §3C-2-03 (obligations of the commission); §3C-2-04 (regulations; enforcement); §3C-2-05 (prohibited activities); §3C-2-06 (referral and review); §3C-2-07 (advisory committees); §3C-2-08 (reports); §3C-2-09 (condemnation proceedings); §3C-2-10 (meetings, hearings and records); §3C-2-11 (tort liability); §3C-3-01 (coordination and cooperation); §3C-3-02 (project costs and evaluation standards); §3C-3-03 (projects of the signatory parties); §3C-3-04 (cooperative services); §4C-1-01 (joint exercise of authority); §4C-1-02 (interrelationship of water resources); §4C-1-03 (comprehensive water management plan); §4C-1-04 (purpose and objectives of comprehensive water management plan); §4C-1-05 (conditions of comprehensive water management plan); §4C-1-06 (deviation from comprehensive water management plan); §4C-1-07 (allocation during flood conditions); §4C-1-08 (allocation during drought conditions); §4C-1-09 (non-impairment of comprehensive water management plan by state action); §4C-2-01 (water allocation, generally); §4C-2-02 (waters not subject to allocation); §4C-2-03 (basin water sources); §4C-2-04 (basin water demands and needs); §4C-2-05 (allocation to equitable and reasonable uses); §4C-2-06 (watershed management); §4C-3-01 (existing rights recognized); §4C-3-02 (flood protection works); §4C-3-03 (minimum flows); §4C-3-04 (withdrawals and diversion; protected areas); §4C-3-05 (water levels protected); §4C-3-06 (augmentation of supply); §4C-3-07 (water quality); §4C-3-08 (underground water; limit on withdrawals); §4C-3-09 (atmospheric water); §4C-3-10 (recreation); Article 5C (financing); §6C-1-01 (good faith implementation); §6C-1-02 (modification of agreement); §6C-1-03 (material breach); §6C-1-04 (alternative dispute resolution); §6C-1-05 (negotiations and consultations); §6C-1-06 (consultations and mediation); §6C-1-07 (arbitration); §6C-1-08 (right to litigate).

Similar Agreements: *Klamath River Basin Compact*, 71 Stat. 497 (1957); *Delaware River Basin Compact*, Pub. L. 87-328, 75

Stat. 688-716 (1961); *Susquehanna River Basin Compact*, Pub. L. 91-575, 84 Stat. 1509-1541 (1970); *Alabama–Coosa–Tallapoosa River Basin Compact*, O.C.G.A. 12-10-110 (1997); *Apalachicola–Chatahoochee–Flint River Basin Compact*, O.C.G.A. 12-10-100 (1997); *Convention on Wetlands of International Importance Especially as Waterfowl Habitat,* UN Treaty Series 14583 (1971), amended by the Paris Protocol (1982), Regina Amendments (1987); *Convention on Biological Diversity*, 1760 UNTS 79; 31 ILM 818 (1992); *Rio Declaration on Environment and Development*, 31 ILM 874 (1992); *Declaration of Principles on Interim Self-Government Arrangements*, 32 ILM 1525 (1993); *Convention on the Law of the Non-Navigational Uses of International Watercourses*, United Nations Document A/51/869 (United Nations 1997).

§1C-1-03 Objectives of Agreement (Optional)
The Parties agree to the following objectives:

(a) To cooperate in all fields of sustainable development, utilization, management, and conservation of the water and related resources of the waters of the _____ River Basin including, but not limited to, agricultural irrigation, domestic and commercial use, industrial and mining use, hydropower, navigation, flood control, fisheries, recreation and tourism, and maintenance of natural water environments in a manner to optimize the multiple use and mutual benefits of all riparians and to minimize the harmful effects that might result from natural occurrences and man-made activities.

(b) To promote, support, cooperate, and coordinate in the development of the full potential of sustainable waters of the Basin, with emphasis and preference on joint and/or basin-wide development projects and Basin programs through the formulation of a Basin development plan that would be used to identify, categorize, and prioritize the projects and programs to seek assistance for and to implement at the Basin level.

(c) To protect the environment, natural resources, aquatic life and conditions, and ecological balance of the waters of the Basin from pollution or other harmful effects resulting from any development plans and uses of water and related resources in the Basin.

(d) To cooperate on the basis of equality and territorial integrity in the utilization and protection of water resources of the waters of the Basin.

(e) To utilize the waters of the Basin in a reasonable and equitable manner in their respective territories.

Commentary: This provision provides a framework for the parties in the development of their individual water development planning. It recognizes that there are certain fundamental principles that each party should follow in their rational management of water resources. It would be irrational for one party to agree to "equitable and reasonable utilization" when it does not follow a similar philosophy within its own borders. These general objectives and principles improve the attainment of purposes of the water sharing (Draper 2002b).

Cross References: §1C-1-01 (general policies); §1C-1-02 (purposes of agreement); §1C-1-04 (coordination and cooperation); §1C-1-05 (reservation of federal rights); §1C-1-06 (national security; for use in the international context); §2C-2-07 (flood); §2C-2-09 (party or parties); §4C-1-02 (interrelationship of water resources); §4C-1-03 (comprehensive water management plan); §4C-1-04 (purpose and objectives of comprehensive water management plan); §4C-1-05 (conditions of comprehensive water management plan); §4C-1-06 (deviation from comprehensive water management plan); §4C-1-07 (allocation during flood conditions); §4C-1-08 (allocation during drought conditions); §4C-2-01 (water allocation, generally); §4C-2-02 (waters not subject to allocation); §4C-2-05 (allocation to equitable and reasonable uses); §4C-2-06 (watershed management); §4C-3-01 (existing rights recognized); §4C-3-02 (flood protection works); §4C-3-03 (minimum flows); §4C-3-04 (withdrawals and diversion; protected areas); §4C-3-05 (water levels protected); §4C-3-06 (augmentation of supply); §4C-3-07 (water quality); §4C-3-08 (underground water; limit on withdrawals); §4C-3-09 (atmospheric water); §4C-3-10 (recreation).

Similar Agreements: *Convention between Switzerland and Italy Concerning the Protection of Italo-Swiss Waters Against Pollution*, UNTS, Vol. 957, 277 (1972); *Canada–Mexico–United States: North American Agreement on Environmental Cooperation*, 32 ILM 1480 (1993); *Agreement on the Cooperation for the Sustainable Development of the Mekong River Basin*, 34 ILM 864 (1995).

§1C-1-04 Coordination and Cooperation

(a) Each of the Parties pledges to support implementation of the provisions of this Agreement, and covenants that its officers and agencies will not hinder, impair, or prevent any other Party carrying out any provision or recommendation of this Agreement.

(b) The Parties shall at all times endeavor to agree on the interpretation and application of this Agreement, and shall make every attempt through cooperation and consultations to arrive at a mutually satisfactory resolution of any matter that might affect its operation.

(c) The Parties acknowledge that their respective governmental organizations shall provide the information necessary to assist in the equitable and reasonable utilization of the waters of the Basin. Such information shall include, but not be limited to, all planning and management activities and water projects affecting their common boundary water resources.

(d) The Parties further acknowledge that all states are expected to conduct themselves with an absence of malice and with no intention to seek unconscionable advantage, or otherwise be deceitful.

Commentary: A party normally enters into any international agreement with a position of self-interest. In the negotiations, each party seeks the rights and authorities critical to certain political, economic, or social objectives while ceding less critical rights and authorities to the other nations. While accepting this fact, all parties have a duty to cooperate and negotiate in good faith. This principle is the foundation of international law, and it applies in all relations between states (Draper 2002b).

Cross References: §1C-1-01 (general policies); §1C-1-02 (purposes of agreement); §1C-1-06 (national security); §2C-1-01 (activation of agreement); §2C-1-03 (duration of agreement); §2C-1-04 (powers of signatory parties; withdrawal); §2C-1-05 (amendments and supplements); §2C-1-06 (existing agencies); §2C-1-07 (limited applicability); §2C-1-08 (annexes); §2C-2-09 (party or parties); §3C-1-02 (jurisdiction of the commission); §3C-1-03 (commissioners); §3C-1-04 (status, immunities, and privileges); §3C-1-05 (commission organization and staffing); §3C-1-06 (rules of procedure); §3C-1-07 (commission administration); §3C-2-01 (general powers and duties); §3C-2-02 (powers and duties reserved to the commissioners); §3C-2-03 (obligations of the commission); §3C-2-04 (regulations; enforcement); §3C-2-05 (prohibited activities); §3C-2-06 (referral and review); §3C-2-07 (advisory committees); §3C-2-08 (reports); §3C-2-09 (condemnation proceedings); §3C-2-10 (meetings, hearings and records); §3C-2-11 (tort liability); §3C-3-01 (coordination and cooperation); §3C-3-02 (project costs and evaluation standards); §3C-3-03 (projects of the signatory parties); §3C-3-04 (cooperative services); §4C-1-01 (joint exercise of authority); §4C-1-02 (interrelationship of water resources); §4C-1-03 (comprehensive water management plan); §4C-1-04 (purpose and objectives of comprehensive water management plan); §4C-1-05 (conditions of comprehensive water management plan); §4C-1-06 (deviation from comprehensive water management plan); §4C-1-07 (allocation during flood conditions); §4C-1-08 (allocation during drought conditions); §4C-1-09 (non-impairment of comprehensive water management plan by state action); §4C-2-01 (water allocation, generally); §4C-2-02 (waters not subject to allocation); §4C-2-05 (allocation to equitable and reasonable uses); §4C-2-06 (watershed management); §4C-3-01 (existing rights recognized); §4C-3-02 (flood protection works); §4C-3-03 (minimum flows); §4C-3-04 (withdrawals and diversion; protected areas); §4C-3-05 (water levels protected); §4C-3-06 (augmentation of supply); §4C-3-07 (water quality); §4C-3-08 (underground water; limit on withdrawals); §4C-3-09 (atmospheric water); §4C-3-10 (recreation); Article 5C (financing); §6C-1-01 (good faith implementation); §6C-1-02 (modification of agreement); §6C-1-03 (material breach); §6C-1-04 (alternative dispute resolution); §6C-1-05 (negotiations and consultations); §6C-1-06 (consultations and mediation); §6C-1-07 (arbitration); §6C-1-08 (right to litigate).

Similar Agreements: *Agreement Between the People's Republic of Bulgaria and the Republic of Turkey Concerning Co-operation in the Use of the Waters of Rivers Flowing through the Territory of Both Countries*, UNTS, Vol. 807, 117 (1968); *Stockholm Declaration of the United Nations Conference on the Human Environment*, 11 ILM 1416 (United Nations 1972); *Treaty for Amazonian Cooperation*, 17 ILM 1046 (1978); *Convention on Long-Range Transboundary Air Pollution*, 18 ILM 1442 (1979); *Vienna Convention for the Protection of the Ozone Layer*, 26 ILM 1529 (1985); *Convention between the Federal Republic of Germany and the Czech and Slovak Federal Republic and the European Economic Community on the International Commission for the Protection of the Elbe*, International Environmental Law, Multilateral Agreements, 976:90/1 (1990); *Convention on the Protection and Use of Transboundary Watercourses and International Lakes*, 31 ILM 1312 (1992); *Canada–Mexico–United States: North American Agreement on Environmental Cooperation*, 32 ILM 1480 (1993); *Convention on the Law of the Non-Navigational Uses of International Watercourses*, United Nations Document A/51/869 (United Nations 1997); *Treaty of Peace between the State of Israel and the Hashemite Kingdom of Jordan*, 34 ILM 43 (1994).

§1C-1-05 Preservation of Federal Rights (For U.S. use only)

Nothing in this Agreement shall be deemed:

(a) To impair or affect any rights or powers of the United States, its agencies, or instrumentalities, in and to the use of the waters of the _____ River Basin nor its capacity to acquire rights in and to the use of said waters;

(b) To subject any property of the United States, its agencies, or instrumentalities to taxation by either Party, nor to create an obligation on the part of the United States, its agencies, or instrumentalities, by reason of the acquisition, construction, or operation of any property or works of whatsoever kind, to make any payments to any State or political subdivision thereof, State agency, municipality, or entity whatsoever in reimbursement for the loss of taxes;

(c) To subject any property of the United States, its agencies, or instrumentalities, to the laws of any Party to an extent other than the extent to which these laws would apply without regard to the Agreement.

Commentary: These articles may be included in agreements between states of the United States. They are probably unnecessary to preserve federal rights, but inasmuch as Congress should approve the agreement, the inclusion of these provisions may make it easier to obtain that approval (Draper 2002b).

Cross References: §1C-1-01 (general policies); §1C-1-02 (purposes of agreement); §1C-1-03 (objectives of agreement); §2C-1-03 (duration of agreement); §2C-1-04 (powers of signatory parties; withdrawal); §2C-1-05 (amendments and supplements); §4C-1-01 (joint exercise of authority).

Similar Agreements: *Republican River Compact*, 57 Stat. 86 (1943); *Belle Fourche River Compact*, 58 Stat. 94 (1944); *Pecos River Compact*, 63 Stat. 159 (1948); *Snake River Compact*, 64 Stat. 29 (1949); *Yellowstone River Compact*, 65 Stat. 663 (1950); *Canadian River Compact*, 66 Stat. 74 (1952); *Klamath River Basin Compact*, 71 Stat. 497 (1957); *Bear River Compact*, 72 Stat. 38 (1955), amended 94 Stat. 4, Art. XIII(2) (1980); *Upper Colorado River Basin Compact*, 63 Stat. 31 (1949).

§1C-1-06 National Security (For international use)

(a) Nothing in this Agreement shall be construed to require any Party to make available or provide access to information the disclosure of which it determines to be contrary to its essential security interests.

(b) Nothing in this Agreement shall be construed to prevent any Party from taking any actions that it considers necessary for the protection of its essential security interests relating to a formal declaration of war.

Commentary: National security concerns necessarily take precedence over any program of water management and the exchange of data.

Cross References: §1C-1-01 (general policies); §1C-1-02 (purposes of agreement); §1C-1-03 (objectives of agreement); §1C-1-04 (coordination and cooperation); §2C-1-04 (powers of signatory parties; withdrawal); §2C-2-09 (party or parties).

Similar Agreement: *Canada–Mexico–United States: North American Agreement on Environmental Cooperation*, 32 ILM 1480 (1993).

C.2 ARTICLE 2C: GENERAL

§2C-1-01 General Provisions

<u>Part 1 General Obligations</u>

§2C-1-01 Effective Date

Alternative 1: (For international use)

This Agreement shall become operative when approved by the appropriate governing authorities of all Parties. The Agreement will go into full force and effect at 12:01 a.m. [time zone] on the day immediately following the final act necessary for approval of the Agreement, as defined by the domestic law of each Party, by the last Party to give such approval.

Alternative 2: (For U.S. use)

This Agreement shall become operative when, subsequent to adoption of each of the States, the Congress of the United States adopts legislation providing, among other things, that:

(a) **Any equitable and reasonable utilization hereafter made by the United States, or those acting by or under its authority, within a State, of the waters allocated by this Agreement, shall be within the allocations hereinabove made for use in that State and shall be taken into account in determining the extent of use within that State.**

(b) **The United States, or those acting by or under its authority, in the exercise of rights or powers arising from whatever jurisdiction the United States has in, over, and to the waters of the __________ River and all its tributaries, shall recognize, to the extent consistent with the best utilization of the waters for multiple purposes, that equitable and reasonable use of the waters within the __________ Basin is of paramount importance to development of the Basin, and no exercise of such power or right by the United States government or those acting under its authority that would interfere with the full equitable and reasonable use of the waters shall be made except upon a determination, giving due consideration to the objectives of this Agreement and after consultation with all interested Federal agencies and the State officials charged with the administration of this Agreement, that such exercise is in the interest of the best utilization of such waters for multiple purposes.**

(c) **The United States or those acting by or under its authority will recognize any established use, for domestic and irrigation purposes, of the apportioned waters which may be impaired by the exercise of Federal jurisdiction in, over, and to such waters; provided, that such use is being exercised equitably and reasonably, is valid under the laws of the appropriate State and in conformity with this Agreement at the time of the impairment thereof and was validly initiated under State law prior to the initiation or authorization of the Federal program or project which causes such impairment.**

Commentary: Any agreement of this nature should specify the date or conditions upon which it takes effect. In the case of agreements between states of the United States, the conditions with respect to Congress are designed to provide some measure of protection against subsequent federal action that might disturb the allocation system agreed upon by the contracting parties. Despite the requirement of federal approval of interstate compacts, the federal government is not normally a party to those agreements and might not be bound by the provisions of those agreements unless there is specific legislation committing the federal government to be so bound. The provisions of Alternative 2, modeled after the *Republican River Compact*, 57 Stat. 86 (1943), and the *Belle Fourche River Compact*, 58 Stat. 94 (1944), condition the effectiveness of the agreement on passage of such legislation by Congress and also establish a basis for compensation for takings under the Fifth Amendment should a subsequent Congress decide to take action contrary to that commitment (a later Congress has the power to set aside the actions of an earlier Congress, but the question of takings and just compensation then arises). If these conditions are not incorporated, the states making the agreement may later find that federal actions render their agreement ineffective (Draper 2002b).

Cross References: §2C-1-03 (duration of agreement); §2C-1-04 (powers of signatory parties; withdrawal); §2C-1-05 (amendments and supplements); §2C-2-09 (party or parties); §6C-1-02 (modification of agreement).

Similar Agreements: *Republican River Compact*, 57 Stat. 86 (1943); *Belle Fourche River Compact*, 58 Stat. 94 (1944); *Delaware River Basin Compact*, Pub. L. 87-328, 75 Stat. 688-716 (1961); *Susquehanna River Basin Compact*, Pub. L. 91-575, 84 Stat. 1509-1541 (1970); *Convention on the Protection and Use of Transboundary Watercourses and International Lakes*, 31 ILM 1312 (1992); *Convention on the Law of the Non-Navigational Uses of International Watercourses*, United Nations Document A/51/869 (United Nations 1997).

§2C-1-02 Consent to Jurisdiction (For U.S. use)

This Agreement shall be effective upon the United States Congress giving its consent for the United States to be named and joined as a Party defendant or otherwise in any case or controversy involving the construction or application of this Agreement in which one or more of the States is a plaintiff, without regard to any requirement as to the sum or value in controversy or diversity of citizenship of Parties to the case or controversy.

Commentary: The predominance of federal interests in water resources makes it likely that any litigation concerning the agreement between states may involve federal interests. The doctrine of sovereign immunity could prevent joinder of the federal interests as parties to the suit absent a waiver of sovereign immunity. The discretionary decision not to join federal parties led to dismissal of a suit filed by Texas against New Mexico in 1951 to enforce certain provisions of the *Rio Grande Compact of 1938*, 53 Stat. 785, 938 (1938). The U.S. Supreme Court dismissed the case because the federal government was not joined as a party but had important interests that would be affected by any such suit (*Texas v. New Mexico*, 352 U.S. 991, 1957). In 1952, Congress enacted the *McCarran Amendment*, 73 U.S.C. 66 (1952), which waived federal sovereign immunity to be joined in general stream adjudications. However, in most cases, the agreement covers management and development issues that reach beyond general stream adjudication. The parties should consider including such a waiver of sovereign immunity as a condition to effectiveness of the agreement. They may also wish to add a provision granting jurisdiction over any such cases to the district courts, which may be preferable to the U.S. Supreme Court as the initial forum for resolving certain types of disputes. The *Red River Compact*, 94 Stat. 3305 (1980) takes this approach (Draper 2002b).

Cross References: §1A-1-04 (preservation of federal right); §1A-1-05 (national security); §2C-2-09 (party or parties).

Similar Agreements: *Kansas–Nebraska Big Blue River Compact*, 86 Stat. 193 (1972); *Red River Compact*, 94 Stat. 3305 (1980).

§2C-1-03 Duration of Agreement (Optional)

(a) **The Parties intend that the duration of this Agreement shall be for an initial period of [__] years from its effective date.**

(b) **In the event that this Agreement should be terminated by operation of paragraph (a) above, the management structure for the Agreement shall be dissolved, its assets and liabilities transferred, and its corporate affairs wound up, in such manner as may be provided by agreement of the signatory Parties.**

Commentary: The parties may prefer to establish no duration and rely on later provisions to modify or terminate the agreement. However, two significant principles are established by this provision. First, setting duration for an extended period of time, recommended at 10–50 years, allows for predictability in terms of water resources development; it also allows sufficient time to recover capital costs in the financing of projects. Second, establishing duration ensures that the parties reconsider the agreement only after a sufficient hydrologic record is established. It should be noted, however, that this provision greatly affects the exercise of authority of the parties involved (Draper 2002b).

Cross References: §2C-1-01 (activation of agreement); §2C-1-02 (consent to jurisdiction); §2C-1-04 (powers of signatory parties; withdrawal); §2C-1-05 (amendments and supplements); §2C-2-09 (party or parties); §6C-1-02 (modification of agreement).

Similar Agreements: *Delaware River Basin Compact*, Pub. L. 87-328, 75 Stat. 688-716 (1961); *Susquehanna River Basin Compact*, Pub. L. 91-575, 84 Stat. 1509-1541 (1970).

§2C-1-04 Powers of Signatory Parties; Withdrawal (Optional, international use only)

(a) Nothing in this Agreement shall be construed to relinquish the functions, powers, or duties of the government of any signatory Party with respect to the control of any navigable waters within the Basin, nor shall any provision hereof be construed in derogation of any of the powers of the parties to regulate commerce within their borders. The power and right of any party to withdraw from this Agreement or to revise or modify the terms, conditions, and provisions under which it may remain a party is recognized by the signatory Parties. Notification of the withdrawal should be made (___) months in advance of the prospective withdrawal.

(b) Inasmuch as the other Parties may have committed resources to the comprehensive joint management of the waters of the Basin, the withdrawing Party shall equitably compensate the other Parties who have relied on the term of the Agreement to justify their capital costs for facilities and other appurtenances.

Commentary: This provision acknowledges the inherent authority of the individual parties and recognizes that any relinquishment of authority is limited solely to the purposes of this agreement. It does, however, recognize that some parties may have committed financial and other resources to capital projects and that the early withdrawal of one party does not allow them to recover those costs. Consequently, part (b) of this provision provides for equitable compensation for the unrealized recovery of capital expenditures (Draper 2002b).

Cross References: §2C-2-09 (party or parties); §2C-2-11 (waters of the basin); §6C-1-01 (good faith implementation); §6C-1-02 (modification of agreement).

Similar Agreements: *Delaware River Basin Compact*, Pub. L. 87-328, 75 Stat. 688-716 (1961); *Susquehanna River Basin Compact*, Pub. L. 91-575, 84 Stat. 1509-1541 (1970); *Apalachicola–Chattahoochee–Flint River Basin Compact*, O.C.G.A. 12-10-100 (1997); *Alabama–Coosa–Tallapoosa River Basin Compact*, O.C.G.A. 12-10-110 (1997); *Convention on the Protection and Use of Transboundary Watercourses and International Lakes*, 31 ILM 1312 (1992); *Canada–Mexico–United States: North American Agreement on Environmental Cooperation*, 32 ILM 1480 (1993); *Agreement on the Cooperation for the Sustainable Development of the Mekong River Basin*, 34 ILM 864 (1995).

§2C-1-05 Amendments and Supplements (Optional)

The provisions of this Agreement shall remain in full force and effect until amended by action of the governing bodies of the Parties and consented to and approved by any other necessary authority in the same manner as this Agreement is required to be ratified to become effective.

Commentary: Agreements may, over time, cease to operate as well as initially hoped. Some amendment process should be specified. In some cases, the approval of another institution may be required. If, for example, the agreement is between states of the United States, the U.S. Constitution probably requires congressional approval of any amendment as well as approval of the original agreement, unless the agreement provides for a different method of amendment. In this latter case, the congressional approval of the initial agreement would implicitly grant consent to modify the agreement in accordance with the terms of the agreement. If the agreement is between nations, the references to other "necessary authority" would probably be omitted, but the particular circumstances of each case should be considered (Draper 2002b).

Cross References: §2C-1-07 (limited applicability); §2C-2-09 (party or parties); §6C-1-02 (modification of agreement).

Similar Agreements: *Delaware River Basin Compact*, Pub. L. 87-328, 75 Stat. 688-716 (1961); *Susquehanna River Basin Compact*, Pub. L. 91-575, 84 Stat. 1509-1541 (1970); *Convention on the Protection and Use of Transboundary Watercourses and International Lakes*, 31 ILM 1312 (1992); *Agreement on the Cooperation for the Sustainable Development of the Mekong River Basin*, 34 ILM 864 (1995).

§2C-1-06 Existing Agencies

It is the purpose of the signatory Parties to preserve and utilize the functions, powers, and duties of existing offices and agencies of government to the extent not inconsistent with the Agreement, and the institution established to enforce this Agreement is authorized and directed to utilize and employ such offices and agencies for the purpose of this Agreement to the fullest extent it finds feasible and advantageous.

Commentary: The use of existing offices and agencies prevents duplication, and consequently costs, of effort in the data collection and management of the water resource subject to the agreement (Draper 2002b).

Cross References: §2C-2-09 (party or parties); §3C-1-05 (commission organization and staffing); §3C-1-06 (rules of procedures); §3C-1-07 (commission administration); §3C-3-04 (cooperative services).

Similar Agreements: *Delaware River Basin Compact*, Pub. L. 87-328, 75 Stat. 688-716 (1961); *Susquehanna River Basin Compact*, Pub. L. 91-575, 84 Stat. 1509-1541 (1970); *Convention on the Use of Transboundary Watercourses and International Lakes*, 31 ILM 1312 (1992).

§2C-1-07 Severability

Should a tribunal of competent jurisdiction hold any part of this Agreement to be void or unenforceable, it shall be considered severable from those portions of the Agreement capable of continued implementation in the absence of the voided provisions. All other severable provision capable of continued implementation shall continue in full force and effect.

Commentary: The drafters of the agreement should consider whether they wish this clause to be included. The advantage of such a clause is that it avoids the possibility of having the entire agreement become null and void if any part is found to be void or unenforceable (Draper 2002b).

Cross Reference: §2A-1-04 (amendments and supplements).

Similar Agreements: *Yellowstone River Compact,* 65 Stat. 663 (1950); *Sabine River Compact*, 68 Stat. 690 (1953), amended 76 Stat. 34 (1962), 91 Stat. 281 (1977), 106 Stat. 4661 (1992); *Klamath River Basin Compact*, 71 Stat. 497 (1957); *Delaware*

River Basin Compact, Pub. L. 87-328, 75 Stat. 688-716 (1961); *Susquehanna River Basin Compact*, Pub. L. 91-575, 84 Stat. 1509-1541 (1970).

§2C-1-08 Annexes

The Annexes to this Agreement to the extent consistent with the objectives and intent of the Agreement constitute an integral part of the Agreement.

Commentary: An effective water management agreement necessarily contains detailed information and data of a procedural nature. Although such information may be essential to the effectiveness of the particular agreement, its inclusion in the main body of the agreement may take away from the essence of the contractual nature of the agreement. The use of annexes minimizes this effect.

Cross References: Art. 6C (financing); §6C-1-07 (arbitration).

Similar Agreements: *Agreement on the Conservation of Nature and Natural Resources* (ASEAN 1985); *Canada–Mexico–United States: North American Agreement on Environmental Cooperation*, 32 ILM 1480 (1993).

<u>Part 2 Definitions</u>

§2C-2-01 Atmospheric Water

"Atmospheric water" means all available moisture above the surface of the earth, including underground and surface water and all forms of precipitation but not including water projecting from irrigation systems.

Commentary: This definition is consistent with the definitions usually used in state and federal laws on weather modification (Beck 2001, §3.04; Davis 1987; Gochis 2001; Keyes 2006).

Cross References: §1B-1-01 (purposes and scope of agreement); §2B-2-09 (waters of the basin).

§2C-2-02 __________ Basin

"__________ Basin" means the area of drainage into the _______ River and its tributaries, [and] aquifers underlying the drainage, or only the aquifers themselves.

Commentary: The agreement could include the total surface area of drainage throughout the basin and contain aquifers underlying the surface drainage. Some tributaries can be connected to the underlying aquifers holding the underground water. Some of the aquifers could be connected to more than one of the surface water basins. The geographic scope of the agreement should be defined to ensure that there are no future disagreements about what lands are or are not covered by the agreement. A map may be incorporated, but care should be taken that the map is cartographically accurate. Because the map is likely to be at a scale too small for precise delineation of boundaries, it should be made clear that it is for general reference only. In the event of a dispute over land or within the defined ________ River and its tributaries, the actual limits of the watershed as determined on the ground should be controlling.

Cross References: §3C-1-01 (commission created); §3C-1-02 (jurisdiction of the commission); §3C-2-01 (general powers and duties); §3C-2-02 (powers reserved to the commissioners); §3C-2-03 (obligations of the commission); §3C-2-04 (regulations; enforcement); §3C-3-01 (coordination and cooperation); §3C-3-03 (projects of the signatory parties); §4C-1-01 (joint exercise of authority); §4C-1-03 (comprehensive water management plan); §4C-1-04 (purpose and objectives); §4C-1-05 (conditions of comprehensive plan); §4C-1-07 (allocations during flood conditions); §4C-1-09 (non-impairment of comprehensive water plan by state action); §4C-2-01 (basin water sources); §4C-2-03 (basin water demands and needs); §4C-2-06 (watershed management); §4C-3-01 (existing rights recognized); §4C-3-02 (flood protection works); §4C-3-04 (withdrawals and diversions); §4C-3-06 (augmentation of supply); §4C-3-07 (water quality).

§2C-2-03 Comprehensive Water Management Plans

"Comprehensive Water Management Plan" means a plan which considers the interrelationship of transboundary water resources, describes current and prospective water uses, identifies water supplies, and matches these supplies to water uses. It also identifies needed water-related management measures, facility needs and costs, addresses environmental concerns, and offers program and policy recommendations to better manage the Basin's water resources and water quality. This plan is for the long term to produce the highest quality and most efficient use of water resources for the greatest benefit to the public and the environment.

Commentary: Comprehensive water management plans generally are guides for the orderly development and management of water resources. Generally, these plans span a 10- to 50-year horizon and consider population growth, development and availability of new water supplies, water transfers from one basin to another, data sources and methodologies, cost of water, regional or subregional plans, regulatory issues, economic development, specific projects, health and public safety issues, and other concerns. The parties should negotiate the time horizon and the frequency for updating the plan.

Cross References: §4C-1-01 (joint exercise of authority); §4C-1-02 (interrelationship of water resources); §4C-1-03 (comprehensive water management plan); §4C-1-04 (purpose and objectives of comprehensive water management plan); §4C-1-05 (conditions of comprehensive water management plan); §4C-1-06 (deviation from comprehensive water management plan); §4C-1-07 (allocation during flood conditions); §4C-1-08 (allocation during drought conditions); §4C-1-09 (non-impairment of comprehensive water management plan by state action); §4C-2-01 (water allocation, generally); §4C-2-03 (basin water sources); §4C-2-04 (basin water demands and needs); §4C-2-05 (allocation to equitable and reasonable uses); §4C-2-06 (watershed management); §4C-3-01 (existing rights recognized); §4C-3-02 (flood protection works); §4C-3-03 (minimum flows); §4C-3-04 (withdrawals and diversion; protected areas); §4C-3-05 (water levels protected); §4C-3-06 (augmentation of supply); §4C-3-07 (water quality); §4C-3-08 (underground water; limit on withdrawals); §4C-3-09 (atmospheric water); §4C-3-10 (recreation).

§2C-2-04 Conservation Measures

"Conservation measures" refers to any measures adopted by a water right holder, or several water right holders acting in concert pursuant to a conservation agreement reviewed and approved by the Commission, as being appropriate water-saving strategies for purposes of the Comprehensive Water Management Plan, to reduce the withdrawals and/or consumptive uses, including, but not limited to:

- **(a) Improvements in water transmission and water use efficiency;**
- **(b) Reduction in water use;**
- **(c) Enhancement of return flows; and**
- **(d) Reuse of return flows.**

Commentary: Sustainable development requires steps to conserve the waters of the basin. This definition limits the application of the term "conservation measures" to practices that have been reviewed and approved by the commission as being appropriate water-saving strategies for the purposes of the comprehensive water management plan. Specifically excluded from this definition are practices applied to native or naturally occurring waters, return flows from other water rights, or other water sources not associated with the water right holder or sought by the applicant.

Nothing in this model agreement attempts to detail appropriate conservation measures. Such efforts as improved efficiency in manufacturing processes, the substitution of drip irrigation for sprinklers, or the introduction by a public water supply enterprise of a requirement that customers use low-flow toilets or showerheads would all be appropriate examples. The model agreement leaves the precise details regarding the suitability of these or other possible conservation measures to be developed by the regulatory and planning processes prescribed for the state agency.

Cross Reference: §4C-1-08 (allocation under drought conditions).

§2C-2-05 Drought

"Drought" conditions means conditions of abnormal water scarcity in a specific area, resulting from natural conditions.

Commentary: Management action may arise from a drought, or lack of mean annual rainfall, but could arise from other causes as well, such as the collapse of a dam with the resulting draining of a reservoir on which the commission users depend. The definition should be determined, in large measure, by the use intended. Then a "drought management strategy" would be a specific course of conduct planned by the commission as a necessary or appropriate response to the lack of precipitation.

Cross References: §4C-1-03 (comprehensive water management plan); §4C-1-04 (purpose and objectives of comprehensive water management plan); §4C-1-05 (conditions of comprehensive water management plan); §4C-1-06 (deviation from comprehensive water management plan); §4C-1-08 (allocation during drought conditions); §4C-1-09 (non-impairment of comprehensive water management plan by state action); §4C-2-01 (water allocation, generally); §4C-2-03 (basin water sources); §4C-2-04 (basin water demands and needs); §4C-2-05 (allocation to equitable and reasonable uses); §4C-2-06 (watershed management); §4C-3-01 (existing rights recognized); §4C-3-03 (minimum flows); §4C-3-04 (withdrawals and diversion; protected areas); §4C-3-05 (water levels protected); §4C-3-06 (augmentation of supply); §4C-3-07 (water quality); §4C-3-08 (underground water; limit on withdrawals); §4C-3-09 (atmospheric water).

§2C-2-06 Equitable and Reasonable Utilization

Utilization of a transboundary water resource in an equitable and reasonable manner requires taking into account all relevant factors and circumstances, including:

- **(a) Geographic, hydrographic, hydrological, climatic, ecological, and other factors of a natural character;**
- **(b) The social and economic needs of the Parties concerned;**
- **(c) The population dependent on the water resource in each of the Parties;**
- **(d) The effects of the use or uses of the water resources by one Party on other Parties;**
- **(e) Existing and potential uses of the water resource;**
- **(f) Conservation, protection, development, and economy of use of the water resource and the costs of measures taken to that effect;**
- **(g) The availability of alternatives, of comparable value, to a particular planned or existing use;**
- **(h) The sustainability of proposed or existing uses; and**
- **(i) The minimization of environmental harm.**

Commentary: The term "equitable and reasonable utilization" appears in the 1997 UN *Convention on the Law of the Non-Navigational Uses of International Watercourses* (United Nations 1997). Although ratifications of the UN convention have proceeded slowly and it has yet to enter into force, it has become recognized as authoritative on the customary international law governing the issues it addresses. To cope with emerging problems of international or global water management for the 21st century, the International Law Association added the last two factors to those defining the term "equitable and reasonable utilization." This term is wider in scope than the term for "equitable apportionment," providing consideration of many more water uses and programs. For this reason, the committee for the comprehensive water management model has approved the term "equitable and reasonable utilization." For those parties within the United States who wish to maintain allegiance to the term "equitable apportionment," that term should replace the term "equitable and reasonable utilization" (ILA 2004; Dellapenna 2007b).

Cross References: §1C-1-02 (purposes and scope of agreement); §1C-1-04 (coordination and cooperation); §3C-2-01 (general powers and duties); §3C-2-10 (meetings, hearings and records); §4C-3-01 (existing rights recognized).

§2C-2-07 Flood

"Flood" means a rising of water to levels that have detrimental effects on or in one or more basin States with a frequency specified jointly by the Parties.

Commentary: The flood condition is the opposite of a drought. A large amount of water is to be controlled by facilities of the commission. The parties are to agree as to the frequency of the flow of high waters in the basin. Most of the time, these flows occur during periods that exceed the amount of flow that occurs during the years of mean annual precipitation (ILA 2004).

Cross References: §4C-1-03 (comprehensive water management plan); §4C-1-04 (purpose and objectives of comprehensive water management plan); §4C-1-05 (conditions of comprehensive water management plan); §4C-1-06 (deviation from comprehensive water management plan); §4C-1-07 (allocation during flood conditions); §4C-1-09 (non-impairment of comprehensive water management plan by state action); §4C-2-01 (water allocation, generally); §4C-2-03 (basin water sources); §4C-2-06 (watershed management); §4C-3-01 (existing rights recognized); §4C-3-02 (flood protection works).

§2C-2-08 Interbasin Transfer

An "interbasin transfer" is any bulk transfer of water from one water basin to another.

Commentary: This definition merely makes clear that the term "interbasin transfer" is not limited in any fashion but refers to all transfers from one water basin to another. The provisions regarding interbasin transfers allow regulations to exempt certain small transfers. In many states within the United States, 378,571 L per day (100,000 gal. per day) are exempt from regulation.

Cross Reference: §4C-1-09 (non-impairment of comprehensive water management plan by state action).

§2C-2-09 Party or Parties

"Party or Parties" means, unless the text otherwise indicates, those governmental units signatory to this Agreement.

Commentary: Defining the terms in this way avoids the need to include similar language at numerous points throughout the agreement. As a matter of law, it may be unnecessary to state this principle.

Cross References: §1C-1-02 (purposes of agreement); §1C-1-03 (objectives of agreement); §1C-1-04 (coordination and cooperation); §1C-1-06 (national security); §2C-1-01 (activation of agreement); §2C-1-02 (consent to jurisdiction); §2C-1-03 (duration of agreement); §2C-1-04 (powers of signatory parties; withdrawal); §2C-1-05 (amendments and supplements); §2C-1-06 (existing agencies); §3C-1-03 (commissioners); §3C-1-04 (status, immunities, and privileges); §3C-1-05 (commission organization and staffing); §3C-2-01 (general powers and duties); §3C-2-02 (powers and duties reserved to the commissioners); §3C-2-03 (obligations of the commission); §3C-2-05 (prohibited activities); §3C-2-08 (reports); §3C-2-09 (condemnation proceedings); §3C-2-10 (meetings, hearings and records); §3C-2-11 (tort liability); §3C-3-01 (coordination and cooperation); §3C-3-02 (project costs and evaluation standards); §3C-3-03 (projects of the signatory parties); §3C-3-04 (cooperative services); §4C-1-01 (joint exercise of authority); §4C-1-05 (conditions of comprehensive water management plan); §4C-1-07 (allocation during flood conditions); §4C-2-02 (waters not subject to allocation); §4C-2-05 (allocation to equitable and reasonable uses); §4C-3-01 (existing rights recognized); §4C-3-02 (flood protection works); §4C-3-03 (minimum flows); §4C-3-04 (withdrawals and diversion; protected areas;) §4C-3-05 (water levels protected); §4C-3-07 (water quality); §4C-3-08 (underground water; limit on withdrawals); §4C-3-09 (atmospheric water); §4C-3-10 (recreation); §6C-1-01 (good faith implementation); §6C-1-02 (modification of agreement); §6C-1-03 (material breach); §6C-1-04 (alternative dispute resolution); §6C-1-05 (negotiations and consultations); §6C-1-06 (consultations and mediation); §6C-1-07 (arbitration); §6C-1-08 (right to litigate); Annex C-1; Annex C-2.

§2C-2-10 Underground Water

The term "underground water" means water found beneath the ground, regardless of whether flowing through defined channels or percolating through the ground, and regardless of its existence in water table or artesian condition, whether the result of natural or artificial recharge.

Commentary: This definition of "underground water" includes all forms of water in the ground, being equivalent to terms such as "groundwater" or similar expressions. It excludes soil (capillary) moisture that might be drawn upon by plants but cannot practically be withdrawn by direct human activity. A somewhat more precise definition is found in the *Illinois Water Use Act* (1983) water under the ground where the fluid pressure in the pore space is equal to or greater than atmospheric pressure (Dellapenna 2001, §6.04; Murphy 2001, §18.02).

Cross References: §3C-1-02 (jurisdiction of the commission); §3C-2-03 (obligations of the commission); §4C-2-03 (basin water sources); §4C-3-04 (withdrawals and diversions; protected areas); §4C-3-07 (water quality); §4C-3-08 (underground water; limit on withdrawals).

§2C-2-11 Waters of the Basin

"Waters of the Basin" shall include all water found within the Basin, whether surface water, underground water, and/or atmospheric water.

Commentary: This definition should be included to make it clear that underground water and atmospheric water are included within the scope of the agreement, if that is the intent of the parties. With regard to underground water, where current scientific knowledge is available in an aquifer to establish direct interconnectivity between surface and underground water, it is essential that such underground water be included in the agreement. The technological questions relating to atmospheric water may result in uncertainty regarding its allocation, but to the extent the parties wish to reach a complete agreement, the matter should be addressed, or recognition should be given to the fact that the parties have chosen to reserve that issue for later resolution. The parties should also decide whether water imported from other basins should be included within the scope of the agreement. If it is not to be so included, that exclusion should be noted in this article.

Cross References: §1C-1-02 (purposes and scope of agreement); §2C-1-04 (powers of signatory parties; withdrawal); §2C-2-10 (underground water); §3C-1-02 (jurisdiction of the commission); §3C-2-01 (general powers and duties); §4C-1-01 (purpose and objectives of the comprehensive water management plan); §4C-2-01 (water allocation, generally); §4C-3-01 (existing rights recognized); §4C-3-04 (withdrawals and diversions; protected areas); §4C-3-07 (water quality).

C.3 ARTICLE 3C: ADMINISTRATION

__Part 1 Administrative Authority__

§3C-1-01 Commission Created

(a) **The ____________ River Commission (hereinafter called the Commission) is hereby created as a body politic and corporate, with succession for the duration of this Agreement, as an agency and instrumentality of the governments of the respective signatory Parties.**

(b) **The Commission shall develop and effectuate policies for the allocation of the water resources of the ____________ River and its tributaries in accordance with this Agreement and may amend the Agreement should changing conditions or extreme events require alteration of the Agreement terms, but only when such changes are agreed to by all Commissioners.**

Commentary: Almost two-thirds of the interstate water apportionment compacts create a compact commission (Grant 2001, §46.03). Comprehensive management of a water basin shared by two or more political parties requires joint or communal management (Dellapenna 1994). To be effective in managing water and precluding conflict, the institutional structure should not only have to embody concepts of cooperative management, but it should also have to be able to (United Nations 1975; Le Marquand 1978).

1. determine the facts of water use in each nation;
2. resolve disputes across political boundaries;
3. guide responses to unusual temporary water shortages;
4. regulate or design long-term solutions; and
5. enforce its decision.

The organizational structure of the commission should be constituted according to the specific geographical and hydrological characteristics of the shared water resource and the political structures of the parties involved. "What works for wealthy nations may not work for developing countries," and in some shared water situations, cultural differences (e.g., the Jordan River) may require different management structures than those used in situations involving similar cultures (e.g., the Rhine River) (Williams 1993).

Sterner (1994) states that "there should be an appropriate legal structure and set of institutions that define property rights and establish the framework within which an environmental authority can function."

Cross References: §2C-2-02 (basin); §3C-1-02 (jurisdiction of the commission); §3C-1-03 (commissioners); §3C-1-04 (status, immunities and privileges); §3C-1-05 (commission organization and staffing); §3C-1-06 (rules of procedures); §3C-1-07 (commission administration); §3C-2-01 (general powers and duties); §3C-2-02 (powers and duties reserved to the commissioners); §3C-2-03 (obligations of the commission); §3C-2-04 (regulations; enforcement); §3C-2-05 (prohibited activities); §3C-2-06 (referral and review); §3C-2-07 (advisory committees); §3C-2-08 (reports); §3C-2-09 (condemnation proceedings); §3C-2-10 (meetings, hearings and records); §3C-2-11 (tort liability); §3C-3-01 (coordination and cooperation); §3C-3-02 (project costs and evaluation standards); §3C-3-03 (projects of the signatory parties); §3C-3-04 (cooperative services).

Similar Agreements: *Klamath River Basin Compact*, 71 Stat. 497 (1957); *Delaware River Basin Compact*, Pub. L. 87-328, 75 Stat. 688-716 (1961); *Susquehanna River Basin Compact*, Pub. L. 91-575, 84 Stat. 1509-1541 (1970); *Apalachicola–Chattahoochee–Flint River Basin Compact*, O.C.G.A. 12-10-100 (1997); *Alabama–Coosa–Tallapoosa River Basin Compact*, O.C.G.A. 12-10-110 (1997); *Treaty between the United States and Great Britain Relating to Boundary Waters*, 36 Stat. 2451 (1909); *Treaty between the United States and Mexico. Utilization of Waters of the Colorado and Tijuana Rivers and of the Rio Grande*, 59 Stat. 1219 (1944); *The Indus Waters Treaty 1960*, 419 UNTS 126 (1960); *Canada–Mexico–United States: North American Agreement on Environmental Cooperation*, 32 ILM 1480 (1993); *Agreement on the Cooperation for the Sustainable Development of the Mekong River Basin*, 34 ILM 864 (1995).

§3C-1-02 Jurisdiction of the Commission

(a) The Commission shall have, exercise and discharge its functions, powers, and duties within the limits of the Basin, except that it may in its discretion act outside the Basin whenever such action may be necessary or convenient to effectuate its powers or duties within the Basin. The Commission shall exercise such power outside the Basin only upon the consent of the Party in which it proposes to act.

(b) Those sources of underground water that have a direct connection and recharge capability to surface waters of the Basin shall be included in the water allocations of this Agreement. In the event such underground water aquifers are or have the potential of being used by parties not signatory to this Agreement, the Commission will enter into consultations and negotiations with such parties to reach an agreement as to the allocation of the interconnected underground water.

Commentary: This provision describes the geographic and hydrologic jurisdiction of the commission. Similar provisions appear in most agreements to clarify and define the limits of supranational authority, although conjunctive management has not been the norm (Hayton and Utton 1989).

Cross References: §2C-2-02 (basin); §2C-2-11 (waters of the basin); §3C-1-01 (commission created); §3C-1-03 (commissioners); §3C-1-04 (status, immunities and privileges); §3C-1-05 (commission organization and staffing); §3C-1-06 (rules of procedures); §3C-1-07 (commission administration); §3C-2-01 (general powers and duties); §3C-2-02 (powers and duties reserved to the commissioners); §3C-2-03 (obligations of the commission); §3C-2-04 (regulations; enforcement); §3C-2-05 (prohibited activities); §3C-2-06 (referral and review); §3C-2-07 (advisory committees); §3C-2-08 (reports); §3C-2-09 (condemnation proceedings); §3C-2-10 (meetings, hearings and records); §3C-2-11 (tort liability); §3C-3-01 (coordination and cooperation); §3C-3-02 (project costs and evaluation standards); §3C-3-03 (projects of the signatory parties); §3C-3-04 (cooperative services).

Similar Agreements: *Delaware River Basin Compact*, Pub. L. 87-328, 75 Stat. 688-716 (1961); *Susquehanna River Basin Compact*, Pub. L. 91-575, 84 Stat. 1509-1541 (1970); *Treaty between the United States and Great Britain Relating to Boundary Waters*, 36 Stat. 2451 (1909); *Treaty between the United States and Mexico. Utilization of Waters of the Colorado and Tijuana Rivers and of the Rio Grande*, 59 Stat. 1219 (1944); *Agreement on the Cooperation for the Sustainable Development of the Mekong River Basin*, 34 ILM 864 (1995).

§3C-1-03 Commissioners

(a) The Commission shall be governed by a Board of Commissioners, consisting of the Governor or other Chief Executive Officer of the signatory Parties.

(b) Each Commissioner may appoint an alternate to act in his place and stead, with authority to attend all meetings of the Commission, and with power to vote in the absence of the member. Unless otherwise provided by law of the signatory Party for which he is appointed, each alternate shall serve during the term of the member appointing him, subject to removal at the pleasure of the member. In the event of a vacancy in the office of alternate, it shall be filled in the same manner as an original appointment for the unexpired term only.

(c) Any and all changes to the timing or amount of water allocations or changes to the financial or enforcement provisions of the Agreement shall be decided upon by the Commissioners and unanimity shall be required. For all other matters, the rule of decision shall be simple majority. Each member shall be entitled to one vote on all matters that may come before the Commission. No action by the Commissioners shall be taken at any meeting unless a majority of the membership shall vote in favor thereof. The federal representative shall have no vote.

Commentary: The membership of the typical commission in the United States includes one or more members from each party plus a federal representative. The federal representative usually has no vote (Grant 2001, §46.03). The *Delaware River Basin Compact* and the *Susquehanna River Basin Compact* are notable exceptions. This provision recognizes that the responsibility (and authority) for all policy decisions or decisions affecting the substance of the agreement remains with the principal executive offices of the respective parties.

Cross References: §2C-2-09 (party or parties); §3C-1-01 (commission created); §3C-1-02 (jurisdiction of the commission); §3C-1-04 (status, immunities and privileges); §3C-1-05 (commission organization and staffing); §3C-1-06 (rules of procedures); §3C-1-07 (commission administration); §3C-2-01 (general powers and duties); §3C-2-02 (powers and duties reserved to the commissioners); §3C-2-03 (obligations of the commission); §3C-2-04 (regulations; enforcement); §3C-2-05 (prohibited activities); §3C-2-06 (referral and review); §3C-2-07 (advisory committees); §3C-2-08 (reports); §3C-2-09 (condemnation proceedings); §3C-2-10 (meetings, hearings and records); §3C-2-11 (tort liability); §3C-3-01 (coordination and cooperation); §3C-3-02 (project costs and evaluation standards); §3C-3-03 (projects of the signatory parties); §3C-3-04 (cooperative services).

Similar Agreements: *Klamath River Basin Compact*, 71 Stat. 497 (1957); *Delaware River Basin Compact*, Pub. L. 87-328,

75 Stat. 688-716 (1961); *Susquehanna River Basin Compact*, Pub. L. 91-575, 84 Stat. 1509-1541 (1970); *Apalachicola–Chattahoochee–Flint River Basin Compact*, O.C.G.A. 12-10-100 (1997); *Alabama–Coosa–Tallapoosa River Basin Compact*, O.C.G.A. 12-10-110 (1997); *Convention on the Protection and Use of Transboundary Watercourses and International Lakes*, 31 ILM 1312 (1992).

§3C-1-04 Status, Immunities and Privileges (Optional)
To enable the Commission to fulfill its purpose and the functions with which it is entrusted, the status, immunities and privileges set forth in this Article shall be accorded to the Commission in the territories of each Party.

(a) The Commission, its property and its assets, wherever located, and by whomsoever held, shall enjoy the same immunity from suit and every form of judicial process as is enjoyed by the Parties, except to the extent that the Commission may expressly waive its immunity for the purposes of any proceedings or by the terms of any contract.
(b) Property and assets of the Commission, wheresoever located and by whomsoever held, shall be considered public property and shall be immune from search, requisition, confiscation, expropriation, or any other form of taking or foreclosure by executive or legislative action.
(c) To the extent necessary to carry out the purpose and functions of the Commission and to conduct its operations in accordance with this Agreement, all property and other assets of the Commission shall be free from restrictions, regulations, controls, and moratoria of any nature affecting the implementation of this Agreement, except as may otherwise be provided in this Agreement.
(d) The official communications of the Commission shall be accorded by each Party the same treatment that it accords to the official communications of the other Parties.
(e) (Optional, for international use only) The Commissioners and other personnel engaged directly in the affairs of the Commission shall have the following privileges and immunities:
   (1) Immunity from legal process with respect to acts performed by them in their official capacity except when the Commission expressly waives this immunity.
   (2) When not citizens of one of the signatory Parties, the same immunities from immigration restrictions, alien registration requirements and national service obligations and the same facilities as regards exchange provisions as are accorded by each Party to the representatives, officials, and employees of comparable rank of the other Party; and
   (3) The same privileges in respect of traveling and facilities as are accorded by each Party to representatives, officials, and employees of comparable rank of the other Party.
(f) The Commission, its property, other assets, income, and the operations it carries out pursuant to this Article shall be immune from all state taxation. The Commission shall also be immune from any obligation relating to the payment, withholding, or collection of any tax or customs duty. No state tax shall be levied on or in respect of salaries and benefits paid by the Commission to officers or staff of the Commission who are not local citizens.
(g) Each Party, in accordance with its juridical system, shall take such action as is necessary to make effective in its own territories the principles set forth in this Article, and shall inform the Commission of the action which it has taken on the matter.

Commentary: This provision provides the commissioners and their personnel with the same legal protections that normally exist for governmental officials of the parties.

Cross References: §1C-1-01 (general policies); §1C-1-02 (purposes of agreement); §1C-1-03 (objectives of agreement); §2C-2-09 (party or parties); §3C-1-01 (commission created); §3C-1-02 (jurisdiction of the commission); §3C-1-03 (commissioners); §3C-1-05 (commission organization and staffing); §3C-1-06 (rules of procedures); §3C-1-07 (commission administration); §3C-2-01 (general powers and duties); §3C-2-02 (powers and duties reserved to the commissioners); §3C-2-03 (obligations of the commission).

Similar Agreement: *Agreement between the Governments of the United States of America and the Government of the United Mexican States Concerning the Establishment of a Border Environment Cooperation Commission and a North American Development Bank*, 19 U.S.C. §3473 (1993).

§3C-1-05 Commission Organization and Staffing

(a) The Commission shall meet and organize at _______ promptly after the members thereof are appointed, and when organized the Commission may fix such times and places for its meetings as may be necessary, subject at all times to special call or direction by the Parties. Each Commissioner, upon the first joint meeting of the Commission after his appointment, shall, before proceeding with the work of the Commission, make and subscribe a solemn declaration in writing that he will faithfully and impartially perform the duties imposed upon him under this Agreement, and such declaration shall be entered on the records of the proceedings of the Commission.
(b) The respective Commissioners may each appoint a secretary, and these shall act as joint secretaries of the Commission at its joint sessions, and the Commission may employ professional and administrative personnel from time to time as it may deem advisable. The salaries and personal expenses of the Commission and of the secretaries shall be paid by their respective Governments, and all reasonable and necessary joint expenses of the Commission, incurred by it, shall be paid in equal portions by the Parties unless otherwise stipulated in this Agreement.

Commentary: This provision provides the authority and instructions for the organization and initiation of commission undertakings. The parties may prefer to apportion joint expenses according to benefits received or some other economic algorithm.

Cross References: §1C-1-01 (general policies); §1C-1-02 (purposes of agreement); §1C-1-03 (objectives of agreement); §2C-2-09 (party or parties); §3C-1-01 (commission created).

Similar Agreements: *Treaty between the United States and Great Britain Relating to Boundary Waters*, 36 Stat. 2451 (1909); *Convention between the Federal Republic of Germany and the Czech and Slovak Federal Republic and the European Economic Community on the International Commission for the Protection* *of the Elbe*, International Environmental Law, Multilateral Agreements, 976:90/1 (1990).

§3C-1-06 Rules and Procedures

(a) The Commission shall adopt its own Rules of Procedure.
(b) The Commission may seek technical advisory services as it deems necessary.

Commentary: Except when necessary for policy reasons, the agreement should not bind the commission to specific procedural requirements.

Cross References: §3C-1-01 (commission created); §3C-1-02 (jurisdiction of the commission); §3C-1-03 (commissioners); §3C-1-04 (status, immunities and privileges); §3C-1-05 (commission organization and staffing); §3C-1-07 (commission administration); §3C-2-01 (general powers and duties); §3C-2-02 (powers and duties reserved to the commissioners); §3C-2-03 (obligations of the commission); §3C-2-04 (regulations; enforcement); §3C-2-05 (prohibited activities); §3C-2-06 (referral and review); §3C-2-07 (advisory committees); §3C-2-08 (reports); §3C-2-09 (condemnation proceedings); §3C-2-10 (meetings, hearings and records); §3C-2-11 (tort liability); §3C-3-01 (coordination and cooperation); §3C-3-02 (project costs and evaluation standards); §3C-3-03 (projects of the signatory parties); §3C-3-04 (cooperative services).

Similar Agreements: *Klamath River Basin Compact*, 71 Stat. 497 (1957); *Delaware River Basin Compact*, Pub. L. 87-328, 75 Stat. 688-716 (1961); *Susquehanna River Basin Compact*, Pub. L. 91-575, 84 Stat. 1509-1541 (1970); *Canada–Mexico–United States: North American Agreement on Environmental Cooperation*, 32 ILM 1480 (1993); *Agreement on the Cooperation for the Sustainable Development of the Mekong River Basin*, 34 ILM 864 (1995); *Convention on the Law of the Non-Navigational Uses of International Watercourses*, United Nations Document A/51/869 (United Nations 1997).

§3C-1-07 Commission Administration

(a) Supervision and management of the routine and customary affairs of the Commission shall be vested in a Commission Administration consisting of an Executive Director and such additional officers, deputies, and assistants as the Commissioners may determine. The Executive Director shall be appointed and may be removed by the affirmative vote of a majority of the full membership of the Commissioners. All other officers and employees shall be appointed by the Executive Director and confirmed by the Commissioners under such rules of procedure as the Commission may determine.
(b) Except for the express limitations provided in this Agreement, implementation and administration of the Agreement terms shall be vested in the Commission Officers appointed by the Commissioners. The Commission Officers shall report directly to the Commission.
(c) In the appointment and promotion of Commission Officers and employees, no political, racial, gender, religious, or residence test or qualification shall be permitted or given consideration, but all such appointments and promotions shall be solely on the basis of merit and fitness. Any officer or employee of the Commission who is found by the Commissioners to be guilty of a violation of this Article shall be removed from office.

Commentary: The commission administration provides the day-to-day administrative management of the comprehensive plan for the commission. Although all substantive policy-making decisions involving the agreement rightly remain with the commissioners, the commission's purpose is to carry out the technical facets involved with agreement implementation.

The parties may wish to establish a commission membership that equitably allocates officers and employees among citizens of the various governments. In such a case, the references to political and residency tests can be eliminated. Nondiscrimination on the basis of race, gender, or religion is strongly encouraged, however.

Cross References: §3C-1-01 (commission created); §3C-1-02 (jurisdiction of the commission); §3C-1-03 (commissioners); §3C-1-04 (status, immunities and privileges); §3C-1-05 (commission organization and staffing); §3C-1-06 (rules of procedures); §3C-2-01 (general powers and duties); §3C-2-02 (powers and duties reserved to the commissioners); §3C-2-03 (obligations of the commission); §3C-2-04 (regulations; enforcement); §3C-2-05 (prohibited activities); §3C-2-06 (referral and review); §3C-2-07 (advisory committees); §3C-2-08 (reports); §3C-2-09 (condemnation proceedings); §3C-2-10 (meetings, hearings and records); §3C-2-11 (tort liability); §3C-3-01 (coordination and cooperation); §3C-3-02 (project costs and evaluation standards); §3C-3-03 (projects of the signatory parties).

Similar Agreements: *Klamath River Basin Compact*, 71 Stat. 497 (1957); *Delaware River Basin Compact*, Pub. L. 87-328, 75 Stat. 688-716 (1961); *Susquehanna River Basin Compact*, Pub. L. 91-575, 84 Stat. 1509-1541 (1970); *Apalachicola–Chattahoochee–Flint River Basin Compact*, O.C.G.A. 12-10-100 (1997); *Alabama–Coosa–Tallapoosa River Basin Compact*, O.C.G.A. 12-10-110 (1997); *Agreement on the Cooperation for the Sustainable Development of the Mekong River Basin*, 34 ILM 864 (1995).

<u>Part 2 Powers and Duties</u>

§3C-2-01 General Powers and Duties

The Commission as a corporate body shall:

(a) Allocate the waters of the ______ River Basin among the Parties signatory to this Agreement and their respective political subdivisions, in accordance with the doctrine of equitable and reasonable utilization, and impose conditions, obligations and release requirements related thereto;
(b) Develop, implement, and effectuate plans and projects for the use of the water of the Basin for the purposes of economical growth and development, maintenance of adequate public health and safety, and environmental protection;
(c) Provide administration and coordination of the Agreement implementation;
(d) Conduct activities consistent with its functions so as to further the conservation and enhancement of natural beauty and the sustainability of the environment;
(e) Coordinate the collection, compilation, and analysis of data and information of hydrologic, environmental, and economic importance within the Basin;
(f) Respond on an emergency basis to changing conditions when time constraints do not allow review and analysis by the Commissioners;
(g) Report (<u>quarterly/semi-annually/annually</u>) to the Parties on the implementation of the Agreement;
(h) Conduct active public involvement programs to ensure the needs of all stakeholders are considered;

(i) **Prepare annual budgets for implementation of the Agreement;**
(j) **Identify and recommend to the Commissioners appointments to the Commission administration and any review panel;**
(k) **Establish standards for planning, design, and operation of all water resources projects and facilities in the Basin;**
(l) **Conduct and sponsor research on water resources issues that may arise during the life of the Agreement;**
(m) **Prepare, publish, and disseminate information and reports with respect to the water problems of the Basin and for the presentation of the needs, resources, and policies of the Basin to the Commissioners and the executive and legislative branches of the signatory Parties;**
(n) **Plan, manage, budget, and allocate financial resources necessary to effectuate the purposes of this Agreement; and**
(o) **Exercise such other powers as may be delegated to it by this Agreement or otherwise pursuant to law, and have and exercise all powers necessary or convenient to carry out its express powers or powers which may be reasonably implied therefrom.**

Commentary: This list of powers and duties should be sufficient to adequately manage the implementation of the agreement. Additionally, the stated powers should not be so detailed as to restrict the management of the inherently variable conditions of water resources.

Cross References: §1C-1-01 (general policies); §1C-1-02 (purposes of agreement); §1C-1-03 (objectives of agreement); §2C-2-02 (basin); §2C-2-06 (equitable and reasonable apportionment); §2C-2-09 (party or parties); §2C-2-11 (waters of the basin); §3C-2-02 (powers and duties reserved to the commissioners); §3C-2-03 (obligations of the commission); §3C-2-04 (regulations; enforcement); §3C-2-05 (prohibited activities).

Similar Agreements: *Klamath River Basin Compact*, 71 Stat. 497 (1957); *Delaware River Basin Compact*, Pub. L. 87-328, 75 Stat. 688-716 (1961); *Susquehanna River Basin Compact*, Pub. L. 91-575, 84 Stat. 1509-1541 (1970); *Apalachicola–Chattahoochee–Flint River Basin Compact*, O.C.G.A. 12-10-100 (1997); *Alabama–Coosa–Tallapoosa River Basin Compact*, O.C.G.A. 12-10-110 (1997); *Treaty between the United States and Mexico. Utilization of Waters of the Colorado and Tijuana Rivers and of the Rio Grande*, 59 Stat. 1219 (1944); *Canada–Mexico–United States: North American Agreement on Environmental Cooperation*, 32 ILM 1480 (1993); *Declaration of Principles on Interim Self-Government Arrangements,* 32 ILM 1525 (1993); *Agreement on the Cooperation for the Sustainable Development of the Mekong River Basin*, 34 ILM 864 (1995).

§3C-2-02 Powers and Duties Reserved to the Commissioners
The Commissioners shall supervise and control the implementation of the provisions of the Agreement. The Commissioners may delegate certain powers but shall retain the following:

(a) **To appoint the Executive Director of the Commission and confirm the appointment of other Commission Officers designated to administer the Agreement provisions;**
(b) **To adopt annual budgets prepared for Agreement administration;**
(c) **To approve all plans and capital projects developed by the signatory Parties relating to the water resources of the Basin;**
(d) **(Optional) To approve all fees and assessments levied by the Commission;**
(e) **To adopt and promote uniform and coordinated policies for water conservation, control, use, and management in the Basin and to approve the planning, development, and financing of water resources projects according to such plans and policies;**
(f) **To require the maintenance of faithful records of all meetings and decision-making activities of the Commission and provide public access to those records (within [___] calendar days) of the meeting or activities taking place;**
(g) **To conduct public meetings (once/twice) yearly concerning the activities of the Commission and provide opportunity for all interest groups and the general public to express their views;**
(h) **To determine the character of and the necessity for its obligations and expenditures and the manner in which they shall be incurred, allowed, and paid subject to any provisions of law specifically applicable to agencies or instrumentalities created by the Agreement;**
(i) **To provide for the internal organization and administration of the Commission;**
(j) **To appoint the principal executive officers of the Commission and delegate to and allocate among them administrative functions, powers, and duties;**
(k) **To create and abolish offices, employment opportunities, and positions as it deems necessary for the purposes of the Commission, and subject to the provisions of this Article, fix and provide for the qualification, appointment, removal, term, tenure, compensation, pension, and retirement rights of its officers and employees; and**
(l) **(Optional) To let and execute contracts over ($500,000) to carry out the purposes of this Agreement.**

Commentary: These powers and duties of the commissioners are the minimum necessary to effectively administer a comprehensive water management plan while ensuring that all stakeholders have a channel to express their needs. Although the list can be expanded, care should be taken to limit any infringement of national autonomy that may cause the agreement to be breached.

Subparagraph (d) contains an optional provision for approval of a power granted below to the commission itself to levy fees and assessments to cover costs of specific operations and maintenance. Its inclusion is appropriate only when the power to levy fees and assessments (*Delaware River Basin Compact*, Pub. L. 87-328, 75 Stat. 688-716, 1961) has been given to the Commission (§3C-2-03(e)).

The list of powers also includes the requirement for approval by the commissioners of contracts in excess of a certain amount, to be determined by the parties.

Cross References: §1C-1-01 (general policies); §1C-1-02 (purposes of agreement); §1C-1-03 (objectives of agreement); §2C-2-02 (basin); §2C-2-09 (party or parties); §2C-2-10 (underground water); §3C-2-03 (obligations of the commission); §3C-2-04 (regulations; enforcement); §3C-2-05 (prohibited activities).

Similar Agreements: *Delaware River Basin Compact*, Pub. L. 87-328, 75 Stat. 688-716 (1961); *Susquehanna River*

Basin Compact, Pub. L. 91-575, 84 Stat. 1509-1541 (1970); *Apalachicola–Chattahoochee–Flint River Basin Compact*, O.C.G.A. 12-10-100 (1997); *Alabama–Coosa–Tallapoosa River Basin Compact*, O.C.G.A. 12-10-110 (1997); *Treaty between the United States and Mexico. Utilization of Waters of the Colorado and Tijuana Rivers and of the Rio Grande*, 59 Stat. 1219 (1944); *Canada–Mexico–United States: North American Agreement on Environmental Cooperation*, 32 ILM 1480 (1993).

§3C-2-03 Powers and Duties of the Commission
The Commission, for the purposes of this Agreement, may:

(a) Enter into contracts, sue, and be sued in all courts of competent jurisdiction;
(b) Receive and accept such payments, appropriations, grants, gifts, loans, advances, and other funds, properties and services as may be transferred or made available to it by any signatory Party or by any other public or private corporation or individual, and enter into agreements to make reimbursement for all or part thereof;
(c) Provide for, acquire, and adopt detailed engineering, administrative, financial, and operating plans and specifications to effectuate, maintain, or develop any facility or project;
(d) Control and regulate the use of facilities owned or operated by the Commission;
(e) (Optional) Assess on an annual basis or otherwise the cost of operations and maintenance upon water users or any classification of them specially benefited thereby to a measurable extent, provided that no such assessment shall exceed the actual benefit to any water user and such assessment shall follow the procedure prescribed by law for the assessment of fees for governmental services and shall be subject to judicial review in any court of competent jurisdiction;
(f) Acquire, own, operate, maintain, control, sell, and convey real and personal property and any interest therein by contract, purchase, lease, license, mortgage, or otherwise as it may deem necessary for any project or facility, including any and all appurtenances thereto necessary, useful, or convenient for such ownership, operation, control, maintenance or conveyance;
(g) Provide for, construct, acquire, operate, and maintain dams, reservoirs, and other facilities for utilization of surface and water resources, and all related structures, appurtenances, and equipment in the Basin and its tributaries and at such off-river sites as it may find appropriate, and may regulate and control the use thereof.
(h) Have and exercise all corporate powers essential to the declared objects and purposes of the Commission.

Commentary: This provision allows the commission to operate without the day-to-day decision making by the commissioners themselves. However, although the powers and duties listed in this article assist in effective implementation of the agreement, the provision is optional and may be deleted or expanded, according to the specific geopolitical situation.

Subparagraph (e) contains an optional provision for the levy of fees and assessments to cover costs of specific operations and maintenance (*Delaware River Basin Compact*, Pub. L. 87-328, 75 Stat. 688-716, 1961). Its inclusion is appropriate only if this power is subject to the approval of the commissioners (§3C-2-02(d)).

Cross References: §1C-1-01 (general policies); §1C-1-02 (purposes of agreement); §1C-1-03 (objectives of agreement); §2C-2-02 (basin); §2C-2-09 (party or parties); §2C-2-10 (underground water); §3C-2-01 (general powers and duties); §3C-2-02 (powers and duties reserved to the commissioners); §3C-2-04 (regulations; enforcement); §3C-2-05 (prohibited activities).

Similar Agreements: *Delaware River Basin Compact*, Pub. L. 87-328, 75 Stat. 688-716 (1961); *Susquehanna River Basin Compact*, Pub. L. 91-575, 84 Stat. 1509-1541 (1970).

§3C-2-04 Regulations; Enforcement
The Commission may:

(a) Make and enforce reasonable rules and regulations for the initiation, application, and enforcement of this Agreement; and it may adopt and enforce practices and schedules for or in connection with the use, maintenance, and administration of projects and facilities it may own or operate and any product or service rendered thereby, provided that any rule or regulation, other than one which deals solely with the internal management of the Commission, shall be adopted only after public hearing and shall not be effective unless and until filed in accordance with the law of the respective signatory Parties applicable to administrative rules and regulations;
(b) Provide regulations for water management in the Basin; and
(c) Designate any officer, agent or employee of the Commission to be an investigator or watchman, and such person shall be vested with the police powers of the jurisdiction in which he is duly assigned to perform his duties.

Commentary: Any successful comprehensive plan requires detailed regulations. This article provides the commission with the power to establish and enforce regulations necessary for implementation of the agreement. The provision was modeled on a similar provision in the *Delaware River Basin Compact* and the *Susquehanna River Basin Compact*, with several changes.

Cross References: §1C-1-01 (general policies); §1C-1-02 (purposes of agreement); §1C-1-03 (objectives of agreement); §2C-2-06 (equitable and reasonable apportionment); §3C-2-01 (general powers and duties); §3C-2-02 (powers and duties reserved to the commissioners); §3C-2-03 (obligations of the commission); §3C-2-05 (prohibited activities).

Similar Agreements: *Delaware River Basin Compact*, Pub. L. 87-328, 75 Stat. 688-716 (1961); *Susquehanna River Basin Compact*, Pub. L. 91-575, 84 Stat. 1509-1541 (1970).

§3C-2-05 Prohibited Activities

(a) No Commissioner or Commission Officer or employee shall:
  (1) Be financially interested, either directly or indirectly, in any contract, sale, purchase, lease, or transfer of real or personal property in which the Commission is involved;
  (2) Solicit or accept money or any other thing of value, either directly or indirectly, in addition to the compensation or expenses paid him by the Commission for services performed within the scope of his official duties; or
  (3) Offer money or any thing of value for or in consideration of obtaining an appointment, promo-

tion or privilege in his employment with the Commission.

(b) **Any officer or employee who shall willfully violate any of the provisions of this Article shall forfeit his office or employment.**

(c) **Any contract or agreement knowingly made in contravention of this Article is void.**

(d) **Officers and employees of the Commission shall be subject to such criminal and civil sanctions for misconduct in office as may be imposed by [federal law and] the law of the signatory Party in which such misconduct occurs.**

Commentary: Although optional, inclusion of this article provides professionalism of the administration of the agreement and guards against corruption in its enforcement.

Cross References: §1C-1-01 (general policies); §1C-1-02 (purposes of agreement); §1C-1-03 (objectives of agreement); §2C-2-09 (party or parties); §3C-2-01 (general powers and duties); §3C-2-02 (powers and duties reserved to the commissioners); §3C-2-03 (obligations of the commission); §3C-2-04 (regulations; enforcement).

Similar Agreements: *Delaware River Basin Compact*, Pub. L. 87-328, 75 Stat. 688-716 (1961); *Susquehanna River Basin Compact*, Pub. L. 91-575, 84 Stat. 1509-1541 (1970).

§3C-2-06 Commission Approval of Water Resource Projects

No project having a substantial effect on the water resources of the Basin shall hereafter be undertaken by any person, corporation or governmental authority unless it shall have been first submitted to and approved by the Commission. The Commission shall approve a project whenever it finds and determines that such project would not substantially impair or conflict with the comprehensive plan and may modify and approve as modified, or may disapprove any such project whenever it finds and determines that the project would substantially impair or conflict with such plan. The Commission shall provide by regulation for the procedure of submission, review, and consideration of projects, and for its determinations pursuant to this Article.

Commentary: This article establishes the primacy of the commission in water resource planning, providing it with the sole authority to approve projects that substantially affect the basin's water resources. Without this restriction, the purposes of the plan may be thwarted. The determination of what constitutes "substantial impairment or conflict" is case specific. Although optional, its inclusion is strongly advised.

Cross References: §2C-1-06 (existing agencies); §2C-2-07 (flood); §3C-3-02 (project costs and evaluation standards); §3C-3-03 (projects of the signatory parties); §3C-3-04 (cooperative services); §4C-1-01 (joint exercise of authority); §4C-1-02 (interrelationship of water resources); §4C-1-03 (comprehensive water management plan); §4C-1-04 (purpose and objectives of comprehensive water management plan); §4C-1-05 (conditions of comprehensive water management plan); §4C-1-06 (deviation from comprehensive water management plan); §4C-1-07 (allocation during flood conditions); §4C-1-08 (allocation during drought conditions); §4C-1-09 (non-impairment of comprehensive water management plan by state action); §4C-2-01 (water allocation, generally); §4C-2-02 (waters not subject to allocation); §4C-2-03 (basin water sources); §4C-2-04 (basin water demands and needs); §4C-2-05 (allocation to equitable and reasonable uses); §4C-2-06 (watershed management); §4C-3-01 (existing rights recognized); §4C-3-02 (flood protection works); §4C-3-03 (minimum flows); §4C-3-04 (withdrawals and diversion; protected areas); §4C-3-05 (water levels protected); §4C-3-06 (augmentation of supply); §4C-3-07 (water quality); §4C-3-08 (underground water; limit on withdrawals); §4C-3-09 (atmospheric water); §4C-3-10 (recreation).

Similar Agreements: *Delaware River Basin Compact*, Pub. L. 87-328, 75 Stat. 688-716 (1961); *Susquehanna River Basin Compact*, Pub. L. 91-575, 84 Stat. 1509-1541 (1970); *Treaty between the United States and Mexico. Utilization of Waters of the Colorado and Tijuana Rivers and of the Rio Grande*, 59 Stat. 1219 (1944).

§3C-2-07 Advisory Committees (Optional)

The Commission may constitute and empower advisory committees, which may comprise representatives of the public and of agencies and Officials of the Parties or any sub-division thereof, water-using industries, water-interest groups, and academic experts in related fields.

Commentary: Although optional, inclusion of this article allows the commission to seek independent advice and counsel regarding water issues within the basin, thereby ensuring that maximum technical competence is applied to the comprehensive water management.

Cross References: §3C-1-05 (commission organization and staffing); §3C-1-06 (rules of procedures); §3C-1-07 (commission administration); §3C-2-01 (general powers and duties); §3C-2-02 (powers and duties reserved to the commissioners); §3C-2-03 (obligations of the commission).

Similar Agreements: *Delaware River Basin Compact*, Pub. L. 87-328, 75 Stat. 688-716 (1961); *Susquehanna River Basin Compact*, Pub. L. 91-575, 84 Stat. 1509-1541 (1970).

§3C-2-08 Reports

The Commission shall make and publish an Annual Report to the Commissioners and the legislative bodies of the signatory Parties and to the public reporting on its programs, operations, and finances. It may also prepare, publish, and distribute such other public reports and informational materials as it may deem necessary or desirable.

Commentary: Although optional, inclusion of this article is strongly advised. It reinforces the public nature of the agreement administration. Perhaps more importantly, such reporting provides for public accountability of the commission and ensures maximum effective management of the water sources and uses.

Cross References: §1C-1-01 (general policies); §1C-1-02 (purposes of agreement); §1C-1-03 (objectives of agreement); §2C-2-09 (party or parties); §3C-1-01 (commission created); §3C-1-02 (jurisdiction of the commission); §3C-3-02 (project costs and evaluation standards); §3C-3-03 (projects of the signatory parties); §4C-1-03 (comprehensive water management plan); §4C-1-04 (purpose and objectives of comprehensive water management plan); §4C-1-05 (conditions of comprehensive water management plan); §4C-1-06 (deviation from comprehensive water management plan); §4C-1-07 (allocation during flood conditions); §4C-1-08 (allocation during drought conditions); §4C-1-09 (non-impairment of comprehensive water management plan by state action); §4C-2-01 (water allocation, generally); §4C-2-02 (waters not subject to allocation); §4C-2-03 (basin water sources); §4C-2-04 (basin water demands and needs); §4C-2-05 (allocation to equitable and reasonable uses); §4C-2-06 (watershed management); §4C-3-01 (existing rights recognized); §4C-3-02 (flood protection works); §4C-3-03 (minimum flows); §4C-3-04 (withdrawals and diversion; protected areas); §4C-3-05 (water levels protected); §4C-3-06 (augmentation of supply); §4C-3-07 (water quality); §4C-3-08 (underground water; limit on withdrawals); §4C-3-09 (atmospheric water); §4C-3-10 (recreation).

Similar Agreements: *Delaware River Basin Compact*, Pub. L. 87-328, 75 Stat. 688-716 (1961); *Susquehanna River Basin Compact*, Pub. L. 91-575, 84 Stat. 1509-1541 (1970).

§3C-2-09 Condemnation Proceedings (Optional)

(a) The Commission shall have the power to acquire by condemnation the fee or any lesser interest in lands, lands lying under water, development rights in land, riparian rights, water rights, waters and other real or personal property within the Basin for any project or facility authorized pursuant to this Agreement. This grant of power of eminent domain includes but is not limited to the power to condemn for the purposes of this Agreement any property already devoted to a public use, by whomsoever owned or held, other than property of a signatory Party. Any condemnation proceeding of any property or franchises owned or used by a municipal or privately owned public utility, unless the affected public utility facility is to be relocated or replaced, shall be subject to the authority of such state board, Commission or other body as may have regulatory jurisdiction over such public utility.

(b) Such power of condemnation shall be exercised in accordance with the provisions of any law applicable to the jurisdiction in which the property is located.

(c) Nothing in this Agreement authorizes the taking of any existing vested property right in the use of water except for just compensation, in accordance with the internal laws of the Party in which the property or usufructuary right exists.

Commentary: This provision is primarily concerned with the condemnation of real property that may be required for projects and facilities needed to support comprehensive water management, as opposed to the usufructuary rights provided by any water withdrawal permitting system of the individual parties. Naturally, however, the acquisition of real property has associated with it the potential for a "taking" of water rights that may be associated with the land so condemned. This association is especially true in jurisdiction, especially those parties in which appropriative rights have been separated from the land itself.

Providing the commission with the power of condemnation and eminent domain should be carefully considered. Without such power, the effectiveness of the comprehensive plan might be significantly impaired. However, such power also provides the commission with the capability of interfering with the inherent authority of the parties in ways that were unintended. The key to preventing this interference is in carefully crafting internal laws that address condemnation and invocation of eminent domain.

This provision expressly requires "just compensation" for any taking of property rights. The "just compensation," however, depends largely on the individual internal laws of the parties themselves. In the United States, the recent rulings in regulatory affairs by the U.S. Supreme Court have held that a serious impairment of the value of land by a regulation of its use should be compensated, but they have noted that the state could diminish the value of a water right by as much as 95% without incurring liability, at least when a system of regulated riparian rights exist (Sax 1990; *Lucas v. South Carolina Coastal Council* 1992; Byrne 1995; and Houck 1995).

This policy may not, however, apply to governments with a system of appropriative rights. Appropriative rights are defined in terms of a specific quantity of water applied to a beneficial use and are much more closely related to property rights. Such rights to use water are usually vested property rights that may not be abolished without compensation (*United States v. State Water Resources Control Board* 1986).

Cross References: §2C-2-09 (party or parties); §4C-1-03 (comprehensive water management plan); §4C-1-04 (purpose and objectives of comprehensive water management plan); §4C-1-05 (conditions of comprehensive water management plan); §4C-1-06 (deviation from comprehensive water management plan); §4C-1-07 (allocation during flood conditions); §4C-1-08 (allocation during drought conditions); §4C-1-09 (non-impairment of comprehensive water management plan by state action); §4C-2-01 (water allocation, generally); §4C-2-02 (waters not subject to allocation); §4C-2-03 (basin water sources); §4C-2-04 (basin water demands and needs); §4C-2-05 (allocation to equitable and reasonable uses); §4C-2-06 (watershed management); §4C-3-01 (existing rights recognized); §4C-3-02 (flood protection works); §4C-3-03 (minimum flows); §4C-3-04 (withdrawals and diversion; protected areas); §4C-3-05 (water levels protected); §4C-3-06 (augmentation of supply); §4C-3-07 (water quality); §4C-3-08 (underground water; limit on withdrawals); §4C-3-09 (atmospheric water); §4C-3-10 (recreation).

Similar Agreements: *Delaware River Basin Compact*, Pub. L. 87-328, 75 Stat. 688-716 (1961); *Susquehanna River Basin Compact*, Pub. L. 91-575, 84 Stat. 1509-1541 (1970).

§3C-2-10 Meetings, Hearings and Records

(a) The signatory Parties recognize the importance and necessity of public participation in promoting utilization of the water resource of the _____ River Basin. Consequently, all meetings of the Commission shall be open to the public except with respect to issues of personnel.

(b) The Commission shall conduct at least one public hearing prior to the adoption of the comprehensive plan, water resources program, annual capital and current expense budgets, the letting of any contract for the sale or other disposition by the Commission of hydroelectric energy or water resources to any person, corporation or entity, and in all other cases wherein this Agreement requires a public hearing. Such hearing shall be held upon at least (ten) days public notice given by posting at the offices of the Commission. The Commission shall also provide forthwith for distribution of such notice to the press and by the mailing of a copy thereof to any person who shall request such notices.

(c) The minutes of the Commission shall be a public record open to inspection at its offices during regular business hours.

Commentary: Effective water sharing demands that all stakeholders have information upon which they can rely to make rational decisions about water use. Without sufficient public participation, the parties would be unable to maximize the use of the water resource. Although optional, inclusion of this provision is essential to maximum effectiveness of the comprehensive plan, which cannot be realized without full disclosure of its administration. Incorporation of this article memorializes the public nature of the enterprise. The U.S.–Canada International Joint Commission has recognized the need for "engaging public support" (International Joint Commission 1984). *Agenda 21* (United Nations 1992) recognizes the need for the widest cooperation between governmental and nongovernmental organizations. However, it is recognized that public notice provisions may not conform to the requirements of the specific jurisdictions involved; jurisdictional requirements control.

Cross References: §2C-2-06 (equitable and reasonable apportionment); §2C-2-09 (party or parties); §3C-2-01 (general powers and duties); §3C-2-02 (powers and duties reserved to the commissioners); §3C-2-03 (obligations of the commission).

Similar Agreements: *Delaware River Basin Compact*, Pub. L. 87-328, 75 Stat. 688-716 (1961); *Susquehanna River Basin Compact*, Pub. L. 91-575, 84 Stat. 1509-1541 (1970); *Apalachicola–Chattahoochee–Flint River Basin Compact*, O.C.G.A. 12-10-100 (1997); *Alabama–Coosa–Tallapoosa River Basin Compact*, O.C.G.A. 12-10-110 (1997); *Convention on the Protection and Use of Transboundary Watercourses and International Lakes*, 31 ILM 1312 (1992); *Canada–Mexico–United States: North American Agreement on Environmental Cooperation*, 32 ILM 1480 (1993).

§3C-2-11 Tort Liability

The Commission shall be responsible for claims arising out of the negligent acts or omissions of its Officers, agents, and employees only to the extent and subject to the procedures prescribed by law generally with respect to officers, agents, and employees of the Government of the Party in which the tortious conduct occurred.

Commentary: This provision provides accountability by the Commission for actions by its officers, agents, and employees. To take effect, similar provisions should exist in the national laws of the respective parties.

Cross References: §2C-2-09 (party or parties); §3C-2-01 (general powers and duties); §3C-2-02 (powers and duties reserved to the commissioners); §3C-2-03 (obligations of the commission); §3C-2-09 (condemnation proceedings); §6C-1-01 (good faith implementation).

Similar Agreements: *Delaware River Basin Compact*, Pub. L. 87-328, 75 Stat. 688-716 (1961); *Susquehanna River Basin Compact*, Pub. L. 91-575, 84 Stat. 1509-1541 (1970).

Part 3 Intergovernmental Relations

§3C-3-01 Coordination and Cooperation

The Commission shall promote and aid the coordination of the activities and programs of governmental and private agencies concerned with water resources administration in the Basin. To this end, but without limitation thereto, the Commission may:

(a) Advise, consult, contract, financially assist, or otherwise cooperate with any and all such agencies;
(b) Employ any other agency or instrumentality of any of the signatory Parties or of any political subdivision thereof, in the design, construction, operation, and maintenance of structures, and the installation and management of river control systems, or for any other purpose;
(c) Develop and adopt plans and specifications for particular water resource projects and facilities which so far as consistent with the comprehensive plan incorporate any separate plans of other public and private organizations operating in the Basin, and permit the decentralized administration thereof;
(d) Qualify as a sponsoring agency under any local, national or regional legislation heretofore or hereafter enacted to provide financial or other assistance for the planning, conservation, utilization, development, management or control of water resources.

Commentary: This provision also appears as §1C-1-04. Its exact placement is optional. However, this provision is important because it allows the commission and its officers to coordinate directly with water policy makers and decision makers on matters related to water management. Without this or a similar provision, inordinate bureaucratic entanglement in the maze of administrative and legal institutions within the structure of the parties may frustrate effective management efforts.

Cross References: §2C-1-02 (consent to jurisdiction); §2C-1-06 (existing agencies); §2C-2-02 (basin); §2C-2-09 (party or parties); §3C-1-05 (commission organization and staffing); §3C-1-06 (rules of procedures); §3C-1-07 (commission administration); §3C-2-01 (general powers and duties); §3C-2-02 (powers and duties reserved to the commissioners); §3C-2-03 (obligations of the commission); §3C-2-06 (referral and review); §3C-2-07 (advisory committees); §3C-3-02 (project costs and evaluation standards); §3C-3-03 (projects of the signatory parties); §3C-3-04 (cooperative services).

Similar Agreements: *Delaware River Basin Compact*, Pub. L. 87-328, 75 Stat. 688-716 (1961); *Susquehanna River Basin Compact*, Pub. L. 91-575, 84 Stat. 1509-1541 (1970); *Agreement on the Conservation of Nature and Natural Resources* (ASEAN 1985); *Convention on the Protection and Use of Transboundary Watercourses and International Lakes*, 31 ILM 1312 (1992); *Canada–Mexico–United States: North American Agreement on Environmental Cooperation*, 32 ILM 1480 (1993); *Treaty of Peace between the State of Israel and the Hashemite Kingdom of Jordan*, 34 ILM 43 (1994); *Convention on the Law of the Non-Navigational Uses of International Watercourses*, United Nations Document A/51/869 (United Nations 1997).

§3C-3-02 Project Costs and Evaluation Criteria

Consistent with Article 5C, the Commission shall establish uniform criteria and procedures for the evaluation, determination of benefits, and cost allocations of water projects affecting the Basin, and for the determination of project priorities, pursuant to the requirements of the comprehensive plan and its water resources program. The Commission shall develop equitable cost sharing and reimbursement formulas for the signatory Parties including:

(a) Uniform and consistent procedures for the allocation of project costs among purposes included in multiple-purpose programs;
(b) Procedures for allocation of cost sharing between the signatory Parties and with other public and semi-public entities, private organizations, and groups according to benefit received;
(c) Establishment and supervision of a system of accounts for reimbursable purposes and directing the payments and charges to be made from such accounts.

Commentary: This provision establishes a rational basis for the implementation of comprehensive management of water projects that can survive appropriate economic analysis and cost–benefit scrutiny.

Cross References: §2C-2-09 (party or parties); §3C-3-03 (projects of the signatory parties); §3C-3-04 (cooperative services); Article 5C (financing).

Similar Agreements: *Delaware River Basin Compact*, Pub. L. 87-328, 75 Stat. 688-716 (1961); *Susquehanna River Basin Compact*, Pub. L. 91-575, 84 Stat. 1509-1541 (1970).

§3C-3-03 Projects of the Signatory Parties

For the purposes of avoiding conflicts of jurisdiction and giving full effect to the Commission as a regional agency of the signatory Parties, the following rules shall govern proj-

ects of the individual Parties and their subdivisions that affect the water resources of the Basin:

(a) The planning of all projects related to powers delegated to the Commission by this Agreement shall be undertaken in consultation with the Commission;

(b) Prior to entering upon the execution of any project authorized by this Article, the Commission shall review and consider and integrate into the Comprehensive Water Management Plan insofar as possible all existing rights, plans and programs of the signatory Parties, their political subdivisions, private parties, and water users which are pertinent to such project, and shall hold a public hearing on each proposed project;

(c) No expenditure or commitment shall be made for or on account of the construction, acquisition or operation of any project or facility nor shall it be deemed authorized, unless the Commission in the Comprehensive Water Management Plan shall have first included it;

(d) Each governmental agency otherwise authorized by law to plan, design, construct, operate, or maintain any project or facility in or for the Basin shall continue to have, exercise, and discharge such authority except as specifically provided by this Article.

Commentary: By this provision, the parties agree to refrain from independently developing and operating water management facilities that may affect the comprehensive water management plan. Without its inclusion, the effectiveness of the comprehensive water management plan may be weakened.

Cross References: §2C-2-02 (basin); §2C-2-09 (party or parties); §3C-3-03 (projects of the signatory parties); §3C-3-04 (cooperative services); §4C-1-03 (comprehensive water management plan); §4C-1-04 (purpose and objectives of comprehensive water management plan); §4C-1-05 (conditions of comprehensive water management plan); §4C-1-06 (deviation from comprehensive water management plan); §4C-1-07 (allocation during flood conditions); §4C-1-08 (allocation during drought conditions); §4C-1-09 (non-impairment of comprehensive water management plan by state action); Article 5C (financing).

Similar Agreements: *Delaware River Basin Compact*, Pub. L. 87-328, 75 Stat. 688-716 (1961); *Susquehanna River Basin Compact*, Pub. L. 91-575, 84 Stat. 1509-1541 (1970); *Agreement on the Conservation of Nature and Natural Resources* (ASEAN 1985).

§3C-3-04 Cooperative Services

The Commission shall furnish technical services, advice, and consultation to authorized agencies of the signatory Parties with respect to the water resources of the Basin, and each of the signatory Parties pledges itself to provide technical and administrative services to the Commission upon request, within the limits of available appropriations and to cooperate generally with the Commission for the purposes of this Agreement, and the cost of such services may be reimbursable whenever the Parties deem appropriate.

Commentary: This provision suggests that both the parties and the commission and its officers provide consultation and services to the other in carrying out the provisions of the agreement.

Cross References: §2C-2-09 (party or parties); §3C-3-01 (coordination and cooperation); §3C-3-02 (project costs and evaluation standards); §3C-3-03 (projects of the signatory parties); §4C-1-03 (comprehensive water management plan); §4C-1-04 (purpose and objectives of comprehensive water management plan); §4C-1-05 (conditions of comprehensive water management plan); §4C-1-06 (deviation from comprehensive water management plan); §4C-1-07 (allocation during flood conditions); §4C-1-08 (allocation during drought conditions); §4C-1-09 (non-impairment of comprehensive water management plan by state action).

Similar Agreements: *Delaware River Basin Compact*, Pub. L. 87-328, 75 Stat. 688-716 (1961); *Susquehanna River Basin Compact*, Pub. L. 91-575, 84 Stat. 1509-1541 (1970); *Treaty of Peace between the State of Israel and the Hashemite Kingdom of Jordan*, 34 ILM 43 (1994).

C.4 ARTICLE 4C: COMPREHENSIVE WATER MANAGEMENT

Part 1 Comprehensive Water Management

§4C-1-01 Joint Exercise of Authority

The water resources of the Basin are subject to the rights and responsibilities of the signatory Parties, and it is the purpose of this Agreement to provide for the joint exercise of such powers of authority over the waters of the _______ River Basin in the common interests of the people of the region.

Commentary: All principal stakeholders directly affected by the shared use should be identified and included in the negotiations. This list includes any government that has direct access to surface or underground water. Both governmental and private stakeholders should have a voice in agreement formulation.

Although private water rights holders within the various jurisdictions are not joined as parties to the agreement, it is recommended that the various persons and organizations associated with the various water needs and demands within the basin have a voice in formulating the agreement. Although the extent of consultation with private groups depends largely on the political nature of the governments themselves, some recognition of existing rights is required. For instance, in most situations a conflict exists between demands for water for economic purposes and the needs of environmental protection. Water users frequently need external incentives to accept that some water is reserved for environmental and ecological protection. Unless the governmental entities involved with formulating the agreement actively seek to include the various interest groups, the effectiveness of the agreement may be compromised.

Many international agreements reserve the right of each signatory to exploit its own resources. Some, however, allow for a measure of joint authority but reserve activities that affect industrial and commercial secrecy or national security. However, recent agreements testify to the trend of joint development of shared water resources, and joint development presupposes the joint exercise of authority (Trolldallen 1992).

Cross References: §1C-1-01 (general policies); §1C-1-02 (purposes of agreement); §1C-1-03 (objectives of agreement); §2C-1-05 (powers of sovereign parties; withdrawal); §2C-1-06 (existing agencies); §2C-1-07 (limited applicability); §2C-2-02 (basin); §2C-2-03 (comprehensive water management plan); §2C-2-09 (party or parties); §2C-2-11 (waters of the basin); §3C-1-01 (commission created); §3C-1-02 (jurisdiction of the commission); §3C-1-03 (commissioners); §3C-1-04 (status, immunities and privileges); §3C-1-05 (commission organization and staffing); §3C-1-06 (rules of procedures); §3C-1-07 (commission administration); §3C-2-01 (general powers and duties); §3C-2-02 (powers and duties reserved to the commissioners); §3C-2-03 (obligations of the commission).

Similar Agreements: *Delaware River Basin Compact*, Pub. L. 87-328, 75 Stat. 688-716 (1961); *Susquehanna River Basin*

Compact, Pub. L. 91-575, 84 Stat. 1509-1541 (1970); *Act Regarding Navigation and Economic Co-operation Between the States of the Niger Basin*, 587 UNTS 9 (1963); *Convention and Statutes Relating to the Development of the Chad Basin* (1974); *Nouakchott Convention Establishing the OMVS (Organisation pour la Mise en Valeur du Fleuve Senegal)*, 672 UNTS 251 (1974); *Convention on the Law of the Non-Navigational Uses of International Watercourses*, United Nations Document A/51/869 (United Nations 1997).

§4C-1-02 Interrelationship of Water Resources

The water resources of the Basin are functionally interrelated, and the uses of these resources are interdependent. Joint management and coordination of efforts, programs, and policies within the Basin is essential to effective and efficient use of the water resource.

Commentary: This provision expands the purpose statement by expressing the intent of the parties to seek effective water management. Similar provisions are included in *Agenda 21* (United Nations 1992), which proposed the need for a program of integrated water resource development and management.

Cross References: §1C-1-01 (general policies); §1C-1-02 (purposes of agreement); §1C-1-03 (objectives of agreement); §2C-2-03 (comprehensive water management plan); §4C-1-03 (comprehensive water management plan); §4C-1-04 (purpose and objectives of comprehensive water management plan); §4C-1-05 (conditions of comprehensive water management plan); §4C-1-06 (deviation from comprehensive water management plan); §4C-1-07 (allocation during flood conditions); §4C-1-08 (allocation during drought conditions); §4C-1-09 (non-impairment of comprehensive water management plan by state action); §4C-2-01 (water allocation, generally); §4C-2-02 (waters not subject to allocation); §4C-2-03 (basin water sources); §4C-2-04 (basin water demands and needs); §4C-2-05 (allocation to equitable and reasonable uses); §4C-2-06 (watershed management); §4C-3-01 (existing rights recognized); §4C-3-02 (flood protection works); §4C-3-03 (minimum flows); §4C-3-04 (withdrawals and diversion; protected areas); §4C-3-05 (water levels protected); §4C-3-06 (augmentation of supply); §4C-3-07 (water quality); §4C-3-08 (underground water; limit on withdrawals); §4C-3-09 (atmospheric water); §4C-3-10 (recreation).

Similar Agreements: *Delaware River Basin Compact*, Pub. L. 87-328, 75 Stat. 688-716 (1961); *Susquehanna River Basin Compact*, Pub. L. 91-575, 84 Stat. 1509-1541 (1970).

§4C-1-03 Comprehensive Water Management Plan

The Commission shall develop and adopt, and may from time to time review and revise, a Comprehensive Water Management Plan for the immediate and long-range development and use of the water resources of the _______ River Basin. The Plan may be as general or specific as the Commission may deem appropriate for efficient and sustainable development of the Basin. At a minimum, the Master Plan shall include:

(a) A water allocation program to allocate the waters of the _____________ River Basin with the goal of providing sufficient quantities of quality water resources to satisfy the needs of the Basin for such reasonably foreseeable period as the Commission may determine. This allocation program may be supplemented by existing and proposed projects which may be required to satisfy such needs, including all public and private projects affecting the Basin, together with a separate statement of the projects proposed to be undertaken by the Commission during such period; and

(b) A program of water quality management that shall describe, at a minimum, the maximum amounts of designated pollutants allowable in the ___________ River and its tributaries.

Commentary: This provision directs the commission to consider all water-related activities that may affect efficient and sustainable development of the basin. The parties to this agreement still retain the right, authority, and responsibility to plan and manage their own economic, social, and environmental development. Irrespective of whether the planning is the result of state planning or free-market processes, this agreement suggests that it is the intent of the parties to provide a baseline of existing and projected water demands. The commission integrates the input from the various parties and develops a plan that equitably optimizes water use among the parties for economic development within the basin. Effective basinwide planning normally requires that it be done on a subbasin, or reach of river, basis. An objective of sustainable development, a significant objective of the *North American Free Trade Agreement*, 19 U.S.C. §§3311–3473 (1993), should be considered. The absence of comprehensive management of a shared resource significantly reduces successful conflict resolution. For instance, the lack of coordinated management has been highlighted as a major hindrance to resolution of the water conflicts between India and Pakistan and India and Bangladesh (Clarke 1993).

This article directs the commission to develop and implement a plan to manage the total water resources of the basin, subject to certain restrictions, in an effort to maximize the efficiency of water use according to specific goals established by the agreement. Management of the water allocation can be as detailed as necessary, according to the capabilities and limitations of the basin. Integration of water quality and quantity is essential. The *Convention on the Law of the Non-Navigational Uses of International Watercourses*, United Nations Document A/51/869 (United Nations 1997), mandates "sustainable development of an international watercourse" and the "rational and optimal utilization, protection and control of the watercourse."

Cross References: §1C-1-01 (general policies); §1C-1-02 (purposes of agreement); §1C-1-03 (objectives of agreement); §2C-2-02 (basin); §2C-2-03 (comprehensive water management plan); §2C-2-05 (drought); §4C-1-01 (joint exercise of authority); §4C-1-02 (interrelationship of water resources); §4C-1-04 (purpose and objectives of comprehensive water management plan); §4C-1-05 (conditions of comprehensive water management plan); §4C-1-06 (deviation from comprehensive water management plan); §4C-1-07 (allocation during flood conditions); §4C-1-08 (allocation during drought conditions); §4C-1-09 (non-impairment of comprehensive water management plan by state action); §4C-2-01 (water allocation, generally); §4C-2-02 (waters not subject to allocation); §4C-2-03 (basin water sources); §4C-2-04 (basin water demands and needs); §4C-2-05 (allocation to equitable and reasonable uses); §4C-2-06 (watershed management); §4C-3-01 (existing rights recognized); §4C-3-02 (flood protection works); §4C-3-03 (minimum flows); §4C-3-04 (withdrawals and diversion; protected areas); §4C-3-05 (water levels protected); §4C-3-06 (augmentation of supply); §4C-3-07 (water quality); §4C-3-08 (underground water; limit on withdrawals); §4C-3-09 (atmospheric water); §4C-3-10 (recreation).

Similar Agreements: *Klamath River Basin Compact*, 71 Stat. 497 (1957); *Delaware River Basin Compact*, Pub. L. 87-328, 75 Stat. 688-716 (1961); *Susquehanna River Basin Compact*, Pub.

L. 91-575, 84 Stat. 1509-1541 (1970); *Convention on the Law of the Non-Navigational Uses of International Watercourses*, United Nations Document A/51/869 (United Nations 1997); *Treaty for Amazonian Cooperation*, 17 ILM 1046 (1978); *Agreement on the Cooperation for the Sustainable Development of the Mekong River Basin*, 34 ILM 864 (1995).

§4C-1-04 Purposes and Objectives of Comprehensive Management Plan

The purpose of the Comprehensive Water Management Plan is to facilitate maximum utilization of the waters of the _____ River Basin to meet both existing and potential needs and uses for sustained economic growth while providing for adequate public health and safety, water-based recreation, and environmental protection. A [___]-year planning horizon shall be established. The objectives of the Plan shall include the following:

- **(a) To identify and assess the sources of water available to the users within the Basin, and to coordinate public and private projects and facilities to augment water availability, which will allow for the optimum planning, development, conservation, utilization, management, and control of the water resources of the Basin as later described in §4C-2-03.**
- **(b) To identify, quantify, and value where appropriate all existing and planned water uses within the Basin by sub-Basin and by political jurisdiction as described in §4C-2-04.**
- **(c) To allocate efficiently the available Basin waters according to the needs of the various categories of uses as described in §4C-2-04 and in a manner that ensures equitable and reasonable use of the water.**

Commentary: The parties should agree on a common planning horizon for economic evaluation. In the United States, a 50-year horizon is used by many states for water planning and/or permitting purposes. Although the choice of horizon does not in any way limit the duration of the agreement, equitable partition of the water source requires a common time horizon for the economic evaluation.

Surface and underground water resources are often managed as separate and distinct sources. However, such individualized management does not provide an integrated understanding of either the water sources or their use. Before implementing transboundary water use allocation or transfer, participants should conduct a comprehensive water resources assessment. Certain parameters define the framework of the agreement. The parties should specify the range of hydrologic events that the agreement covers. This specification means that the parties should clearly establish the quantitative measures of drought and flood conditions and establish the levels when special management for drought or flood conditions arise. The parties should also define the levels of water quality degradation that they are willing to accept as a result of meeting the needs and demands for the water.

The assessment team should identify and quantify the existing and planned water demands in the basin according to type of use, as described later. Optimal water sharing may be accomplished when sharing is done by subbasin or by reach, as appropriate. Within each category of use, the team should further classify uses as consumptive or nonconsumptive. Demands are classified as consumptive when they remove water from the water source for further use. Their major effect on the watercourse is to reduce the quantity of flow. They may, however, return to the water source a portion of the water used, in which case the quality of the receiving waters may be affected. The actual amount of water consumption varies both by type of demand and within the general category of particular demands. The efficient allocation of water for consumptive use presents the greatest challenge to the commission.

Efficient allocation to equitable and reasonable uses suggests that the various water uses be compared by the economic benefit each use produces, wherever possible. Economists contend that the best way to allocate water efficiently is to use market mechanisms (Hillel 1994). The accepted procedures for measuring benefits and costs are based on the idea that the benefit from making an increment of water available for a particular use is measured by society's willingness to pay for the increment of production resulting from the additional allocation of water. This valuation methodology can easily be applied to water uses that are involved in the "stream of commerce" (e.g., water used to produce agricultural or silvacultural products, water used by the pulp and paper industry, or water supplied by municipalities). However, other uses are difficult to measure objectively (e.g., water needed to sustain the aquatic ecology or water needed to sustain hunting and fishing). In some uses of water, adequate methodologies for quantifying benefits have not been developed (e.g., the value of maintaining riparian vegetative buffers). Care should be taken not to neglect these less easily valued water uses simply because it is difficult to do so. Additionally, equity is the other broad principle that has guided the allocation of water resources. It is more difficult to articulate a means of achieving equitable allocations than to define rules for maximizing returns.

Although an objective of optimum allocation was not introduced into this model agreement, that objective exists in a number of international agreements [e.g., *Council on European Communities Third Action Programme on the Environment* 1983; *Convention on the Law of the Non-Navigational Uses of International Watercourses*, United Nations Document A/51/869 (United Nations 1997)].

Cross References: §1C-1-01 (general policies); §1C-1-02 (purposes of agreement); §1C-1-03 (objectives of agreement); §2C-2-02 (basin); §2C-2-03 (comprehensive water management plan); §2C-2-05 (drought); §4C-1-01 (joint exercise of authority); §4C-1-02 (interrelationship of water resources); §4C-1-03 (comprehensive water management plan); §4C-1-05 (conditions of comprehensive water management plan); §4C-1-06 (deviation from comprehensive water management plan); §4C-1-07 (allocation during flood conditions); §4C-1-08 (allocation during drought conditions); §4C-1-09 (non-impairment of comprehensive water management plan by state action); §4C-2-01 (water allocation, generally); §4C-2-02 (waters not subject to allocation); §4C-2-03 (basin water sources); §4C-2-04 (basin water demands and needs); §4C-2-05 (allocation to equitable and reasonable uses); §4C-2-06 (watershed management); §4C-3-01 (existing rights recognized); §4C-3-02 (flood protection works); §4C-3-03 (minimum flows); §4C-3-04 (withdrawals and diversion; protected areas); §4C-3-05 (water levels protected); §4C-3-06 (augmentation of supply); §4C-3-07 (water quality); §4C-3-08 (underground water; limit on withdrawals); §4C-3-09 (atmospheric water); §4C-3-10 (recreation).

Similar Agreements: *Delaware River Basin Compact*, Pub. L. 87-328, 75 Stat. 688-716 (1961); *Susquehanna River Basin Compact*, Pub. L. 91-575, 84 Stat. 1509-1541 (1970); *Agreement on the Cooperation for the Sustainable Development of the Mekong River Basin*, 34 ILM 864 (1995); *Convention on the Law of the Non-Navigational Uses of International Watercourses*, United Nations Document A/51/869 (United Nations 1997).

§4C-1-05 Conditions of Comprehensive Management Plan

(a) **The plan shall allocate waters under all hydrologic conditions, with the exception of floods and droughts, as defined herein.**

(b) **The plan shall include consideration of all public and private projects and facilities which are required, in the judgment of the Commission, for the optimum planning, development, conservation, utilization, management, and control of the water resources of the Basin to meet present and future needs, provided that the plan shall include any projects required to conform with any present or future decree or judgment of any court of competent jurisdiction.**

(c) **The basis of the management plan shall be the integration of economic development, maintenance of adequate public health and safety, and environmental protection plans and policies established and submitted by the signatory Parties. However, before the adoption of the plan or any part or revision thereof, the Commission shall consult with water users and interested public bodies and public utilities and shall consider and give due regard to the findings and recommendations of the various agencies of the signatory Parties and their political subdivisions.**

(d) **The Commission shall conduct public hearings with respect to the comprehensive management plan prior to the adoption of the plan or any part of the revision thereof.**

Commentary: This Article provides additional framework for the comprehensive plan. Certain provisions are optional, however. For instance, the parties may wish the comprehensive plan to include the procedures to be used in floods and droughts.

Cross References: §2C-2-02 (basin); §2C-2-03 (comprehensive water management plan); §2C-2-05 (drought); §2C-2-07 (flood); §2C-2-09 (party or parties); §4C-1-06 (deviation from comprehensive water management plan); §4C-1-07 (allocation during flood conditions); §4C-1-08 (allocation during drought conditions); §4C-1-09 (non-impairment of comprehensive water management plan by state action); §4C-2-01 (water allocation, generally); §4C-2-02 (waters not subject to allocation); §4C-2-03 (basin water sources); §4C-2-04 (basin water demands and needs); §4C-2-05 (allocation to equitable and reasonable uses); §4C-2-06 (watershed management); §4C-3-01 (existing rights recognized); §4C-3-02 (flood protection works); §4C-3-03 (minimum flows); §4C-3-04 (withdrawals and diversion; protected areas); §4C-3-05 (water levels protected); §4C-3-06 (augmentation of supply); §4C-3-07 (water quality); §4C-3-08 (underground water; limit on withdrawals); §4C-3-09 (atmospheric water); §4C-3-10 (recreation).

Similar Agreements: *Delaware River Basin Compact*, Pub. L. 87-328, 75 Stat. 688-716 (1961); *Susquehanna River Basin Compact*, Pub. L. 91-575, 84 Stat. 1509-1541 (1970); *Agreement on the Cooperation for the Sustainable Development of the Mekong River Basin*, 34 ILM 864 (1995); *Convention on the Law of the Non-Navigational Uses of International Watercourses*, United Nations Document A/51/869 (United Nations 1997).

§4C-1-06 Deviation from Comprehensive Management Plan
The Commission may authorize deviations from the Comprehensive Water Management Plan during extreme hydrologic conditions, either flood or drought.

Commentary: The comprehensive water management plan can be crafted to address both flood and drought conditions. However, because the nature and characteristics of extreme hydrologic events are difficult to predict, and because no plan is capable of adequately reacting to all such events, the plan should have provisions that provide for the unexpected, unplanned situation.

The issue of climate change has surfaced as a potential impediment to effective long-range policies and management of water resources. At the present time, the accuracy and precision of prediction of the effects is not yet sufficient to provide specific planning capability; the parties should recognize that the comprehensive water management plan may require some modifications as predictability of climate changes become more accurate and precise. The American Society of Civil Engineers, in *Policy Statement 360* (ASCE 2010a), recognized the problems associated with water resource planning and climate change:

> Global climate change could pose a potentially serious effect on worldwide water resources, energy production and use, agriculture, forestry, coastal development and resources, flood control, and public infrastructure. . . . Such effects could require modified agricultural practices and could require measures to deal with rising sea levels with associated effects on estuaries, coastal shoreline land uses, and infrastructure. Early participation of the engineering community in the debate and decisions on appropriate solutions are desirable because the practice of engineering deals with the development of cost-effective and environmentally acceptable methods of providing for human needs in relation to the natural environment.

Because the effects of climate change extend across political boundaries, the parties should recognize that the comprehensive water management plan may need to be modified as continued international research develops more accurate and precise predictability on those changes.

Inland aquatic ecosystems are influenced by climate change through altered water temperatures, flow regimes, and water levels. In lakes and streams, warming would have the greatest biological effects at high latitudes, where biological productivity would increase, and at the low-latitude boundaries of cold- and cool-water species ranges, where extinctions would be greatest. Warming of larger and deeper temperate zone lakes would increase their productivity; although in some shallow lakes and in streams, warming could increase the likelihood of anoxic conditions. Increases in flow variability, particularly the frequency and duration of large floods and droughts, would tend to reduce water quality and biological productivity and habitat in streams. Water-level declines are most severe in lakes and streams in dry evaporative drainages and in basins with small catchments. The geographical distribution of wetlands is likely to shift with changes in temperature and precipitation. There is an effect from climate change on greenhouse gas release from nontidal wetlands, but there is uncertainty regarding the exact effects from site to site.

Policy makers have to decide to what degree they want to take precautionary measures by mitigating greenhouse gas emissions and enhancing the resilience of vulnerable systems by means of adaptation. Uncertainty does not mean that a nation or the world community cannot position itself better to cope with the broad range of possible climate changes or protect against potentially costly future outcomes. Delaying such measures may leave a nation or the world poorly prepared to deal with adverse changes and may increase the possibility of irreversible or costly consequences. Options for adapting to change or mitigating change that can be justified for other reasons today (e.g., abatement of air and water pollution) and for making society more flexible or resilient to anticipated adverse effects of climate change appear particularly desirable.

To restate the problem for the parties, it is important to recognize the potential problems that climate change may cause to the long-range effectiveness of the comprehensive water management plan. However, predictability of the specific effect of climate change on any specific region is not yet sufficiently accurate or precise to be included in a plan at this time.

Cross References: §2C-2-03 (comprehensive water management plan); §2C-2-05 (drought); §2C-2-07 (flood); §4C-1-07 (allocation during flood conditions); §4C-1-08 (allocation during drought conditions); §4C-1-09 (non-impairment of comprehensive water management plan by state action); §4C-2-01 (water allocation, generally); §4C-2-02 (waters not subject to allocation); §4C-3-06 (augmentation of supply); §4C-3-07 (water quality); §4C-3-08 (underground water; limit on withdrawals).

Similar Agreement: *Convention on the Law of the Non-Navigational Uses of International Watercourses*, United Nations Document A/51/869 (United Nations 1997).

§4C-1-07 Allocation During Flood Conditions

(a) Retention or release of waters for flood protection operations reasonably needed to prevent loss of life, substantial public disruption, or major property damage shall supersede any other allocation provided in this Agreement. The Commission may plan, design, construct, and operate and maintain projects and facilities, as it may deem necessary or desirable for flood damage reduction. It shall have power to operate such facilities and to store and release waters on the ________ River Basin, in such manner, at such times, and under such regulations as the Commission may deem appropriate to meet flood conditions as they may arise.

(b) The Commission shall have power to adopt, amend, and repeal recommended standards, in the manner provided by this Article, relating to the nature and extent of the uses of land in areas subject to flooding by waters of the Basin and its tributaries. Such standards shall not be deemed to impair or restrict the power of the signatory Parties or their political subdivisions to adopt zoning and other land use regulations not inconsistent therewith.

(c) The Commission may study and determine the nature and extent of the flood plains of the _______ and its tributaries. Upon the basis of such studies, it may establish encroachment lines and delineate the areas subject to flood, including a classification of lands with reference to relative risk of flood and the establishment of standards for flood plain use which will safeguard the public health, safety, and property. Prior to the adoption of any standards delineating such area or defining such use, the Commission shall hold public hearings with respect to the substance of such standards. At or before such public hearings the proposed standards shall be available, and all interested persons shall be given an opportunity to be heard thereon at the hearing. Upon the adoption and promulgation of such standards, the Commission may enter into agreements to provide technical and financial aid to any political subdivision for the administration and enforcement of any local land use ordinances or regulations giving effect to such standards.

(d) The Commission shall have power to acquire the fee or any lesser interest in lands and improvements thereon within the area of a flood plain for the purpose of restricting the use of such property so as to minimize the flood hazard, converting property to uses appropriate to flood plain conditions, or preventing unwarranted constrictions that reduce the ability of the river channel to carry flood water.

(e) The Commission may cause lands particularly subject to flood to be posted with flood hazard warnings, and may from time to time cause flood advisory notices to be published and circulated as conditions may warrant.

Commentary: Some preparation can be made to respond to flood conditions. These provisions provide the minimum necessary to prepare for and react to flood conditions.

Cross References: §2C-2-02 (basin); §2C-2-03 (comprehensive water management plan); §2C-2-07 (flood); §2C-2-09 (party or parties); §4C-1-04 (purpose and objectives of comprehensive water management plan); §4C-1-05 (conditions of comprehensive water management plan); §4C-1-06 (deviation from comprehensive water management plan); §4C-1-09 (non-impairment of comprehensive water management plan by state action); §4C-2-01 (water allocation, generally); §4C-2-02 (waters not subject to allocation); §4C-3-01 (existing rights recognized); §4C-3-02 (flood protection works); §4C-3-03 (minimum flows); §4C-3-06 (augmentation of supply).

Similar Agreements: *Delaware River Basin Compact*, Pub. L. 87-328, 75 Stat. 688-716 (1961); *Susquehanna River Basin Compact*, Pub. L. 91-575, 84 Stat. 1509-1541 (1970); *Convention on the Law of the Non-Navigational Uses of International Watercourses*, United Nations Document A/51/869 (United Nations 1997).

§4C-1-08 Allocation Under Drought Conditions

The Commission shall, within (____) years of the coming into force of this Agreement, complete the preparation of a Drought Management Plan applicable to the ______ River Basin. The Drought Management Plan shall:

(a) Specify the hydro-meteorological preconditions of a Drought Alert and, thereunder, the conservation measures to be observed by all water users or by individual classes of water users according to a predetermined priority classification in the ______ River Basin;

(b) Specify the hydro-meteorological preconditions of a Drought Emergency and, thereunder, the specific additional measures to be observed by all water users in the ______ River Basin;

(c) Specify the priorities of water demands within the Basin; and

(d) Modify or suspend the conservation and other specific measures provided in the Drought Management Plan in order to meet the specific requirements of the individual drought circumstances;

(e) Undergo analysis and revision on a five-year cycle, as drought conditions occur, or as circumstances arise.

Commentary: Droughts of any duration, but especially extended droughts, interpose conditions that may require extensive alteration of the allocation prescribed by the comprehensive water allocation management plan. During the most extreme droughts, it may be necessary to allocate all water solely for human survival, meaning that water for human consumption must be made available along with water for subsistence farming. In other situations, water may be reserved solely for those uses necessary to sustain long-term economic viability. In any event, the Commission should be authorized to deviate from the plan in a manner that ensures equity among the parties.

Cross References: §2C-2-03 (comprehensive water management plan); §2C-2-04 (conservation measures); §2C-2-05 (drought); §4C-1-04 (purpose and objectives of comprehensive water management plan); §4C-1-05 (conditions of comprehensive water management plan); §4C-1-06 (deviation from comprehensive water management plan); §4C-1-09 (non-impairment of comprehensive water management plan by state action); §4C-2-01 (water allocation, generally); §4C-2-02 (waters not subject to allocation); §4C-2-05 (allocation to equitable and reasonable uses); §4C-2-06 (watershed management); §4C-3-04 (withdrawals and diversion; protected areas); §4C-3-05 (water levels protected); §4C-3-06 (augmentation of supply); §4C-3-07 (water quality); §4C-3-08 (underground water; limit on withdrawals); §4C-3-09 (atmospheric water); §4C-3-10 (recreation).

Similar Agreements: *Delaware River Basin Compact*, Pub. L. 87-328, 75 Stat. 688-716 (1961); *Susquehanna River Basin Compact*, Pub. L. 91-575, 84 Stat. 1509-1541 (1970); *Convention on the Law of the Non-Navigational Uses of International Watercourses*, United Nations Document A/51/869 (United Nations 1997).

§4C-1-09 Non-Impairment of Comprehensive Management Plan by State Action

(a) State laws which may allow interbasin transfer of surface water, the transfer of surface water from one part of the Basin to another, or the permitting of water withdrawal rights for private sale, shall be integrated into the Comprehensive Plan by the Commission to the greatest extent possible but only to the degree such integration is consistent with the rights of the State under the terms of the Agreement.

(b) No state laws authorizing interbasin transfer from the _____ River Basin will take effect without the certification by the Commission that such interbasin transfer will not adversely affect the Comprehensive Water Management Plan.

Commentary: Large transfers of water from one water basin to another, or from one part of a single water basin to another, have become common. Often these transfers involve the transfer of water from a rural to an urban area (Dellapenna 2001, §7.05(c)(2); Tarlock 2001, §10.04(2)(c)). Protection for the water basin of origin is now a well-established part of the law of many states, particularly in the western states (e.g., *Ariz. Rev. Stat. Ann.* 2012, §45-172; *California Water Code* 2012; *Texas Water Code Ann.* 2011). Such transfers can be contemplated by the state itself, authorizing the transfer of water resources as part of its economic planning initiatives. Such transfers may also result from the institution of private permitting of water withdrawal for commercial purposes that may include interbasin transfer for private sale. In any event, it is necessary that the agreement acknowledge that the individual parties may not adopt policies that reduce the available water supply for the comprehensive water management plan purposes.

Cross References: §2C-2-02 (basin); §2C-2-08 (interbasin transfer); §4C-1-04 (purpose and objectives of comprehensive water management plan); §2C-2-05 (drought); §4C-1-05 (conditions of comprehensive water management plan); §4C-1-06 (deviation from comprehensive water management plan); §4C-1-08 (allocation during drought conditions); §4C-2-01 (water allocation, generally); §4C-2-02 (waters not subject to allocation); §4C-2-03 (basin water sources); §4C-2-04 (basin water demands and needs); §4C-2-05 (allocation to equitable and reasonable uses); §4C-2-06 (watershed management); §4C-3-04, (withdrawals and diversion; protected areas); §4C-3-05 (water levels protected); §4C-3-06 (augmentation of supply); §4C-3-08 (underground water; limit on withdrawals); §4C-3-09 (atmospheric water); §4C-3-10 (recreation).

Similar Agreements: None.

Part 2 Water Allocation

§4C-2-01 Water Allocation, Generally

(a) The Comprehensive Water Management Plan, updated on an annual basis, shall allocate water resources according to the needs of the various equitable and reasonable uses described in §4C-2-04. The water demands based on application of reasonable conservation standards shall be determined for each sub-Basin or reach, as determined by the Commission, and compared to the water available for that sub-Basin or reach.

(b) If the available water in all sub-Basins exceeds the demand, all reasonable demands, as defined by the Commission, within the Basin shall be fully satisfied.

(c) Allocation shall be first made to existing water rights which have been perfected under applicable law of the individual parties prior to the creation of this Agreement. Allocation shall then be made to the needs of environmental protection according to individual environmental policies of the signatory Parties. Allocation thereafter shall be made according to those equitable and reasonable uses which provide maximum economic benefit (whether direct or indirect) to the Parties or which are necessary to enhance specific quality of life objectives of the Parties.

(e) No allocation of waters hereafter made pursuant to this Article shall constitute a prior appropriation of the waters of the Basin or confer any superiority of right in respect to the use of those waters, nor shall any such action be deemed to constitute an apportionment of the waters of the Basin among the Parties hereto. This paragraph shall not be deemed to limit or restrict the power of the Commission to enter into covenants with respect to water supply, with a duration not exceeding the life of this Agreement, as it may deem necessary for a benefit or development of the water resources of the Basin.

(f) No signatory Party shall permit any augmentation of flow by release from storage under the direction of the Commission to be diminished by the diversion of any water of the Basin during any period, except in cases in which such diversion is duly authorized by this Agreement, or by the Commission pursuant thereto, or by the judgment, order, or decree of a court of competent jurisdiction.

(g) Available water in excess of demand will be reserved for use as the Commission may direct.

Commentary: This article describes the fundamental objective function of a water allocation management plan. The *Convention on the Law of the Non-Navigational Uses of International Watercourses*, United Nations Document A/51/869 (United Nations 1997), advises that "in the absence of agreement or custom to the contrary, no use of an international watercourse enjoys inherent priority over other uses." If sufficient water is available to meet all equitable and reasonable uses, all demands will be met and no need for prioritization exists. However, if sufficient water is not available, it is recommended that a priority of allocation be established as follows: vested water rights, minimum instream flow for environmental protection and water quality purposes, followed by allocations that balance demands that provide maximum economic benefit with demands that have a direct influence on quality of life issues but cannot meet the economic threshold. For sustainable development in both transitional and developing countries, there should be an appropriate legal structure and a set of institutions that define property rights (Sterner 1994).

Cross References: §1C-1-01 (general policies); §1C-1-02 (purposes of agreement); §1C-1-03 (objectives of agreement); §2C-2-02 (basin); §2C-2-03 (comprehensive water management plan); §2C-2-05 (drought); §2C-2-11 (waters of the basin); §4C-1-02 (interrelationship of water resources); §4C-1-03 (comprehensive water management plan); §4C-1-04 (purpose and objectives of comprehensive water management plan); §4C-1-05 (conditions of comprehensive water management plan); §4C-1-06 (deviation from comprehensive water management plan); §4C-1-07 (allocation during flood conditions); §4C-1-08 (allocation during drought conditions); §4C-1-09 (non-impairment of comprehensive water management plan by state action); §4C-2-02 (waters not subject to allocation); §4C-2-03 (basin water sources); §4C-2-04 (basin water demands and needs); §4C-2-05 (allocation to equitable and reasonable uses); §4C-2-06 (watershed management); §4C-3-01 (existing rights recognized); §4C-3-02 (flood protection works); §4C-3-03 (minimum flows); §4C-3-04 (withdrawals and diversion; protected areas); §4C-3-05 (water levels protected); §4C-3-06 (augmentation of supply); §4C-3-07 (water quality); §4C-3-08 (underground water; limit on withdrawals); §4C-3-09 (atmospheric water); §4C-3-10 (recreation).

Similar Agreements: *Treaty between the United States and Mexico. Utilization of Waters of the Colorado and Tijuana Rivers and of the Rio Grande*, 59 Stat. 1219 (1944); *Delaware River Basin Compact*, Pub. L. 87-328, 75 Stat. 688-716 (1961); *Susquehanna River Basin Compact*, Pub. L. 91-575, 84 Stat. 1509-1541 (1970); *Agreement on the Cooperation for the Sustainable Development of the Mekong River Basin*, 34 ILM 864 (1995).

§4C-2-02 Waters Not Subject to Allocation

(a) Existing water rights that have been perfected under applicable law of the individual parties prior to the creation of this Agreement have priority over other water rights and allocations established by this Agreement. Non-perfected water rights have no such priority and will be subject to implementation of this Agreement.

(b) The Parties to this Agreement will provide the necessary information to ensure the Commission can allocate sufficient water within the Comprehensive Water Allocation Management Plan to meet the needs of vested water rights holders.

Commentary: This article acknowledges that certain water rights have been vested by internal water laws and that these vested water rights are acknowledged and included within the commission's planning. Initial identification of vested water rights, and knowledge of laws that may vest such rights in the future, is necessary to prevent extremely complex water problems, such as those experienced by Los Angeles as Arizona and Nevada exercise their legal rights to the waters of the Colorado River. It should be noted, however, that if all waters within one or more of the parties are already vested, either the agreement is meaningless or the internal laws of the particular parties should be changed.

Cross References: §2C-2-09 (party or parties); §4C-2-01 (water allocation, generally); §4C-2-03 (basin water sources); §4C-2-04 (basin water demands and needs); §4C-2-05 (allocation to equitable and reasonable uses); §4C-2-06 (watershed management); §4C-3-01 (existing rights recognized); §4C-3-02 (flood protection works); §4C-3-03 (minimum flows); §4C-3-04 (withdrawals and diversion; protected areas); §4C-3-05 (water levels protected); §4C-3-06 (augmentation of supply); §4C-3-07 (water quality); §4C-3-08 (underground water; limit on withdrawals); §4C-3-09 (atmospheric water); §4C-3-10 (recreation).

Similar Agreements: *Delaware River Basin Compact*, Pub. L. 87-328, 75 Stat. 688-716 (1961); *Susquehanna River Basin Compact*, Pub. L. 91-575, 84 Stat. 1509-1541 (1970).

§4C-2-03 Assessment and Enhancement of Basin Water Sources

(a) The Comprehensive Water Allocation Management Plan will identify and examine the factors that influence the availability of water resources in the Basins through an assessment of climatology, physiography, geology, and existing underground and surface water resources, including reservoirs, and the interaction between underground and surface water resources. The assessment will determine the existing and potential future availability and quality of underground and surface water resources within the Basins, the gains and losses of potential out-of-Basin transfers and the costs associated with making such waters available to users.

(b) The Commission shall have the power to develop and implement plans and projects for augmenting the supply of water, both geographically and temporally, for equitable and reasonable uses of water in the Basin. To this end, without limitation thereto, it may provide for, construct, acquire, operate, and maintain dams, reservoirs, and other facilities for the utilization of surface and underground water resources, and all related structures, appurtenances, and equipment in the river Basin and its tributaries and at such off-river sites as it may find appropriate, and may regulate and control use thereof.

(c) The Commission may contract for the transfer of water from outside the Basin when and if such transfer is made for equitable and reasonable purposes.

(d) The Comprehensive Water Management Plan shall include a water availability model that will allow the water availability in specific reaches or sub-Basins to be determined according to planned or programmed water usage.

Commentary: Proper water resource assessment is essential (*Council of European Communities Second Action Program on the Environment* 1977). An assessment of surface waters should include the existing quantity and quality of the available surface water as well as the potential losses resulting from anthropogenic changes to the physiographic and climatic factors within the river basin. The assessment should also consider transfers into and out of the river basin under study. Water demands by a variety of users can dramatically alter both the quantity and quality, as well the temporal and spatial availability of water. Physiographic and climatic factors determine the rate and distribution of runoff within a basin.

In the underground water assessment, the parties should develop an ongoing aquifer assessment program. Periodically, on a seasonal or annual basis, the parties should estimate the total water balance and the "safe yield" for the aquifers. This would entail, in part: (a) identifying the location and quantifying the rate of natural recharge to the aquifer; (b) determining the quantity and rate of diversion, the consumptive use of water, and the rate of natural discharge from the aquifer; (c) estimating changes in underground water storage or flow due to withdrawals; and (d) quantifying the relationships between underground water recharge, water table elevations, and aquifer discharges.

Cross References: §2C-2-02 (basin); §2C-2-03 (comprehensive water management plan); §2C-2-05 (drought); §2C-2-10 (underground water); §4C-2-01 (water allocation, generally); §4C-2-02 (waters not subject to allocation); §4C-2-06 (watershed management); §4C-3-01 (existing rights recognized); §4C-3-02 (flood protection works); §4C-3-03 (minimum flows); §4C-3-04 (withdrawals and diversion; protected areas); §4C-3-05 (water levels protected); §4C-3-06 (augmentation of supply); §4C-3-08 (underground water; limit on withdrawals); §4C-3-09 (atmospheric water); §4C-3-10 (recreation).

Similar Agreements: *Delaware River Basin Compact*, Pub. L. 87-328, 75 Stat. 688-716 (1961); *Susquehanna River Basin Compact*, Pub. L. 91-575, 84 Stat. 1509-1541 (1970); *Treaty between the United States and Mexico. Utilization of Waters of the Colorado and Tijuana Rivers and of the Rio Grande*, 59 Stat. 1219 (1944); *Convention on the Law of the Non-Navigational Uses of International Watercourses*, United Nations Document A/51/869 (United Nations 1997).

§4C-2-04 Assessment and Classification of Basin Water Demands and Needs

Existing and planned water uses shall be identified and quantified according to the categories described below by sub-Basin or by reach of the _____ River. Within each category, uses will be furthered classified as consumptive or nonconsumptive. Water use shall be determined based on existing and projected usage for the following classes of use:

(a) Agricultural water uses;
(b) Public water supply;
(c) Environmental water needs;
(d) Hydropower demands;
(e) Industrial demands;
(f) Navigation water needs;
(g) Recreation and scenic beauty water requirements;
(h) Thermoelectric and nuclear power needs;
(i) Waste assimilation; and
(j) Out-of-Basin transfers.

Commentary: These categories have been used in effective water resource management in the United States for many years. Other categorizing methods may be used, however, at the discretion of the parties.

Agriculture is the most significant consumer of water. Approximately 85% of total global water use is in irrigation (Clarke 1993). In the United States, agricultural use accounts for more than 42% of all freshwater withdrawals. Within the United States, approximately 80% of the agricultural use can be considered consumptive (U.S. Geological Survey 1987). Approximately 90% of western water consumption is for irrigation (Kneese 1993). Accurate projection of water needed for agricultural programs in developing countries is especially critical. Crop irrigation should be quantified to reflect the quantity of water required using existing irrigation techniques and the quantity required using the most efficient techniques, as determined by the commission.

Public supply includes water provided by utilities for public functions, including domestic and commercial uses. Domestic uses include water for normal household purposes, such as drinking, food preparation, bathing, washing clothes, and watering lawns and gardens. Commercial uses include water for the service industry and for retail facilities. It does not include water for industry. That portion of the water withdrawn for public supply and not returned to the water source from which it is drawn is consumptive. Public supply, also know as domestic/commercial use, consumes approximately 20% of the water removed from water sources. Quantification usually may be determined according to governmental water supply agency records. A minimum usage per person, often set at 100 gal. per day (379 L per day), may be established as a minimum allocation for public supply use.

Environmental needs, defined as the need for water by natural biological systems, includes environmental protection uses but excludes recreational and scenic beauty uses. Estimation of the need is unique because determining the water needed for ecological sustainability goes beyond simple analysis of the quantity and quality of water needed in the short term to prevent significant harm to the environment. Rather, environmental demand should include analysis of the value society places on specific aspects of the environment and the trade-offs it is willing to make. Unlike the other water demands, environmental demand should also consider the long-term effects of not meeting the water needs of natural systems. The political process should apportion environmental sustainability needs by imposing minimum stream flows at selected stations. Special consideration will be given to the habitat requirements of endangered species. Sufficient instream flow of sufficient quality shall be set aside for these purposes. These minimum flows should be established in the economic development and environmental protection plans established and submitted by the signatory parties. Such minimum stream flows should be sufficient to ensure that other uses or development do not degrade existing fish and game resources. Special consideration should be given to the habitat requirements of endangered species. Instream flow of sufficient quality should be set aside for these purposes. Consideration of quantity and quality, including temperature, are important. Maintenance and restoration of the natural environment are stressed in *Council on European Communities Third Action Programme on the Environment* (1983). It is recommended that the demand be apportioned according to the minimum stream flow at selected stations as determined by the comprehensive water management plan. However, as it has been learned in Great Britain, it should be recognized that effective application of the minimum flow concept requires reliable long-term flow data to determine flows accurately, as well as analysis based on the scientific method that determines the flows necessary to maintain habitats (Howarth 1990).

Hydropower uses can be identified and quantified according to the existing hydroelectric generating capacity and the flow necessary to support that capacity. The use shall be considered nonconsumptive. It usually may be quantified in terms of stable reservoir levels, river flow regimes, and water quality. For instance, hydropower demand may be quantified according to the seasonal minimum flows necessary to maintain the necessary generator head and navigation demand according to the flow requirements necessary to maintain the appropriate channel depths. Although nonconsumptive with regard to water quantity, hydropower use should be classified as consumptive within the particular subbasin or reach because the thermal effects cause significant water quality concerns for certain other uses.

Industrial demand includes process water for industrial, manufacturing, and mining purposes. On average, approximately 15% of the water withdrawn is consumed. Industrial demand does not include cooling water and steam for fossil fuel and nuclear power plants since the use of cooling water may be considered nonconsumptive.

Navigation uses, as a nonconsumptive and an instream flow demand, shall be quantified according to the flow requirements necessary to maintain the appropriate channel depth and, therefore, necessary minimum river flows.

Recreation and scenic beauty water use shall be quantified according to the need for stable reservoir levels, river flow regimes, and water quality.

Thermoelectric and nuclear power demand, when withdrawn for evaporative cooling purposes, consumes less than 3% of the water removed from the water source. Under these conditions, waters used for evaporative cooling purposes are nonconsumptive with regard to water quantity. However, this demand has thermal effects on the returned water that may cause significant water quality concerns for certain other uses. Water withdrawn for closed water-cooling systems may be considered 100% consumptive.

Waste assimilation is an important use of water. In the United States, current Environmental Protection Agency regulations prohibit waste assimilation as a valid category of water use. However, the reality around the world is that waste assimilation is arguably a significant purpose of surface waters. It is suggested that the demand be quantified according to the minimum instream flows necessary for waste assimilation at various levels of treatment (e.g., primary, secondary, and tertiary treatment). Such minimum flow quantities should include the water necessary to adequately nourish wetland and riparian buffers whose ability to filter nitrates and phosphorus can clearly assist in waste assimilation.

Out-of-basin transfers should be recognized as a demand on the waters available to users in the basin. Both existing and potential out-of-basin demands should be determined. This demand may significantly reduce the future availability of water caused by either existing transfers already recognized in internal water laws or in those cases wherein national water laws allow private transfers under a free-market approach to water allocation. As vested under national water laws, out-of-basin transfers should be specified in national economic development plans provided to the commission.

Cross References: §2C-2-03 (comprehensive water management plan); §2C-2-05 (drought); §2C-2-10 (underground water); §4C-2-01 (water allocation, generally); §4C-2-02 (waters not subject to allocation); §4C-2-04 (basin water demands and needs); §4C-2-05 (allocation to equitable and reasonable uses); §4C-2-06 (watershed management); §4C-3-01 (existing rights recognized); §4C-3-02 (flood protection works); §4C-3-03 (minimum flows); §4C-3-04 (withdrawals and diversion; protected areas); §4C-3-05 (water levels protected); §4C-3-06 (augmentation of supply); §4C-3-07 (water quality); §4C-3-08 (underground water; limit on withdrawals); §4C-3-09 (atmospheric water); §4C-3-10 (recreation).

Similar Agreements: *Delaware River Basin Compact*, Pub. L. 87-328, 75 Stat. 688-716 (1961); *Susquehanna River Basin Compact*, Pub. L. 91-575, 84 Stat. 1509-1541 (1970); *Convention on the Law of the Non-Navigational Uses of International Watercourses*, United Nations Document A/51/869 (United Nations 1997).

§4C-2-05 Criteria for Water Allocation

(a) Allocation of the water among the various equitable and reasonable uses shall be made on the basis of an analysis of the benefit provided to the user as compared to the detriment suffered by other uses that do not receive an allocation. The benefit evaluation shall include at a minimum (1) economic benefit, (2) environmental protection needs and minimum flows as determined by state laws, and (3) water uses which the Parties determine are required to support social and public policies and purposes.

(b) The Commission shall develop and publish a standard methodology for benefit evaluation of each category, and sub-category where applicable, of water demand to be used by all Parties to this Agreement.

Commentary: The essence of current international law establishes "equitable and reasonable utilization" as the basis for water policy. To implement the policy, however, standards should be established that describe the conditions under which "equitable and reasonable utilization" occur. The definition of the value of water has been one of the important sources of controversy with respect to water policy. Natural resource economists tend to focus on water as a commodity and argue that water should be allocated to those uses that bring the highest economic return. Governmental subsidies are discouraged as causing distortions of price. Others support subsidies as important to issues of national interest, equity, prior commitments, and regional and local development. Current discussion concerning valuation of water as a commodity has met resistance from some economists and political scientists who argue that some persons in society, specifically the poor, are materially disadvantaged in the bargaining process and thus may be willing to give up their water rights for a short-term, and potentially damaging, advantage. They claim that water allocation should be a matter of environmental ethics and justice (Frederick 1993; Mann 1993). In the United States, the claim has been made that it is not possible to include relative economic benefit as an allocation criteria because the "relative value of water uses can be determined only through an interactive process of voluntary exchange among thousands of water users" (Saarinen and Lynn 1993).

Those comments notwithstanding, economic evaluation of allocation is mandatory in some form. The benefit evaluation techniques later described are relevant to free-market economies. Where such economies do not exist, other means should be undertaken to value the different categories of water use.

The economic assessment should include both intra- and interparty economic return for existing and potential water use, as well as the potential for commercial transfers as a means for resolving conflicts between the parties. The allocation process should involve trade-offs between the various uses, in terms of their individual benefit–cost ratio, as well as trade-offs with the minimum flows established for environmental sustainability. To begin the assessment, the parties should estimate the intrastate demand water that provides optimal economic return for each party over a specific planning horizon based on a standard rate of return. The parties may then make trade-offs among themselves for regional allocation of the water resource.

Using the concept of marginal willingness to pay is of limited effectiveness in evaluating environmental sustainability demand. Whereas many economists have proposed that a marginal willingness to pay may be estimated using techniques similar to those used for recreational demand, such techniques may be grossly inexact because the consumer in this case can be classified as society at large. Contingent valuation has become the most widely used approach to value public goods. However, there are a number of problems associated with a contingent valuation study, such as a vague or unclear description of the good, lack of key information about context and substitutes, lack of attention to ways respondents could misperceive the good, implausible or overly hypothetical scenarios (particularly with respect to provision of the good and payment for it), willingness-to-pay responses that cannot be predicted by the available covariates, inadequate sampling procedures, poor response rates, and sample sizes that are too small for the purpose for which they were intended (Mitchell and Carson 1995). The direct approach to the estimation of willingness-to-pay is to ask individuals how much they are willing to pay for a risk reduction (Johansson et al. 1995). It has been suggested that the damages could be approximated by the cost for reaching an environmental goal (Dasgupta et al. 1995). Thus, if the officially stated goal is to reduce sulfur emissions by 30%, then one can approximate current sulfur damages by the cost of reaching this goal. The willingness to sacrifice some degree of environmental sustainability relates directly to the political will of the national and/or international community. We therefore recommend allocating environmental sustainability demand according to an established minimum flow determined (as earlier discussed).

Certain water uses are associated with quality-of-life issues or have social and/or civic purposes that cannot be appropriately quantified in a valid benefit evaluation. In a manner similar to minimum flows established for environmental protection purposes, these equitable and reasonable uses should be established by the parties within their internal political process and vested in agreement with the other parties.

Cross References: §1C-1-01 (general policies); §1C-1-02 (purposes of agreement); §1C-1-03 (objectives of agreement); §2C-2-03 (comprehensive water management plan); §2C-2-05 (drought); §2C-2-09 (party or parties); §4C-1-03 (comprehensive water management plan); §4C-1-04 (purpose and objectives of comprehensive water management plan); §4C-1-05 (conditions of comprehensive water management plan); §4C-1-06 (deviation from comprehensive water management plan); §4C-1-07 (allocation during flood conditions); §4C-1-08 (allocation during drought conditions); §4C-1-09 (non-impairment of comprehensive water management plan by state action); §4C-2-01 (water allocation, generally); §4C-2-02 (waters not subject to allocation); §4C-2-03 (basin water sources); §4C-2-04 (basin water demands and needs); §4C-2-06 (watershed management); §4C-3-01 (existing rights recognized); §4C-3-02 (flood protection works); §4C-3-03 (minimum flows); §4C-3-04 (withdrawals and diversion; protected areas); §4C-3-05 (water levels protected); §4C-3-06 (augmentation of supply); §4C-3-07 (water quality); §4C-3-08 (underground water; limit on withdrawals); §4C-3-09 (atmospheric water); §4C-3-10 (recreation).

Similar Agreements: *Klamath River Basin Compact*, 71 Stat. 497 (1957); *Delaware River Basin Compact*, Pub. L. 87-328, 75 Stat. 688-716 (1961); *Susquehanna River Basin Compact*, Pub. L. 91-575, 84 Stat. 1509-1541 (1970); *Agreement on the Cooperation for the Sustainable Development of the Mekong River Basin*, 34 ILM 864 (1995); *Convention on the Law of the Non-Navigational Uses of International Watercourses*, United Nations Document A/51/869 (United Nations 1997).

§4C-2-06 Watershed Management (Optional)

(a) The Commission shall promote sound practices of watershed management in the Basin, including projects and facilities to inhibit uncontrolled runoff and prevent soil erosion. Natural solutions, such as riparian vegetated buffers, will be proposed whenever possible.

(b) The Commission may acquire, sponsor, or operate facilities and projects to encourage soil conservation, prevent and control erosion, and to promote land reclamation and sound forestry practices.

(c) The Commission may acquire, sponsor, or operate projects and facilities for the maintenance and improvement of fish and wildlife habitats related to the water resources of the Basin.

(d) The Commission shall not operate any such project or facility unless it has first found and determined that no other suitable unit or agency of government is available to operate the same upon reasonable conditions, in accordance with the intent and purpose expressed in Article 2C-1-06 of this Agreement.

Commentary: This article reinforces the philosophy that efficient and effective allocation can be best established by use of watershed management techniques.

Cross References: §1C-1-01 (general policies); §1C-1-02 (purposes of agreement); §1C-1-03 (objectives of agreement); §2C-2-02 (basin); §2C-2-05 (drought); §4C-1-03 (comprehensive water management plan); §4C-1-04 (purpose and objectives of comprehensive water management plan); §4C-1-05 (conditions of comprehensive water management plan); §4C-1-06 (deviation from comprehensive water management plan); §4C-1-07 (allocation during flood conditions); §4C-1-08 (allocation during drought conditions); §4C-1-09 (non-impairment of comprehensive water management plan by state action); §4C-2-01 (water allocation, generally); §4C-2-02 (waters not subject to allocation); §4C-2-03 (basin water sources); §4C-2-04 (basin water demands and needs); §4C-2-05 (allocation to equitable and reasonable uses); §4C-3-01 (existing rights recognized); §4C-3-02 (flood protection works); §4C-3-03 (minimum flows); §4C-3-04 (withdrawals and diversion; protected areas); §4C-3-05 (water levels protected); §4C-3-06 (augmentation of supply); §4C-3-07 (water quality); §4C-3-08 (underground water; limit on withdrawals); §4C-3-09 (atmospheric water); §4C-3-10 (recreation).

Similar Agreements: *Convention on Wetlands of International Importance Especially as Waterfowl Habitat*, UN Treaty Series 14583 (1971), amended by the Paris Protocol (1982), Regina Amendments (1987); *Rio Declaration on Environment and Development*, 31 ILM 874 (1992); *Agreement on the Protection of the (River) Meuse*, 34 ILM 854 (1995); *Agreement on the Cooperation for the Sustainable Development of the Mekong River Basin*, 34 ILM 864 (1995).

Part 3 General Provisions

§4C-3-01 Existing Rights Recognized

Rights to the use of the Waters of the Basin, whether instream use or based on direct diversion or storage, are hereby recognized as of the date of this Agreement to the extent these rights are valid under the law of the Party in which the use is made, and shall remain unimpaired hereby. These rights, together with the additional allocations made under §4C-2-05, are agreed to be an equitable and reasonable utilization of the waters of the Basin among the Parties.

Commentary: This provision is recognition of the inherent sovereignty of the individual parties and the inherent rights and privileges that the individual party may have granted for use of water within its borders. The agreement acknowledges the continuation of previously granted property rights. The parties do not cede to the comprehensive water management plan the right to withdraw those water rights existing under state law before the agreement.

Cross References: §2C-2-02 (basin); §2C-2-03 (comprehensive water management plan); §2C-2-05 (drought); §2C-2-06 (equitable and reasonable apportionment); §2C-2-09 (party or parties); §2C-2-11 (waters of the basin); §4C-1-01 (joint exercise of sovereignty); §4C-1-02 (interrelationship of water resources); §4C-1-03 (comprehensive water management plan); §4C-1-04 (purpose and objectives of comprehensive water management plan); §4C-1-05 (conditions of comprehensive water management plan); §4C-1-06 (deviation from comprehensive water management plan); §4C-1-07 (allocation during flood conditions); §4C-1-08 (allocation during drought conditions); §4C-1-09 (non-impairment of comprehensive water management plan by state action); §4C-2-01 (water allocation, generally); §4C-2-02 (waters not subject to allocation); §4C-2-03 (basin water sources); §4C-2-04 (basin water demands and needs); §4C-2-05 (allocation to equitable and reasonable uses); §4C-2-06 (watershed management); §4C-3-02 (flood protection works); §4C-3-03 (minimum flows); §4C-3-04 (withdrawals and diversion; protected areas); §4C-3-05 (water levels protected); §4C-3-06 (augmentation of supply); §4C-3-07 (water quality); §4C-3-08 (underground water; limit on withdrawals); §4C-3-09 (atmospheric water); §4C-3-10 (recreation).

Similar Agreements: *Red River Compact*, 94 Stat. 3305 (1980); *Sabine River Compact*, 68 Stat. 690 (1953), amended 76 Stat. 34 (1962), 91 Stat. 281 (1977), 106 Stat. 4661 (1992).

§4C-3-02 Flood Protection Works

(a) As a general concept, the use of the channels of the waters of the _____ River Basin for the discharge of flood or other excess waters shall be free and not subject to limitation by any Party, and no Party shall have any claim against the other in respect of any damage caused by such use. However, each of the signatory Parties declares its intention to manage flood control programs and activities in such manner, consistent with the normal operations of its hydraulic systems, so as to avoid, as far as feasible, material damage in the territory of the other.

(b) The signatory Parties agree to furnish the Commission with complete documentation of existing flood protection programs and works. The Commission shall analyze the documentation to determine the potential flood damages that may therefrom arise and consult with the signatory Parties concerning the findings of such analysis.

Commentary: Flood control policies and works can have a dramatic effect on the timing and elevation of water levels and thus may become a major contentious issue between the parties. The issue often eclipses the boundaries of "equitable and reasonable utilization" and should be addressed as an individual area of coordination. This provision recognizes the sovereign right of each party to make efforts to safeguard its people and economic assets from flood damages but also establishes an avenue for the sharing of data on flood control efforts as well as an independent analysis of the effects of those efforts on other parties.

Cross References: §1C-1-01 (general policies); §1C-1-02 (purposes of agreement); §1C-1-03 (objectives of agreement); §2C-2-02 (basin); §2C-2-03 (comprehensive water management plan); §2C-2-07 (flood); §2C-2-09 (party or parties); §3C-3-03 (projects of the signatory parties); §3C-3-04 (cooperative services); §4C-1-01 (joint exercise of sovereignty); §4C-1-02 (interrelationship of water resources); §4C-1-03 (comprehensive water management plan); §4C-1-04 (purpose and objectives of comprehensive water management plan); §4C-1-05 (conditions of comprehensive water management plan); §4C-1-06 (deviation from comprehensive water management plan); §4C-1-07 (allocation during flood conditions).

Similar Agreements: *Treaty between the United States and Mexico. Utilization of Waters of the Colorado and Tijuana Rivers and of the Rio Grande*, 59 Stat. 1219 (1944); *Agreement between the People's Republic of Bulgaria and the Republic of Turkey Concerning Co-operation in the Use of the Waters of Rivers Flowing through the Territory of Both Countries*, UNTS, Vol. 807, 117 (1968).

§4C-3-03 Minimum Flows

Alternative 1

Any requirements by either Party for water to maintain minimum flows or for instream uses shall be satisfied from the allocation provided to that Party.

Alternative 2

Any allocation formula shall apply after sufficient allowance is made to satisfy requirements for maintaining minimum flow or for instream use.

Commentary: For ecological reasons and for wastewater discharge purposes, a minimum flow should be maintained in all surface waters. This protected minimum flow or level for each water source, normally established by state law, is not subject to allocation for other purposes. In the transboundary context, conflict may occur because the parties may differ on the quantification of this protected level.

The alternatives presented in this provision provide a model for two different eventualities. Alternative 1 is provided for those cases in which the parties cannot agree on the specific minimum flow or level to be maintained in the watercourse. Alternative 2 provides a model for the case in which an agreed-upon minimum flow is set in the agreement.

Cross References: §1C-1-01 (general policies); §1C-1-02 (purposes of agreement); §1C-1-03 (objectives of agreement); §2C-2-02 (basin); §2C-2-03 (comprehensive water management plan); §2C-2-05 (drought); §2C-2-09 (party or parties); §2C-2-10 (underground water); §4C-1-02 (interrelationship of water resources); §4C-1-03 (comprehensive water management plan); §4C-1-04 (purpose and objectives of comprehensive water management plan); §4C-1-05 (conditions of comprehensive water management plan); §4C-1-06 (deviation from comprehensive water management plan); §4C-1-08 (allocation during drought conditions); §4C-1-09 (non-impairment of comprehensive water management plan by state action); §4C-2-01 (water allocation, generally); §4C-2-02 (waters not subject to allocation); §4C-2-03 (basin water sources); §4C-2-04 (basin water demands and needs).

Similar Agreements: *Red River Compact*, 94 Stat. 3305 (1980); *Sabine River Compact*, 68 Stat. 690 (1953), amended 76 Stat. 34 (1962), 91 Stat. 281 (1977), 106 Stat. 4661 (1992).

§4C-3-04 Withdrawals and Diversions; Protected Areas (Optional)

(a) The Commission may regulate and control withdrawals and diversions from surface waters and underground waters of the _____ River Basin, as provided by this Article. The Commission may enter into agreements with the signatory Parties relating to the exercise of such power or regulation or control and may delegate to any of them such powers of the Commission as it may deem necessary or desirable.

(b) The Commission may from time to time, after public hearing upon due notice, determine and delineate such areas within the Basin wherein the demands upon supply made by water users have developed or threaten to develop to such a degree as to create a water shortage or to impair or conflict with the requirements or effectuation of the Comprehensive Water Management Plan, and any such areas may be designated as protected areas. The Commission, whenever it determines that such shortage no longer exists, shall terminate the protected status of such area and shall give public notice of such termination.

(c) In any protected areas so determined and delineated, no person, firm, corporation, or other entity shall

divert or withdraw water for domestic, municipal, agricultural, or industrial uses in excess of such quantities as the Commission may prescribe by general regulation, except (i) pursuant to a permit granted under this Article, or (ii) pursuant to a permit or approval heretofore granted under the laws of any of the signatory Parties.**

(d) Permits shall be granted, modified, or denied as the case may be so as to avoid such depletion of the natural stream flows and underground waters in the protected area or in an emergency area as will adversely affect the Comprehensive Water Management Plan or the just and equitable interests and rights of other lawful users of the same source, giving due regard to the need to balance and reconcile alternative and conflicting uses in the event of an actual or threatened shortage of water of the quality required.

(e) Each Party shall provide for the maintenance and preservation of such records of authorized diversions and withdrawals and the annual volume thereof as the Commission shall prescribe. Such records and supplementary reports shall be furnished to the Commission at its request.

(f) Whenever the Commission finds it necessary or desirable to exercise the powers conferred by this Article, any diversion or withdrawal permits authorized or issued under the laws of any of the signatory Parties shall be superseded to the extent of any conflict with the control and regulation exercised by the Commission.

Commentary: Optimum management of water resources requires that a central authority regulate withdrawals and diversions from surface waters and underground waters of the basin. However, this optional article has a significant effect on the sovereign integrity of the individual state's control over internal affairs.

Cross References: §2C-1-05 (powers of sovereign parties; withdrawal); §2C-2-03 (comprehensive water management plan); §2C-2-05 (drought); §2C-2-09 (party or parties); §2C-2-10 (underground water); §2C-2-11 (waters of the basin); §3C-2-01 (general powers and duties); §3C-2-02 (powers and duties reserved to the commissioners); §3C-2-03 (obligations of the commission); §3C-2-04 (regulations; enforcement); §3C-2-05 (prohibited activities); §4C-1-01 (joint exercise of sovereignty); §4C-1-02 (interrelationship of water resources); §4C-1-03 (comprehensive water management plan); §4C-1-04 (purpose and objectives of comprehensive water management plan); §4C-1-05 (conditions of comprehensive water management plan); §4C-1-09 (non-impairment of comprehensive water management plan by state action); §4C-2-04 (basin water demands and needs); §4C-2-05 (allocation to equitable and reasonable uses); §4C-2-06 (watershed management).

Similar Agreements: *Delaware River Basin Compact*, Pub. L. 87-328, 75 Stat. 688-716 (1961); *Susquehanna River Basin Compact*, Pub. L. 91-575, 84 Stat. 1509-1541 (1970); *Treaty between the United States and Great Britain Relating to Boundary Waters*, 36 Stat. 2451 (1909); *Agreement on the Conservation of Nature and Natural Resources* (ASEAN 1985).

§4C-3-05 Augmentation of Supply

(a) Any importation of water from outside the Basin shall be excluded from any allocation set forth elsewhere in this Agreement or the Comprehensive Water Management Plan, and the Party importing such water shall have the right to full and complete use and consumption of such imported water.

(b) Any Party implementing a conservation program with respect to water supplies shall be entitled to full and complete use and consumption of all increased supplies resulting from such conservation program. The burden of showing such increase shall rest on the Party claiming such increase.

Commentary: (1) Article 4C-3-05(a) makes clear that if a party arranges to increase supplies by importing water, it need not share those additional supplies.

(2) Article 4C-3-05(b) provides encouragement for conservation by rewarding the party that undertakes that effort. Caution should be exercised in incorporating this provision, however, inasmuch as the level of conservation efforts between the parties may be unequal at the time the agreement is negotiated. A party that has already made significant efforts should not be placed at a disadvantage relative to a party that, before the agreement, made little effort to conserve.

Cross References: §2C-1-05 (powers of sovereign parties; withdrawal); §2C-2-02 (basin); §2C-2-03 (comprehensive water management plan); §2C-2-05 (drought); §3C-3-03 (projects of the signatory parties); §3C-3-04 (cooperative services); §4C-1-01 (joint exercise of sovereignty); §4C-1-02 (interrelationship of water resources); §4C-1-03 (comprehensive water management plan); §4C-1-04 (purpose and objectives of comprehensive water management plan); §4C-1-05 (conditions of comprehensive water management plan); §4C-1-06 (deviation from comprehensive water management plan); §4C-1-08 (allocation during drought conditions); §4C-1-09 (non-impairment of comprehensive water management plan by state action); §4C-2-01 (water allocation, generally); §4C-2-02 (waters not subject to allocation); §4C-2-03 (basin water sources); §4C-2-04 (basin water demands and needs).

Similar Agreements: None.

§4C-3-06 Water Quality
Alternative 1
The Commission shall:

(a) Manage the waters of the _____ River Basin to maintain ecosystem integrity, preserve and protect aquatic ecosystems effectively from any form of (significant) degradation on a drainage Basin or sub-Basin basis;

(b) Publish biological, health, physical, and chemical quality criteria for all water bodies (surface and underground water), according to Basin capacities and needs, with a view to an ongoing improvement of water quality;

(c) Establish standards for the discharge of effluents and for the receiving waters, to include standards for land use management that relate to best management practices for water quality;

(d) Establish minimum flow criteria to maintain the desired instream environmental conditions and ensure nourishment of wetlands and riparian buffers as necessary to properly filter nitrates and phosphorus arising from nonpoint runoff;

(e) Maintain the quality of the waters of the Basin at or above water quality standards as may be adopted, now or hereafter, by the water pollution control agencies of the respective Parties in compliance with the provisions of the Federal Water Quality Act of 1965, and amendments thereto.

Alternative 2

(a) **The Parties mutually agree to the principle of individual Party efforts to control natural and man-made water pollution within each Party and to the continuing support of both Parties in active water pollution control programs.**

(b) **The Parties agree to cooperate, through their appropriate Party agencies, in the investigation, abatement, and control of sources of alleged interparty pollution within the Basin.**

(c) **(Alternative 1). The Parties agree to cooperate in maintaining the quality of the waters of the _____ River Basin at or above water quality standards as may be developed and agreed to by the Parties.**

(d) **(Alternative 2). The Parties agree to cooperate in maintaining the quality of the waters of the Basin at or above water quality standards as may be adopted, now or hereafter, by the water pollution control agencies of the respective Parties in compliance with the provisions of the Clean Water Act of 1965, and amendments thereto.**

Commentary: The quality of the water allocated is as important as the quantity of water allocated. Poor-quality water imposes risks that the parties should consider. First, there is the health risk to the population that uses the water for domestic purposes. Second, if the available water does not meet the standards for certain industrial purposes, there is the risk that economic growth may be impaired. Finally, there is the risk that quality degradation may have a severe effect on the ecology of the basin, resulting in long-term sustainability complications. Integration of water quality and quantity is essential. *Agenda 21* (United Nations 1992) obligated all signatories to develop a program of water and sustainable development (Ahlander 1994). The *Convention on the Law of the Non-Navigational Uses of International Watercourses*, United Nations Document A/51/869 (United Nations 1997), establishes the criterion that "watercourse States shall, individually and, where appropriate, jointly, protect and preserve the ecosystems of international watercourses."

Surface and underground water may be degraded by a variety of factors. Major problems affecting the quality of these water resources arise, for instance, from inadequate domestic sewage treatment, inadequate controls on the discharge of industrial waste and effluent, the diversion of waters resulting in insufficient water to assimilate waste, the loss and destruction of catchment areas, the improper siting of industrial plants, deforestation, and poor agricultural practices, which cause leaching of nutrients and pesticides. Transboundary water sharing should include effective plans and programs that eliminate, or at least minimize, the possible sources of water quality degradation. In most cases, water quality can be significantly advanced by the use of best management practices. Such practices are forms of land use management that limit nonpoint source pollution from entering water bodies. Examples include vegetative riparian buffers and limitations on contiguous impervious surfaces, among others. Best management practices are normally voluntary in nature but are complemented with tax incentives. Alternative 1(c) provides the commission with the power to establish such standards, although the parties may choose to make the standards voluntary.

The complex, interconnected nature of freshwater systems suggests that freshwater management should be systemically integrated, taking a catchment management approach that balances the needs of people and the environment. The parties should manage the waters of the basin to maintain ecosystem integrity, preserve aquatic ecosystems, and protect them effectively from any form of degradation on a drainage basin or subbasin basis.

The parties should establish biological, health, physical, and chemical quality criteria for all significant water bodies in the basin to continually improve water quality. The parties should establish minimum standards both for discharging effluents and for receiving waters. We recommend that the parties institute standards for land use management, such as limits on agrochemical use, deforestation, and wasteful irrigation practices. Such rational land use standards should prevent land degradation, erosion, and siltation of lakes and other water bodies.

Cross References: §1C-1-01 (general policies); §1C-1-02 (purposes of agreement); §1C-1-03 (objectives of agreement); §2C-2-02 (basin); §2C-2-03 (comprehensive water management plan); §2C-2-05 (drought); §2C-2-09 (party or parties); §2C-2-10 (underground water); §2C-2-11 (waters of the basin); §3C-3-03 (projects of the signatory parties); §4C-1-01 (joint exercise of sovereignty); §4C-1-02 (interrelationship of water resources); §4C-1-03 (comprehensive water management plan); §4C-1-04 (purpose and objectives of comprehensive water management plan); §4C-1-05 (conditions of comprehensive water management plan); §4C-1-06 (deviation from comprehensive water management plan); §4C-1-09 (non-impairment of comprehensive water management plan by state action); §4C-2-01 (water allocation, generally); §4C-2-02 (waters not subject to allocation); §4C-2-03 (basin water sources); §4C-2-04 (basin water demands and needs); §4C-2-05 (allocation to equitable and reasonable uses); §4C-2-06 (watershed management).

Similar Agreements: *Rio Grande Compact of 1938*, 53 Stat. 785, 938 (1938); *Klamath River Basin Compact*, 71 Stat. 497 (1957); *Arkansas River Basin Compact of 1965*, 80 Stat. 1409 (1966); *Kansas–Nebraska Big Blue River Compact*, 86 Stat. 193 (1972); *Arkansas River Basin Compact of 1970*, 87 Stat. 569 (1973); *Delaware River Basin Compact,* Pub. L. 87-328, 75 Stat. 688-716 (1961); *Susquehanna River Basin Compact*, Pub. L. 91-575, 84 Stat. 1509-1541 (1970); *Red River Compact*, 94 Stat. 3305 (1980); *Oregon–California Goose Lake Interstate Compact*, 98 Stat. 291 (1984); *Treaty between the United States and Great Britain Relating to Boundary Waters,* 36 Stat. 2451 (1909); *Agreement between the United States and Canada on Great Lakes Water Quality*, 1153 UNTS 187 (1978); *Convention on the Protection and Use of Transboundary Watercourses and International Lakes*, 31 ILM 1312 (1992); *Convention on the Law of the Non-Navigational Uses of International Watercourses*, United Nations Document A/51/869 (United Nations 1997).

§4C-3-07 Underground Water; Limit on Withdrawals

Alternative 1

When such action is necessary [to maintain an allocation set out elsewhere], [Upstream Party] shall regulate, in the same manner that surface flow is regulated, withdrawal of water from irrigation wells located within ___ miles of the river or its tributaries.

Alternative 2

When such action is necessary [to maintain an allocation set out elsewhere], [Upstream Party] shall regulate, in the same manner that surface flow is regulated, withdrawal of water from irrigation wells that may be hydrologically connected to the river or its tributaries.

Commentary: If underground water is subject to the allocation provisions of the agreement, it may be useful to specifically address the steps to be taken with respect to withdrawals. Alternative 1, adapted from the *Kansas–Nebraska Big Blue River Compact*, 86 Stat. 193 (1972), uses a distance limitation to determine which wells fall within the scope of the agreement. If it is possible to establish the hydrological connection between all wells and the surface flow, the mileage limitation may be replaced with references to wells with such connection. In the absence of definitive hydrologic information, the mileage limitation may make administration easier, if less precise.

No further specific allocation systems for underground water are provided because it is assumed that if underground water is allocated by agreement, that allocation will be in conjunction with allocation of related surface water sources and the allocation of underground water will be incorporated as part of the overall allocation of water. If underground water is allocated independently from surface water, the parties might use the surface models as a guide with respect to types of allocations (proportional, guaranteed minimum, etc.).

Cross References: §1C-1-01 (general policies); §1C-1-02 (purposes of agreement); §1C-1-03 (objectives of agreement); §2C-2-03 (comprehensive water management plan); §2C-2-05 (drought); §2C-2-09 (party or parties); §2C-2-10 (underground water); §4C-1-01 (joint exercise of sovereignty); §4C-1-02 (interrelationship of water resources); §4C-1-03 (comprehensive water management plan); §4C-1-04 (purpose and objectives of comprehensive water management plan); §4C-1-05 (conditions of comprehensive water management plan); §4C-1-06 (deviation from comprehensive water management plan); §4C-1-07 (allocation during flood conditions); §4C-1-08 (allocation during drought conditions); §4C-1-09 (non-impairment of comprehensive water management plan by state action); §4C-2-01 (water allocation, generally); §4C-2-02 (waters not subject to allocation); §4C-2-03 (basin water sources); §4C-2-04 (basin water demands and needs); §4C-2-05 (allocation to equitable and reasonable uses); §4C-2-06 (watershed management); §4-3-004 (withdrawals and diversion; protected areas); §4C-3-05 (water levels protected); §4C-3-06 (augmentation of supply).

Similar Agreement: *Kansas–Nebraska Big Blue River Compact*, 86 Stat. 193 (1972).

§4C-3-08 Atmospheric Water

Any Party which augments precipitation within the Basin shall be entitled to full and exclusive use of additional water supplies resulting from such augmentation, notwithstanding any other standard of allocation set forth in this agreement or the Comprehensive Water Management Plan. In the event the Parties cannot agree on whether or to what extent precipitation has been augmented, the Party asserting the right to such increased supplies shall bear the burden of proving that the increase, if any, was the result of the Party's augmentation efforts and not simply the result of natural variation in precipitation amounts.

Commentary: As technology advances, precipitation augmentation may become more widely practiced and accepted. This provision would allow a party that financed such efforts to retain the benefits, provided it could meet the burden of proof. Absent inclusion of this alternative, the augmented supply will be treated in the same manner as any other water within the basin.

Cross References: §1C-1-01 (general policies); §1C-1-02 (purposes of agreement); §1C-1-03 (objectives of agreement); §2C-2-01 (atmospheric water); §2C-2-03 (comprehensive water management plan); §2C-2-05 (drought); §2C-2-09 (party or parties); §4C-1-01 (joint exercise of sovereignty); §4C-1-02 (interrelationship of water resources); §4C-1-03 (comprehensive water management plan); §4C-1-04 (purpose and objectives of comprehensive water management plan); §4C-1-05 (conditions of comprehensive water management plan); §4C-1-06 (deviation from comprehensive water management plan); §4C-1-07 (allocation during flood conditions); §4C-1-08 (allocation during drought conditions); §4C-1-09 (non-impairment of comprehensive water management plan by state action); §4C-2-01 (water allocation, generally); §4C-2-02 (waters not subject to allocation); §4C-2-03 (basin water sources); §4C-2-04 (basin water demands and needs); §4C-2-05 (allocation to equitable and reasonable uses); §4C-2-06 (watershed management); §4-3-04 (withdrawals and diversion; protected areas); §4C-3-05 (water levels protected); §4C-3-06 (augmentation of supply).

Similar Agreement: *Rio Grande Compact of 1938*, 53 Stat. 785, 938 (1938).

§4C-3-09 Recreation (Optional)

(a) The Commission shall provide for the development of water-related public sports and recreational facilities. The Commission on its own account or in cooperation with a signatory Party, political subdivision or any agency thereof, may provide for the construction, maintenance, and administration of such facilities.

(b) The Commission shall not operate any such project or facility unless it has first found and determined that no other suitable unit or agency of government is available to operate the same upon reasonable conditions, in accordance with the intent and purpose expressed in §2C-1-06 of this Agreement.

(c) The Commission shall encourage activities of other public agencies having water-related recreational interests and assist in the coordination thereof.

(d) The Commission shall recommend standards for the development and administration of water-related recreational facilities.

(e) The Commission shall provide for the administration, operation, and maintenance of recreational facilities owned or controlled by the Commission and for the letting and supervision of private concessions in accordance with this Article.

(f) The Commission shall, after notice and public hearing, provide by regulation for the award of contracts for private concessions in connection with recreational facilities, including any renewal or extension thereof, upon sealed competitive bids after public advertisement therefor.

Commentary: This article may be appropriate for agreements involving developed nations. However, the issue may not be material to agreements involving developing nations, or those in transition.

Cross References: §1C-1-01 (general policies); §1C-1-02 (purposes of agreement); §1C-1-03 (objectives of agreement); §2C-2-03 (comprehensive water management plan); §2C-2-09 (party or parties); §4C-1-01 (joint exercise of sovereignty); §4C-1-02 (interrelationship of water resources); §4C-1-03 (comprehensive water management plan); §4C-1-04 (purpose and objectives of comprehensive water management plan); §4C-1-05 (conditions of comprehensive water management plan); §4C-1-06 (deviation from comprehensive water management plan); §4C-1-07 (allocation during flood conditions); §4C-1-08 (allocation during drought conditions); §4C-1-09 (non-impairment of comprehensive water management plan by state action); §4C-2-01 (water allocation, generally); §4C-2-02 (waters not subject to allocation); §4C-2-03 (basin water sources); §4C-2-04 (basin water demands and needs); §4C-2-05 (allocation to equitable and reasonable uses); §4C-2-06 (watershed management); §4C-3-01 (existing rights recognized); §4C-3-02 (flood protection works); §4C-3-03 (minimum flows); §4C-3-04 (withdrawals and diversion; protected areas); §4C-3-05 (water levels protected).

Similar Agreements: *Delaware River Basin Compact*, Pub. L. 87-328, 75 Stat. 688-716 (1961); *Susquehanna River Basin Compact*, Pub. L. 91-575, 84 Stat. 1509-1541 (1970).

C.5 ARTICLE 5C: FINANCING

Alternative 1 (Limited)
The budget of the Commission shall be prepared by the Executive Director and approved by unanimous vote of the Commissioners.

Alternative 2 (Comprehensive)
The Commission shall annually adopt a current expense budget and a capital budget that includes any and all capital projects it proposes to undertake or continue during the budget period. The amount required to balance the current expense budget and the capital budget will be financed according to the arrangements described in Annex C-1.

Commentary: Comprehensive planning and management require dependable sources of finances. If a program of financing is uncertain, or if the program is not sufficiently detailed to provide certainty, comprehensive planning and management often fail. However, in some situations, the parties may wish to minimize complexities and maximize flexibility on the part of the commission. Therefore, two alternative financing methodologies are provided for consideration.

Alternative 1 is a simple methodology that leaves funding of the agreement implementation to predetermined shares from the parties. The provision earlier provided suggests equality of contribution. Equally valid would be contributions on an equitable basis, where "equitable" is determined according to the economic benefit received as the result of the comprehensive water management. The contributions could also be based on the percentage of basin within each of the parties. Other possibilities include the relative wealth of the parties as expressed in GNP or the relative wealth of various reaches of the watercourses within the parties. It should be recognized that such simplified funding exposes the comprehensive management to the vagaries of political forces within the various sovereign parties.

Alternative 2 provides a more comprehensive methodology that provides a great degree of independence from political forces that may arise in one or more of the parties. It is, however, detailed and appropriate only where the parties to the agreement do not impose significant control over economic policies and individual and corporate decision making. This alternative is drawn, largely unmodified, from the *Delaware River Basin Compact* (1961) and its amendments. They are not appropriate in all situations because they are oriented toward a federal political system.

Cross References: §1C-1-01 (general policies); §1C-1-02 (purposes of agreement); §1C-1-03 (objectives of agreement); §1C-1-04 (coordination and cooperation); §1C-1-05 (reservation of federal rights); §1C-1-06 (national security); §2C-1-01 (activation of agreement); §2C-1-02 (consent to jurisdiction); §2C-1-03 (duration of agreement); §2C-1-04 (amendments and supplements); §2C-1-05 (powers of sovereign parties; withdrawal); §2C-1-06 (existing agencies); §2C-1-07 (limited applicability); §2C-1-08 (annexes); §2C-2-02 (basin); §2C-2-03 (conservation measures); §2C-2-04 (consumptive use); §2C-2-05 (domestic use); §2C-2-06 (drought management strategies); §2C-2-07 (equitable apportionment); §2C-2-08 (equitable and reasonable utilization); §2C-2-09 (interbasin transfer); §2C-2-10 (municipal uses); §2C-2-11 (nonconsumptive uses); §2C-2-12 (party); §2C-2-13 (underground water); §2C-2-14 (waters of the basin); §3C-1-01 (commission created); §3C-1-02 (jurisdiction of the commission); §3C-1-03 (commissioners); §3C-1-04 (status, immunities and privileges); §3C-1-05 (commission organization and staffing); §3C-1-06 (rules of procedures); §3C-1-07 (commission administration); §3C-2-01 (general powers and duties); §3C-2-02 (powers and duties reserved to the commissioners); §3C-2-03 (obligations of the commission); §3C-2-04 (regulations; enforcement); §3C-2-05 (prohibited activities); §3C-2-06 (referral and review); §3C-2-07 (advisory committees); §3C-2-08 (reports); §3C-2-09 (condemnation proceedings); §3C-2-10 (meetings, hearings and records); §3C-2-11 (tort liability); §3C-3-01 (coordination and cooperation); §3C-3-02 (project costs and evaluation standards); §3C-3-03 (projects of the signatory parties); §3C-3-04 (cooperative services).

Similar Agreements: *Delaware River Basin Compact*, Pub. L. 87-328, 75 Stat. 688-716 (1961); *Klamath River Basin Compact*, 71 Stat. 497 (1957); *Agreement on the Cooperation for the Sustainable Development of the Mekong River Basin*, 34 ILM 864 (1995); *Susquehanna River Basin Compact*, Pub. L. 91-575, 84 Stat. 1509-1541 (1970).

C.6 ARTICLE 6C: DISPUTE RESOLUTION

Commentary: Disputes inevitably arise as an agreement is implemented. These disputes may involve differences in interpretation of the agreement's provisions or noncompliance with the agreement itself. The disputes may also arise because of changing conditions that alter the effectiveness of the agreement for one or more of the parties. Although a speedy and equitable process of dispute resolution serves all parties well, some sovereign entities do not wish to enter into an obligatory process. In such a case, Article 6C may be omitted. In other instances, the parties may recognize the need to institutionalize a dispute resolution process.

§6C-1-01 Good Faith Implementation
Each of the Parties pledges to support implementation of all provisions of this Agreement, and covenants that its officers and agencies will not hinder, impair, or prevent any other Party carrying out any provision of this Agreement.

Commentary: In the negotiations, each party seeks the rights and authorities critical to certain political, economic, or social objectives while ceding less critical rights and authorities to the other nations. While accepting this fact, under the *Convention on the Law of the Non-Navigational Uses of International Watercourses*, United Nations Document A/51/869 (United Nations 1997), all parties have a duty to cooperate and negotiate in good faith. This principle is the foundation of international law, and it applies in all relations between sovereign states. All states are expected to conduct themselves with an absence of malice and with no intention to seek unconscionable advantage or otherwise be deceitful.

It should be noted, however, that good faith misinterpretation of compact obligations does not excuse a party from damage liability (*Texas v. New Mexico*, 482 U.S. 124, 1987). In that case, the U.S. Supreme Court reasoned that a compact is a contract, and standard contract law does not allow a defense based on misinterpretation of contract obligations (Grant 2001, §45.07(c), §46.05(d)).

Cross References: §1C-1-01 (general policies); §1C-1-02 (purposes of agreement); §1C-1-03 (objectives of agreement); §1C-1-04 (coordination and cooperation); §1C-1-05 (reservation of federal rights); §1C-1-06 (national security); §2C-1-01 (activation of agreement); §2C-1-02 (consent to jurisdiction); §2C-1-03 (duration of agreement); §2C-1-04 (amendments and supplements); §2C-1-05 (powers of sovereign parties; withdrawal); §2C-1-06 (existing agencies); §2C-1-07 (limited applicability); §2C-1-08 (annexes); §2C-2-09 (party or parties); §3C-1-01 (commission created); §3C-1-02 (jurisdiction of the commission); §3C-1-03 (commissioners); §3C-1-04 (status, immunities and privileges); §3C-1-05 (commission organization and staffing); §3C-1-06 (rules of procedures); §3C-1-07 (commission administration); §3C-2-01 (general powers and duties); §3C-2-02 (powers and duties reserved to the commissioners); §3C-2-03 (obligations of the commission); §3C-2-04 (regulations; enforcement); §3C-2-05 (prohibited activities); §3C-2-06 (referral and review); §3C-2-07 (advisory committees); §3C-2-08 (reports); §3C-2-09 (condemnation proceedings); §3C-2-10 (meetings, hearings and records); §3C-2-11 (tort liability); §3C-3-01 (coordination and cooperation); §3C-3-02 (project costs and evaluation standards); §3C-3-03 (projects of the signatory parties); §3C-3-04 (cooperative services); §4C-1-01 (joint exercise of sovereignty); §4C-1-02 (interrelationship of water resources); §4C-1-03 (comprehensive water management plan); §4C-1-04 (purpose and objectives of comprehensive water management plan); §4C-1-05 (conditions of comprehensive water management plan); §4C-1-06 (deviation from comprehensive water management plan); §4C-1-07 (allocation during flood conditions); §4C-1-08 (allocation during drought conditions); §4C-1-09 (non-impairment of comprehensive water management plan by state action); §4C-2-01 (water allocation, generally); §4C-2-02 (waters not subject to allocation); §4C-2-03 (basin water sources); §4C-2-04 (basin water demands and needs); §4C-2-05 (allocation to equitable and reasonable uses); §4C-2-06 (watershed management); §4C-3-01 (existing rights recognized); §4C-3-02 (flood protection works); §4C-3-03 (minimum flows); §4C-3-04 (withdrawals and diversion; protected areas); §4C-3-05 (water levels protected); §4C-3-06 (augmentation of supply); §4C-3-07 (water quality); §4C-3-08 (underground water; limit on withdrawals); §4C-3-09 (atmospheric water); §4C-3-10 (recreation); Article 5C (financing); §6C-1-01 (good faith implementation).

Similar Agreements: *Convention on the Law of the Non-Navigational Uses of International Watercourses*, United Nations Document A/51/869 (United Nations 1997); "Charter of Economic Rights and Duties of States" (United Nations General Assembly 1975); *Helsinki Rules on the Uses of the Waters of International Rivers*, 52 ILA 484 (1966); *Stockholm Declaration of the United Nations Conference on the Human Environment*, 11 ILM 1416 (United Nations 1972).

§6C-1-02 Modification of Agreement

In the event that any Party is substantially hindered or prevented from performing any obligation or implementing any provision under this Agreement, by reasons of circumstances beyond the control of the Party (including but not limited to Acts of God, natural disasters, or labor disputes), the Parties agree to meet and negotiate an appropriate modification of the applicable provisions of the Agreement to reflect the effect of such force majeure. Such modifications may include extensions of applicable schedules and timetables, or agreements on substitute actions to fulfill the objectives and spirit of this Agreement.

Commentary: This provision provides a remedy for unintentional breaches of the agreement that may occur because of unforeseen situations or changed conditions.

Cross References: §1C-1-01 (general policies); §1C-1-02 (purposes of agreement); §1C-1-03 (objectives of agreement); §1C-1-04 (coordination and cooperation); §1C-1-05 (reservation of federal rights); §1C-1-06 (national security); §2C-1-01 (activation of agreement); §2C-1-02 (consent to jurisdiction); §2C-1-03 (duration of agreement); §2C-1-04 (amendments and supplements); §2C-1-05 (powers of sovereign parties; withdrawal); §2C-1-06 (existing agencies); §2C-1-07 (limited applicability); §2C-1-08 (annexes); §2C-2-09 (party or parties); §3C-1-01 (commission created); §3C-1-02 (jurisdiction of the commission); §3C-1-03 (commissioners); §3C-1-04 (status, immunities and privileges); §3C-1-05 (commission organization and staffing); §3C-1-06 (rules of procedures); §3C-1-07 (commission administration); §3C-2-01 (general powers and duties); §3C-2-02 (powers and duties reserved to the commissioners); §3C-2-03 (obligations of the commission); §3C-2-04 (regulations; enforcement); §3C-2-05 (prohibited activities); §3C-2-06 (referral and review); §3C-2-07 (advisory committees); §3C-2-08 (reports); §3C-2-09 (condemnation proceedings); §3C-2-10 (meetings, hearings and records); §3C-2-11 (tort liability); §3C-3-01 (coordination and cooperation); §3C-3-02 (project costs and evaluation standards); §3C-3-03 (projects of the signatory parties); §3C-3-04 (cooperative services); §4C-1-01 (joint exercise of sovereignty); §4C-1-02 (interrelationship of water resources); §4C-1-03 (comprehensive water management plan); §4C-1-04 (purpose and objectives of comprehensive water management plan); §4C-1-05 (conditions of comprehensive water management plan); §4C-1-06 (deviation from comprehensive water management plan); §4C-1-07 (allocation during flood conditions); §4C-1-08 (allocation during drought conditions); §4C-1-09 (non-impairment of comprehensive water management plan by state action); §4C-2-01 (water allocation, generally); §4C-2-02 (waters not subject to allocation); §4C-2-03 (basin water sources); §4C-2-04 (basin water demands and needs); §4C-2-05 (allocation to equitable and reasonable uses); §4C-2-06 (watershed management); §4C-3-01 (existing rights recognized); §4C-3-02 (flood protection works); §4C-3-03 (minimum flows); §4C-3-04 (withdrawals and diversion; protected areas); §4C-3-05 (water levels protected); §4C-3-06 (augmentation of supply); §4C-3-07 (water quality); §4C-3-08 (underground water; limit on withdrawals); §4C-3-09 (atmospheric water); §4C-3-10 (recreation); §5C (financing); §6C-1-01 (good faith implementation).

Similar Agreements: *Agreement on the Full Utilization of the Nile Waters*, UNTS Vol. 453, 51 (1959); *Vienna Convention for the Protection of the Ozone Layer*, 26 ILM 1529 (1985); *Delaware River Basin Compact*, Pub. L. 87-328, 75 Stat. 688-716 (1961); *Susquehanna River Basin Compact*, Pub. L. 91-575, 84 Stat. 1509-1541 (1970).

§6C-1-03 Material Breach

The Parties consider this Agreement to be complete and an integral whole. Each recommendation and provision of this Agreement is considered material to the entire Agreement, and failure to implement or adhere to any recommendation or provision may be considered a material breach.

Commentary: This provision is standard to most contractual agreements, indicating that all provisions of the agreement are interrelated and that breach of any provision by one party may void the agreement.

Cross References: §1C-1-01 (general policies); §1C-1-02 (purposes of agreement); §1C-1-03 (objectives of agreement); §1C-1-04 (coordination and cooperation); §1C-1-05 (reservation of federal rights); §1C-1-06 (national security); §2C-1-01 (activation of agreement); §2C-1-02 (consent to jurisdiction); §2C-1-03 (duration of agreement); §2C-1-04 (amendments and supplements); §2C-1-05 (powers of sovereign parties; withdrawal); §2C-1-06 (existing agencies); §2C-1-07 (limited applicability); §2C-1-08 (annexes); §2C-2-09 (party or parties); §3C-1-01 (commission created); §3C-1-02 (jurisdiction of the commission); §3C-1-03 (commissioners); §3C-1-04 (status, immunities and privileges); §3C-1-05 (commission organization and staffing); §3C-1-06 (rules of procedures); §3C-1-07 (commission administration); §3C-2-01 (general powers and duties); §3C-2-02 (powers and duties reserved to the commissioners); §3C-2-03 (obligations of the commission); §3C-2-04 (regulations; enforcement); §3C-2-05 (prohibited activities); §3C-2-06 (referral and review); §3C-2-07 (advisory committees); §3C-2-08 (reports); §3C-2-09 (condemnation proceedings); §3C-2-10 (meetings, hearings and records); §3C-2-11 (tort liability); §3C-3-01 (coordination and cooperation); §3C-3-02 (project costs and evaluation standards); §3C-3-03 (projects of the signatory parties); §3C-3-04 (cooperative services); §4C-1-01 (joint exercise of sovereignty); §4C-1-02 (interrelationship of water resources); §4C-1-03 (comprehensive water management plan); §4C-1-04 (purpose and objectives of comprehensive water management plan); §4C-1-05 (conditions of comprehensive water management plan); §4C-1-06 (deviation from comprehensive water management plan); §4C-1-07 (allocation during flood conditions); §4C-1-08 (allocation during drought conditions); §4C-1-09 (non-impairment of comprehensive water management plan by state action); §4C-2-01 (water allocation, generally); §4C-2-02 (waters not subject to allocation); §4C-2-03 (basin water sources); §4C-2-04 (basin water demands and needs); §4C-2-05 (allocation to equitable and reasonable uses); §4C-2-06 (watershed management); §4C-3-01 (existing rights recognized); §4C-3-02 (flood protection works); §4C-3-03 (minimum flows); §4C-3-04 (withdrawals and diversion; protected areas); §4C-3-05 (water levels protected); §4C-3-06 (augmentation of supply); §4C-3-07 (water quality); §4C-3-08 (underground water; limit on withdrawals); §4C-3-09 (atmospheric water); §4C-3-10 (recreation); §5C (financing); §6C-1-01 (good faith implementation).

Similar Agreements: *Delaware River Basin Compact*, Pub. L. 87-328, 75 Stat. 688-716 (1961); *Susquehanna River Basin Compact*, Pub. L. 91-575, 84 Stat. 1509-1541 (1970).

§6C-1-04 Negotiations and Consultations

(a) If any Party believes another Party has violated or failed to carry out any provision of this Agreement, it shall notify such Party, and all other Parties, in writing specifying the alleged violation or failure.

(b) The complaining Party shall notify the Commission of the dispute and the intention to enter into negotiations and consultations.

(c) Within (___) days of notice provided under paragraph (a), all Parties will meet to discuss the alleged violation or failure and to negotiate an appropriate settlement, including actions to correct such violation or failure. Such discussions and negotiations shall be pursued in good faith for not less than (___) days after original notice.

(d) The Parties shall seek to avoid any resolution that adversely affects the interests under this Agreement of any other Party.

Commentary: Negotiation is a process in which the conflicting parties engage in face-to-face discussions to develop a mutually satisfactory agreement on the issues or problems at hand. No outside, independent party or individual is involved. If negotiations between the parties themselves are not effective, the process evolves into mediation.

Cross References: §1C-1-01 (general policies); §1C-1-02 (purposes of agreement); §1C-1-03 (objectives of agreement); §1C-1-04 (coordination and cooperation); §2C-1-02 (consent to jurisdiction); §2C-1-03 (duration of agreement); §2C-1-04 (amendments and supplements); §2C-1-05 (powers of sovereign parties; withdrawal); §2C-2-09 (party or parties); §4C-1-01 (joint exercise of sovereignty).

Similar Agreements: *Convention on the Law of the Sea*, 33 ILM 1309 (1982); *Convention on the Law of the Non-Navigational Uses of International Watercourses*, United Nations Document A/51/869 (United Nations 1997); *Convention on the Protection and Use of Transboundary Watercourses and International Lakes*, 31 ILM 1312 (1992); *North American Free Trade Agreement*, 19 U.S.C. §§3311–3473 (1993); *Treaty of Peace between the State of Israel and the Hashemite Kingdom of Jordan*, 34 ILM 43 (1994); *Apalachicola–Chattahoochee–Flint River Basin Compact*, O.C.G.A. 12-10-100 (1997); *Alabama–Coosa–Tallapoosa River Basin Compact*, O.C.G.A. 12-10-110 (1997); *Delaware River Basin Compact*, Pub. L. 87-328, 75 Stat. 688-716 (1961); *Susquehanna River Basin Compact*, Pub. L. 91-575, 84 Stat. 1509-1541 (1970).

Alternative 1 (Right to Litigate)
§6C-1-05 Right to Litigate (Optional)

Nothing in this Agreement shall be construed to limit or prevent either Party from instituting or maintaining any action or proceeding, legal or equitable, in any tribunal of competent jurisdiction for the protection of any right under this Agreement or the enforcement of any of its provisions.

Commentary: If the parties choose not to adopt a provision that states that the decision of the arbitral panel shall be considered final and shall not be appealable to any court of law or equity (§6C-1-05 et seq), the agreement should provide for appropriate legal or equitable action or proceeding. The existence of an appropriate tribunal may pose a problem in cases not involving an entity like the United States or the European Union. It may be advisable to specify the tribunal in the agreement itself to avoid dispute over jurisdictional questions at a later date.

Cross References: §1C-1-01 (general policies); §1C-1-02 (purposes of agreement); §1C-1-03 (objectives of agreement); §1C-1-04 (coordination and cooperation); §2C-1-02 (consent to jurisdiction); §2C-1-03 (duration of agreement); §2C-1-04 (amendments and supplements); §2C-1-05 (powers of sovereign parties; withdrawal); §2C-2-09 (party or parties); §4C-1-01 (joint exercise of sovereignty).

Similar Agreements: *Belle Fourche River Compact*, 58 Stat. 94 (1944); *Colorado River Compact*, 45 Stat. 1057 (1928); *Snake River Compact*, 64 Stat. 29 (1949).

Alternative 2 (Alternative Dispute Resolution)
§6C-1-05 Alternative Dispute Resolution

(a) Desiring that this Agreement be carried out in full, the Parties agree that disputes between the Parties regarding interpretation, application, and implementation of this Agreement shall be settled by alternative dispute resolution and agree to forswear litigation.

(b) The dispute settlement provisions of this Article shall apply with respect to the avoidance or settlement of all disputes between the Parties regarding the interpretation or application of this Agreement or wherever a Party considers that an actual or proposed measure of another Party is or would be inconsistent with the obligations of this Agreement or cause nullification or impairment of the application of the Agreement.

Commentary: Disputes inevitably arise as an agreement is implemented and enforced. They may involve differences in interpretation of the agreement's provisions or noncompliance with the agreement itself. The disputes may also arise because of changing conditions that alter the effectiveness of the agreement for one or more of the parties. Therefore, the institutional provisions should provide for a process to resolve disputes quickly, effectively, and permanently. The mechanism should emphasize a streamlined process of dispute resolution that minimizes costly, time-consuming litigation. Alternate dispute resolution (ADR) is a major force in the resolution of disputes about the terms of an international agreement. This force is necessarily so because of the lack of effective supranational mechanisms for jurisprudence. Even where judicial remedies are available, however, ADR may be a preferable way to resolve disputes.

Judicial conflict resolution holds significant disadvantages. (a) Judges are generalists who are, in most cases, dependent on the testimony presented before them. (b) The judicial process, with its manifold procedural safeguards, is too slow for effective natural resource management. (c) Judicial decrees are retrospective, geographically limited, and quite fact-specific. Unlike administrative agencies, courts are incapable of issuing prospective, uniform regulations of general applicability. (d) Courts lack the ability to consistently monitor and evaluate solutions they have devised. (e) Water is a public trust resource that should be managed by institutions that are politically responsive to the public (Goldfarb 1993).

ADR processes are designed to resolve disputes as quickly as possible and at the lowest cost to the parties involved. The process consists of a successive series of techniques that become increasingly time-consuming and expensive. These techniques are negotiation, mediation, arbitration, and litigation. With each successive step, the parties spend more time and more money for a result over which the parties have less and less control.

The agreement to forswear litigation appearing in §6C-1-04 is optional. If this phrase is included and §6C-1-08 is not included, judicial intervention will not obtain (Grant 2001, §46.05, discussing the U.S. Supreme Court's statements in *Texas v. New Mexico*, 462 U.S. 554 (1983)).

Cross References: 1-1-01 (general policies); §1C-1-02 (purposes of agreement); §1C-1-03 (objectives of agreement); §1C-1-04 (coordination and cooperation); §2C-1-02 (consent to jurisdiction); §2C-1-03 (duration of agreement); §2C-1-04 (amendments and supplements); §2C-1-05 (powers of sovereign parties; withdrawal); §2C-2-09 (party or parties); §4C-1-01 (joint exercise of sovereignty).

Similar Agreements: *Agreement on the Cooperation for the Sustainable Development of the Mekong River Basin*, 34 ILM 864 (1995); *Convention on the Law of the Non-Navigational Uses of International Watercourses*, United Nations Document A/51/869 (United Nations 1997); *Convention on the Law of the Sea*, 33 ILM 1309 (1982); *Convention on the Protection and Use of Transboundary Watercourses and International Lakes*, 31 ILM 1312 (1992); *North American Free Trade Agreement*, 19 U.S.C. §§3311–3473 (1993); *Delaware River Basin Compact*, Pub. L. 87-328, 75 Stat. 688-716 (1961); *Susquehanna River Basin Compact*, Pub. L. 91-575, 84 Stat. 1509-1541 (1970).

§6C-1-06 Conciliation and Mediation

(a) **If the Parties are unable to reach agreement on a settlement, after good faith discussions and negotiations within the period provided in §6C-1-05(c), any aggrieved Party may request that the Commission institute measures of conciliation and mediation. The requesting Party shall state in the request the nature of the dispute and indicate the provisions of the Agreement that it considers relevant, and shall deliver the request to the other Parties, and to the Commission. Unless it decides otherwise, the Commission shall convene within (___) days of delivery of the request and shall endeavor to resolve the dispute promptly.**

(b) **The Commission may call on such technical advisers or create such working groups or expert groups as it deems necessary to act as mediators and assist in the resolution of the dispute and make recommendations as may assist the consulting Parties to reach a mutually satisfactory resolution of the dispute.**

Commentary: Mediation is the intervention of a "third party" in a dispute between two other parties in an attempt to reconcile their differences, usually upon their request. The qualifications of the mediator require his or her knowledge, experience, and background in the water resource issues themselves, as well as an understanding and background in the legal issues involved. The mediation moves through three stages:

(a) The mediator identifies and develops a factual discussion of the disputed and undisputed issues with the parties, both individually and collectively. The purpose is to ensure that all parties understand the strengths and weaknesses of their cases and the perceived weaknesses and strengths of the opposing parties.

(b) The mediator explores with the parties their goals and interests, attempting to create alternative solutions to their perceived concerns. This portion of the process also involves discussions with the parties individually and collectively.

(c) After the mediator intervenes, the parties themselves may then reassume a negotiating posture and possibly agree on a mutually acceptable alternative solution, or they may proceed to arbitration.

Cross References: §1C-1-01 (general policies); §1C-1-02 (purposes of agreement); §1C-1-03 (objectives of agreement); §1C-1-04 (coordination and cooperation); §2C-1-02 (consent to jurisdiction); §2C-1-03 (duration of agreement); §2C-1-04 (amendments and supplements); §2C-1-05 (powers of sovereign parties; withdrawal); §2C-2-09 (party or parties); §4C-1-01 (joint exercise of sovereignty).

Similar Agreements: *Agreement on the Cooperation for the Sustainable Development of the Mekong River Basin*, 34 ILM 864 (1995); *Convention on the Law of the Non-Navigational Uses of International Watercourses*, United Nations Document A/51/869 (United Nations 1997); *Convention on the Protection and Use of Transboundary Watercourses and International Lakes*, 31 ILM 1312 (1992); *Treaty of Peace between the State of Israel and the Hashemite Kingdom of Jordan*, 34 ILM 43 (1994); *Apalachicola–Chattahoochee–Flint River Basin Compact*, O.C.G.A. 12-10-100 (1997); *Alabama–Coosa–Tallapoosa River Basin Compact*, O.C.G.A. 12-10-110 (1997); *Delaware River Basin Compact*, Pub. L. 87-328, 75 Stat. 688-716 (1961); *Susquehanna River Basin Compact*, Pub. L. 91-575, 84 Stat. 1509-1541 (1970).

§6C-1-07 Arbitration

(a) **If the Commission has instituted mediation efforts as described in §6C-1-06 and the matter has not been resolved within 60 days thereafter any consulting Party may request in writing the establishment of an**

arbitral panel. The requesting Party shall deliver the request to the other Parties and the Commission.

(b) On delivery of the request, the Commission shall establish an arbitral panel.

(c) Unless otherwise agreed by the disputing Parties, the panel shall be established and perform its functions in a manner consistent with the provisions of Annex C-2.

(d) (Optional) The decision of the Arbitral Panel shall be considered final and shall not be appealable to any court of law or equity.

Commentary: Essentially arbitration is an informal trial, with the parties choosing the judge in a process that has a less formal evidentiary process. Arbitration differs from mediation only in the rules of decision as to the solution chosen to resolve the dispute. Whereas the mediator seeks to persuade the parties to agree on a mutually acceptable solution, the parties agree to allow the arbitrator to make decisions that are binding on the parties.

It should be recognized that implementation and enforcement of the agreement is as important as the actual development. Most dispute resolution is most likely accomplished through arbitral panels. Unless the parties ensure effective dispute resolution, the agreement may well become meaningless. If the parties decide to eliminate this detail, it is recommended that a side agreement be formed that describes the dispute resolution mechanism. The main agreement should make reference to the side agreement. In any event, the standards presented for the arbitral panel should ensure that the members of the panel are both technically qualified in the subject of water resources and law and independent of the parties involved in the dispute. It is preferable that the commissioners adopt panel selection criteria established by recognized national arbitral organizations. An optional framework for the arbitral panel is presented at Annex C-2.

A five-member panel is recommended, although that number is somewhat arbitrary. An odd number was chosen to ensure that a decision can be reached by majority vote. The use of a single panel member is discouraged because it is unlikely that all parties could agree on the selection. Also a decision by a single arbitrator is usually difficult to accept by all parties, especially those for whom the decision is negative. The number five has been chosen as a means of reducing a conscious or unconscious bias that may affect the decision.

A major consideration of the parties is whether the decision of the arbitral panel is final or whether to allow further conflict resolution by litigation. On the one hand, arbitration finality allows for faster and more equitable conflict resolution; if the framework of the panel is properly established, arbitration finality provides for conclusive resolution by persons acknowledged as experts in the field. However, the parties may consider that the use of arbitration finality lacks sufficient judicial competence for such judgments to be politically acceptable in their particular domestic circumstances. If the decision is made to include ADR with binding arbitration in the agreement, §6C-1-07(d) should be included and §6C-1-08 (right to litigate) should be omitted. If the parties choose to include ADR in the agreement but do not choose to include binding arbitration, §6C-1-07(d) should be omitted and §6C-1-08 (right to litigate) should be included.

Cross References: §1C-1-01 (general policies); §1C-1-02 (purposes of agreement); §1C-1-03 (objectives of agreement); §1C-1-04 (coordination and cooperation); §2C-1-02 (consent to jurisdiction); §2C-1-03 (duration of agreement); §2C-1-04 (amendments and supplements); §2C-1-05 (powers of sovereign parties; withdrawal); §2C-2-09 (party or parties); §4C-1-01 (joint exercise of sovereignty).

Similar Agreements: *Convention on the Law of the Sea*, 33 ILM 1309 (1982); *Convention on the Protection and Use of Transboundary Watercourses and International Lakes*, 31 ILM 1312 (1992); *North American Free Trade Agreement*, 19 U.S.C. §§3311–3473 (1993); *Treaty of Peace between the State of Israel and the Hashemite Kingdom of Jordan*, 34 ILM 43 (1994); *Delaware River Basin Compact*, Pub. L. 87-328, 75 Stat. 688-716 (1961); *Susquehanna River Basin Compact*, Pub. L. 91-575, 84 Stat. 1509-1541 (1970).

§6-1-08 Right to Litigate

Nothing in this Agreement shall be construed to limit or prevent either Party from instituting or maintaining any action or proceeding, legal or equitable, in any tribunal of competent jurisdiction for the protection of any right under this Agreement or the enforcement of any of its provisions.

Commentary: If the parties choose not to adopt a provision that states that the decision of the arbitral panel shall be considered final and may not be appealed to any court of law or equity (§6C-1-04), the agreement should provide for appropriate legal or equitable action or proceeding. The existence of an appropriate tribunal may pose a problem in cases not involving an entity like the United States or the European Union. It may be advisable to specify the tribunal in the agreement itself to avoid dispute over jurisdictional questions at a later date.

Cross References: §1C-1-01 (general policies); §1C-1-02 (purposes of agreement); §1C-1-03 (objectives of agreement); §1C-1-04 (coordination and cooperation); §2C-1-02 (consent to jurisdiction); §2C-1-03 (duration of agreement); §2C-1-04 (amendments and supplements); §2C-1-05 (powers of sovereign parties; withdrawal); §2C-2-09 (party or parties); §4C-1-01 (joint exercise of sovereignty).

Similar Agreements: *Belle Fourche River Compact*, 58 Stat. 94 (1944); *Colorado River Compact*, 45 Stat. 1057 (1928); *Snake River Compact*, 64 Stat. 29 (1949).

C.7 SIGNATURES

IN WITNESS WHEREOF, and in evidence of the adoption and enactment into law of this Agreement by the signatory Parties, the representatives of the sovereign States of ____________, ____________, ________ do hereby, in accordance with authority conferred by law, sign this Agreement in (six) duplicate original copies, as attested by the appropriate authorities of the respective sovereign Parties, and have caused the seals of the respective States to be hereunto affixed this ____ day of ______.

C.8 FINANCING (EXAMPLE)

Annex C-1: Financing (Example)

1. Annual Current Expense and Capital Budgets

(a) The Commission shall annually adopt a capital budget that includes any and all capital projects it proposes to undertake or continue during the budget period. This budget shall contain a statement of the estimated cost of each project and the method of financing thereof. Revenues accruing to the capital projects will be credited to the Commission's capital account, and will be used to defray annualized capital expenses.

(b) The Commission shall annually adopt a current expense budget for each fiscal year. Such budget shall include the Commission's estimated expenses for administration, operation, maintenance, and

repairs. It shall include a separate statement for each project, together with its cost allocation. The total of such expenses shall be balanced by the Commission's estimated revenues from all sources, including the cost allocations undertaken by any of the signatory Parties in connection with any project.

(c) The Parties agree to include the amounts so apportioned for the support of the current expense budget in their respective budgets next to be adopted, subject to such review and approval as may be required by their respective budgetary processes. Such amounts shall be due and payable to the Commission in quarterly installments during its fiscal year, provided that the Commission may draw upon its working capital to finance its current expense budget pending remittances by the signatory Parties.

2. Capital Financing by Signatory Parties; Guarantee

The Parties will provide such capital funds required for projects of the Commission as may be authorized by their respective statutes in accordance with a cost sharing plan prepared by the Commissioners; but nothing in this Article shall be deemed to impose any mandatory obligation on any of the Parties other than such obligations as may be assumed by a Party in connection with a specific project or facility.

3. Grants, Loans, or Payments by Parties

(a) The Commission may receive and accept, and the Parties may make, loans, grants, appropriations, advances, and payments of reimbursable or non-reimbursable funds or property in any form for the capital or operating purposes of the Commission.

(b) Any funds which may be loaned to the Commission either by a Party or a political subdivision thereof shall be repaid by the Commission through the issuance of bonds or out of other income of the Commission, such repayment to be made within such period and upon such terms as may be agreed upon between the Commission and the signatory Party or political subdivision making the loan.

4. Rates and Charges

The Commission may from time to time after public notice and hearing fix, alter, and revise rates, rentals, charges and tolls, and classifications thereof, for the use of facilities which it may own or operate and for products and services rendered thereby, without regulation or control by any department, office, or agency of any signatory Party.

5. Borrowing Power

(a) The Commission may borrow money for any of the purposes of this Agreement, and may issue its negotiable bonds and other evidences of indebtedness in respect thereto.

(b) All such bonds and evidences of indebtedness shall be payable solely out of the properties and revenues of the Commission without recourse to taxation. The bonds and other obligations of the Commission, except as may be otherwise provided in the indenture under which they were issued, shall be direct and general obligations of the Commission and the full faith and credit of the Commission are hereby pledged for the prompt payment of the debt service thereon and for the fulfillment of all other undertakings of the Commission assumed by it to or for the benefit of the holders thereof.

6. Credit Excluded

The Commission shall have no power to pledge the credit of any signatory Party or to impose any obligation for payment of the bonds upon any Party or sub-division thereof. Neither the Commission nor any person executing the bonds shall be liable personally on the bonds of the Commission or be subject to any personal liability or accountability by reason of the issuance thereof.

7. Bonds; Authorization

(a) Bonds and other indebtedness of the Commission shall be authorized by resolution of the Commissioners.

(b) The Commission may issue bonds in one or more series and may provide for one or more consolidated bond issues, in such principal amounts and with such terms and provisions as the Commission may deem necessary. The bonds may be secured by a pledge of all or any part of the property, revenues, and franchises under its control.

(c) Bonds may be issued by the Commission in such amount, with such maturities and in such denominations and form or forms, whether coupon or registered, as to both principal and interest, as may be determined by the Commission.

(d) The Commission may provide for redemption of bonds prior to maturity on such notice and at such time or times and with such redemption provisions, including premiums, as the Commission may determine.

8. Funding and Refunding

Whenever the Commission deems it necessary, it may fund and refund its bonds and other obligations, whether or not such bonds and obligations have matured.

9. Remedies for Default on Bonds (Optional)

(a) The holder of any bond may for the equal benefit and protection of all holders of bonds similarly situated require and compel the performance of any of the duties imposed upon the Commission or assumed by it, its officers, agents, or employees under the provisions of any indenture, in connection with the acquisition, construction, operation, maintenance, repair, reconstruction, or insurance of the facilities, or in connection with the collection, deposit, investment, application, and disbursement of the rates, rents, tolls, fees, charges, and other revenues derived from the operation and use of the facilities, or in connection with the deposit, investment, and disbursement of the proceeds received from the sale of bonds.

(b) In the alternative, the holder of any bond may, by action or suit in a court of competent jurisdiction of any signatory Party, require the Commission to account as if it were the trustee of an express trust, or enjoin any acts or things which may be unlawful or in violation of the rights of the holders of the bonds.

(c) The enumeration of such rights and remedies does not, however, exclude the exercise or prosecution of any other rights or remedies available to the holders of bonds.

10. Annual Independent Audit
   (a) A yearly audit shall be made of the financial accounts of the Commission. The audit shall be made by qualified certified public accountants, selected by the Commissioners, who have no personal interest direct or indirect in the financial affairs of the Commission or any of its officers or employees. The report of audit shall be prepared in accordance with accepted accounting practices and shall be filed with the chairman and such other Officers as the Commission shall direct. Copies of the report shall be made available for public distribution.
   (b) Each Party shall be entitled to examine and audit at any time all of the books, documents, records, files, and accounts and all other papers, things, or property of the Commission. The representatives of the Parties shall have access to all books, documents, records, accounts, reports, files, and all other papers, things, or property belonging to or in use by the Commission and necessary to facilitate the audit and they shall be afforded full facilities for verifying transactions with the balances or securities held by depositaries, fiscal agents, and custodians.

C.9 ARBITRAL PANEL (OPTIONAL)

Annex C-2: Arbitral Panel (Optional)

1. Roster and Qualifications of Panelists
   (a) The Commission shall establish and maintain a roster of up to 30 individuals who are willing and able to serve as panelists. The roster membership shall be approved by the Parties to this Agreement. The roster members shall be appointed by consensus for terms of three years, and may be reappointed.
   (b) Roster members shall have expertise or experience in law and in water resources engineering, environmental engineering, ecology or water resources economics, or other disciplines related to matters covered by this Agreement and shall be chosen strictly on the basis of objectivity, reliability, and sound judgment.
   (c) Panel members shall comply with a code of conduct to be established by the Commission.

2. Panel Selection
   (a) Where there are two disputing Parties, the following procedures shall apply: (i) The panel shall comprise five members. (ii) The disputing Parties shall endeavor to agree on the chair of the panel within 15 days of the delivery of the request for the establishment of the panel. If the disputing Parties are unable to agree on the chair within this period, the disputing Party chosen by lot shall select within five days as chair an individual who is not a citizen of that Party. (iii) Each disputing Party, within 15 days of selection of the chair, shall select two panelists who are citizens of the other disputing Party. (iv) If a disputing Party fails to select its panelists within such period, such panelists shall be selected by lot from among the roster members who are citizens of the other disputing Party.
   (b) Where there are more than two disputing Parties, the following procedures shall apply: (i) The panel shall comprise five members. (ii) The disputing Parties shall endeavor to agree on the chair of the panel within 15 days of the delivery of the request for the establishment of the panel. If the disputing Parties are unable to agree on the chair within this period, the Party or Parties on the side of the dispute chosen by lot shall select within 10 days a chair that is not a citizen of such Party or Parties. (iii) Within 15 days of selection of the chair, the Party complained against shall select two panelists, one of whom is a citizen of a complaining Party, and the other of whom is a citizen of another complaining Party. The complaining Parties shall select two panelists who are citizens of the Party complained against. (iv) If any disputing Party fails to select a panelist within such period, such panelist shall be selected by lot in accordance with the citizenship criteria of subparagraph (iii).
   (c) Panelists shall normally be selected from the roster. Any disputing Party may exercise a peremptory challenge against any individual not on the roster who is proposed as a panelist by a disputing Party within 15 days after the individual has been proposed.
   (d) If a disputing Party believes that a panelist is in violation of the Model Code of Conduct for Arbitration, the disputing Parties shall consult and if they agree, the panelist shall be removed and a new panelist shall be selected in accordance with this Article.

3. Model Rules of Arbitral Procedure
   (a) The Commission shall establish within one year of the entry into force of this Agreement Model Rules of Procedure, in accordance with the following principles: (i) the procedures shall ensure a right to at least one hearing before the panel as well as the opportunity to provide initial and rebuttal written submissions; and (ii) the panel's hearings, deliberations, and initial report, and all written submissions to and communications with the panel shall be open to the public.
   (b) Unless the disputing Parties otherwise agree, the panel shall conduct its proceedings in accordance with the Model Rules of Arbitral Procedure.
   (c) Unless the disputing Parties otherwise agree within 20 days from the date of the delivery of the request for the establishment of the panel, the terms of reference shall be:

"To examine, in the light of the relevant provisions of the Agreement, the matter referred to the Commission (as set out in the request for a Commission meeting) and to make findings, determinations, and recommendations."

4. Panel Decision; Initial Report
   (a) Unless the disputing Parties otherwise agree, the panel shall base its report on the submissions and arguments of the Parties and its independent analysis of the dispute.
   (b) Unless the disputing Parties otherwise agree the panel shall, within (_____) days after the last pan-

elist is selected, present to the disputing Parties its recommendations, if any, for resolution of the dispute.

(c) Panelists may furnish separate opinions regarding the initial report on matters not unanimously agreed.

(d) A disputing Party may submit written comments to the panel regarding the initial report within 14 days of presentation of the report.

(e) In such an event, and after considering such written comments, the panel, on its own initiative or on the request of any disputing Party, may: (i) request the views of any participating Party; (ii) reconsider its report; and (iii) make any further examination that it considers appropriate.

5. Panel Decision; Final Report

(a) The panel shall present to the disputing Parties a final report, including any separate opinions on matters not unanimously agreed, within 30 days of presentation of the initial report, unless the disputing Parties otherwise agree.

(b) No panel may, either in its initial report or its final report, disclose which panelists are associated with majority or minority opinions.

(c) The disputing Parties shall transmit to the Commission the final report of the panel, including the independent analysis of the panel, as well as any written views that a disputing Party desires to be appended, within a reasonable period of time after it is presented to them.

(d) Unless the Commission decides otherwise, the final report of the panel shall be published 15 days after it is transmitted to the Commission.

6. Implementation of Final Report

(a) On receipt of the final report of a panel, the disputing Parties shall agree on the resolution of the dispute, which normally shall conform to the determinations and recommendations of the panel, and shall notify the Commission of any agreed resolution of any dispute.

(b) There shall be no appeal from the determinations in the final report.

GLOSSARY

This glossary contains items that are used extensively in this Guideline and most of them came out of Draper (2006).

Allocation: The act of distributing water by allotting or apportioning a share of the resource among users according to a rational plan or program.

Aquaculture: The commercial raising of fish.

Aquatic environment: All underground and surface waters, the lands and subsurface geological formations connected to those waters, and the atmosphere related to those waters and lands.

Aquifer: A subsurface layer or layers of geological strata of sufficient porosity and permeability to allow either a flow of or the withdrawal of usable quantities of underground water.

Atmospheric water: All available moisture above the surface of the Earth, including underground and water surface and all forms of precipitation, but not including water projecting from irrigation systems.

Availability of water: The quality or state of source water being present or ready for immediate use.

Beneficial use: Water use that has value. This value can be derived economically, as in the use of water to produce some financial benefit, or the use may have been specifically acknowledged by the political process as having public value, as in wetland replenishment for ecological reasons.

Climatology: The study of the Earth's atmosphere over extended periods of time.

Comprehensive water management plan: A plan that considers the interrelationship of transboundary water resources, describes current and prospective water uses, identifies water supplies, and matches these supplies to water uses. It also identifies needed water-related management measures, facility needs, and costs; addresses environmental concerns; and offers program and policy recommendations to better manage the basin's water resources and water quality.

Conjunctive water management: The management of hydraulically connected underground and surface waters in a coordinated way such that the total benefits of integrated management exceed the sum of the benefits that would result from independent management of the surface and underground water components.

Conservation measures: Any measures adopted by a water right holder, or several water right holders acting in concert, pursuant to a conservation agreement reviewed and approved by the commission, as being appropriate water-saving strategies for purposes of the comprehensive water management plan, to reduce the withdrawals and/or consumptive uses, including, but not limited to the following:

a. Improvements in water transmission and water use efficiency;
b. Reduction in water use;
c. Enhancement of return flows; and
d. Reuse of return flows.

Consumptive use: That portion of the water withdrawn from the shared resource that is not available for use to meet water needs immediately downstream or in the immediate area downgradient, either because water is not discharged (by return flow) into a water source at or near the point of withdrawal or because the biological, chemical, or physical quality of the water is sufficiently impaired to prevent further use.

Depletion: Withdrawal of water from an aquifer at a rate faster than its recharge rate.

Drainage basin: An area determined by the geographic limits of a system of interconnected waters the surface waters of which normally share a common terminus.

Drought: Condition of abnormal water scarcity in a specific area, resulting from natural conditions.

Ecology: Scientific study of the interrelationships of plants, animals, and the environment.

Efficiency: 1. Economic efficiency refers to production of a good or service in the least costly manner and distributed or provided to those who value them most.

2. Water use efficiency refers to
   a. the quantity of water used for a specific beneficial purpose compared to the amount commonly used for that same purpose within a defined community;
   b. the quantity of water consumed proportional to the amount of water diverted, thus excluding return flows; or
   c. the quantity of water used for a specific beneficial purpose proportional to the amount of water allocated to that purpose according to a water right or withdrawal permit.

Environment: The surroundings of an individual organism or a community of organisms, ranging up to the entire biosphere, the zone of Earth that is able to sustain life.

Equitable: Fair; just; according to the principles of justice. An equitable settlement of a dispute is fair to both sides.

Equitable and reasonable utilization, consumption, or diversion: Utilization, consumption, or diversion of a transboundary water resource in an equitable and reasonable manner that takes into account all relevant factors and circumstances, including the following:

a. Geographic, hydrographic, hydrological, climatic, ecological, and other factors of a natural character;
b. The social and economic needs of the parties concerned;
c. The population dependent on the water resource in each of the parties;
d. The effects of the use or uses of the water resources by one party on other parties;
e. Existing and potential uses of the water resource;
f. Conservation, protection, development, and economy of use of the water resource and the costs of measures taken to that effect;
g. The availability of alternatives of comparable value to a particular planned or existing use; and
h. The potential or actual material injury or harm to other parties utilizing the shared water resource.

Extraction of atmospheric water: Removal of water from the atmosphere; precipitation induced by nonnatural causes.

Flood: Inundation of normally dry land resulting from the rising and overflowing of a body of water, a rising of water to levels that have detrimental effects on or in one or more basin states.

Geology: Study of the composition of Earth materials and the various geological processes to locate and exploit Earth's mineral resources.

Hydrology: The study of water moving over and through the land and its temporary storage on or within the ground.

Instream use: Use that does not withdraw water from the watercourse. Examples include navigation, generation of hydroelectric power, maintaining high water quality, provision for fish and wildlife habitat, and fulfillment of aesthetic goals.

Integrated: Coordinated.

Integrated water resource management: A water management process that integrates the assessment, management, protection, and reasonable utilization of all water resources within a basin to meet the needs of humans on a sustainable basis but still protects the integrity of the resource and its associated ecosystems.

Interbasin transfer: Transfer of bulk quantities of water from one water basin to another.

Interstate compacts: Agreements between or among sovereign political entities.

Legal duty: Legal obligation.

Meteorology: Study of the Earth's atmosphere and the variations in temperature and moisture patterns that produce different weather conditions.

Marine waters: Fully saline, brackish, or almost fresh water. Marine habitats include those below spring high tide limit (or below mean water level in non-tidal waters) and enclosed coastal saline or brackish waters.

Nonconsumptive use: Use that either does not withdraw water from the water source (see instream use) or that returns a portion of the water to its point of origin after use.

Obligation: Activity or conduct a person is bound to do or bound not to do; a moral or legal duty. Penalties may be imposed upon people who fail in their obligations.

Physiology: The branch of biology dealing with the functions of living organisms and their components.

Political entities: Sovereign government, or any political subdivision thereof.

Reasonable use: Rational use of water, which may have an effect on the quantity or quality of a watercourse but does not cause unreasonable harm to another's reasonable use.

Renewable water: Water continuously replenished in the same quantity and quality within reasonable time spans by the hydrological cycle, such as that in streams, reservoirs, or aquifers that refill from precipitation, runoff, or underground water recharge.

Riparian: Geographically adjacent to a watercourse.

Runoff: Water from precipitation that moves across the surface; the difference between the precipitation rate and evapotranspiration and infiltration rates.

Safe yield: The condition in an aquifer in which the withdrawal rate equals the rate of water replenishment by natural means.

Source water: Water resources in their natural state, in underground and surface waters.

Sovereign: An independent entity that governs itself.

Spatial: Having to do with geographical space.

State: Depending upon the context, a state may be a national government or may be a sovereign political subdivision of a nation organized on a federal system.

Surface water: The water from all sources that occurs on the Earth's surface either as diffused water or as water in natural channels, artificial channels, or other surface water bodies.

Sustainable: The integrated management of resources to ensure efficient use of and equitable access to waters for the benefit of current and future generations while preserving renewable resources and maintaining nonrenewable resources to the maximum extent reasonably possible.

Temporal: Having to do with time, such as the time of day, month, or season.

Transboundary: Moving across or along political boundaries separating two or more sovereign governments.

Underground water: Water beneath the surface of the ground located in a saturated zone and in direct contact with the ground or soil.

Waste assimilation: Dilution of effluents from sewage by water.

Water demands and uses: Activities that use water to fulfill their purposes.

Waters of the basin: All water found within the basin, whether surface, underground, or atmospheric water, excluding marine waters.

REFERENCES

Abdalla, C., Drohan, J. R., Becker, J. C., and Saacke Blink, K. (2007). *A primer on access and allocation of water in Pennsylvania*, Penn State Cooperative Extension, College of Agricultural Sciences, University Park, PA.

Accord for the prohibition of bulk water removal from drainage basins. (2004). Canadian Intergovernmental Conference Secretariat, Ottawa, Ontario.

Act regarding navigation and economic co-operation between the states of the Niger Basin. (1963). 587 UNTS 9.

Agreement between the government of the United States of America and the government of the United Mexican States concerning the establishment of a border environment cooperation commission and a North American development bank. (1993). 19 U.S.C. §473.

Agreement between the People's Republic of Bulgaria and the Republic of Turkey concerning co-operation in the use of the waters of rivers flowing through the territory of both countries. (1968). UNTS, Vol. 807, 117.

Agreement between the United States and Canada on Great Lakes water quality. (1978). 1153 UNTS 187.

Agreement on the cooperation for the sustainable development of the Mekong River Basin. (1995). 34 ILM 864.

Agreement on the full utilization of the Nile Waters. (1959). UAR–Sudan, UNTS Vol. 453, 51.

Agreement on the Protection of the (River) Meuse. (1995). 34 ILM 854.

Ahlander, A. S. (1994). "Environmental policies in the former Soviet Union." *Economic Policies for Sustainable Development,* Sterner, T., ed., Kluwer Academic Publishers, Dordrecht, Netherlands, 68–81.

Alabama–Coosa–Tallapoosa River Basin Compact. (1997). O.C.G.A. 12-10-110.

Albert, J. (2000). "Rethinking the management of transboundary freshwater resources: A critical examination of modern international law and practice." *National Resources Forum,* 24(1).

Alley, W. M., Reilly, T. E., and Franke, O. L. (1999). *Sustainability of ground-water resources,* Circular 1186, U.S. Geological Survey, Denver.

Anderson, D. L. (2002). *Utah's perspective: The Colorado River,* Utah Division of Water Resources, Salt Lake City, UT, <www.water.utah.gov/interstate/ thecoloradoriverart.pdf>

Anderson, O. L. (2001). "Reallocation." 2 Water and Water Rights, §16.02(c)(2), Lexis Law Publishing, Charlottesville, VA.

Apalachicola–Chattahoochee–Flint River Basin Compact. (1997). O.C.G.A. 12-10-100.

Arizona v. California. (1963). 373 U.S. 546.

Arkansas River Basin Compact of 1965. (1966). 80 Stat. 1409.

Arkansas River Basin Compact of 1970. (1973). 87 Stat. 569.

Arkansas River Compact. (1949). 63 Stat. 145.

ASCE. (2001). "Standard guidelines for artificial recharge of ground water." *EWRI/ASCE 34-01*, Reston, VA.

ASCE. (2004a). "Regulated riparian model water code." *EWRI/ASCE 40-03*, Reston, VA.

ASCE. (2004b). "Standard practice for the design and operation of precipitation enhancement projects." *EWRI/ASCE 42-04*, Reston, VA.

ASCE. (2006). *Guidelines for cloud seeding to augment precipitation*, Manual of Practice 81, 2d Ed., Reston, VA.

ASCE. (2009a). "Alternative dispute resolution." *ASCE Policy Statement 258,* Reston, VA, <http://www.asce.org/Public-Policies-and-Priorities/Public-Policy-Statements/Policy-Statement-258---Alternative-Dispute-Resolution/> (Apr. 25, 2012).

ASCE. (2009b). "Atmospheric water resources management." *ASCE Policy Statement 275*, Reston, VA, <http://www.asce.org/Public-Policies-and-Priorities/Public-Policy-Statements/Policy-Statement-275---Atmospheric-Water-Resources-Management/> (Apr. 25, 2012).

ASCE. (2009c). "Comprehensive transboundary water quality management agreement." *ASCE/EWRI 33-09*, Reston, VA.

ASCE. (2010a). "Impact of climate change." *ASCE Policy Statement 360*, Reston, VA. <http://www.asce.org/Public-Policies-and-Priorities/Public-Policy-Statements/Policy-Statement-360---Impact-of-Climate-Change/> (Apr. 25, 2012).

ASCE. (2010b). "The role of the civil engineer in sustainable development." *ASCE Policy Statement 418*, Reston, VA. <http://www.asce.org/Public-Policies-and-Priorities/Public-Policy-Statements/Policy-Statement-418---The-Role-of-the-Civil-Engineer-in-Sustainable-Development/> (Apr. 25, 2012).

Asheesh, M. (2000). "Management approach to Middle Eastern water resources and services." *18th International Peace Research Association Conf.*, International Peace Research Association, Tampere, Finland. < http://www.copri.dk/ipra/papers/html > (Jun. 4, 2012).

Association of Southeast Asian Nations (ASEAN). (1985). *Agreement on the Conservation of Nature and Natural Resources*, Kuala Lumpur.

Bandaragoda, D. J. (2000). "A framework for institutional analysis for water resources management in a river basin context." *Working Paper 5*, International Water Management Institute, Sri Lanka.

Barlow, M., and Clarke, T. (2002). *Blue Gold*, New Press, New York.

Barlow, P. M., Ahlfeld, D. P., and Dickerman, D. C. (2003). "Conjunctive-management models for sustained yield of stream-aquifer systems." *J. Water Resour. Plann. Manage.*, 129(1), 35–48.

Basagaolu, H., Mariño, M. A., and Shumway, R. H. (1999). "δ-Form approximating problem for a conjunctive water resource management model." *Advances in Water Resources*, 23(1), 69–81.

Basin Advisory Council. (2002). *June Meetings, Wastewater Briefing*, Metropolitan North Georgia Water Planning District, Atlanta.

Bear River Compact. (1955). 72 Stat. 38. Amended (1980). 94 Stat. 4, Art. XIII(2).

Beck, R. E. (2001). "Riparian rights." *Waters and Water Rights*, Vol. 1, R. E. Beck, ed., Lexis Law Publishing, Charlottesville, VA.

Belle Fourche River Compact. (1944). 58 Stat. 94.

Bennett, J., and Lawrence, P. (2002). "Adaptive management framework for catchment and coastal management and decision making." *Proc., Coast to Coast 2002: Source to Sea, National Coastal Conf.*, Tweed Heads, New South Wales, Australia.

Berlin Recommendations for the International Roundtable on Transboundary Water Management. (1998). Berlin Roundtable: A Consultation on the World Bank Strategy, Berlin, June.

Berman, I., and Wihbey, P. M. (1999). "The new water politics of the Middle East." *Strategic Review*, Summer.

Bernauer, T. (2001). *Explaining success and failure in international river management*, Center for International Studies, Swiss Federal Institute of Technology (ETH), Zurich.

Bomar, G. W. (2006). "Legal aspects of weather modification operations." *Guidelines for Cloud Seeding to Augment Precipitation*, Manual of Practice 81, 2d Ed., Keyes et al., eds., ASCE, Reston, VA, 44–60.

Bonneville Power Administration, U.S. Bureau of Reclamation, and U.S. Army Corps of Engineers. (1999). *Multi-species biological assessment of the Federal Columbia River Power System*, U.S. Bureau of Reclamation, Washington, DC, <www.usbr.gov/pn/programs/fcrps/pdf/1999ba.pdf> (Jun. 4, 2012).

Boulder Canyon Project Act. (1928). Ch. 42, 45 Stat. 1057.

Brundtland, G. H. (1987). "Chairman's foreword." *Our Common Future: Rep. of the World Commission on Environment and Development, A/42/427*, United Nations, Oslo, <http://www.un-documents.net/ocf-cf.htm> (Apr. 28, 2012).

Bush, P. M. (2004). "Transboundary water allocation in the Delaware River Basin—A history of conflict management and growing challenges." *Eastern Water Resources: Law, Policy and Technology,*

American Bar Association, Section of Environment, Energy and Resources.
Byrne, J. P. (1995). "Ten arguments for the abolition of the Regulatory Takings Doctrine." *Ecology Law Quarterly*, 22, 89.
Canada–Mexico–United States: North American Agreement on Environmental Cooperation. (1993). 32 ILM 1480.
Canadian River Compact. (1952). 66 Stat. 74.
Caponera, D. A. (1995). "Shared waters and international law." *The peaceful management of transboundary resources*, Gerald H. Blake et al., eds., London: Graham & Trotman/Martinus Nijhoff, 121–126.
Carter, R. H., and Morehouse, B. J. (2001). "An examination of Arizona water law and policy from the perspective of climate impacts." *CLIMAS Rep. CL2-01*, Institute for the Study of Planet Earth Climate Assessment for the Southwest, Univ. of Arizona, Tucson, AZ.
Cavallo, B., Ligon, F., McBain, S., Stevens, C., Williams, J., and Birk, S. (2002). "Topic area report: Flow manipulation." *CALFED Adaptive Management Workshop*, CALFED, Sacramento, CA.
Cherry, R. N., and Badr, A. W. (1998). *South Carolina Water Plan: Goals and Guidelines for Management of the State's Water Resources*, Dept. of Natural Resources, Columbia, SC.
Christensen, Jon. (1996). "River in Nevada helps its own restoration." *New York Times*, Sept. 24.
Clarke, R. (1993). *Water: The International Crisis*, The MIT Press, Boston.
Clean Water Act. (1987). 33 U.S.C. §§1362 et seq.
Coleman, W. T. (1998). "Legal barriers to the restoration of aquatic systems and the utilization of adaptive management." *Vermont Law Review*, 23, 177.
Colorado River Basin Project Act. (1956). Pub. L. No. 485, <www.usbr.gov/lc/region/pao/pdfiles/crspuc.pdf> (Jun. 4, 2012).
Colorado River Basin Project Act. (1968). Pub. L. No. 90-537, <www.usbr.gov/lc/region/g1000/pdfiles/crbproj.pdf> (Jun. 4, 2012).
Colorado River Compact. (1922). 43 U.S.C. § 6171, <www.usbr.gov/lc/region/g1000/pdfiles/crcompct.pdf> (Jun. 4, 2012).
Colorado River Compact. (1928). 45 Stat. 1057, <www.usbr.gov/lc/region/g1000/pdfiles/bcpact.pdf> (Jun. 4, 2012).
Colorado Weather Modification Permit Program. (2002). Colorado Dept. of Natural Resources, Denver, CO, <cwcb.state.co.us/WeatherModification/weather>
Convention and Statute on the Regime of Navigable Waterways of International Concern. (1921). 7 LNTS 35.
Convention and Statutes Relating to the Development of the Chad Basin. (1974). Reprinted in *Journal Officiel de la Republique Federale du Cameroon* (Yaounde), 4th year, No. 18 (Sept. 15).
Convention between the Federal Republic of Germany and the Czech and Slovak Federal Republic and the European Economic Community on the International Commission for the Protection of the Elbe. (1990). International Environmental Law, Multilateral Agreements, 976:90/1.
Convention between Switzerland and Italy Concerning the Protection of Italo-Swiss Waters Against Pollution. (1972). UNTS, Vol. 957, 277.
Convention on Access to Information, Public Participation in Decision-making and Access to Justice in Environmental Matters. ECE/CEP/43, 1998. United Nations Economic Commission for Europe. New York and Geneva. (1998).
Convention on Biological Diversity. (1992). 1760 UNTS 79; 31 ILM 818.
Convention on the Law of the Sea. (1982). 33 ILM 1309.
Convention on Long-Range Transboundary Air Pollution. (1979). 18 ILM 1442.
Convention on the Protection and Use of Transboundary Watercourses and International Lakes. (1992). 31 ILM 1312, ECE/UNEP Network of Experts on Public Participation and Compliance. Secretariat of the United Nations Economic Commission for Europe and the Regional Office for Europe of the United Nations Environment Programme, Geneva.
Convention on the Protection of the Rhine. (1998). International Law Project, Rotterdam, January 22. <www.internationalwaterlaw.org/documents/regionaldocs/rhine_river> (Jun. 4, 2012).
Convention on Wetlands of International Importance Especially as Waterfowl Habitat. (1971). UN Treaty Series 14583, amended by the Paris Protocol (1982), Regina Amendments (1987).
Cooperative agreement for Platte River research and other efforts relating to endangered species habitats along the central Platte River, Nebraska. (1997). U.S. Dept. of Interior, Washington, DC, <www.platteriverprogram.org/PubsAndData/ProgramLibrary/> (Jun. 4, 2012).
Costilla Creek Compact. (1946). 60 Stat. 246; Amended (1963). 77 Stat. 350.
Council of European Communities Second Action Program on the Environment (1977). Resolution of the Council of the European Communities and of the Representatives of the Governments of the Member States meeting within the Council, of 17 May 1977, on the continuation and implementation of a European Community policy and action programme on the environment, O.J. C139/1.
Council on European Communities Third Action Programme on the Environment. (1983). Resolution of the Council of the European Communities and of the Representativesof the Governments of the Member States meeting within the Council of 7 February 1983 on the continuation of a European Community policy and action programme on the environment (1982 to 1986), OJ. C4611.
Cummings, R. G., Keeler, A., and Thompson, B. (2002). "Georgia water—"A public resource or a commodity": What are the real policy questions?" *Water Policy Working Paper 2002-0081*, Water Resources Policy Program, School of Policy Studies, Georgia State Univ., Atlanta, <www.cviog.uga.edu/water/>
Cummings, R. G., Norton, N. A., and Norton, V. J. (2001a). "Water rights transfers: Options for institutional reform." *Water Policy Working Paper 2001-001*, Water Resources Policy Program, School of Policy Studies, Georgia State Univ. and Albany State Univ., <www2.gsu.edu/~wwwenv/waterPDF/W2001001.pdf> (Apr. 28, 2012).
Cummings, R. G., Norton, N. A., Norton, V. J., and Eigenberg, D. A. (2001b). "Changing rules for agricultural use: Policy options related to metering and forfeiture for non-use." *Water Policy Working Paper 2001-003*, Water Resources Policy Program, School of Policy Studies, Georgia State Univ. and Albany State Univ. Flint River Planning and Policy Center, <www2.gsu.edu/~wwwenv/waterPDF/W2001003.pdf>
Cummings, R. G., Terrebonne, P., and Valdez, G. (1996). *Management principles for ground water with salt-water intrusion: An analysis of alternative policies for Georgia's Upper Floridan Aquifer*, Water Resources Policy Program, School of Policy Studies, Georgia State Univ., Atlanta.
Dasgupta, P., Kristrom, B., and Maler, K.-G. (1995). "Current issues in resource accounting." *Current issues in environmental economics*, P.-O. Johansson, B. Kristrom, and K.-G. Maler, eds., Manchester University Press, Manchester, 117–152.
Davis, R. J. (1987). "Four decades of American weather modification law." *J. Weather Modification*, 19, 102.
de Villiers, M. (2001). *Water: The fate of our most precious resource responses*, Houghton Mifflin, Boston.
Declaration of Principles on Interim Self-Government Arrangements. (1993). 32 ILM 1525.
Delaware River Basin Commission (DRBC). (2003). "Vision statement: Charting the future." *Who we are*, <www.state.nj.us/drbc/vision.htm>
Delaware River Basin Compact. (1961). Pub. L. No. 87-328, 75 Stat. 688-716.
Dellapenna, J. W. (1994). "Treaties as instruments for managing internationally-shared water resources: Restricted sovereignty vs. community of property." *Case Western Reserve Journal of International Law*, 26(1), 27–56.
Dellapenna, J. W. (1997a). "The Nile as a legal and political structure." *The scarcity of water: Emerging legal and policy responses*, E. Brans, E. de Haan, A. Nollkaemper, and J. Rinzema, eds., Kluwer, London.
Dellapenna, J. W., ed. (1997b). "The regulated riparian model water code." *Final rep. of the Water Laws Committee of the Water Resources Planning and Management Division of ASCE*, ASCE, New York.
Dellapenna, J. W. (2000). "The importance of getting names right: The myth of markets for water." *William and Mary Environmental Law and Policy Review*, 25(2), 4, 317. <http://scholarship.law.wm.edu/cgi/viewcontent.cgi?article=1222&context=wmelpr&sei> (Apr. 28, 2012).

Dellapenna, J. W. (2001a). "The customary international law of transboundary fresh waters." *International Journal of Global Environmental Issues,* 1(3/4), 264–395.
Dellapenna, J. W. (2001b). "Riparianism." *Water and Water Rights,* Ch. 6–9, Vol. 1, R. E. Beck, ed., Lexis Law Publishing, Charlottesville, VA.
Dellapenna, J. W. (2002a). "Southeastern water wars." *Water Resources Committee Newsletter*, 6(1).
Dellapenna, J. W. (2002b). "Water issues white paper: Exercising riparian rights and non-riparian land." Prepared for the Joint Comprehensive State Water Plan Study Committee, Atlanta, <www.cviog.uga.edu/water/ whitepaper>
Dellapenna, J. W. (2007a). *Appropriative rights model water code,* ASCE, Reston, VA.
Dellapenna, J. W. (2007b). "Transboundary water sharing and the need for public management." *J. Water Resour. Plann. Manage.,* 133(5), 397–404.
Dellapenna, J. W., and Draper, S. E. (2002a). "Straight talk about markets for water." Prepared for the Comprehensive State Water Plan Joint Study Committee, Atlanta, <www.cviog.uga.edu/water/ whitepaper>
Dellapenna, J. W., and Draper, S. E. (2002b). "Water issues white paper: Property in water in Georgia." Prepared for the Comprehensive State Water Plan Joint Study Committee, Atlanta, <www.cviog.uga.edu/ water/whitepaper>
Dellapenna, J. W., and Draper, S. E. (2002c). "Water issues white paper: Water law doctrines as applied to Georgia water law." Prepared for the Joint Comprehensive State Water Plan Study Committee, Atlanta, <www.cviog.uga.edu/water/whitepaper>
Derthick, M. (1974). *Between state and nation: Regional organizations of the United States,* Brookings Institution, Washington, DC.
Donkers, H. (1997). "Fresh water as a source of international conflicts: The water conflicts between Israel, Jordan and the Palestinians." *The scarcity of water: Emerging legal and policy responses*, E. Brans, E. de Haan, A. Nollkaemper, and J. Rinzema, eds., Kluwer, London.
Draper, S. E. (1997). "International duties and obligations for transboundary water sharing." *J. Water Resour. Plann. Manage.*, 123(6), 344–349.
Draper, S. E. (2001). "Effective Georgia water policy for the 21st century." *Proc., World Water and Environmental Resources Congress*, Environmental and Water Resources Institute, ASCE, Reston, VA.
Draper, S. E. (2002a). "Future water reform legislation in Georgia: Major water rights issues facing Georgia." *Water Law Seminar*, Institute of Continuing Education, Univ. of Georgia, Athens, GA.
Draper, S. E., ed. (2002b). "Model Water Sharing Agreements for the 21st Century." *Final rep. of the Water Laws Committee of the Water Resources Planning and Management Division of the ASCE*, ASCE, Reston, VA.
Draper, S. E. (2002c). "Water issues white paper: Implications for water rights transfers." Prepared for the Joint Comprehensive State Water Plan Study Committee, Atlanta, <www.cviog.uga.edu/water/ whitepaper/>
Draper, S. E. (2002d). "Water issues white paper: Major water rights issues facing Georgia." Prepared for the Joint Comprehensive State Water Plan Study Committee, Atlanta, <www.cviog.uga.edu/water/ whitepaper>
Draper, S. E. (2002e). "Water issues white paper: Modification of permits based on consumptive use." Prepared for the Joint Comprehensive State Water Plan Study Committee, Atlanta, <www.cviog.uga.edu/water/whitepaper>
Draper, S. E. (2003). *Estimated U.S. Population Served by Shared Waters* (unpublished).
Draper, S. E. (2004). "Sharing water through interbasin transfer and basin of origin protection in Georgia." *Georgia State University Law Review*, 20(2), 339–372.
Draper, S. E., ed. (2006). *Sharing water in times of scarcity: Guidelines and procedures in the development of effective agreements to share water across political boundaries,* ASCE, Reston, VA.
Draper, S. E. (2007a). "Administration and institutional provisions of water sharing agreements." *J. Water Resour. Plann. Manage.*, 133(5), 446–455.
Draper, S. E. (2007b). "Introduction to transboundary water sharing." *J. Water Resour. Plann. Manage.*, 133(5), 377–381.
Draper, S. E., and Kundell, J. E. (2007). "Impact of climate change on transboundary water sharing." *J. Water Resour. Plann. Manage.*, 133(5), 405–415.
DuMars, Charles T. (1990). "Consequences of Rio Grande Compact Noncompliance." *WRRI Conf. Proc.,* <wrri.nmsu.edu/publish/ watcon/proc/proc44/ dumars.pdf> (Jun. 4, 2012).
Dwyer, B. N. (2001). "The evolving role of hydrologic computer modeling for adaptive management programs." *4th DHI Software Conf. and DHI Software Courses*, Helsingør, Denmark, <www.dhisoftware.com/uc2001/Abstracts_Proceedings/abstracts01/117.doc>
Eaux partagées. (Shared Water—Working Group of the French Académie de l'Eau). (2002). "Proposals for a Strategic Guide to Assist in the Constitution of International Inter-State Commissions for Shared Water Resources (Draft)." International Network of Basin Organizations, Constitutive Meeting of the Network of Transboundary Basin Organizations, Thonon-Le-Bains, France, <www.inbonews.org/ divers/thonon/ AcademieRiobGuide_gb.PDF>
Eckstein, G. E. (1998). "Hydrologic reality: International water law and transboundary ground-water resources." *Conf. on Water: Dispute Prevention and Development*, American University Center for the Global South, Washington, DC, <www.internationalwaterlaw.org/ Articles/GlobalSouth>.
Eheart, J. W., ed. (2002). "Riparian water regulations: Guidelines for withdrawal limitations and permit trading." *Final rep. of the Water Laws Committee of the Water Resources Planning and Management Division of ASCE*, ASCE, Reston VA.
Eisel, L., and Aiken, J. D. (1997). "Platte River basin study." *Rep. to the Western Water Policy Review Advisory Commission*, McLaughlin Water Engineers, Ltd., Denver.
El-Naser, H. K. (2001) "Beantwortung der Fragen von: Fragebogen der Enquete-Kommission des Deutschen Bundestages," *Globalisierung der Weltwirtschaft – Herausforderungen und Antworten*, zum Thema "Wasser – Ein weltweit immer knapper werdendes Gut," <www .bundestag.de/gremien/welt/ weltto/weltto122_stell> (Jun. 4, 2012).
Endangered Species Act. (1982). 7 U.S.C. § 136, 16 U.S.C. § 1531 et seq.
Espland, W. N. (1998). *The struggle for water: Politics, rationality, and identity in the American southwest*, University of Chicago Press, Chicago.
Evans, J. W., and England, R. H. (1995). *Recommended method to protect instream flows in Georgia*, Georgia Department of Natural Resources Wildlife Resources Division, Atlanta.
Falkenmark, M., and Lindh, G. (1974). "How can we cope with the water resources situation by the year 2050?" *Ambio* 3(3–4), 114–122.
FAO/UNDP/World Bank. (1995). "Water sector policy review and strategy formulation: A general framework." *FAO Land and Water Bulletin*, 3.
Fisher, R., and Ury, W. (1983). *Getting to yes: Negotiating agreement without giving in*, Penguin Books, New York.
Fitch, E. (2003). "Introduction: Social foundations of water management." *Water Resources IMPAC*, 5(6), 3–4, <www.awra.org/impact/ issues/ 0311imp_toc.pdf>
Fort, D. D. (1998). "Water in the west: Challenge for the next century." *Rep. of the Western Water Policy Review Advisory Commission*, Univ. of New Mexico, Albuquerque, <repository.unm.edu/dspace/ bitstream/1928/2788/13/Introductory+Pages.pdf>
Frederick, K. D. (1993). "The economics of risk in water resources planning." *Water resources administration in the United States*, M. Reuss, ed., American Water Resources Association and Michigan State University Press, East Lansing, MI.
Frederick, K. D., and Schwarz, G. E. (1999). "Socioeconomic impacts of climate change on U.S. water supplies." *Journal of the American Water Resources Association*, 35(6), 1563–1583.
Freeman, R. M., and Lesesne, E. H. (1980). "TVA revisited: Implications for long range water resources management in the Tennessee River Valley." *Unified River Basin Management*, 67.
French, D. B. (2000). "The U.S.–Mexican Border Environment: Mexican Policies and NAFTA." <www.angelfire.com/az3/ bluewindsong/mexsem.htm> (Apr. 29, 2012).

Freyfogle, E. T. (1999). "Eight principles for property rights in the anti-sprawl age." *William and Mary Environmental Law and Policy Review*, 23, 777–799.

Gelt, J. (1997). "Sharing Colorado River water: History, public policy and the Colorado River Compact." *Arroyo*, 10(1), <ag.arizona.edu/AZWATER/arroyo/101comm>

Georgeyewsky, V. Y. (1998). "On global climate warming on water resources." *Water: A looming crisis?* H. Zebidi, ed., UNESCO IHP-V Technical Documents in Hydrology, No. 18, 37–46.

Georgia Chamber of Commerce. (2002). *The role of water rights and Georgia law in comprehensive water planning for Georgia*, Prepared for the Joint Comprehensive State Water Plan Study Committee, Atlanta, GA, <www.cviog.uga.edu/water/whitepaper>

Georgia Joint Comprehensive Water Plan Study Committee. (2002). "A Report to the Governor and General Assembly." *Final Rep. of the Joint Comprehensive Water Plan Study*, <www.cviog.uga.edu/water/whitepaper> (Jun. 4, 2012).

Georgia Water Coalition. (2002). "A report of the Georgia Water Coalition." Atlanta, <www.cviog.uga.edu/water/whitepaper>

German Foundation for International Development. (1998). "The Nile: Sharing experiences, sharing visions." *International Round Table, Transboundary Water Management Experience of International River and Lake Commissions, Berlin Recommendations* <www.dse.de/ef/nile/>

Gleick, P. H. (1993). "Water in the 21st century." *Water in crisis: A guide to the world's fresh water resources*, P. H. Gleick, ed., Oxford University Press, New York.

Gleick, P. H. (1996). "Basic water requirements for human activities: Meeting basic needs." *Water International*, 21, 83–92 <www.pacinst.org/basic_water_needs.pdf>

Gleick, P. H. (1998). "Water in crisis: Paths to sustainable water use." *Ecological Applications*, 8(3), 571–579.

Gleick, P. H. (2000a). "The changing water paradigm: A look at twenty-first century water resources development." *Water International*, 25(1), 127–138.

Gleick, P. H. (2000b). *The world's water, 2000–2001, the biennial report on freshwater resources*, Island Press, Chicago.

Gleick, P. H. (2001). "Global water: Threats and challenges facing the United States." *Environment*, March, 18–26.

Glennon, R. (2002). *Water follies*, Island Press, Washington, DC.

Glicken, J. (2000). "Getting stakeholders participation 'right': A discussion of participatory processes and possible pitfalls." *Environmental Science and Policy*, 3(6), 305–310.

Gochis, D. J. (2001). "Emergent precipitation enhancement and the rights to developed water." *Globalization and water resources management: The changing value of water*, AWRA/Univ. of Dundee Int. Specialty Conf., American Water Resources Association, Dundee.

Goldfarb, W. (1993). "The trend toward judicial integration of water quality and quantity management: Facing the new century." *Water resources administration in the United States*, M. Reuss, ed., American Water Resources Association and Michigan State University Press, East Lansing, MI.

Gooch, G. D., Höglund, P., Roll, G., Lopman, E., and Aliakseyeva, N. (2003). "Review of existing structures, models and practices for transboundary water management in Europe—The implementation of transboundary water management—Identification of present problems and a design for future research." *Proc., Int. Conf. on Sustainable Management of Transboundary Waters in Europe*, F. Bernardini, M. Landsberg-Uczciwek, S. Haunia, M. Adriaanse, and R. E. Enderlein, eds., 285–293, <www.unece.org/env/water/meetings/conf2> (Apr. 29, 2012).

Grant, D. L. (2001). "Interstate water allocation." *Waters and Water Rights*, Ch. 43–48, Vol. 4, R. E. Beck, ed., Lexis Law Publishing, Charlottesville, VA.

Haddadin, M. J. (2000). "International waters—Challenges an outside perspective." *International Round Table Transboundary Water Management: A Consultation on a World Bank Strategy*, Villa Borsig, Sweden, <lnweb18.worldbank.org/essd/essdext.nsf/18DocByUnid/D65DD89CD155B2BD85256B5000697EBB/$FILE/Haddadin.pdf>

Hall, S., and Meral, J. (2002). *Meeting Summary*, California Bay-Delta Public Advisory Committee, Water Supply Subcommittee, California Bay-Delta Authority <ca/water.ca.gov/content/Documents/meetings/2003AndLess/WSS_Meeting_Sumary_02-28-02.pdf/> (Apr. 29, 2012).

Hamner, J., and Wolf, A. (1998). "Patterns in international water resource treaties: The transboundary freshwater dispute database." *Colorado Journal of International Environmental Law and Policy*, 1997 Yearbook, 23(1), 3–30.

Hanson, K. M. (2006). "The evolution of interstate water disputes into regional cooperative management regimes: Launching a new model compact for interstate water issues." *Eastern Water Law Conf.*, American Bar Association, Miami.

Hassall & Associates Pty., Ltd., and the NSW Department of Land and Water Conservation, NSW National Parks and Wildlife Service, and CSIRO Division of Atmospheric Research. (1998). "Climate change scenarios and managing the scarce water resources of the Macquarie River: Final report." *Prepared for Environment Australia [Climate Change Impacts and Adaptation Grants Program]*, Sydney, New South Wales, Australia, March.

Hassoun, R. (1998). "Water between Arabs and Israelis." *Water, culture, and power: Local struggles in a global context*, J. M. Donahue and B. R. Johnston, eds., Island Press, Washington, DC.

Hayton, R. D., and Utton, A. E. (1989). *Transboundary groundwater: The Bellagio Draft Treaty*, University of New Mexico International Transboundary Resources Center, Albuquerque, NM.

Helsinki Rules on the Uses of the Waters of International Rivers. (1966). 52 ILA 484.

Helweg, O. J. (2000). "Water for a growing population: Water supply and groundwater issues in developing countries." *Water International*, 25(1), 33–39.

Hey, E. (1995). "Sustainable use of shared water resources: The need for a paradigmatic shift in international watercourses law." *The peaceful management of transboundary resources*, Blake, G. H., et al., eds., Graham and Trotman, London, 127–128.

Hightower, M. (2004). "National concentrate management guidelines." *National Salinity Management and Desalination Summit, Concentrating on Solutions*, Nevada Water Resources Association, Las Vegas, NV, Dec. 13–14.

Hillel, D. (1994). *Rivers of Eden: The struggle for water and the quest for peace in the Middle East*, Oxford University Press, New York.

Hirji, R., and Ibrekk, H. O. (2001). "Environmental and water resources management." *Environment Strategy Paper 2*, World Bank, Washington, DC.

Holling, C. S. (1978). *Adaptive Environmental Assessment and Management*, Wiley, London.

Houck, O. (1995). "Why do we protect endangered species, and what does that say about whether restrictions on private property to protect them constitute 'takings'?" *Iowa Law Review*, 80, 297.

Howarth, W. (1990). "The Law of the National Rivers Authority." *Cambridge Law Journal*, 49, 550–551.

Illinois State Water Survey. (2001). *A plan for scientific assessment of water supplies in Illinois*, ISWS Information/Educational Material 2001–03, Illinois Dept. of Natural Resources, Champaign, IL.

Illinois Water Use Act. (1983). Public Act 096-0222. <www.ilga.gov/legislation/publicacts> (Jun. 4, 2012).

The Indus Waters Treaty 1960. (1960). 419 UNTS 126. Karachi.

International Conference on Water and the Environment (ICWE). (1992). "The Dublin statement on water and sustainable development." <www.un-documents.net/h2o-dub.htm> (Apr. 29, 2012).

International Joint Commission. (1941). "Order of approval, in the matter of the application of the Government of the United States for approval of the construction and operation of the Grand Coolee Dam and Reservoir, 15 December 1941." <www.ijc.org/php/publications/html/columbia/columbiaord.htm> (Jun. 4, 2012).

International Joint Commission. (1984). *Second biennial rep. under the Great Lakes Water Quality Agreement*, Washington and Ottawa.

International Law Association (ILA). (2004). "The Berlin rules on water resources." *Report of the Seventy-First Conference (Berlin)*, London.

International Network of Basin Organizations. (2002). "Constitutive meeting of the network of transboundary basin organizations." Thonon-Les-Bains, France, <www.riob.org/divers/Constita.PDF>

International Souris River Board. (1948). "Reference letter to the International Joint Commission." <www.ijc.org/conseil_board/souris_river/en/souris_mandate_mandat>.
Jackson, R. B., Carpenter, S. R., Dahm, C. N., McKnight, D. M., Naiman, R. J., Postel, S. L., and Running, S. W. (2001). "Water in a changing world." *Ecological Applications*, 11(4), 1027–1045.
Johansson, M., Johansson, P.-O., Jonsson, B., and Soderqvist, T. (1995). "Valuing changes in health: Theoretical and empirical issues." *Current Issues in Environmental Economics*, P.-O. Johansson, B. Kristrom, and K.-G. Male, eds., Manchester University Press, Manchester, 78–97.
Kahan, A. M., Rottner, D., Sena, R., and Keyes, C. G., Jr. (1995). *Guidelines for cloud seeding to augment precipitation*, Manual of Practice 81, ASCE, New York.
Kakebeeke, W., Wouters, P., and Bouman, N. (2000). *Water management: Guidance on public participation and compliance with agreements*, ECE/UNEP Network of Experts on Public Participation and Compliance, Geneva.
Kansas–Nebraska Big Blue River Compact. (1972). 86 Stat. 193.
Kansas v. Colorado. 185 U.S. 143 (1902) and 206 U.S. 46 (1907).
Kaya, I. (1998). "The Euphrates–Tigris basin: An overview and opportunities for cooperation under international law." *Arid Lands Newsletter*, 44 <http://cals.arizona.edu/OALS/ALN/aln44/kaya.html> (Apr. 29, 2012).
Kenney, D. S. (1995). "Resolving interstate water allocation conflicts: Tools, strategies, and administrative options." Paper presented at the annual conference of the Southern Regional Information Exchange Group, Natural Resources Economics Committee, Auburn Univ., Auburn, AL.
Kenney, D., and Lord, W. B. (1994). "Coordination mechanisms for the control of interstate water resources: A synthesis and review of the literature." *Rep. for the ACF-ACT Comprehensive Study*, U.S. Army Corps of Engineers, Mobile District.
Keyes, C. G., Jr. (1977). "Options for new water supplies." *Drought: Our heritage*, 9th Annual Land-Use Symposium Proceedings, Greater Albuquerque Chamber of Commerce, Albuquerque, NM.
Keyes, C. G., Jr. (2006). "Societal, environmental, and economic aspects of precipitation enhancement by cloud seeding." *Guidelines for cloud seeding to augment precipitation*, Manual of Practice 81, 2d Ed., ASCE, Reston, VA.
Keyes, C. G., Jr. (2007). "ASCE/EWRI efforts to establish concentrate management guidelines for desalination and water reuse." *World environmental and water resources congress 2007*, K. C. Kabbes, ed., ASCE, Reston, VA.
Killgore, M. W., and Eaton, D. J. (1995). *NAFTA Handbook for Water Resource Managers and Engineers*, ASCE, New York.
Klamath River Basin Compact. (1957). 71 Stat. 497.
Kliot, N., and Shmueli, D. (2001). "Development of institutional frameworks for the management of transboundary water resources." *International Journal of Global Environmental Issues*, 1(3/4), 306–328.
Kliot, N., Shmueli, D., and Shamir, U. (1998). *Institutional frameworks for the management of transboundary water resources.* Water Research Institute, Haifa, Israel.
Kneese, A. V. (1993). "Economics and water resources." *Water resource administration in the United States*, M. Reuss, ed., American Water Resources Association and Michigan State University Press, East Lansing, MI, 23–35.
Kroes, M. (1997). "Protection of international watercourses as sources of fresh water in the interest of future generations." *The scarcity of water: Emerging legal and policy responses*, E. Brans, E. de Haan, A. Nollkaemper, and J. Rinzema, eds., Kluwer, London, 80–89.
Kundell, J. E., and Tetens, D. (1997). *Whose water is it? Major water allocation issues facing Georgia*, Carl Vinson Institute of Government, Univ. of Georgia, Athens, GA.
La Plata River Compact. (1925). 43 Stat. 796.
Lee, K. (1993). *Compass and gyroscope: Integrating science and politics for the environment*, Island Press, Washington, DC.
Lee, K. N., and Lawrence, J. (1986). "Adaptive management: Learning from the Columbia River Basin Fish and Wildlife Program." *Environmental Law,* 16, 431.
Le Marquand, D. (1978). "Politics of international basin cooperation and management." *Water in a developing world*, A. E. Utton and L. A. Teclaff, eds., Westview Press, Boulder.
Levy, J. K., Hipel, K. W., Kilgour, D. M., Hoque, B. A., and Gopalakrishnan, C. (2003). "Social mental models in water resources management." *Water Resources IMPAC*, 5(6), 14–17.
Libiszewski, S. (1995). "Water disputes in the Jordan Basin region and their role in the resolution of the Arab–Israeli conflict." *ENCOP Occasional Paper 13*, Center for Security Policy and Conflict Research Zurich/Swiss Peace Foundation Berne, Zurich/Berne, Switzerland.
Light, S. (2002). "Resilience—Putting the arcane goal of adaptive management to work." *Symp. at the Sustainable Fisheries Foundation Conf., Toward ecosystem management: Breaking down the barriers in the Columbia River basin and beyond,* Sustainable Fisheries Foundation, Snohomish, WA, <www.iatp.org/AEAM/Proceedings%5CAMinColumbiaBasin.pdf>
Lord, W. B., Booker, J. F., Goetches, D. M., Harding, B. L., Kenney, D. S., and Young, R. A. (1995). "Managing the Colorado River in a severe sustained drought: An evaluation of institutional options." *Journal of the American Water Resources Association*, 31(5), 939–944.
Lubell, M. (2000). "Cognitive conflict and consensus building in the national estuary program." *American Behavioral Scientist*, 44(4), 628–647.
Lucas v. South Carolina Coastal Council. (1992). 112 S. Ct. 2886.
Mann, D. E. (1993). "Political science: The past and future of water resources policy and management." *Water resources administration in the United States: Policy, practices, and emerging issues*, M. Reuss, ed., American Water Resources Association and Michigan State University Press, East Lansing, MI.
Mather, J. R. (1974). *Climatology: Fundamentals and applications*, McGraw-Hill, New York.
Matthews, O. P., and St. Germain, D. (2007). "Boundaries and transboundary water conflicts." *J. Water Resour. Plann. Manage.*, 133(5), 386–396.
McCaffrey, S. C. (1993). "Water, politics, and international law." *Water in crisis: A guide to the world's fresh water resources*, P. H. Gleick, ed., Oxford University Press, New York.
McCaffrey, S. C. (1997). "Middle East water problems: The Jordan River." *The scarcity of water: Emerging legal and policy responses*, E. Brans, E. de Haan, A. Nollkaemper, and J. Rinzema, eds., Kluwer, London.
McCain, Katie L. (2005). "President's comment: Value of water critical message for customers." *Main Stream*, 49(1), American Water Works Association, Denver.
McCarran Amendment. (1952). 43 U.S.C. 66.
McCormick, Z. L. (1994a). "Interstate water allocation compacts in the western United States—Some suggestions." *Water Resources Bulletin*, 30(3), 385–395.
McCormick, Z. L. (1994b). "The use of interstate compacts to resolve transboundary water allocation issues." Ph.D. thesis, Oklahoma State Univ., Stillwater, OK.
McDonald, J. W. (1997). "A deal is not a deal." *Rep. to the Western Water, Policy Review Advisory Commission*, U.S. Department of the Interior, Washington, DC.
McMahon, G. F., and Farmer, M. C. (2002). *Efficiency, equity and sustainability aspects of reallocation.* <www.cviog.uga.edu/services/policy/environmental/policyreports/waterconservation.pdf
Miller, J. A. (1990). "Ground water atlas of the United States: Alabama, Florida, Georgia, South Carolina." *HA 730-G*, U.S. Geological Survey, Washington, DC.
Mitchell, B., ed. (1990). *Integrated water management*, Belhaven, London.
Mitchell, R. C., and Carson, R. T. (1995). "Current issues in the design, administration, and analysis of contingent valuation surveys." *Current issues in environmental economics*, P.-O. Johansson, B. Kristrom, and K.-G. Male, eds., Manchester University Press, Manchester.
Murakami, M. (1995). *Managing water for peace in the Middle East: Alternative strategies*, United Nations University Press, Tokyo, <www.unu.edu/unupress/ unupbooks/80858e>

Murphy, E. F. (2001). "Quantitative groundwater." *Waters and Water Rights*, Vol. 3, §18.02, R. E. Beck, ed., Lexis Law Publishing, Charlottesville, VA.

Muys, J. C. (1971). *Interstate water compacts.* National Water Commission, Washington, DC.

Muys, J. C. (2001). *Testimony before the House of Representatives Committee on the Judiciary*, Dec. 19, <www.house.gov/judiciary/muys_121901>

Muys, J. C., Sherk, G. W., and O'Leary, M. C. (2006). *Utton Transboundary Resources Center Model Interstate Water Compact* <uttoncenter.unm.edu>

Narmada Water Disputes. (1979). "Final order and decision of the tribunal," India, Dec. 12, <www.sscac.gov.in/NWDT.pdf> (Apr. 8, 2012).

National Recreational Lakes Study Commission. (1999). "Reservoirs of opportunity." *Committee rep., submitted to the President, the Senate Committee on Energy and Natural Resources, and the House Committees on Resources and Transportation and Infrastructure*, <www.doi.gov/nrls/ findings/natrec.pdf> (Jun. 4, 2012).

Natural Resources Conservation Service. (2003). *Work plan for adaptive management, Klamath River Basin, Oregon and California*, U.S. Department of Agriculture, Washington, DC.

Nebraska Department of Natural Resources. (1983). "Policy issue study on selected water rights issues." *Report 6, Interstate Water Uses and Conflicts*, Lincoln, NE.

Neuman, J. (2001). "Adaptive management: How water law needs to change." *Environmental Law Reporter*, 31(12), 11432.

Nevada Division of Water Planning. (1999). *Nevada state water plan*, <water.nv.gov/Water%20planning/wat-plan/sum-sec8.pdf>

Nicholls, R. J. (2000). "Changing climate and sea level: How to evaluate?" *Water Resources Impact*, 2(4), 15–19.

Nile Management Roundtable. (2001). "Executive summary." *Roundtable: Nile waters management and linkage to conflict management and food security in the horn of Africa,* Canadian Centre for Foreign Policy Development, Ottawa, <publications.gc.ca/collections/Collection/E2-422-2001E.pdf> (Jun. 4, 2012).

North American Free Trade Agreement. (1993). 19 U.S.C. §§3311–3473.

Nouakchott Convention Establishing the OMVS (Organisation pour la Mise en Valeur du Fleuve Senegal). (1974). 672 UNTS 251.

Okidi, C. O. (1997). "International law and water scarcity in Africa." *The scarcity of water: Emerging legal and policy responses*, E. Brans, E. de Haan, A. Nollkaemper, and J. Rinzema, eds., Kluwer, London.

Olem, H., and Duda, A. M. (1995). "International watercourses: The World Bank looks toward a more comprehensive approach to management." *Water Science Technology*, 31(8), 345–352.

Opie, J. (2000). *Ogallala: Water for a dry land,* 2nd Ed., University of Nebraska Press, Lincoln, NE.

Oregon–California Goose Lake Interstate Compact. (1984). 98 Stat. 291.

Overseas Development Institute. (2001). "Transboundary water management as an international public good." *A study as part of Development Financing 2000*, Swedish Ministry for Foreign Affairs, Stockholm, <www.odi.org.uk/rpeg/transboundary_summary.pdf> (Jun. 4, 2012).

Pagano, T., Hartmann, H., Sorooshian, S., and Bales, R. (1999). "Advances in seasonal forecasting for water management in Arizona: A case study of the 1997–98 El Niño." *HWR No. 99-040*, Dept. of Hydrology and Water Resources, Univ. of Arizona, Tucson, AZ.

Pankratz, T., ed. (2004). "Water desalination and reuse strategies for New Mexico." *Proc., 49th Annual New Mexico Water Conf.*, New Mexico Water Resources Research Institute, WRRI, Las Cruces, NM, <waterdesalreport.com/articles/13250>

Patino-Gomez, C., McKinney, D. C., and Maidment, D. R. (2007). "Sharing water resources data in the Binational Rio Grande/Bravo Basin." *J. Water Resour. Plann. Manage.*, 133(5), 416–426.

Pawley, A. (2000). *Program performance indicators for the CALFED Bay-Delta Ecosystem Restoration Program*, Bay Institute of San Francisco, San Francisco, CA, <www.bay.org/assets/Indrpt02.pdf> (Jun. 4, 2012).

Pearson Education. (2002). *Online almanac, encyclopedia, and homework help*, <www.Factmonster.com/ipka/> (Apr. 29, 2012).

Pecos River Compact. (1948). 63 Stat. 159.

Pielou, E. C. (1998). *Fresh water*, University of Chicago Press, Chicago.

Planning and Management Consultants. (2000). "Regional listening session meeting notes." *Rep. submitted to U.S. Army Corps of Engineers*, Alexandria, VA.

Poff, N. L., Allan, J. D., Bain, M. B., Karr, J. R., Prestegaard, K. L., Richter, B. D., Sparks, R. E., and Stromberg, J. C. (1997). "The natural flow regime: A paradigm for river conservation and restoration." *BioScience,* 47(11), 769–784, <http://floridarivers.ifas.ufl.edu/RiverClass/Papers/Poff%20et%20al.%2097%20natflow_paradigm.pdf> (Jun. 4, 2012).

Postel, S. (1993). "Water and agriculture." *Water in crisis: A guide to the world's fresh water resources*, P. H. Gleick, ed., Oxford University Press, New York.

Postel, S. (1996). *Dividing the waters: Food security, ecosystem health, and the new politics of scarcity*, Worldwatch Institute, Washington, DC.

Puri, S., and Arnold, G. (2002). "Challenges to management of transboundary aquifers." *Proc., Int. Conf. on Sustainable Management of Transboundary Waters in Europe*, F. Bernardini, M. Landsberg-Uczciwek, A. Haunia, and R. Enderlein, eds., United Nations Economic Commission for Europe, Miedzyzdroje, Poland, 21–24, <www.unece.org/fileadmin/DAM/env/water/meetings/conf2/3-transboundaquifers_puri.pdf > (Jun. 4, 2012).

Pyne, D. (2008). *What is ASR?* ASR Systems, Gainesville, FL, <www.asrforum.com/fatestudy/what.html> (Apr. 29, 2012).

Radosevich, G. E., and Olson, D. C. (1999). "Existing and emerging basin arrangements in Asia, Mekong River Commission case study." *Third Workshop on River Basin Institution Development*, MRC Case Study, World Bank, Washington, DC.

Reagan, R. (1981). Executive order 12319, Sept. 30.

Red River Compact. (1980). 94 Stat. 3305.

Reinking, R. F., Berg, N. H., Farhar, B. C., and Foehner, O. H., Jr. (1995). "Economic, environmental, and societal aspects of precipitation enhancement by cloud seeding." *Guidelines for cloud seeding to augment precipitation*, Manual of Practice 81, ASCE, Reston, VA, 9–47.

Report of the United Nations Conference on Environment and Development. (1992). Rio de Janeiro.

Republican River Compact. (1943). 57 Stat. 86.

Rio Declaration on Environment and Development. (1992). 31 ILM 874.

Rio Grande Compact of 1938. (1938). 53 Stat. 785, 938.

Rogers, P. (1993). "Why are widely accepted principles of water management so often not followed in practice?" *Water Resources Update*, 90, Winter.

Roll, G., and Lopman, E. (2001). "EU water policy and implementation of water management regimes on transboundary water bodies in the Baltic Sea basin." *Proc., 2001 Berlin Conf. on the Human Dimensions of Global Environmental Change,* F. Biermann, R. Brohm, and K. Dingwerth, eds., Potsdam Institute for Climate Impact Research, Potsdam, December, <www.pik-potsdam.de/research/publications/pikreports/.files/pr80.pdf> (Jun. 4, 2012).

Rothfeder, J. (2001). *Every drop for sale,* Penguin Group, New York.

Ryan, P. L., et al. (1998). *District water supply assessment: Northwest Florida water management district*, Water Resources Assessment, 98-2.

Saarinen, P. P., and Lynn, G. D. (1993). "Getting the most valuable water supply pie: Economic efficiency in Florida's reasonable-beneficial use standard." *Journal of Land Use and Environmental Law*, 8(2), Summer Supplement.

Sabine River Compact. (1953). 68 Stat. 690, amended 76 Stat. 34 (1962), 91 Stat. 281 (1977), 106 Stat. 4661 (1992).

Sanchez, R. A. (1997). "Water conflicts between Mexico and the United States: Towards a transboundary regional water market?" *The scarcity of water: Emerging legal and policy responses*, E. Brans, E. de Haan, A. Nollkaemper, and J. Rinzema, eds., Kluwer, London.

Sando, R. (2002). "Adaptive management in the Columbia Basin: Will it be implemented or be a good idea on the shelf? Adaptive management in the Columbia Basin and beyond: What is it and where is it happening?" *Symp. at the Sustainable Fisheries Foundation Conf., Toward ecosystem management: Breaking down the barriers in the*

Columbia River basin and beyond, <www.iatp.org/AEAM/Proceedings%5CAMinColumbiaBasin.pdf>

Sax, J. (1990). "The Constitution, property rights and the future of water law." *University of Colorado Law Review*, 61, 257.

Sax, J. L., Thompson, B. H., Leshy, J. D., and Abrams, R. H. (2006). *Legal control of water resources*, 4th Ed., West Publishing, Minneapolis-St. Paul, MN.

Sea River. (2002). *Water News*, <www.sea-river.com/eau/>

Sehlke, G. (2004). "Basin management case studies: Tools, techniques, and methods for managing river basins." *Water Resources Impact*, 6(2).

Sehke, G., and Colosimo, M. (2007). "Adaptive management of water resources in light of future climate uncertainty." *Water Resources Impact*, 9(6).

Sherk, G. W. (2000). *Dividing the waters*, Kluwer Law International, The Hague, Netherlands.

Sherk, G. W. D. (2002). *Protecting instream flows: An economic benefits summary*, <www.cviog.uga.edu/water/whitepaper>

Shiklomanov, I. A. (1993). "World fresh water resources." *Water in crisis: A guide to the world's fresh water resources,* P. H. Gleick, ed., Oxford University Press, New York.

Shiklomanov, I. A. (1998). "Global renewable water resources." *Water: A looming crisis*? H. Zebidi, ed., UNESCO IHP-V Technical Documents in Hydrology 18, 3–13. Paris, France.

Sibanda, P. N. (2002). "Water supply and sanitation: How have African cities managed the sector? What are the possible options?" *Presented at the Urban and City Management Course for Africa*, Uganda Management Institute, Kampala, Uganda.

Snake River Compact. (1949). 64 Stat. 29.

Solley, W. B. (1997). "Estimates of water use in the western United States in 1990 and water-use trends 1960–90." *Rep. to the Western Water Policy Review Advisory Commission,* U.S. Geological Survey, Reston, VA, <http://bioe.orst.edu/Faculty/selker/Oregon%20Water%20Policy%20and%20Law%20Website/Report%20of%20the%20WWPRAC/WATERUSE.PDF> (Jun. 4, 2012).

Sophocleous, M., ed. (1998). "Perspectives on sustainable development of water resources in Kansas." *Kansas Geological Survey Bulletin*, 239.

South Platte River Compact. (1923). 44 Stat. 195.

State of Georgia. (1997). *Interim strategy for managing salt water intrusion in the Upper Floridan Aquifer of southeast Georgia*, Dept. of Natural Resources, Environmental Protection Division, Atlanta.

Statute of the International Court of Justice. (1989). Charter of the United Nations, statute and rules of court, and other documents, KZ6277.U56.

Sterner, T. (1994). *Policy instruments for a sustainable economy, economic policies for sustainable development*, Kluwer Academic Publishers, Dordrecht, Netherlands.

Stockton, W. C., and Jacoby, G. C., Jr. (1976). "Long-term surface-water supply and streamflow trends in the Upper Colorado River basin based on tree-ring analyses." *Lake Powell Research Project Bulletin*, 18.

Strickland, D., Derby, C., and Howlin, S. (2002). "Adaptive resource management and the Platte River Endangered Species Partnership." *Toward Ecosystem-Based Management: Breaking Down the Barriers in the Columbia River Basin and Beyond*, Lake Roosevelt Forum, Spokane, WA.

Sundeen, M., and Runyon, L. C. (1998). "Interstate compacts and administrative agreements." *State Legislative Rep.*, National Conference of State Legislators, <www.ncsl.org/programs/esnr/slr238>

Susquehanna River Basin Compact. (1970). Pub. L. No. 91-575, 84 Stat. 1509-1541.

Tarlock, A. D. (1989). *Law of water rights and resources*, §4.05.

Tarlock, A. D. (2000). "How well can international allocation regimes adapt to global climate change?" *Journal of Land Use and Environmental Law*, 15(3), 423–449.

Tarlock, A. D. (2001). "Water policy adrift." *Forum for Applied Research and Public Policy,* Summer.

Tarlock, A. D., Corbridge, J. N., Getches, D., and Benson, R. (2009). *Water Resource Management, A Casebook in Law and Public Policy*, 6th Ed., §10.04(2)(c), West Law Publishers, Minneapolis-St. Paul, MN.

Teclaff, L. A. (1987). *The river basin in history and law*, Martinus Nijhofl, The Hague, Netherlands.

Tennessee Valley Authority. (1933). 16 U.S.C.S. §83/c.

Texas v. New Mexico. (1957). 352 U.S. 991.

Texas v. New Mexico. (1983). 462 U.S. 554.

Texas v. New Mexico. (1987). 482 U.S. 124.

Texas v. New Mexico. (1991). 111 S. Ct. 2281.

Texas Water Code Ann., <www.statutes.legis.state.tx.us> (June 4, 2012).

Thurow, L. (1983). *Dangerous currents*, Random House, New York.

Tietenberg, T. (1998). *Environmental economics and policy*, 2nd Ed., Addison-Wesley, Reading, MA.

Toole, C. (2002). "The NMFS hydro biological opinion and adaptive management." *Symp. at the Sustainable Fisheries Foundation Conf., Toward ecosystem management: Breaking down the barriers in the Columbia River basin and beyond*, Sustainable Fisheries Foundation, Spokane, WA.

The Transboundary Freshwater Dispute Database. (2012). Oregon State University, Corvallis, OR, <http://www.transboundarywaters.orst.edu/database/DatabaseIntro.html> (Apr. 29, 2012).

Treaty Addendum. (1973). Addendum to *Treaty between the United States and Mexico. Utilization of Waters of the Colorado and Tijuana Rivers and of the Rio Grande.* (1944). Treaty Series 994; 59 Stat. 1219.

Treaty between the United States and Great Britain Relating to Boundary Waters. (1909). 36 Stat. 2451.

Treaty between the United States and Mexico. Utilization of Waters of the Colorado and Tijuana Rivers and of the Rio Grande. (1944). 59 Stat. 1219.

Treaty for Amazonian Cooperation. (1978). 17 ILM 1046.

Treaty of Peace between the State of Israel and the Hashemite Kingdom of Jordan. (1994). 34 ILM 43.

Trelease, F. (1979). *Water law: Cases and materials,* 3rd Ed., West Publishing, Minneapolis, MN.

Trolldalen, J. M. (1992). *International environmental conflict resolution: The role of the United Nations,* United Nations Institute for Training and Research, Washington, DC.

Truckee–Carson–Pyramid Lake Water Rights Settlement Act. (1990). Title II, Pub. L. No. 101-618, 104 Stat. 3289, Oct. 27.

Tsur, Y. (1995). "Uncertainty and irreversibility in groundwater resource management." *Journal of Environmental Economics and Management*, 29, 149–161.

Tsur, Y. (1997). "The economics of conjunctive ground and surface water irrigation systems: Basic principles and empirical evidence from Southern California." *Decentralization and coordination of water resource management,* D. Parker and Y. Tsur, eds., Kluwer Academic Publishers, Boston.

United Nations. (1972). *Stockholm Declaration of the United Nations Conference on the Human Environment*, 11 ILM 1416.

United Nations. (1975). "Management of international water resources: Institutional and legal aspects." *Rep. of the panel of experts on the legal and institutional aspects of international water resources development*, Natural Resources/Water Series No. 1, UN Doc. St/ESA/5, Department of Economic and Social Affairs, New York.

United Nations. (1992). *Agenda 21: The United National Programme of Action from Rio,* Rio de Janerio, <ww.un.org/documents/ga/conf151/aconf15126-1annex1> (Jun. 4, 2012).

United Nations. (1992a). *Convention on the Protection and Use of Transboundary Watercourses and International Lakes,* 31 ILM 1312.

United Nations. (1997). *Convention on the Law of the Non-Navigational Uses of International Watercourses,* United Nations Document A/51/869.

United Nations, Committee on Economic, Social and Cultural Rights. (2002). *General Comment No. 15 on the right to water*, Geneva, July.

United Nations General Assembly. (1975). "Charter of economic rights and duties of states." *GA Res. 3281(XXIX)*, New York.

United Nations General Assembly. (2010). *Resolution A/RES/64/292*, New York.

U.S. Advisory Commission on Intergovernmental Relations. (1991). *Coordinating water resources in the federal system: The groundwater-surface water connection*, Washington, DC.

U.S. Army Corps of Engineers. (1995). *Alabama-Coosa-Tallapoosa and Apalachicola-Chattahoochee-Flint River Basins Comprehensive Water Resources Study: Status Report.*

U.S. Army Corps of Engineers. (1999). "Water resources policies and authorities: Digest of water resources policies and authorities." *EP 1165-2-1*, <www.usace.army.mil/inet/usace-docs/eng-pamphlets/ep1165-2-1/toc>

U.S. Army Corps of Engineers. (2007). "Final environmental impact statement: Summary." *FES-07-05,* U.S. Army Corps of Engineers, U.S. Department of the Interior, New Mexico Interstate Stream Commission, Albuquerque, NM. <www.spa.usace.army.mil/urgwops>.

U.S. Geological Survey. (1987). *National water summary*, Washington, DC.

U.S. House of Representatives. (1998). *Hearing before the Subcommittee on Water and Power of the Committee on Resources, in reference to H.R. 3658, A Bill to Provide for the Settlement of the Water Rights Claims of the Chippewa Cree Tribe of the Rocky Boy's Reservation, and for Other Purposes*, 105th Cong., 2nd sess., Sept. 23, Washington, DC.

U.S. Water Resources Planning Act. (1965). Pub. L. 89-80, 79 Stat. 244.

United States v. State Water Resources Control Board. (1986). 182 Cal. App. 3d 82, 227, California Reporter 161.

Upper Colorado River Basin Compact. (1949). 63 Stat. 31.

Upper Niobrara River Compact. (1969). 83 Stat. 86, Art V.

Utton, A. E., and Atkinson, C. K. (1979). *International groundwater management: The case of the Mexico-United States frontier*, New Mexico Water Resources Research Institute, Las Cruces, NM.

Utton Transboundary Resources Center. (2007). *Model interstate water compact project.* Univ. of New Mexico, Albuquerque, NM, <uttoncenter.unm.edu/model_compacts> (Jun. 4, 2012).

van Edig, A., van de Giesen, N., Andreini, M., and Laube, W. (2001). *Transboundary, institutional, and legal aspects of the Water Resources Commission in Ghana*, Center for Development Research, Univ. of Bonn, Bonn, Germany.

Vienna Convention for the Protection of the Ozone Layer. (1985). 26 ILM 1529.

Virginia v. Tennessee. (1893). 148 U.S. 503.

Vlachos, E. (1997). "Approaching integrated water resources planning and management." *Int. Conf. on Water in the Mediterranean*, International Network of Basin Organizations, Istanbul, Turkey.

Vörösmarty, C. J. (2002). "Global water assessment and potential contributions from earth systems science." *Aquatic Science*, 64(2002), 328–351.

Wade, J. S., Tucker, J. C., and Leitman, S. F. (1994). "Alternative strategies for basinwide management of the ACF Basin: The Florida perspective." *Prepared for the Northwest Florida Water Management District, Contract 9209310*, Havana, FL, <www.law.ufl.edu/cgr/publications/alternative_strategies.pdf> (Jun. 4, 2012).

Walker, W. R., Hrezo, M. S., and Haley, C. J. (1991). "Management of water resources for drought conditions." *1988–89—Hydrologic Events and Floods and Droughts, U.S. Geological Survey Water Supply Paper 2375*, Paulson, Chase, Roberts, and Moody, eds., USGS, Washington, DC, <http://md.water.usgs.gov/publications/wsp-2375/al/index.html> (Jun. 4, 2012).

Water Plan Advisory Committee, Data and Information Management Subcommittee. (2002). *Final report: Water quality in Georgia 1998–1999*, Environmental Protection Division, Georgia Department of Natural Resources, Atlanta.

Water Resources Planning Act of 1965. (1965). Pub. L. No. 89-90, 79 Stat. 244, repealed Exec. Order No. 12319, Sept. 30, 1981.

Water Science and Technology Board. (2001). *Aquifer storage and recovery in the Comprehensive Everglades Restoration Plan: A critique of the pilot projects and related plans for ASR in the Lake Okeechobee and Western Hillsboro areas*, National Research Council, Division on Earth and Life Studies, Washington, DC.

Williams, A. E. (1993). "The role of technology in sustainable development." *Water Resources Administration in the United States: Policy, Practice, and Emerging Issues*, M. Reuss, ed., Michigan State University Press, East Lansing, MI.

Wilson, L. (2006). "Strategic concepts, conjunctive water resource management plan, Central Platte Valley, Nebraska." *Prepared for the Nebraska Dept. of Natural Resources*, Central Platte Natural Resource District, Nebraska Public Power District, Lee Wilson and Associates, Inc., Santa Fe, NM, <http://dnr.ne.gov/PlatteRiver/docs/Strategies_WhitePaper_11-21-06.pdf > (Jun. 4, 2012).

Winter, T. C., Harvey, J. W., Franke, O. L., and Alley, W. M. (1998). "Ground water and surface water: A single resource." *U.S. Geological Survey Circular 1139*, USGS, Denver.

Wiseburgh, P. (2002). "Preface." *Efficient use of limited water resources*, B. Rubun, ed., Begin–Sadat Center, Bar Ilan Univ., Tel Aviv, Israel.

Wolf, A. T. (1998). "Conflict and cooperation along international waterways." *Water policy*, 1(2), 251–265.

Wolf, A. T. (1999). "Criteria for equitable allocations: The heart of international water conflict." *Natural Resources Forum*, 23(1), 3–30.

Wolf, A. T. (2001). "Transboundary waters: Sharing benefits, lessons learned." *Thematic Background Paper for the Int. Ministerial Conf. on Freshwater*, Oregon State Univ., Corvallis, OR, <www.agnos-online.de/inwent1/images/pdfs/transboundary-waters.pdf>

World Bank. (1998). "Strategy for integrated water resources management." *N ENV-125*, Washington, DC.

World Bank. (2001). *World Development Indicators 2001*, Washington, DC.

World Commission on Dams. (2000). *Dams and development: A new framework for decision-making,* Earthscan Publications, London.

Yellowstone River Compact. (1950). 65 Stat. 663.

Zalewski, M., ed. (2002). "Guidelines for the integrated management of the watershed—Phytotechnology and ecohydrology." *Freshwater Management Series No. 5*, United Nations Environment Programme, Division of Technology, Industry and Economics, Internationale Environmental Technology Center.

Zimmerman, A. P. (2002). *Water use in North America*, University of Toronto, Toronto, Canada, <www.cquest.utoronto.ca/env/jie222y/LECTURES/Lecture3/>

Zusman, P. (1997). "Informational imperfections in water resource systems and the political economy of water supply and pricing." *Decentralization and coordination of water resource management*, D. Parker and U. Tsur, eds., Kluwer Academic Publishers, Boston, 133–154.

Zylicz, T. (1994). "Environmental policy reform in Poland." *Economic policies for sustainable development*, T. Sterner, ed., Springer, New York.

INDEX

Proceedings of
2001 ASME International Mechanical Engineering Congress and Exposition
November 11–16, 2001, New York, NY
EPP-Vol. 1

IMECE2001/EPP-24717

EXPERIMENTAL INVESTIGATION ON CHEMICAL-MECHANICAL POLISHING OF WAFERS WITH LOW-CONDUCTIVITY DIELECTRICS

Jhy-Cherng Tsai[1]
Department of Mechanical Engineering
National Chung-Hsing Univ,
250 Koukwang Road
Taichung, Taiwan 402
Republic of China

Charls Liu
Mosel Vitelic Inc.,
No.1, Creation Road 1
Science-Based Industrial Park
Hsinchu, Taiwan 300
Republic of China

Ming-Hsih Tsai, Bao-Tong Dai
Nano Device Laboratories
The National Science Council
1001-1 University Road
Hsinchu, Taiwan 300
Republic of China

ABSTRACT

Low conductivity (low-k) dielectric material is used in the sandwich structure of next-generation semiconductor devices in order to reduce the RC time delay. While global flatness of wafer surface becomes critical for deep sub-micro semiconductor fabrication process, chemical-mechanical polishing (CMP) becomes one of the key technologies for planarization of wafer surface. This paper investigated the effect of the low-k material on the CMP of the SiO_2 cap layer of such a sandwiched wafer. Two types of wafers, blanket wafer and wafer with circuit pattern, are designed and conducted to investigate the effects of the thickness of the low-k layer under different polishing pressures and velocities. Material removal rate (RR) and non-uniformity (NU) are used as indices of the CMP process performance. The results show that the RR and NU of wafers with low-k layer, either blanket or with circuit pattern, become better when the pressure or velocity increases. The thickness of the low-k layer, however, has only tiny effect on the RR and NU.

Keywords: chemical-mechanical polishing (CMP), low-k dielectric, polishing pressure, polishing velocity, removal rate, non-uniformity.

INTRODUCTION

Integrated circuit (IC) is rice of electronic industry to serve as the most fundamental components for modern products, in particular computer, communication and consumer (3C) products. The IC industry is one of the fast growing

[1] Corresponding author. Email: jctsai@mail.nchu.edu.tw

and most important high-technology industries because ICs have been used in many different applications. As the device density per area and performance are the two most common indices for measuring the progress of the IC technology, technologies have been pushed to increase these indices.

The most common approach to increase device density, however, is to reduce feature sizes of devices and to increase the layer of circuits with inter-layer interconnects. This leads to the requirement of high flatness of wafer surface in order to focus the mask in the lithography process. That is, surface flatness and roughness of the wafer must be well controlled in the IC fabrication process. It is therefore requires a high-efficiency process that can quickly remove the surface peaks and to produce a new surface with good surface flatness and roughness. Among many other polishing technologies for wafer polishing, chemical-mechanical polishing (CMP) has been recognized as the most important process in IC fabrication for global planarization of wafer surface required for small feature size and large wafer size. CMP is a chemo-mechanical polishing process that combines both chemical etching and mechanical polishing processes. It has been widely used in polishing brittle material such as glass and ceramics [Steigerwald et al., 1997; Towery & Fury, 1998].

The other important index of an IC, performance, is commonly measured by its operating speed, or frequency. The common rule is the higher the operating frequency, the better the performance. One of the bottlenecks to increase the operating frequency is the resistance-capacitance (RC) delay due to conductivities in multi-layer interconnects. Studies showed that RC time delay per unit length of wire overrides intrinsic gate delay when feature size of IC is less than 0.25μm [Lin & Pramanik, 1998]. To reduce the time delay, one can either use low-resistance interconnects, such as copper rather than aluminum, or replace the insulating layer between circuits by low conductivity (low-k) material. While copper processes are still under developing, processes for ICs with low-k layers need fewer modifications therefore it has been adcpted to increase the performance of an IC. Studies on process technologies for such ICs are important for process performance and production quality [Lee at al., 1996].

This paper is an investigation of the effect of low-k layer on the CMP process. Experiments are designed and conducted to measure the performance of CMP process using wafers with low-k layers. Polishing pressure and velocity are also studied as process parameters. Process performance is evaluated by process efficiency, the material removal rate, and by process quality, the non-uniformity of wafer surface.

CMP OF WAFERS WITH LOW-K LAYER

Wafer with low-k layer is basically the same as traditional one but insulating layers are replaced by low-k material. An insulating layer is first grown on a silicon substrate. The low-k layer, such as polyimide, is then coated on the surface. Another cap layer, often oxide, is then platting on top of the surface [Fury, 1999].

Figure 1 is the sketch of a typical CMP process for a wafer with low-k layer. The wafer is detained on the carrier with face pressed against the polishing pad, which is attached to a rotating polishing platen. Both the wafer carrier and the platen are rotating, either in the same or in the opposite direction, to generate relative motion between the wafer and the pad. Slurry, which contains abrasive particles in chemical solution, is supplied to form a liquid film between the wafer and the pad. The pad is made of porosity material such that abrasive particles, size in between 10 to 3000 nanometers (nm), can be carried into the gap between the wafer and the pad. The slurry chemically etches the wafer surface thus weaken or break molecular bonds on the wafer surface so that molecules on wafer can be mechanically removed by abrasives [Tsai et al., 1999].

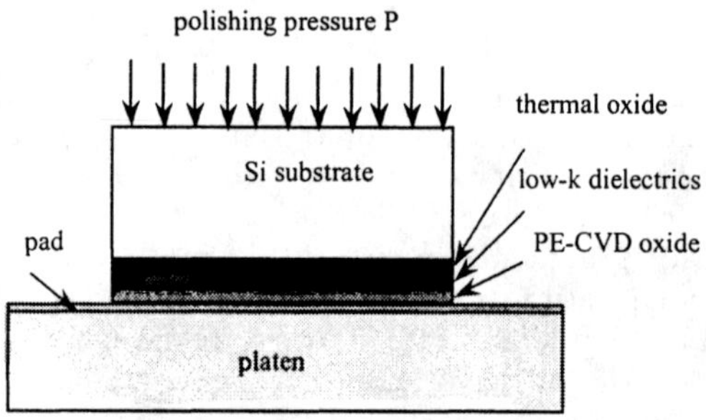

Figure 1. Sketch of a typical CMP process for a wafer with low-k layer.

EXPERIMENT DESIGN AND SET-UP

In order to investigate the effects of thickness of the low-k layer as well as the polishing pressure and velocity, experiments are designed and conducted. As CMP has been used for polishing both insulating and circuit layers [Chen et al.,

1999], two sets of experiments, one with blanket wafer with insulating layer and the other one with circuit pattern, are designed. The thickness of low-k layer are set as 0, 8500Å(10^{-10}m) and 13500Å. Polishing pressures, referring to the range in practice, are set as 5psi, 7psi, 9psi and 11psi($3.45\times10^4 \sim 7.58\times10^4 N/m^2$). Polishing velocities are set as 20/20, 40/40 and 60/60 (rpm/rpm) as it is important to choose the same rotation speed for platen and pad in order to keep uniform relative speed across the wafer. Circuit pattern used in the experiment employs the NDL SKW-11N mask as shown in Figure 2. The mask contains circuit patterns for different feature sizes from 0.35μm to 500μm. Each circuit pattern is allocated in a 4mm by 2mm area as shown in the figure.

Experimental specimens are 6-inch P(100) silicon wafers that were sent to thermal process to grow a layer of 1000Å-thick oxide. A layer of low-k material (polyimide) is then coated on the thermal oxide by spinning. Thickness of this layer, either 0, 8500Å or 13500Å, depends on experiment design. Finally, a 3000Å-thick cap layer of oxide by PE-CVD is then deposited on top of the low-k layer.

Experiments are conducted on an IPEC 372M with Cobat SS25 slurry and Rodel IC1400 pad. The Thickness of oxide layer and low-k layer are measured by a NanoSpec 210UV and an n&k 1200 analyzer respectively. Profile of wafer is measured by a KLA Tencor P-10 surface profiler.

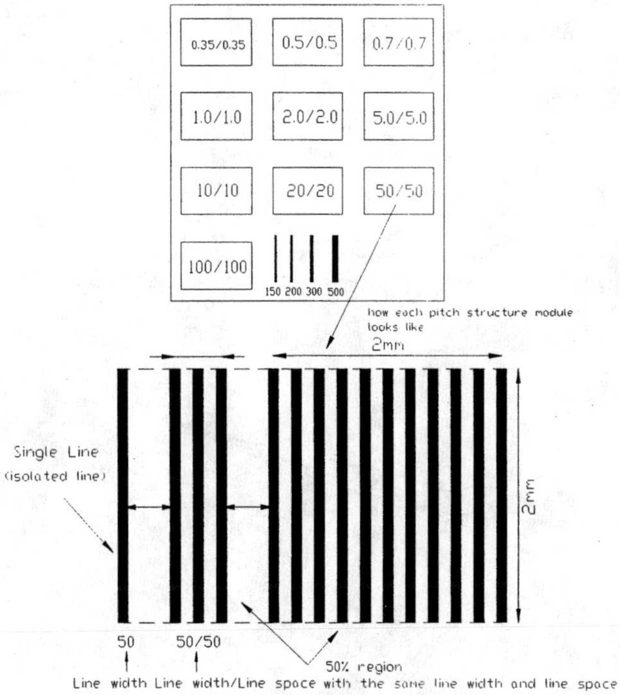

Figure 2: Circuit pattern of the NDL SKW-11N mask used in the experiment

As throughput and quality are the major considerations in production, the material removal rate (RR) and non-uniformity (NU) are used to evaluate the performance in the experiment. RR is an important index of the CMP process to measure the process efficiency. It measures the thickness of wafer removed in unit time and is calculated based equation (1), where T_{pre} and T_{post} are thickness of wafers before and after polishing; and t is the polishing time.

$$RR = (T_{pre} - T_{post})/t \text{ (Å/min)} \quad \text{.............(1)}$$

As the wavelength of exposition light used in lithography is getting shorter in order to obtain higher resolution for smaller feature size, global flatness of wafer is important for focusing the circuit pattern on mask to produce high quality circuit on substrate. Therefore, uniformity of thickness is another important metric of quality for the CMP process. Thickness across the wafer are measured from the nine points as shown in Figure 3, where the distance between points is 30mm. NU is calculated based on equation (2) that indicates the ratio of thickness variation with respect to the mean value of thickness.

$$NU = \frac{\sigma}{\overline{X}} \times 100\% \quad \text{.................................} (2)$$

Where $\sigma = \sqrt{\frac{\sum_{i=1}^{n}(X_i - \overline{X})^2}{N}}$ .... (3)

is the standard variation of the measurement and $\overline{X}$ is the mean value of measured thickness *Xi*.

DATA ANALYSIS AND RESULTS

Data are measured from both sets of experiments. Figure 4 is the results for blanket wafers under different polishing pressures at 60/60 rpm rotating speed. It shows that RR linearly increases and NU becomes better when polishing pressure increases, no matter how thick the low-k layer is. Same results are also observed at 40/40 rpm and 20/20 rpm.

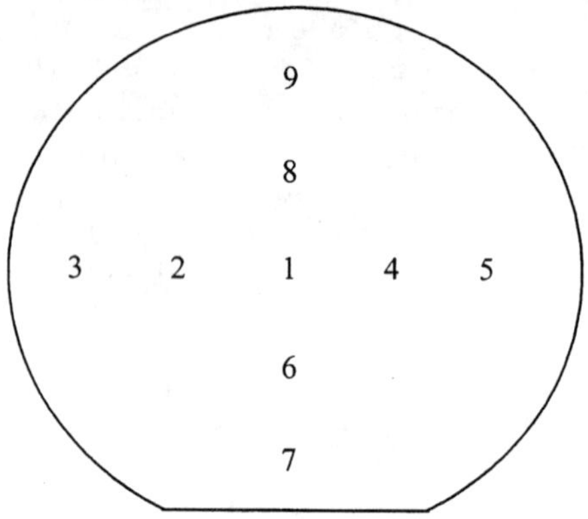

Figure 3: Points on a wafer for thickness measurement.

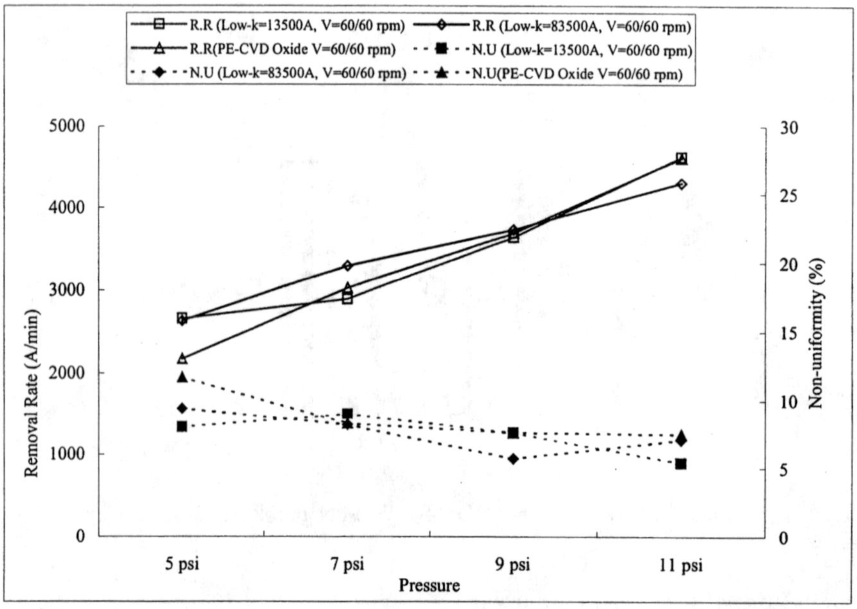

Figure 4. RR and NU for blanket wafers under different polishing pressures at 60/60 rpm.

For the effect of polishing velocity on RR and NU, figure 5 shows the results wafers with various thickness of low-k layer at different polishing velocities under polishing pressure 11psi. It indicates higher RR and better NU can be achieved at higher polishing velocity. RR increases linearly with polishing velocity while NU improved slightly at the same time. It also shows there is no much difference for wafers with different thickness of low-k layer. Wafers under other polishing pressures also showed the same trend. This behavior is the same as those without low-k dielectrics [Tsai & Tsai, 2000].

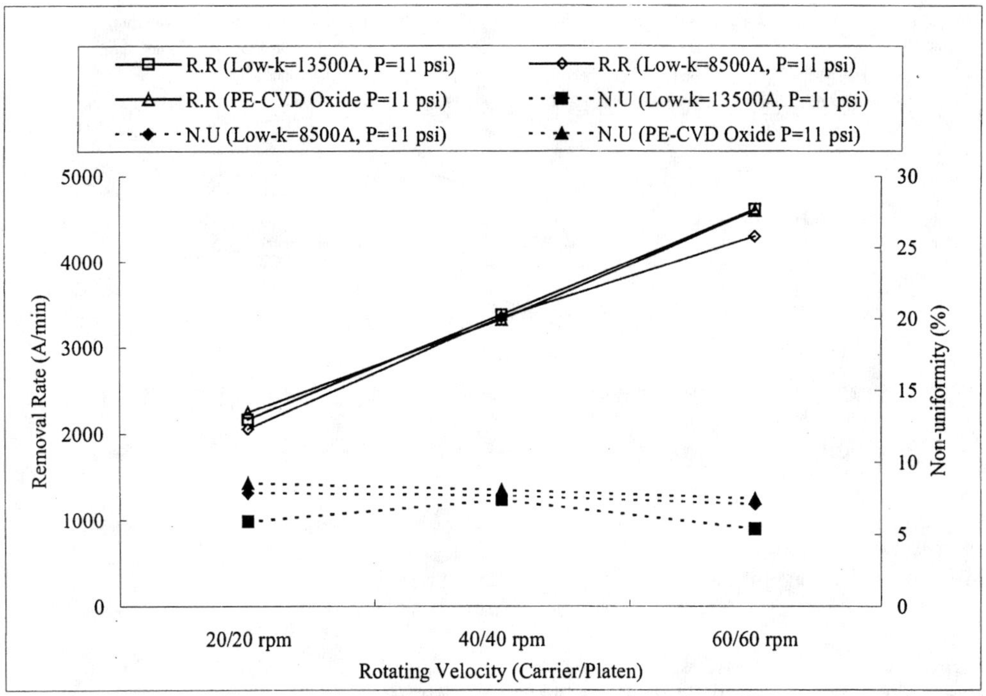

Figure 5. RR and NU for blanket wafers under different polishing velocities with polishing pressure at 11psi.

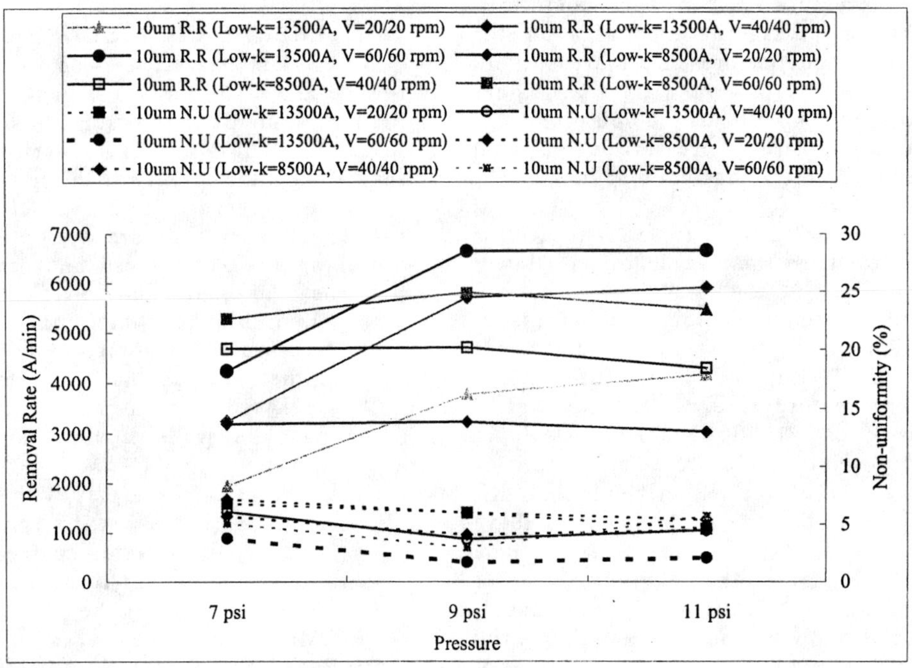

Figure 6. RR and NU on 10μm region under different polishing pressures with various thickness of low-k and different polishing velocities.

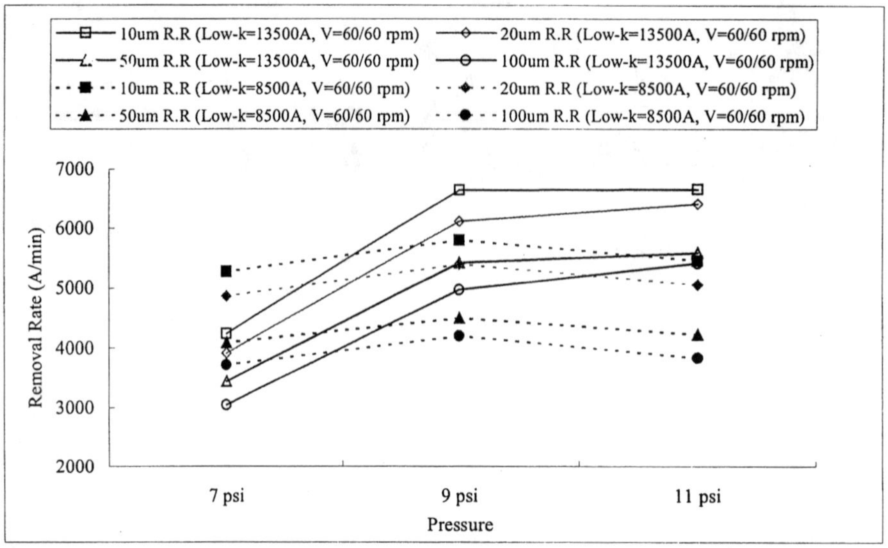

Figure 7. Effect of polishing pressure on RR for different thickness of low-k layer and feature size at 60/60 rpm.

For experiments on wafers with circuit pattern, Figure 6 indicates the effects of polishing pressure on RR and NU on the 10μm region for wafers with different thickness of low-k layer at different polishing velocities. It shows RR increases sharply with polishing pressure below 9psi and then saturated at higher pressure. Uniformity on the wafer gets better when the pressure increases but becomes worse under 11psi for wafers at higher polishing velocities (40/40 and 60/60 rpm). It suggests the optimal polishing pressure is 9psi as it produces better surface flatness without losing RR. It is also shown in the figure that NU is better at higher polishing velocity. For 20, 50 and 100μm regions, we observed the same tendency with lower RR.

For the same polishing velocity, thickness of low-k layer and polishing pressure also affect the performance of CMP. Figure 7 shows the removal rate on different feature size regions for wafers with 8500Å-thick and 13500Å-thick low-k layer under different polishing pressure at 60/60 rpm. It can be observed from the experiment that RR increases, in most cases, at higher polishing pressure. For wafers with low-k layer of the same thickness, RR is higher at the region with wider feature. The effect of thickness of low-k layer, however, does not show a specific rule. Compared with the results on blanket wafers, RR does not linearly increase with polishing pressure. It shows that CMP at 9psi has the optimal performance. Similar results are also observed at different rotation speeds. This comparison indicates that the low-k layer does affect the performance of CMP process.

CONCLUSION AND DISCUSSION

Low-k material has been used in ICs as it can reduce the RC-delay in signal processing. While planarization of wafer surface becomes a critical issue for sub-micro IC fabrication, this paper investigates the effects of the low-k layer as well as polishing pressure and velocity on the CMP process. Experiments are design and conducted for polishing blanket wafers and wafers with circuit patterns. The results show that

i. RR of blanket wafers increases as polishing pressure and velocity increase. This tendency is the same as that of wafers without low-k layer. However, removal rate of wafers with low-k layer is lower than that of regular wafers.
ii. RR, for wafers either with or without low-k layer, increases with polishing velocity. Better NUs of wafer surface are observed at the same time. It indicates that higher polishing velocity helps the performance of CMP process.
iii. Based on RR and NU, the thickness of low-k layer, however, does not show significant

influence on the performance of CMP of blanket wafers.

iv. RR and NU for wafers with circuit pattern does not increase linearly with the polishing pressure or velocity. Experiment shows that best result is obtained when polishing at 9psi.

In addition to CMP on wafers with low-k layer, we observed that polishing rate on the rim is higher than that at the center. From material removal mechanism standpoint, RR and NU are related to pressure and stress distributions across the wafer in the CMP process [Tseng, 1998]. Further investigation should be conducted to have a better understand of the CMP process.

ACKNOWLEDGMENTS

The authors would like to thank Mr. Da-Chung Liu who has helped the experiments. This research was supported by the National Nano Device Laboratories under contract NDL-C89095 and by the National Science Council of Taiwan, R.O.C. under contract NSC 88-2212-E-005-015.

REFERENCES

1. Chen, W. C.; Lin, S. C.; Dai, B. T.; Tsai, M. S.; 1999, "Chemical Mechanical Polishing of Low-Dielectric-Constant Polymers: Hydrogen Silsesquioxane and Methyl Silsesquioxane", *Journal of The Electrochemical Society*, Vol. 146(8), pp. 3004-3008.
2. Fury, M. A., 1999 "CMP Processing with Low-k Dielectrics," *Solid State Technology*, pp.87-96, July.
3. Lee, M. G.; Larson, T.; Beilin, S.; 1996, "Planarization of Copper/Polyimide for Multilevel Interconnects by Chemical Mechanical Polishing," *Proceedings of VMIC*, pp.395-397, June.
4. Lin, X. W.; Pramanik, D; 1998, "Future Interconnect Technologies and Copper Metallization", *Solid State Technology*, pp.63-79, October.
5. Steigerwald, J. M.; Murarka, S. P.; Gutmann, R. J., *Chemical Mechanical Planarization of Microelectronic Materials*, 1997, John Wiley & Sons.
6. Towery, D.; Fury, M. A.; 1998, "Chemical Mechanical Polishing of Polymer Film", *Journal of Electronic Materials*, Vol. 27, No. 10, pp.1088-1094.
7. Tsai, J.-C.; Tsai, M.; 2000, "Experimental Investigation of Pressure And Velocity on Chemical Mechanical Planarization," *Proceedings of The 2000 ASME International Mechanical Engineering Congress and Exhibition, EEP-Vol.28, Packaging of Electronic and Photonic Devices*, pp.221-226.
8. Tsai, M.-Y.; Tsai, J.-C.; Liu, C.; 1999, "Mechanical Wear Models and Experimental Investigation for The CMP Process," *Proceedings of the 6th Symposium on Nano Device Technology*, pp.20-23.
9. Tseng, W. T.; 1998, "Distribution of Pressure and Its Effect on the Removal Rate During CMP Process", *Proceedings of the 1998 VMIC*, pp. 87-94.

Proceedings of
2001 ASME International Mechanical Engineering Congress and Exposition
November 11–16, 2001, New York, NY
EPP-Vol. 1

IMECE2001/EPP-24718

A METHODOLOGY FOR EVALUATING THE MECHANICAL RELIABILITY OF ALTERNATIVES TO TIN-LEAD SOLDER ELECTRONIC INTERCONNECTION SYSTEMS (FOCUS ON ELECTRICALLY CONDUCTIVE ADHESIVES)

Anthony J. Rafanelli
Raytheon Company
Naval & Maritime Integrated Systems
1847 West Main Road-MS 369
Portsmouth, RI 02871
401-842-4850
FAX: 401-842-5326
E-mail: Anthony_J_Rafanelli@raytheon.com

ABSTRACT

A methodology for evaluating suitable alternative materials to traditional tin-lead solder, as an electronic component interconnection media, is presented. This methodology was developed as part of prior research that addressed the mechanical reliability of alternative materials using approaches and methods, which were traditionally unique to tin-lead, solder.

The original research included evaluation and subsequent tabulation of basic mechanical, fatigue, creep, and fracture mechanics properties. Basic mechanical data was acquired either through the literature or laboratory test. Fatigue data was limited to thermal shock/cycling. Creep data consisted of observing material behavior under temperature and thermo-mechanical loads. The fracture mechanics study features the validation and subsequent use of the J-Integral as applied to solder and electrically conductive adhesives (ECAs).

This paper suggests a framework for establishing a methodology for evaluating mechanical reliability of electronic interconnection media based on the acquired materials data.

INTRODUCTION

The electronics industry predominantly utilizes eutectic (63-37) and near-eutectic (60-40) tin-lead solder for component interconnection to printed wiring boards (PWB's). Lead provides many advantages. It is very workable and durable. It provides excellent mechanical strength and reliability as an electronic interconnection medium. As of 1995, world production was approaching an annual amount of 3.5 million metric tons with U. S. consumption being about 1.25 million metric tons [1].

Approximately 20,000 tons of lead in solder, each year, is used globally among electronics manufacturers [2]. This amount is approximately 7% of total lead consumption worldwide [3]. While tin-lead solder has established itself as a more than adequate interconnection material, world-wide legislation continues to surface in attempts to rid lead usage for the betterment of the environment. In the United States, the last several years have seen legislative attempts to ban all use of solder [4-9]. However, realizing the magnitude of effects on the electronics industry, legislation has been limited to lead-based paints and subsequent abatement. In Europe, the Waste Electrical and Electronic Equipment directive dominates the European Council "green" agenda. This directive has asked for a ban on lead by January 2004 [10]. As expected, opposition from Europe's main electronics houses was strong and, to date, suggestions of extending the date to 2009 are now being considered [11]. Nevertheless, some kind of restrictions would seem inevitable. Japan, similarly, has introduced legislation to ban lead having originally looked to ban lead by 2003. Surprisingly, Japanese electronics firms have taken their own initiatives in reducing and eliminating lead usage. Matsushita (Panasonic) and Sony have challenged themselves to remove lead from a large number of their own products by 2001. Hitachi is working on eliminating all lead by 2001. NEC has challenged itself to reduce its lead usage by 50% by 2002. Toshiba's plans are to remove lead from all mobile phones. Fujitsu is tasking itself to make all products lead-free by December 2002.

Subsequently, several initiatives have sought to improve the technology for the purposes of eliminating lead in the environment, saving processing costs, and facilitating production methods [12-14]. Most of these initiatives include non-lead solder alloys and electrically conductive adhesives (ECAs).

However, evaluation and selection of alternatives represents a formidable effort. Several major areas of study are necessary to properly and accurately qualify alternative bonding systems [13]. These areas include basic materials properties, fatigue, creep, fracture, ancillary technologies, functionality, and manufacturability. Final selection must be based on optimization of the results of these studies.

This paper provides a summary-level report of research [13] in which the first four areas of study were addressed by implementing a material selection methodology which included the use of fatigue, creep, and fracture properties as a means of comparatively evaluating mechanical reliability. The methodology also included a rating scheme for determining the most promising alternative candidates.

THE EXPERIMENTAL PLAN

The outline, shown in Table 1 below, illustrates a set of tasks that would provide a comprehensive and sufficient program to suitably evaluate alternatives [12]. However, to accommodate logistical and funding constraints, a subset of these tasks was created and executed as shown in Table 2. Emphasis was placed on deriving a plan that would include the least amount of tests necessary to characterize the mechanical reliability of candidate materials. The overall plan, originally proposed, was based on historical work performed on 63-37 Sn-Pb and 60-40 Sn-Pb. Alternate materials were limited to electrically conductive polymeric (epoxy) adhesives. Extending similar approaches to the adhesives was felt to be appropriate since these tests represent baseline parameters for interconnection media in the industry.

Table 1. Technical Approach to Selection and Optimization of Alternatives to Tin-Lead Solder Component Attachment Systems

I. BASIC MATERIAL PROPERTIES
   - A. Ultimate Tensile Strength
   - B. Ultimate Elongation
   - C. Poisson's Ratio
   - D. Young's Modulus
   - E. Shear Strength
   - F. Stress/Strain
   - G. Fatigue Strength

II. FATIGUE, CREEP, AND FRACTURE
   - A. Fatigue Crack Growth Rate vs. Thermal Cycle
   - B. Coffin-Manson Equation
   - C. Engelmaier Equation
   - D. Creep
   - E. J-Integral

III. ANCILLARY TECHNOLOGIES
   - A. Experimental Strain Measuring Techniques for Solder Joints
     1. foil resistance strain gages
     2. fiber optic strain sensors
     3. photoelastic
     4. other
   - B. Joint Fillet Boundary Conditions
   - C. Interconnection Geometry, Volume, and Dimensional Criteria

IV. FUNCTIONALITY (BEFORE AND AFTER THERMAL CYCLE)
   - A. Bond Strength
   - B. Electrical Conductivity

V. MANUFACTURABILITY
   - A. Processability
     1. printability
     2. pot life/shelf life
     3. surface preparation
   - B. Compatibility
     1. cleaners
     2. rework
   - C. Toxicity/Hazards

VI. OPTIMIZATION/SELECTION

Table 2. Streamlined Set of Tasks Addressing Mechanical Reliability of Tin-Lead Solder Alternate Component Attachment Materials

I. BASIC MATERIAL PROPERTIES
   - A. Ultimate Tensile Strength
   - B. Modulus of Elasticity
   - C. Poisson's Ratio
   - D. Stress/Strain

II. FATIGUE, CREEP, AND FRACTURE
   - A. Fatigue Crack Growth Rate
   - B. Creep
   - C. Fracture
     1. Strain gage approach
     2. Photomechanics approach

III. EVALUATION/OPTIMIZATION

Basic materials properties are very well documented for eutectic and near-eutectic solders for the following: tensile/yield strength, elongation, elastic modulus, shear strength, stress/strain curves, and fatigue strength. Despite some variations, mainly due to effects of temperature and strain rate, a set of baseline values were obtained from this literature. Data on Poisson's ratio was not plentiful but was obtained from testing. At the time of the research, data was not as plentiful for non-lead solder and even more lacking for conductive adhesives. A key area of focus was determining effects of environment (temperature) on the non-lead solders and polymers.

While the majority of solder fatigue studies has resulted in life (i.e. number of cycles) prediction, little work was performed in characterizing crack propagation. Solomon reported on work using load drop criteria which was correlated to crack growth [15]. The ability to determine/measure fatigue crack growth rate (FCGR) is desirable for varying reasons. Primarily, FCGR data would provide a direct measurement of the "degree" of failure. While load drop, for instance, may correlate to crack growth, loads could vary due to creep-fatigue interaction, thereby leading to differing values or load ranges. Industry has expressed interest in measuring FCGR of solder as well as other interconnection media [16, 17]. Thus, thermal cycling was performed to monitor fatigue crack propagation

Creep data, while prevalent for tin-lead, needed to be acquired through test for other polymeric (hereto referred to as epoxies) materials. Tests were performed using temperature and/or dead weights as the loading mechanisms since previous work on solder dealt with room temperature and thermal creep. Focus was directed to the epoxies to determine if creep was prevalent, to what degree, and its interaction with fatigue, if any. Creep measurements were be made at selected temperatures, within the range of -55C to +85C, since these environments represent most military and commercial requirements.

The J-integral was determined using the HRR power law approach [18-20], considered valid since solder is a hardening material. The methodology, including specimen geometry and procedure, was based upon previous work performed by Nigam [21] in his dynamic and elastic-plastic fracture studies of Homalite® 100 polymer, 4340 steel, and 7075-T6 aluminum.

The photoelastic experiment was pursued as an experimental stress analysis technique to Mode I stress intensity factors using a scaled up version of a typical electronic component, i.e., a chip capacitor, bonded to a printed wiring board with simulated solder joints exhibiting cracks of varying lengths and directions. Results of these tests are considered beyond the scope of this particular report.

Optimization was performed by evaluating conductive adhesive data against the 63-37 or 60-40 eutectic and near-eutectic benchmarks utilizing a ranking and rating system to determine which of the alternate materials was comparable to the reliability of the solder.

Upon performing a top-level review of various engineering and processability properties, the electrically conductive adhesives, shown in Table 3, were selected for study.

Depending upon availability, samples of all or some of the materials shown in Table 3 were subjected to each set of tests (Basic Materials Properties, Fatigue, Creep, Fracture).

Table 3. Schedule of Material Samples Tested

material	one or two part	filler	cure temp. & time	volume resist. (ohm-cm)	thermal conduc-tivity
63-37 tin-lead solder	n/a	n/a	n/a	2.83	0.121 Btu/ft^2-cm-C-s
Silver 1	1	silver <80%	150C 1 hr.	0.0001 to 0.0004	1.7 watt/m°K 12.0 Btu-in/ft^2 -hr-F
Silver 2	2	silver 80%	150C 30 min 125C 1 hr.	0.00008 to 0.00030	11.5 Btu-in/ft^2 -hr-F
Silver 3	2	silver <90%	25C 24 hr. 65C 2 hr.	0.0050	n/a
Silver 4	2	silver 50-90%	130C 30 min. 150C 15 min.	<0.1	n/a
Silver 5	2	silver <15%	22C 8 hr. 100C 30 min.	0.018	3.2 Btu -in/ft^2 -hr-F
Nickel 1	2	nickel 70-90%	25C 72 hr. to 160C 5 min.	0.020 to 0.003	37x10^{-4} cal-cm /cm^2-sec^2-C
Silver 6	2	silver 70-90%	25C 24 hr. to 160C 5 min.	0.0010 to 0.0003	37x10^{-4} cal-cm/ cm^2-sec^2-C
Silver 7	2	silver 70-90%	25C 24 hr. to 110C 1 hr.	0.0010 to 0.0008	49x10^{-4} cal-cm/ cm^2-sec^2 -C

RESULTS

Basic Materials Property Tests. Samples were tested on a tensile-compression machine at a pull speed between 7.6 and 12.7 mm per minute. All tests were performed at room temperature. Unless otherwise specified, true strains (i.e. $\varepsilon = \ln [e + 1]$ where e is the engineering strain and ε is the true strain) are used. Table 4 provides a summary of the results.

Table 4. Summary of Materials Property Test Results

material	ultimate tensile strength (MPa)	mean[a] modulus of elasticity (MPa)	mean[a] Poisson's ratio
63-37 Sn-Pb	44.8	$2.00(10)^4$	0.33
Silver 1	21.5	$0.60(10)^4$	0.32
Silver 2	3.1	$0.09(10)^4$	0.30
Silver 3	2.0	$0.04(10)^4$	n/a[c]
Silver 4	6.5	$0.42(10)^4$	0.22
Silver 5	18.3	$1.20(10)^4$	n/a[c]
Nickel 1	15.6	$0.20(10)^4$	0.50
Silver 6	16.6	$0.43(10)^4$	0.20
Silver 7	6.9[b]	n/a[c]	n/a[c]

[a] statistical average of readings at incremental loadings
[b] estimated as twice that of the published shear strength
[c] could not be experimentally determined

Having obtained material properties, the data was rearranged to indicate how each material fared against the tin-lead eutectic (hereon in noted as the benchmark) solder. Table 5, below, shows a ranking of these materials against the benchmark with "1" indicating closest to the benchmark value. For brevity, Poisson's ratio was not listed in this paper although ranking data was included in the original research [13].

Table 5. Ranking of Materials per Material Property Results

Ultimate Tensile Strength (MPa)		Modulus of Elasticity (10^4MPa)	
rank	Benchmark: 44.8	rank	Benchmark: 2.00
1	Silver 1: 21.5	1	Silver 5: 1.22
2	Silver 5: 18.3	2	Silver 1: 0.60
3	Silver 6: 16.6	3	Silver 6: 0.43
4	Nickel 1: 15.6	4	Silver 4: 0.42
			Cont'd
Cont'd 5	Silver 7: 6.9	5	Nickel 1: 0.21
6	Silver 4: 6.5	6	Silver 2: 0.10
7	Silver 2: 3.1	7	Silver 3: 0.04
8	Silver 3: 2.0	8	Silver 7: n/a

Fatigue (Thermal) Test Results. Thermal fatigue tests were performed by subjecting each sample to temperature cycling as follows: -57C to +125C, ramp time of ten (10) seconds and dwell time of ten (10) minutes. These specific limits were requirements of another special electronic assembly, and in the interest of expediency, the samples were "piggy-backed" with the special unit. This decision was validated on the basis that the worst case temperature cycle/shock requirement for military printed circuit assembly hardware is -55C to +125C [22]. Due to logistical constraints, only 1000 cycles were applied resulting in no noticeable increase in crack length for any of the samples. However, it should be noted that the accuracy of the crack measurements was within ± 0.8 mm. Thus, micro-cracking (which was not monitored) may have been present. Past experience on tin-lead solder has shown that significant propagation can be on the order of 0.8 to 1.6 mm. Therefore, for comparative purposes, these results are useful although it would have been more desirable to continue up to the 10000 cycle low-cycle limit. Since all results were identical, no ranking table was derived for fatigue.

Thermal Creep (Thermal Stability). This test measured strain while several samples were exposed to different air temperatures. The objective of this effort was to determine if any of the materials creeped, over long periods of time, when only exposed to temperature. Any significant "thermal" creep could greatly influence failure mechanisms. Consequently, designers would be required to account for material strains at different temperatures without any mechanical loading.

Samples were instrumented with strain gages such axial and lateral data could be obtained. Temperatures selected were room temperature (20C), 0C, -55C, and +85C. These temperatures represent key points of concern in designing for the military environment.

Evaluation of thermal creep was performed by examining strain versus time at various temperatures. A simplistic analysis consisted of inspecting each curve for minimum creep change.

Table 6 is a summary of qualitative evaluations of material behavior at each temperature. (Due to logistical constraints, not all of the candidate materials could be evaluated.)

Table 6. Results of Qualitative Evaluation of Thermal Stability			
Material	**Temp. (C)**	**Thermal Stability Rating**	
		axial creep	**lateral creep**
63-37 Sn-Pb	room	modest variation	unstable
"	0C	unstable	modest variation
"	-55C	stable	modest variation
"	85C	modest variation	exponential
Silver 2	room	stable	exponential
"	0C	stable	unstable
"	-55C	stable	stable
"	85C	stable	exponential
Silver 4	room	modest variation	unstable
"	0C	stable	modest variation
"	-55C	stable	modest variation
"	85C	modest variation	unstable
Silver 5	room	modest variation	exponential
"	0C	modest variation	modest variation
"	-55C	stable	stable
"	85C	modest variation	stable
Nickel 1	room	exponential	exponential
"	0C	stable	stable
"	-55C	stable	stable
"	85C	stable	stable

The materials were then ranked as to how stable they were at each temperature (in the axial mode only). Ranking was established by tallying the total "favorable" ratings each material exhibited at each temperature. While the data of Table 6 is subjective in nature, it still serves a purpose in illustrating the methodology. One would suitably define these qualitative descriptions and then incorporate the data in the final ranking scheme. Ratings were as follows; 1: "Stable" was the most desirable, 2: "Modest variation" was the next best, and 3: "large" or "exponential" or "unstable" were the least desired. Table 7 shows the rankings.

Table 7. Ranking of the Thermal Creep Results						
Material	**Rankings**				**Total Ranking**	**Rating**
	RT.	**0C**	**-55C**	**85C**		
63-37 Sn-Pb	2	3	1	2	8	5
Silver 2	1	1	1	1	4	1
Silver 4	2	1	1	2	6	2
Silver 5	2	2	1	2	7	4
Nickel 1	3	1	1	1	6	3

As shown in Table 8, materials were performance rated (1 = best) based upon their total ranking (the lower the rank, the better the material). Note that the eutectic solder benchmark behaved the worst.

The significance of thermal stability would be manifested in situations where electronic equipment would be subjected to large down times in between operational or power cycling sequences at various locations around the earth. Operation and power cycling tend to thermally stress component interconnections resulting in significant strains. Moreover, if these components were subjected to long periods of "down" time, the interconnections (i.e. solder joints) would thermally return to ambient temperature. Depending on which sector of the world one is located, ambient temperature can range from sub-freezing to room temperature to the tropics. Thus, the creep behavior of interconnection materials at these various ambients should be considered when determining the overall strains that the interconnection will experience during service.

Thermo-mechanical Creep. Samples were subjected to thermo-mechanical creep by applying tensile (dead weight) loads to each sample and subjecting them to the following temperatures: room (20C), 0C, -55C, and +85C. For this test, all eight alternate materials, as well as eutectic solder, were evaluated. An arbitrary decision was made to limit tensile loads to 138 KPa maximum. Table 8 provides a summary of mean creep data rates obtained for each material at each temperature.

In the original research, attempts were made to fit the creep data to various models such as Arrhennius, Dorn, and the linear approach. It was reported that some agreement occurred with the Arrhennius and Dorn models. Agreement with the linear model was limited by critical effects at room temperature. For brevity, details on the modeling efforts are omitted from this report but a full discussion is provided in the dissertation [13].

Table 8. Summary of Mean Thermo-mechanical Creep Rates

Material	Load (KPa)	Mean Creep Rate (microstrain/hour)			
		RT	0C	-55C	+85C
63-37 Sn-Pb	48.3	0.5228	-2.860	-2.6065	3.8843
Silver 1	111.7	2.1625	1.9431	-0.5232	1.5968
Silver 2	90.3	13.7381	23.516	-2.6377	-50.0862
Silver 3	94.5	14.2163	16.668	-6.2792	2.7833
Silver 4	88.3	1.6821	1.1365	-1.9149	2.1275
Silver 5	80.7	2.5711	1.8641	-1.2546	-30.5220
Nickel 1	93.1	1.7990	0.0574	0.3548	1.0588
Silver 6	96.5	0.8810	0.1906	-0.2507	-1.1756
Silver 7	103.4	1.1100	1.0041	-1.5579	-3.9615

Evaluation of materials was performed by reviewing the mean creep rate data and ranking this data within each temperature category. The ranking is defined as follows: 1 = lowest creep rate, 9 = highest creep rate. Table 9 summarizes the materials and rankings.

In evaluating thermo-mechanical creep performance, the material exhibiting the lowest creep ranking earns the best performance rating, i.e., 1 being the best and 9 being the worst. From the data of Table 6, the best five materials concerning thermo-mechanical creep performance are Silver 6, Nickel 1, Silver 7, Silver 4, and Silver 1. Moreover, Table 6 shows that Silver 6, Nickel 1, and Silver 7 were superior to the eutectic solder benchmarks in terms of creep rate.

In closing, the data suggests that , for the majority of materials tested, conductive adhesives exhibited comparable or better creep resistance than the benchmark.

Table 9. Rankings of Thermo-mechanical Creep Performance

Material	Ranking of Creep Rate at Temperature				Total Creep Ranking	Over-all rating
	RT	0C	-55C	85C		
63-37 Sn-Pb	1	7	8	6	22	6
Silver 1	6	6	3	3	18	5
Silver 2	9	9	7	9	34	9
Silver 3	8	8	9	5	30	8
Silver 4	4	4	6	4	18	4
Silver 5	7	5	4	8	24	7
Nickel 1	5	1	2	1	9	2
Silver 6	2	2	1	2	7	1
Silver 7	3	3	5	7	18	3

Fracture Properties (J-Integral). The J-Integral [20] was used to describe the fracture properties of the solder and the candidate materials. The dissertation provides a detailed discussion supporting the use of the J-Integral. Issues and concerns with the J-Integral were identified as summarized in Table 10. Key to the justification was resolving these issues as shown in Table 11.

Table 10 Issues and Concerns With J-Integral

J-Integral Criteria	Concern
Path independence	J can be applied to a stationary crack only.
Small scale yielding	In plastic zone, stress field cannot be dominated by J. Plastic region must be small w/r to size of "elastic" region, i.e. specimen width and ligament length. J dominance for $b \geq 200\ J/\sigma_{ys}$ where b is ligament size [25]
Crack extension	Must be small w/r to region D (area between contours)
Singularities	No singularities (e.g. voids) between the contours (i.e. region D)
Energy dissipation	For elastic-plastic materials, energy is dissipated within, thus the Eq's below may not be equivalent; $J = \int_{\Gamma} (Wdy - T(\delta u/\delta x)ds)$ $J = {}_0\int^{\zeta} (-\delta P/\delta a)_{\zeta} d\zeta$ and $J = {}_0\int^{P} (\delta\zeta/\delta a)_P dP$

Cont'd

Table 10 Issues and Concerns With J-Integral	
J-Integral Criteria	**Concern**
Cont'd Elastic-plastic behavior	An approximation based upon stress/strain curves

Table 11. Resolution of J-Integral Concerns	
CONCERN	**RESOLUTION**
Path independence	Ensure homogeneity
Small-scale yielding	Limit applied stress
Crack extension	Limit applied stress
No singularities	Ensure homogeneity (process control)
Energy dissipation	Limit applied stress
Solder as elastic-plastic material	Limit applied stress and load rate

It appears that use of the J-Integral may require qualification based upon the particular material at hand and conditions surrounding its processing. For example, the path independence and no singularities criteria could be met by ensuring material homogeneity which is achieved by suitable process control. Small-scale yielding, crack extension, and energy dissipation criteria could be met by limiting applied stresses. The idea of bounding J-Integral analysis within applied stress levels may not be unfeasible. When one considers that the bond area of one solder joint is on the order of square mils in magnitude, stress limits of 20.7 MPa (3000 psi), for example, indicate that large forces can be withstood. Finally, the elastic-plastic criteria is satisfied provided uniaxial tension tests prove that the material displays traditional stress/strain characteristics. Other issues included morphology effects. Morphology effects, while possibly a factor for solder, may not be as critical in epoxies assuming optimum homogeneity can be achieved.

Therefore, an argument was presented for use of the J-Integral as a fracture mechanics property for the solder and candidate materials.

The J-Integral was determined using experimental and analytical approaches. Interest, in utilizing these approaches, was inspired by the work of Nigam [21], concerning his dynamic and elastic-plastic fracture studies of HOMALITE®100 polymer, 4340 steel, and 7075-T6 aluminum. In his work, he attempted evaluation of the J-Integral using a strain gage method incorporating the HRR power law for hardening material. Using strain data, expressions for "theoretical" (J_{th}) and "experimental" (J_{exp}) J values were obtained.

In determining J_{exp}, it can be shown that, for small strain formulation, the asymptotic stress σ_{ij} and strain ε_{ij} distribution in the crack tip region can be expressed as follows, in terms of polar coordinates [24-25]

$$\sigma_{ij} = \sigma_o[J/(\alpha\sigma_o\varepsilon_o I_n r)]^{(1/[n+1])} S_{ij}(n,\Theta) \quad (1)$$

$$\varepsilon_{ij} = \alpha\varepsilon_o[J/(\alpha\sigma_o\varepsilon_o I_n\, r)]^{(n/[n+1])} E_{ij}(n,\Theta) \quad (2)$$

where I_n is a dimensionless constant, E_{ij} and S_{ij} are dimensionless functions, α is a material constant, n is the hardening exponent, σ_o and ε_o are the Ramberg-Osgood secant yield strength and strain, respectively [26]. Through a series of derivations, Nigam showed that Eqs. (1) and (2) eventually lead to the following expression:

$$J_{exp} = \alpha\, \sigma_o\, \varepsilon_o\, I_n r[2\varepsilon_g/ (\alpha\, \varepsilon_o\{E_{rr}(1+\cos 2\psi)+E_{\Theta\Theta}(1 - \cos 2\psi)+2E_{r\Theta}\sin 2\psi\}] \quad (3)$$

where ε_g is the measured strain at the angular location ψ. Suggested values of I_n, E_{ij}, and S_{ij} have been determined by [27-28] for values of n.

The expression for J_{th}, derived by Kumar, German, and Shih [29], is as follows:

$$J_{th} = [\pi a_e F^2 P^2/w^2 E] + \alpha\sigma_o\varepsilon_o\, c(a/w)h_1(a/w,n)[P/P_o]^n + 1 \quad (4)$$

where $P = 1.072\delta c\sigma_o$,$c=w-a$, $a=a_e + \phi r_p$, $\delta = \{[1 + (a/c)^2]^{(1/2)} - (a/c)\}$, $r_p = \{(1/2\pi) \times [(n-1)/(n+1)](K/\sigma_o)^2\}$, $\phi = \{1/(1+(P/P_o)^2)\}$, P is the load per unit thickness, P_o is the load per unit thickness for the perfectly plastic case, w is the specimen width, and a is the crack length. K is stress intensity factor. F is a stress intensity geometrical parameter and h_1 is a function representing fully plastic crack solutions [29].

Values for h and F were taken from Kumar et.al., [23] and Tada et.al., [30]. Key to the data, conveyed in this research, was the use of the Ramberg-Osgood relation to calculate the secant yield strength σ_o and yield strain ε_o, as well as the material constant α for strain hardening exponent n [26]. Description of the Ramberg-Osgood method and justification for its use have been previously discussed [31-32]. In short, it was determined that, since the material did not truly conform to Hooke's law, Young's modulus and the yield strength would be insufficient to accurately define a stress-strain curve and, consequently, a third parameter (a secant yield strength) would be required.

The connection between the Ramberg-Osgood relation and the HRR power law is illustrated as follows: For a hardening material, such as solder [33], the HRR power law is expressed as follows [21]:

$$[\varepsilon/\varepsilon_o] = \alpha[\sigma/\sigma_o]^n \quad (5)$$

Since the Ramberg-Osgood relation is:

$$[\varepsilon/\varepsilon_o] = [\sigma/\sigma_o] + \alpha[\sigma/\sigma_o]^n \quad (6)$$

one can see that Eq.(5) is a simplified form of Eq.(6). This special case is the condition of small non-linear strain, i.e., $\sigma/\sigma_o \rightarrow 0$, which is true in most practical applications [34]. This case explains why deflection testing, and in many cases, power cycling, is more representative of "actual life" than thermal cycling [35].

A comprehensive experimental program was executed to determine various material properties in order to use the J_{exp} approach. Samples of both eutectic tin-lead solder and conductive adhesives were fabricated and instrumented using traditional foil resistance strain gages and subjected to tensile tests to measure the required Ramberg-Osgood parameters. Samples were tested for different crack lengths. Sample materials, subjected to J_{exp} and J_{th} analysis, are listed in Table 12.

Table 12. Schedule of Samples Subjected to J-integral Analysis

Material	Material
63-37 tin-lead	Silver 5
Silver 1	Nickel 1
Silver 2	Silver 6
Silver 3	Silver 7
Silver 4	

Table 13. J-Integral Values

Material	J_{exp} (N/m)[1]	J_{th} (N/m)[1]
63-37 tin-lead	9670	2675
Silver 1	637	313
Silver 2	94	336
Silver 3	5352	888
Silver 4	457	358
Silver 5	235	306
Nickel 1	7820	561
Silver 6	488	650
Silver 7	4001	1328

[1] Critical values

In general, one would expect the experimental approach to be more conservative than the theoretical analysis. The variability in the data of Table 13 is probably attributed to three factors regarding experimentation:

1. Use of actual strain gage data in Eq. (3).
2. Calculation of a plastic zone size using stress intensity factor K in Eq. (4).
3. Size of strain gage.

Concerning the first factor, use of actual strain data to validate analysis can be difficult since the data can be slightly influenced by environment and test procedure. Next, a calculated plastic zone size may not be realistic for some of the epoxies since their content (i.e. per cent resin versus per cent metal) could result in plastic zones larger or smaller than expected by analysis. Finally, the size of the strain gage may produce some variability in the value of angle θ due to small specimen dimensions.

Despite the wide disagreement in the actual magnitude of the data, rearrangement of Table 13 into a ranking file is beneficial in identifying those materials exhibiting a J-integral value that meets or exceeds (the most desired situation) the eutectic solder benchmark. Rating numbers are assigned to each material with "1" being the most desired and "8" the least desired situation.

Table 14. Ratings and Rankings of Materials for J-Integral Values

Mate-rial	J_{exp} (N/m)	Rating	J_{th} (N/m)	Rating	Total Rating	Rank
63-37 Sn-Pb	9670	bench-mark	2675	bench-mark		
Silver 1	637	4	313	7	11	4
Silver 2	94	8	336	6	14	5
Silver 3	5352	2	888	2	4	1
Silver 4	457	6	358	5	11	4
Silver 5	235	7	306	8	15	6
Nickel 1	7820	1	561	4	5	2
Silver 6	488	5	650	3	8	3
Silver 7	4001	3	1328	1	4	1

From Table 14, both Silver 3 and Nickel 1 are the best "performers" of J-Integral comparability to the eutectic tin-lead.

Comprehensive Ranking. Having completed several physical/mechanical tests, the data for the alternative materials was compiled into a major matrix (Table 15, located at the end of this paper due to size constraints) highlighting their performance ranking and subsequent overall ranking for comparability to the eutectic solder benchmarks. For each property, the ranking is shown along with a weighted rating which is equal to the product of the material's rank and a weighted percentage for the property. The weighted percentage may be considered somewhat arbitrary. However, since thermo-mechanical creep and fracture are considered the prime areas of concern in interconnection reliability, an equally large weighted percent (40%) was assigned to each. The remaining twenty percent was divided among the four remaining properties. In rating each material overall, the lowest "weighted rank" corresponds to the best candidate material.

The better performing materials are rated from lowest to highest weighted rank. Those materials, marked by footnote "1" in Table 15, were not tested for some of the less significant properties (i.e. those properties with lower weighted percentage rankings) due to material limitations. In all likelihood, there is still enough data to include these materials in the overall comparison since the "missing" data would probably not be factors in the total ratings.

Therefore, in reviewing Table 15, the following materials, as listed in Table 16, are rated in accordance with their near comparability to the mechanical properties of eutectic solder.

Table 16. Ratings of the Alternate Materials as Compared to 63-37 Sn-Pb (eutectic) Solder

Rating	Material
1	Silver 7 (70-90%)
2	Silver 6 (70-90%)
3	Nickel 1 (70-90%)
4	Silver 1 (<80%)
5	Silver 4 (50-90%)
6	Silver 3 (<90%)
7	Silver 2 (80%)
8	Silver 5 (<15%)

A review of Table 16 also provides some insight concerning the relationship of the metal properties and their amounts in each epoxy with overall performance. The top three performing epoxies (Silver 7, 6, and Nickel 1) all contain the highest lower and upper limits of filler metals (i.e. 70 to 90%). The next four materials (Silver 1, 4, 3, and 2) while showing comparable or higher upper limits, did not perform as well as the top three probably due to their containing far less metal filler than the first three, i.e. the first three materials (Silver 7, 6, and Nickel 1) "guarantee" a minimum of 70% metal where the next four materials, citing upper limits only, could have well had far less metal filler than the top three materials. Finally, the last material, Silver 5, was the worst performer, not surprisingly, due to an upper limit metal filler of only 15%, the least of all the epoxies.

Limited morphology analysis also provides some additional insight regarding the behavior and ratings exhibited in Table 16. Appendix A contains a summary of the morphological features of the ECAs. Upon reviewing this data, it can be shown that the three silver epoxies, filled with both flakes and nodules, were among the top five performers per Table 16. Moreover, these three silver epoxies (Silver 7, 6, and 4) contained fillers with the largest flake/nodule sizes. Furthermore, Nickel 1 which was the third best performer, shows smaller filler sizes than the silver materials. Yet, the performance of Nickel 1 as one of the five best materials is probably attributed to its bulk properties, i.e. a higher tensile strength (345 MPa) and elastic modulus ($20.7(10)^4$ MPa) than silver (152MPa, $7.6(10^4)$MPa). Therefore, the limited morphological data of Appendix F provides some correlation to the overall performance (Table16) of these epoxies.

The reliability of these alternate materials was defined as to how well they compare to the tin-lead solder benchmark for all tests performed (Table 17, at end of this paper due to size constraints). In other words, Silver 7 is the most reliable of the alternate materials as compared to tin-lead solder.

CONCLUSIONS

A methodology for evaluating the mechanical reliability of several alternative materials (conductive epoxies) as electronic component interconnection media has been developed based upon a review of several "pockets" of research performed on tin-lead solder technology and subsequent compilation of this research into a sequential method of evaluation. Data from tin-lead research has been used as the benchmark for comparison of the alternative materials, which was performed by employing a ranking/rating system. Reliability was defined as the degree of comparison to the tin-lead benchmark.

The key contribution of this paper is not just in the magnitudes of the data, but in the methodology presented. Despite any questions with absolute values of physical properties, the methodology provides for relative comparisons between a benchmark and alternative materials. The procedure allows the experimenter to determine which properties he/she feels are critical and, subsequently, allows for corresponding weighted allowances in the rating scheme. Such an approach saves time by focusing on relative comparisons against a known baseline thereby accommodating industrial constraints. The methodology also compliments the "drop-in" approach to evaluating tin-lead substitutes that various industry groups are addressing.

Since new materials will have a large-scale impact on future electronic product performance, a disciplined approach to alternative selection is of the utmost importance.

ACKNOWLEDGMENTS

The author would like to thank Mr. William Robinson, Mr. Ernest Swenson, and the Raytheon Company for their support in the writing of this paper. Appreciation is also expressed to Mr. Frank Liotine, Jr. and Dr. Arun Shukla of the University of Rhode Island. Finally, the author would like to thank his wife Karen and sons John and Matthew for their patience and support.

REFERENCES

[1] Reese, Ken, "Getting the Lead Out", **TODAY'S CHEMIST AT WORK**, March 1995, pp. 61-62.

[2] Trumble, Bill, "Get the Lead Out", **IEEE SPECTRUM**, May 1998, pp. 55-60.

[3] Kwoka, Mark A. and Foster, Dawn M., "A Comparison of Lead-Free vs. Eutectic Solders", **CIRCUITS ASSEMBLY**, October 1995, pp. 32-35.

[4] S. 391, "Lead Exposure Reduction Act of 1991", United States Senate, 7 February 1991.

[5] H.R. 1750, "Lead Exposure Reduction Act of 1991, United States House of Representatives, 11 April 1991.

[6] H.R. 2840, "Lead Contamination Control Act Amendments of 1991", United States House of Representatives, 10 July 1991.

[7] H.R. 5730, "Lead Exposure Control Act of 1992", United States House of Representatives, 31 July 1992.

[8] S. 729, "Lead Exposure Reduction Act of 1993", United States Senate, 1 April 1993.

[9] S. 1347, "Lead Abatement Trust Fund Act of 1993", United States Senate, 3 August 1993.

[10] Karim, Zaheed and Martin, Jim, "LEAD-FREE SOLDER BUMP TECHNOLOGIES FOR FLIP-CHIP PACKAGING APPLICATIONS", **ADVANCED PACKAGING TECHNOLOGIES IN THE ELECTRONICS INDUSTRY**, SMTA Proceedings, NEPCON East 2001, June 12-13, 2001, pp. 87-92.

[11] Robins, Mark, "The History of Pb-Free", Special Report to **ELECTRONIC PACKAGING AND PRODUCTION**, June 2000, URL ysig:/ 939/ http://www.epp.com/special/ default0700_2as

[12] Rafanelli, Anthony J., "An Overview of Alternatives to Tin-Lead Solder Component Attachment Systems", **ASME Publication 94-WA/EEP-4**, American Society of Mechanical Engineers, International Mechanical Engineering Congress and Exposition, ASME, November 6-11, 1994.

[13] Rafanelli, Anthony J., "A Mechanical Reliability Approach for the Optimization of Alternatives to Tin-Lead Solder Interconnection Systems for Electronic Packaging", Ph.D. Dissertation, University of Rhode Island, 1995.

[14] Crum, Susan, "Targeting Lead-Free Solutions", **ELECTRONIC PACKAGING AND PRODUCTION**, June 1999, pp. 28-36.

[15] Solomon, H. D., "Strain-Life Behavior in 60/40 Solder", ASME Technical Paper 88-WA/EEP-10, ASME Winter Annual Meeting, 1988.

[16] Dally, James W., University of Maryland, private communication, April 1992.

[17] Read, D., NIST, Boulder, CO, private communication, February, 1992.

[18] Hutchinson, J. W., 'Plastic Stress and Strain Fields at a Crack Tip", **JOURNAL OF MECHANICS AND PHYSICS OF SOLIDS**, Volume 16, 1968, pp. 337-347.

[19] Rice, J. R. and Rosengren, G. F., "Plane Strain Deformation Near a Crack Tip in a Power Law Hardening Material", **JOURNAL OF MECHANICS AND PHYSICS OF SOLIDS**, Volume 16, 1968, pp. 1-12.

[20] Rice, J. R., A Path Independent Integral and the Approximate Analysis of Strain Concentration by Notches and Cracks", **TRANSACTIONS OF THE ASME JOURNAL OF APPLIED MECHANICS**, June 1968, p. 379.

[21] Nigam, Himanshu, **EXPERIMENTAL STUDIES IN DYNAMIC AND ELASTIC PLASTIC FRACTURE**, Ph.D. Dissertation, University of Rhode Island, Kingston, RI, 1988.

[22] MIL-M-28787D, "Military Specification, Modules, Standard Electronic", Naval Sea Systems Command, Washington, D.C., 30 March 1989.

[23] Kumar, V., German, M. D., and Shih, C. F., "An Engineering Approach for Elastic-Plastic Fracture Analysis", EPRI Report NP-1931, July 1981.

[24] Hutchinson, J. W., "Singular Behavior at the End of a Tensile Crack in a Hardening Material", **JOURNAL OF THE MECHANICS AND PHYSICS OF SOLIDS**, Volume 16, 1986, pp. 13-31.

[25] Hutchinson, J. W., "Plastic Stress and Strain Fields at a Crack Tip", **JOURNAL OF THE MECHANICS AND PHYSICS OF SOLIDS**, Volume 16, 1986, pp. 337-347.

[26] Ramberg, Walter and Osgood, William R., "Description of Stress-Strain Curves by Three Parameters", Technical Notes National Advisory Committee for Aeronautics No. 902, Washington, DC, 1943.

[27] Shih, C. F., **"ELASTIC-PLASTIC ANALYSIS OF COMBINED MODE CRACK PROBLEMS"**, Ph.D. Thesis, Harvard University, Cambridge, MA, 1973.

[28] Shih, C. F., "Tables of Hutchinson- Rice-Rosengren Singular Field Quantities", Brown University Report MRL E-147, Brown University, Providence, RI, June 1983.

[29] Kumar, V., German, M. D., and Shih, C. F., "An Engineering Approach for Elastic-Plastic Fracture Analysis", EPRI Report NP-1931, July 1981.

[30] Tada, H., Paris, P. C., and Irwin, G. R., **THE STRESS ANALYSIS OF CRACKS HANDBOOK,** Del Research Corporation, Hellertown, PA., 1973.

[31] Rafanelli, Anthony J., "Ramberg-Osgood and HRR Power Law Relations for 63-37 SN-PB Solder", ASME Technical Paper 91-WA-EEP-21, ASME Winter Annual Meeting, December 1991.

[32] Rafanelli, A. J., "Ramberg-Osgood Parameters for 63-37 Sn-Pb Solder", **TRANSACTIONS OF THE ASME JOURNAL OF ELECTRONIC PACKAGING,** Volume 114, June 1992, pp. 234-238.

[33] Lau, J., Harkings, G., Rice, D., and Kral, J., "Thermal Fatigue Reliability of SMT Packages and Interconnections", **1987 IEEE INTERNATIONAL RELIABILITY PHYSICS SYMPOSIUM.**

[34] Broek, David, **ELEMENTARY ENGINEERING FRACTURE MECHANICS,** Martinus Nijhoff Publishers, Boston, MA, 1987, page 79.

[35] Liotine, F. J., Jr., private communication, May 1993

APPENDIX A. MORPHOLOGICAL DATA

Material	Rating[1]	Filler Geometry	Range in Size (mils)	Range in Size (microns)
Silver 1	4	nodules	0.3 to 1.5	7 to 38
Silver 2	7	nodules	0.3 to 3.0	7 to 76
Silver 3	6	flakes	0.1 to 1.0	2.5 to 25
Silver 4	5	flakes & nodules	0.5 to 1.4 0.4 to 1.2	12 to 36 12 to 32
Silver 5	8	nodules	0.3 to 2.0	8 to 56
Nickel 1	3	nodules	0.1 to 0.6	4 to 16
Silver 6	2	flakes & nodules	0.3 to 1.2 0.8 to 1.5	8 to 32 20 to 40
Silver 7	1	flakes & nodules	0.4 to 1.7 0.3 to 2.0	12 to 44 8 to 56

[1] Refer to Table 16.

Table 15. Rankings/Ratings Summary of the Alternative Materials by Parameter													
Material	Ultimate Tensile Strength (5%)		Modulus of Elasticity (5%)		Fatigue (Thermal) (5%)		Thermal Creep (5%)		Thermo-mech. Creep (40%)		J-Int (40%)		Tot.Wt. Rank
	rank	wt.%	rank	wt %	rank	wt%	rank	wt%	rank	wt%	rank	wt%	
Silver 1	1	0.05	2	0.10	1	0.05	1	0.05	5	2.00	4	1.60	3.85
Silver 2	7	0.35	6	0.30	1	0.05	n/a	n/a	8	3.20	5	2.40	5.10[1]
Silver 3	8	0.40	7	0.35	1	0.05	n/a	n/a	7	2.80	1	0.40	4.00[1]
Silver 4	6	0.30	4	0.20	1	0.05	2	0.10	4	1.60	4	2.00	3.85
Silver 5	2	0.10	1	0.05	1	0.05	4	0.20	6	2.40	6	2.40	5.20
Nickel 1	4	0.20	5	0.25	1	0.05	3	0.15	2	0.80	2	0.40	2.25
Silver 6	3	0.15	3	0.15	1	0.05	n/a	n/a	1	0.40	3	1.60	1.95[1]
Silver 7	5	0.25	n/a	n/a	1	0.05	n/a	n/a	3	1.20	1	0.80	1.90[1]

[1] Based on available data.

Table 17. Summary and Comparison[1] of All Data							
Material	Ultimate Tensile Strength (MPa)	Modulus of Elasticity (10^4MPa)	Poisson's Ratio	Thermal Mechanical Creep Rates (με/hr)	Thermal Creep (με/hr)	Critical J_{exp} (N/m)	Critical J_{th} (N/m)
63-37 Sn-Pb	44.8	2.00	0.33	RT: 0.52 0C: -2.86 -55C:-2.61 85C: 3.88	RT: 0.11 0C: 0.34 -55C:-50.34 85C: 0.38	9670	2675
Silver 1	21.5 (48%)	0.60 (30%)	0.32 (97%)	RT: 2.16 (415%) 0C: 1.94 (-68%) -55C:-0.52 (20%) 85C: 1.60 (41%)	N/A for all temperatures	637 (7%)	313 (12%)
Silver 2	3.1 (7%)	0.09 (5%)	0.30 (91%)	RT: 13.74 (2642%) 0C: 23.52 (-822%) -55C:-2.64 (101%) 85C:-50.09 (-1291%)	RT: -0.10 (-91%) 0C: -0.03 (-9%) -55C: -0.72 (1%) 85C: -0.30 (-70%)	94 (1%)	336 (13%)
Silver 3	2.0 (4%)	0.04 (2%)	N/A	RT: 14.21 (2732%) 0C: 16.67 (-582%) -55C: -6.28 (241%) 85C: 2.78 (72%)	N/A for all temperatures	5352 (55%)	888 (33%)
Silver 4	6.5	0.42	0.22	RT: 1.68 (323%)	RT: -0.20 (-1.81%)	457	358

Table 17. Summary and Comparison[1] of All Data

Material	Ultimate Tensile Strength (MPa)	Modulus of Elasticity (10^4MPa)	Poisson's Ratio	Thermal Mechanical Creep Rates (με/hr)	Thermal Creep (με/hr)	Critical J_{exp} (N/m)	Critical J_{th} (N/m)
	(15%)	(21%)	(67%)	0C: 1.14 (-40%) -55C:-1.91 (730%) 85C: 2.12 (55%)	0C: -2.26 (-664%) -55C: -0.37 (-0.7%) 85C: 0.76 (2%)	(5%)	(13%)
Silver 5	18.3 (41%)	1.22 (61%)	N/A	RT: 2.57 (494%) 0C: 1.86 (-65%) -55C:-1.25 (48%) 85C:-30.52 (-787%)	RT: -0.14 (-127%) 0C: -1.33 (-391%) -55C: -0.20 (0.4%) 85C: 1.93 (508%)	235 (2%)	306 (11%)
Nickel 1	15.6 (35%)	0.21 (10%)	0.50 (152%)	RT: 1.80 (346%) 0C: 0.06 (-2%) -55C: 0.35 (-13%) 85C: 1.06 (27%)	RT: -1.46 (-1327%) 0C: -5.40 (-1588%) -55C: -0.40 (0.8%) 85C: -2.90 (-763%)	7820 (81%)	561 (21%)
Silver 6	16.6 (37%)	0.43 (22%)	0.20 (61%)	RT: 0.88 (169%) 0C: 0.19 (-7%) -55C:-0.25 (10%) 85C: -1.18 (-30%)	N/A for all temperatures	488 (5%)	600 (24%)
Silver 7	6.9 (15%)	N/A	N/A	RT: 1.11 (213%) 0C: 1.00 (35%) -55C:-1.56 (60%) 85C: -3.96 (-102%)	N/A for all temperatures	4001 (41%)	1328 (5%)

[1] Numbers in parentheses indicate percentage comparison to 63-37 Sn-Pb.

Proceedings of
2001 ASME International Mechanical Engineering Congress and Exposition
November 11–16, 2001, New York, NY
EPP-Vol. 1

IMECE2001/EPP-24719

COMPUTATIONAL AND EXPERIMENTAL STUDY OF THERMO-MECHANICS IN A NEW DESIGN OF A TRANSCEIVER FOR HIGH-SPEED TELECOMMUNICATIONS

Ryszard J. Pryputniewicz, **Steven A. Weller***, **Sam R. Shaw***, and **William L. Herb***
NEST – NanoEngineering, Science, and Technology
CHSLT – Center for Holographic Studies and Laser micro-mechaTronics
Mechanical Engineering Department
Worcester Polytechnic Institute
Worcester, MA 01609

* Tyco Electronics Corporation, Harrisburg, PA 17105

ABSTRACT

High-speed telecommunications require ever increasing data rates, pushing the limits of existing hardware and driving the development of new data handling hardware. This paper addresses thermo-mechanical issues in a new Input/Output (I/O) device for high-speed telecommunications. This device is a surface mount-compatible fiber optic transceiver developed as the Small Form-Factor Pluggable (SFP) System. The development of the SFP System is supported by virtually every leading fiber optic transceiver manufacturer in the industry in accordance with the Multi Source Agreement (MSA). We describe the new MSA-compliant SFP System, develop a FEM model to simulate its thermal behavior under a variety of operating conditions, and present results of the simulation which are validated by selected experimental measurements.

Keywords: I/O device, fiber optics, transceiver, Small Form-Factor Pluggable (SFP) System, Multi Source Agreement (MSA), Gigabit Ethernet, Fibre Channel, thermo-mechanics, computational modeling and simulation, validation of a computational model.

1. INTRODUCTION

Increasing demands for information transfers require greater data rates, pushing the limits of the existing hardware and driving the development of new data handling hardware [1-3]. One way to achieve this objective is to place larger number of data transmission lines in parallel. This, however, unless accompanied by a reduction in physical size of the I/O devices, results in bulky systems, which inherently operate at higher temperatures creating severe thermomechanical issues. To overcome this limitation, next generation I/O devices are being developed. Although these "forward looking" devices will offer reduced size and increased functionality, thermomechanical issues associated with their use will not diminish, but may become an unsurpassable challenge, unless concurrently considered at the beginning of the product development process.

Thermomechanical issues in one of such I/O devices are addressed in this paper. More specifically, an overwhelming market demand for a surface mount-compatible fiber optic transceiver has led to the development of the SFP System [4]. In its 10 Megabit implementation, the SFP System extends Ethernet capabilities providing higher bandwidth, distributed processing, imaging, medical, CAD/CAM, and pre-press applications by improving the performance of: (1) LAN connectivity, (2) switch aggregation, (3) regional area network (RAN), and (4) storage area network (SAN).

In this paper, we describe the SFP System, define its functional characteristics, develop a FEM model to simulate its thermal behavior under a variety of operating conditions, present results of the simulation, and compare the simulation results with preliminary experimental results.

2. THE SFP SYSTEM

The SFP System provides a platform to replace fixed front panel copper or fiber optic connections with a flexible and removable transceiver, which may be inserted or removed from a system without powering off the system.

The SFP System consists of a guide rail EMI cage, a blind-mateable host connector, and a module – SFP Transceiver, Fig. 1. The guide rail cage, and blind-mateable connector are mounted to a printed circuit board (PCB) of the host. The open end of the guide rail cage extends through the host box enclosure, and the module is inserted into the guide rail cage where it is aligned and mated with the host connector inside.

The SFP system currently supports data rates of 3Gb/s or more and is suitable for box-to-box or uplink scenarios. Since the SFP is removable and hot pluggable, the user can customize the interface to fiber optic and/or copper cable by simply changing the SFP transceiver and interface connection, Fig. 2. In addition, due to the modularity of the SFP Transceiver, the device is also upgradeable for future technological advances and customer needs.

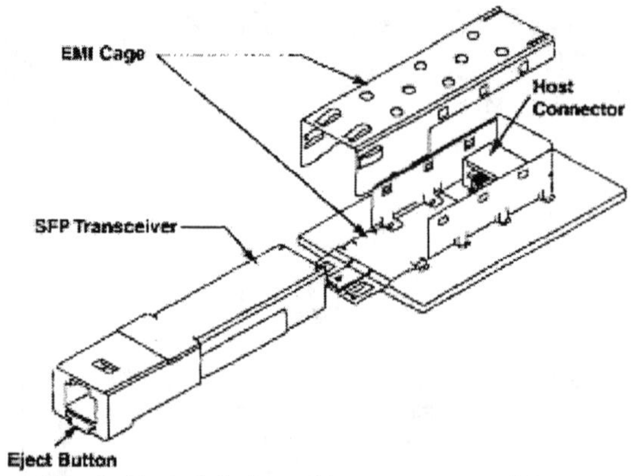

Fig. 1. Exploded view of the components of the SFP System.

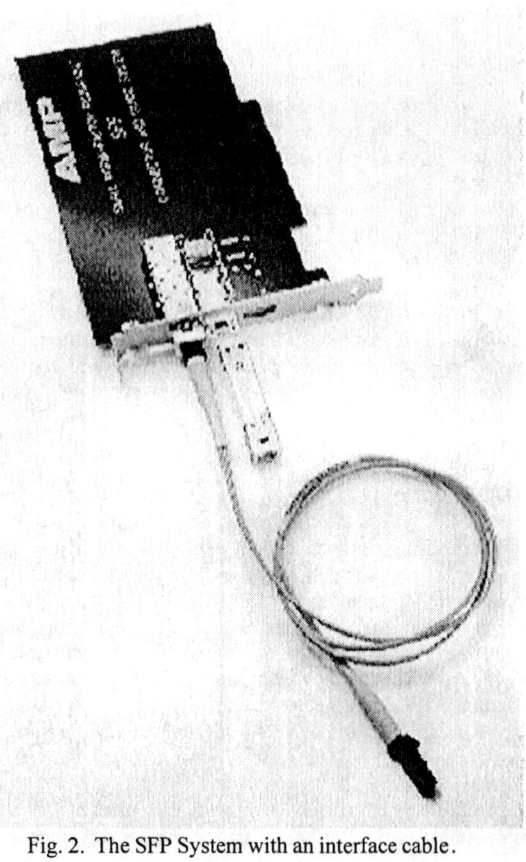

Fig. 2. The SFP System with an interface cable.

3. FEM MODEL OF THE SFP SYSTEM

The SFP System can be configured to consist of either a single or multiple modules, depending on the specific application. However, to facilitate development of a FEM model of the SFP System, a two-module configuration (used in the initial experimental measurements) was selected, Figs 2 and 3. In this configuration, the modules were assembled parallel to each other, in a "brickwalled" arrangement, on the PCB of the SFP System. Description of the FEM model is made with reference to this twin-module configuration. It should be realized, however, that FEM model was developed in such a way that it could be reformulated to perform analyses of the SFP System with either a single or multiple module configurations.

Fig. 3. Twin-module SFP System used in the initial experimental part of this study.

Airflow and heat transfer affecting the SFP System are governed by the principles of conservation of mass, momentum, and energy. These conservation laws may be expressed in terms of partial differential equations, the solution of which forms the bases of computational fluid dynamics (CFD). A finite element based approach [5] was used requiring the region modeled, Fig. 4, to be subdivided into a number of small elements, Fig. 5.

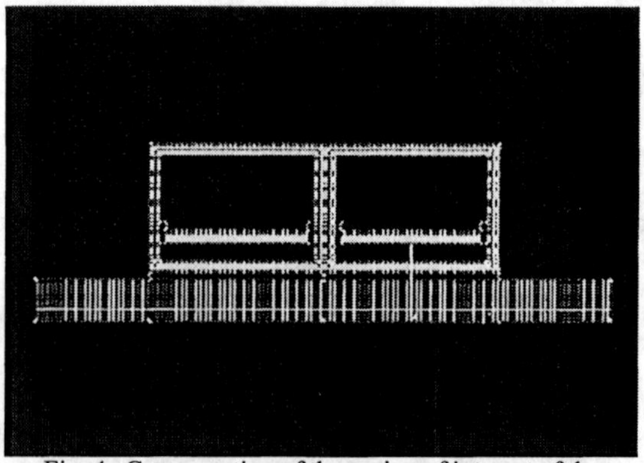

Fig. 4. Cross-section of the region of interest of the twin-module SFP System.

Figure 4 shows two modules each with its own local-PCB where heat is generated and must be removed from during functional application of the SFP System. Using the FEM model developed during this study, rates of heat generation could be set up for each local-PCB independent of each other. With this model we were able to simulate thermal effects within the SFP System when either left or right local-PCBs were active at a given time, as well as when both of them were on simultaneously. Figure 5 shows the FEM mesh of the twin-module SFP System in the case when only the right module is active.

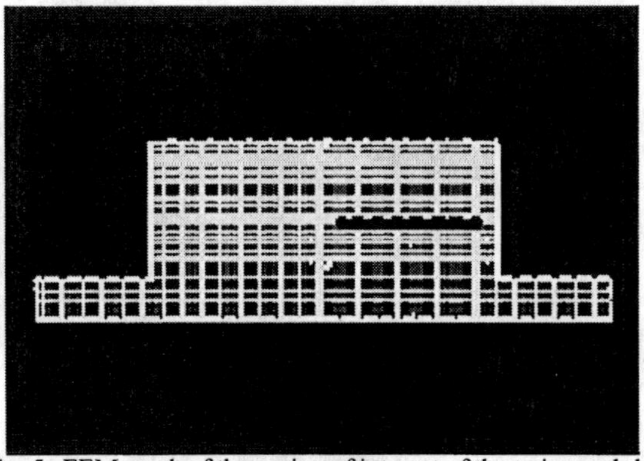

Fig. 5. FEM mesh of the region of interest of the twin-module SFP System when only the right module is active.

The flexibility of individually activating specific modules in the FEM model of the SFP System was particularly useful when simulating thermal performance of the multi-module configurations considered in this study.

The heat transfer from the SFP System would occur over the entire structure. Because of the typical operating conditions, the SFP System would normally be subjected to convective heat transfer from its surface(s).

In our FEM model, the convective heat transfer from a surface is governed by the equation [6]

$$Q_c \quad h_c A_c (T_c \quad T_a) \ , \tag{1}$$

where Q_c is the local convective heat flux, h_c is the local convective heat transfer coefficient at the surface, A_c is the local surface area of the SFP System subjected to convective heat transfer, T_c is the local temperature at the surface of the System, and T_a is the ambient (air) temperature. In definition of Eq. 1, the word "local" refers to the specific surface(s) of the SFP System. This distinction is necessary because the convective heat transfer coefficient is a complex function of geometry, location, and orientation of the surface considered, properties of the fluid (in this case air) in contact with the surfaces, characteristics of the flow of the fluid, the nature of the boundary layer at the surfaces, as well as specific temperatures, all of which play an important role in evaluating convection. This functional dependence of the convective heat transfer coefficient can be expressed by the following fundamental relationship [7]:

$$h_c \quad h_c(geometry, location, orientation, \\ properties, flow, boundary, T_c, T_a \quad . \tag{2}$$

In our FEM simulation of the SFP System, local heat transfer coefficients were determined using the explicit equations relating to the relationship given by Eq. 2, for each surface from which heat was removed by convection. Also, Eqs 1 and 2 were the basis for determination of uncertainties in h_c and in the corresponding temperatures.

4. REPRESENTATIVE RESULTS

Testing and simulation of the SFP System has been performed for various configurations: (1) horizontal, (2) vertical, (3) single-module, (4) multiple-modules, (5) various airflow rates, and (6) various power levels. Representative results of the testing and simulation are shown in Figs 6 to 11 and are summarized in Table 1.

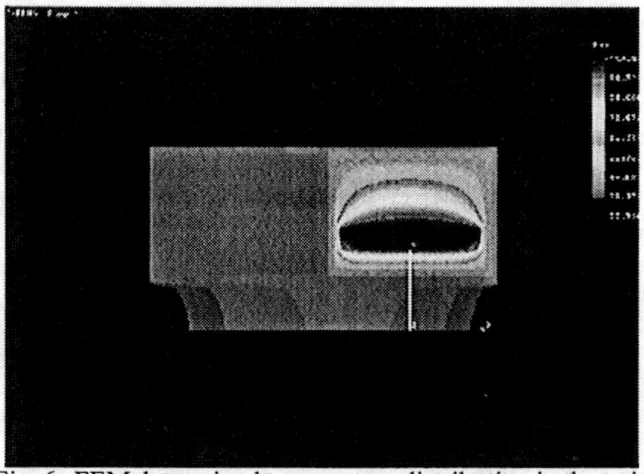

Fig. 6. FEM determined temperature distribution in the twin-module SFP System corresponding to the horizontal configuration shown in Fig. 5, when only the right module is active.

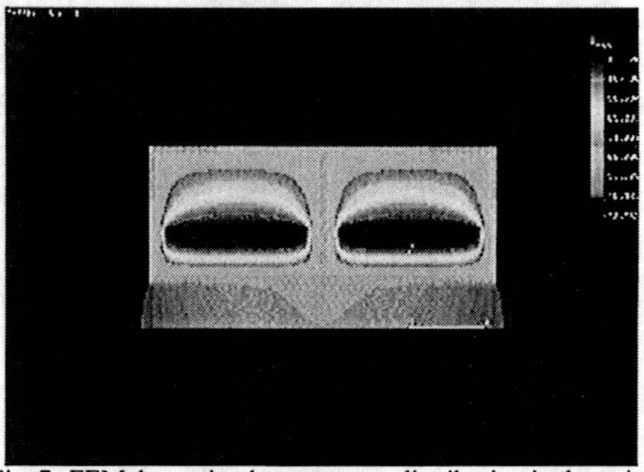

Fig. 7. FEM determined temperature distribution in the twin-module SFP System corresponding to the horizontal configuration shown in Fig. 5, when both modules are active.

Figures 6 and 7 show temperature distributions in the twin-module SFP System. These results are for the local-PCB dissipation of 2 W per module, where applicable, and free convection from the outside surfaces of the EMI cages to the ambient air at 25°C.

Clearly, temperature distributions are affected by the specific configuration. For example, if only one module is active in a twin-module SFP System, lower temperatures of the air, above the local-PCB inside the module, will be obtained (Fig. 6) than if both modules were active (Fig. 7). The reason for this is that the "unpowered" left module, in the configuration shown in Fig. 6, is in a "brickwalled' contact with the right module that is "powered" and, as a result, heat is transferred by conduction from the right cage to the left cage. Therefore, the left module effectively increases external area that is engaged in convective cooling of the SFP System effectively lowering temperature of the entire assembly.

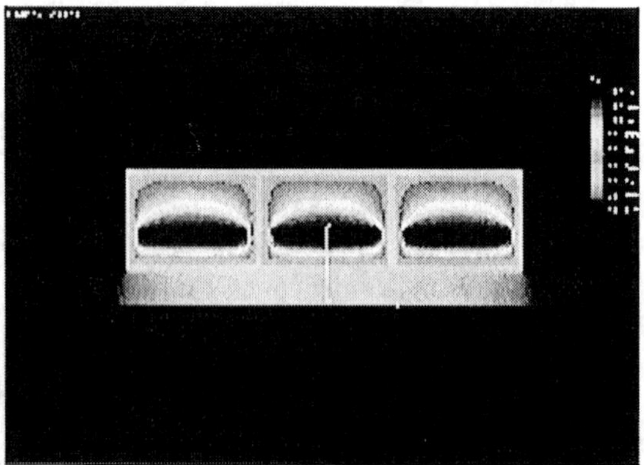

Fig. 8. FEM determined temperature distribution in the three-module SFP System, horizontal configuration, when all modules are active, for the case of free convection.

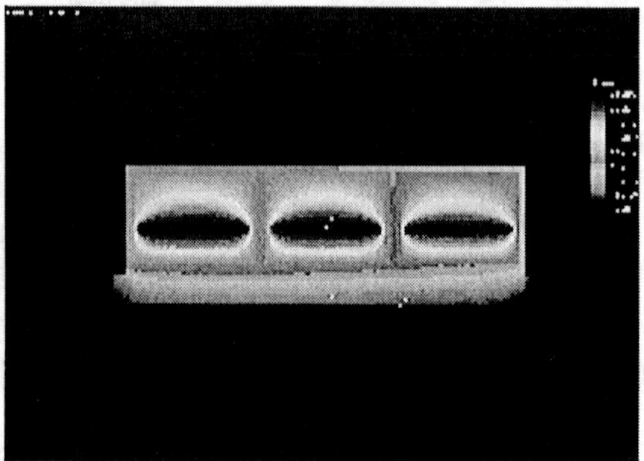

Fig. 9. FEM determined temperature distribution in the three-module SFP System, horizontal configuration, when all modules are active, for the case of forced convection.

Representative FEM results obtained for the three-module configuration of the SFP System are shown in Figs 8 to 11. Figures 8 and 9 display temperatures for the horizontal configuration of the SFP System when all modules are active for the cases of free convection and forced convection, respectively. Figures 10 and 11 show the results for the vertical configuration of the System for the case of free convection, with Fig. 10 displaying temperature distribution when only center module is active, while Fig. 11 displaying the temperature when all three modules are active.

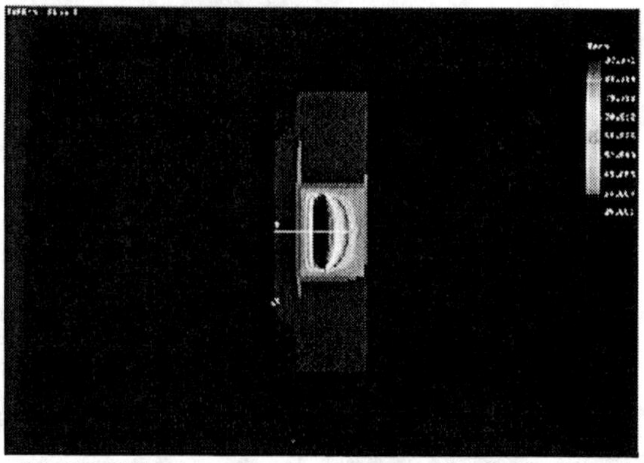

Fig. 10. FEM determined temperature distribution in the three-module SFP System, vertical configuration, when only center module is active, for the case of free convection.

Fig. 11. FEM determined temperature distribution in the three-module SFP System, vertical configuration, when all three modules are active, for the case of free convection.

Experimental measurements of temperatures inside of the three-module configuration of the SFP System in horizontal and vertical configurations, subjected to free and forced convection conditions, were made using thermocouples.

For the experimental setup, three-module guides were soldered to a PCB. The module guides were "brickwalled" on 13.97 mm (0.55 in.) centerlines. The center guide was loaded with an 11.6 mm σ 8.5 mm σ 44.8 mm hollow module made of 0.396 mm (0.0156 in.) thick brass. The hollow module was loaded with a 2 W resistor and a thermocouple and capped at both ends. The thermocouple was located in the hollow module away from the resistor element to measure the air temperature inside the module, above the resistor. The adjacent guides were loaded with 11.6 mm σ 8.5 mm σ 44.8 mm dummy modules. The dummy modules were drilled out and loaded with 2 W resistors.

The experimental setup was "sandwiched" between 2 PCBs on 20.32 mm (0.8 in.) centerlines. The "sandwiched" assembly was then placed in an enclosure and tested under the conditions described in Table 1.

Table 1. Comparison of measured and computed temperatures for the SFP System.

TEST No.	CONFIG	T_a (°C)	AIR FLOW (cfm)	POWER DISSIP IN TEST MODULE (W)	POWER DISSIP IN ADJACENT MODULES (W)	MOLULE VENT HOLE SIZE (mm)	NUMBER OF MODULE VENT HOLES	AIR TEMP INSIDE MODULE (°C)	
								MEAS	SIMUL
1	Vertical	24	0	2	2	0	0	93	85
2	Horiz	26	0	2	2	0	0	103	95
3	Vertical	24	0	2	0	0	0	66	67
4	Horiz	25	0	2	0	0	0	71	70
5	Horiz	25	15	2	2	0	0	67	72
6	Horiz	25	10	2	2	0	0	73	78
7	Horiz	40	15	2	2	3	8	62	65
8	Horiz	40	10	2	2	3	8	67	73

The 3 mm module vent holes in tests 7 and 8 were aligned with 3 mm vent holes in the module guide.

Comparison of the selected measured temperatures with the FEM determined temperatures, for a number of test conditions, is shown in Table 1. According to this table measured and computed temperatures correlated well within the uncertainties determined for the conditions considered in this study.

The experimentally measured values verified that the FEM model presented an accurate simulation of the temperatures within the SFP System.

5. CONCLUSIONS

A new MSA-compliant SFP System, currently supporting data rates of 3 Gb/s or more, suitable for box-to-box and uplink scenarios, as well as its functional characteristics, were described. A FEM model to simulate thermal behavior of the SFP System for a variety of its configurations and operating conditions was developed. Selected experimental measurements of the SFP System were made to validate the FEM model.

The results presented in this paper indicate that careful consideration must be given to physical geometry of the SFP System, materials and components used, their orientation in the array, thermal operational loads, and boundary conditions. It should be noted that because of specific operational functionality of the System, in addition to thermomechanical issues, the corresponding electromagnetic issues should be considered concurrently. In this way, development and optimization of the SFP System will lead to its reliable and durable design.

The cage and module design of the SFP System allow the flexibility to fill specific ports now and add or hot-swap others later, which facilitates interchangeability and/or adaptability to different protocols. As such, designers can easily address both singlemode and multimode requirements on the same board. Also, the Gigabit SFP Transceiver offers a through-hole solutions for Gigabit Ethernet and Fibre Channel applications. In addition, the 10 Megabit Transceiver for 10Base Ethernet applications provides the first surface-mountable solution with the smallest pitch.

All in all, results of this study give a better understanding of the thermo-mechanics in the SFP System. It is imperative that proper thermal management is exercised in the system integration so that quality and reliability of the system is not compromised.

Use of validated model for simulation of the operational functionality of the SFP System facilitates development and optimization of the System configurations for specific high-speed telecommunication applications.

6. REFERENCES

1. Comerford, R., 2000, "The Internet," *IEEE Spectrum*, pp. 40-44.
2. Geppert, L., 2000, "Devices & circuits," *IEEE Spectrum*, pp. 63-69.
3. Cosgrove, K., 2000, "Telecom technology challenges manufacturers," SMT, pp. 60-64.
4. *Small form factor transceivers*, 2001, Tyco Electronics, Harrisburg, PA.
5. *COSMOS/M user's guide*, 1995, Structural Research and Analysis Corporation, Santa Monica, CA.
6. Holman, J. P., 1997, *Heat transfer*, 8th ed., McGraw-Hill, New York.
7. Pryputniewicz, R. J., 1992, *Engineering experimentation*, Worcester Polytechnic Institute, Worcester, MA.

Proceedings of
2001 ASME International Mechanical Engineering Congress and Exposition
November 11–16, 2001, New York, NY
EPP-Vol. 1

IMECE2001/EPP-24720

DSC, TMA, DMA, and TGA of Epoxies for Fiber-Optic Transceiver Applications

John Lau, Steve Erasmus, Fred Sporon-Fiedler,
Agilent Technologies, Inc. 350 West Trimble Road,
MS 90LJ, San Jose, CA 95131

Chris Chang
NSG Technology, Inc. 1705 Junction Ct.,
San Jose, CA 95112

ABSTRACT

The curing conditions (such as temperature and time) and material properties such as the TCE (thermal coefficient of expansion), T_g (glass transition temperature), storage modulus, tangent delta (tan δ), and moisture absorption of two different epoxy materials from two different vendors are measured. Emphasis is placed on specimen preparations, experimental procedures, and data analyses. The methods and results presented herein should be useful for evaluating epoxies for fiber-optic transceiver applications.

(1) INTRODUCTION

Epoxy materials [1-5] are frequently used in electronic packages and fiber-optic modules such as transceivers. Usually, they are used for: (1) gluing two or more pieces of parts of the module together; (2) enhancing the strength of certain part(s) of the module; and (3) blocking the moisture to the moisture-sensitive part(s) of the module.

The most desirable features of epoxy materials for fiber-optic transceiver applications are:

- Low curing temperature/fast curing time, which can reduce manufacturing cost and be less harmful to other components.
- Low TCE, which can reduce thermal expansion mismatch of the structure and enhance its thermal reliability.
- High modulus which leads to good mechanical properties.
- High T_g which can endure higher temperature environments.
- Low moisture absorption, which can effectively block the moisture and extend shelf life.
- Good adhesion which can improve product lifetime.

Usually, the data of these parameters are in the Product Data Sheet, Technical Product Data, or Technical Bulletin provided by the vendors. Very often, however, the vendors don't have all the data. Furthermore, some of their published data are wrong. Thus, it is necessary to measure and analysis these parameters, especially for evaluating new epoxies.

In this study, two different epoxies, A and B, from two different vendors are studied. Their color is black. At room temperature, their specific gravity (mixed), viscosity (mixed), and ASTM D2240 Shore "D" hardness are, respectively, 1.6 ~ 1.7, 10,000 ~ 13,5000 cps, and 90 ~ 92.Their curing conditions such as time and temperature are measured by a differential scanning calorimeter (DSC) unit. Their material properties such as the TCE, T_g, storage modulus, tan δ, and moisture content are carried out using the thermal mechanical analysis (TMA), dynamic mechanical analysis (DMA), and thermal gravimetric analysis (TGA).

(2) CURING CONDITIONS OF EPOXIES

For determining the curing conditions of epoxy materials, A and B, one drop of each is put into an aluminum pan (which will form a disc sample with dimensions: 6.4 ± 0.2 mm in diameter and 1.6 ± 0.1 mm in height), weighed, and then put in a DSC equipment. At this moment, the epoxies are not cured. The objective of DSC is to measure the amount of energy (heat) absorbed or released by a sample as it is heated, cooled or held at a constant (isothermal) temperature.

The instrument consists of two independent furnaces, one for the sample and one for the reference as shown in Figure 1. When an exothermic or endothermic change occurs in the sample material, energy is applied to or removed from one or both furnaces to compensate for the energy change occurring in the sample. Since the system is always directly measuring energy flow to or from the samples, DSC can directly measure melting temperature, temperature onset of crystallization, and temperature at onset of curing. The kinetic software enables analysis of a DSC peak to obtain specific kinetic parameters that characterize a reaction process.

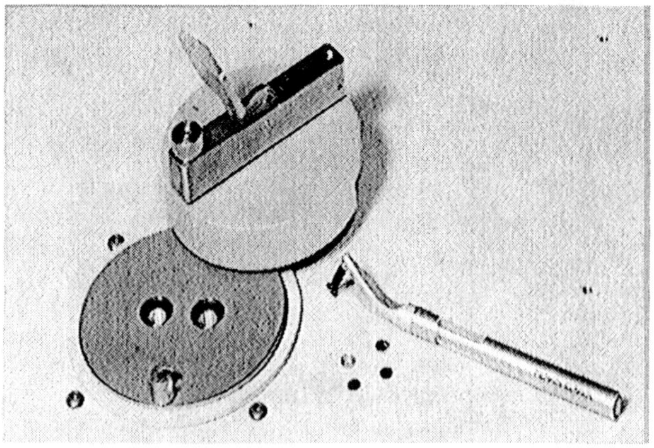

Figure 1 Specimen setup for DSC (differential scanning calorimeter)

Any material reaction can be represented by the following equation:

$$X1 \xrightarrow{k} X2 + \Delta H$$

where X1 is the material before reaction, X2 is the material after reaction, Δ H is the heat absorbed or released, and k is the Arrhenius rate constant.

The Arrhenius equation is given by :

$$k = Z\exp(-E_a/RT)$$

where Z is the pre-exponential constant, E_a is the activation energy of the reaction, R is the universal gas constant (8.314 j/°C/mole), and T is the absolute temperature in degree Kelvin.

The rate of reaction (dx/dt) can be directly measured by the DSC and can be expressed as:

$$dx/dt = k(1-x)^n$$

where dx/dt is the rate of reaction, x is the fraction reacted, t is the time, k is the Arrhenius rate constant, and n is the order of reaction. Combining all the preceding equations and assuming nth-order reaction kinetics and a constant program rate, activation energy, and pre-exponential constant yields:

$$dx/dt = Z\exp(-E_a/RT)\,(1-x)^n$$

The fraction reacted (x) is directly related to the fractional area of the DSC reaction peak. The kinetic parameters Z, E_a, and n are determined by using an advanced multilinear regression method (MLR). Thermal scan is carried out at a 5 °C/min heating rate ranging from 40 to 200 °C.

Figures 2 and 4 show the DSC thermal scan curves and Figures 3 and 5 show the typical degree of conversion (reaction) verses time curves of Epoxy A and Epoxy B, respectively. It can be seen that: (1) the average curing temperature of Epoxy A (129.5 °C) is larger than that of Epoxy B (106.7 °C) as shown in Figures 2 and 4; (2) for both epoxies, in general, the higher the curing temperature the faster the curing time; and (3) for the same curing temperature, Epoxy B cures faster than Epoxy A as shown in Figures 3 and 5.

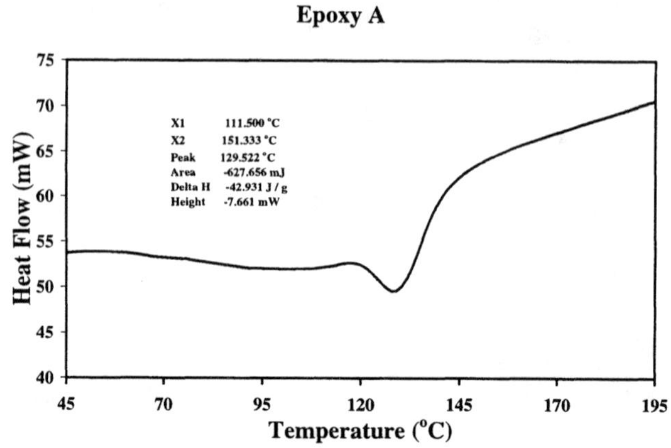

Figure 2 DSC thermal scan curve for epoxy A

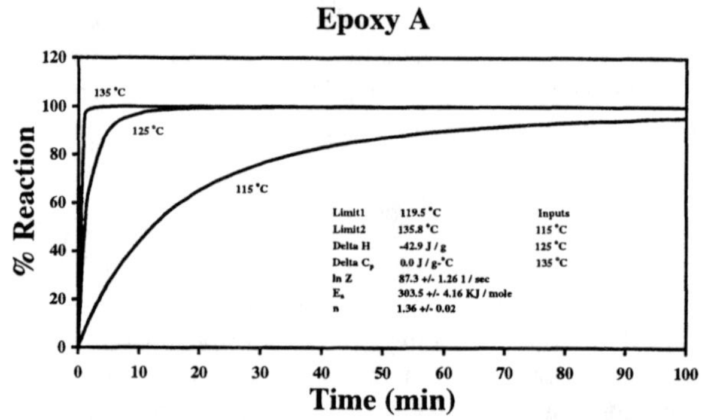

Figure 3 Typical degree of conversion verse time curve of epoxy A

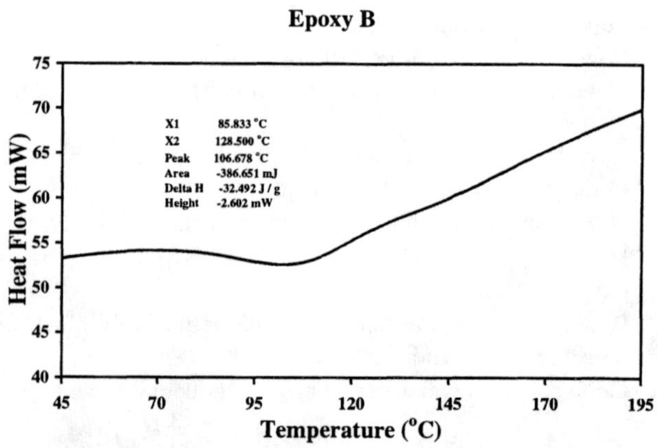

Figure 4 DSC thermal scan curve for epoxy B

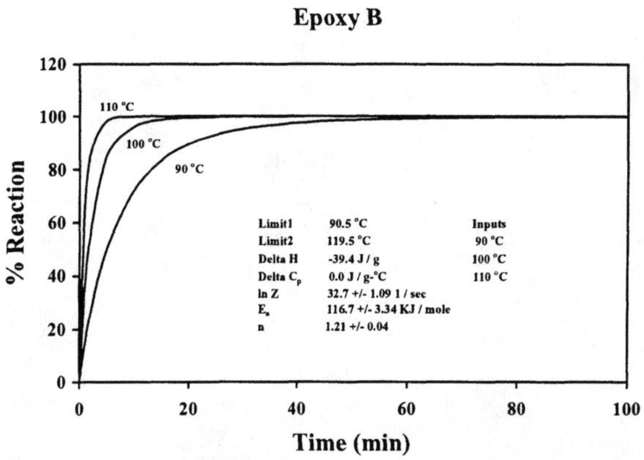

Figure 5 Typical degree of conversion verse time curve of epoxy B

(3) MATERIAL PROPERTIES OF EPOXIES

After the curing conditions of Epoxy A and Epoxy B are determined, they are put into the fixtures for making the specimen, and then cured in an oven at 130 °C for 30 minutes. In order to ensure the epoxies are 100% fully cured, they are intentionally slightly over cured.

(3a) TCE

The TCE of Epoxy A and Epoxy B (with a sample dimensions: 6.4 ± 0.2 mm in diameter and 1.6 ± 0.1 mm in height) is determined by the TMA in an expansion quartz system (55 to 200 °C) at a 5 °C/min heating rate as shown in Figures 6 and 7. The objective of TMA is to measure the change in dimension of a sample (expansion or contraction) as the sample is heated, cooled or held at a constant temperature.

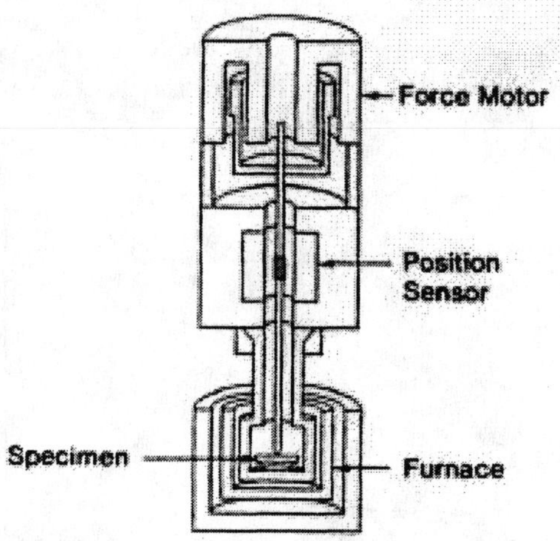

Figure 6 Schematic diagram of a TMA (thermal mechanical analysis) and DMA (dynamic mechanical analysis). TMA force motor applies static load, while DMA force motor applies dynamic load on specimen.

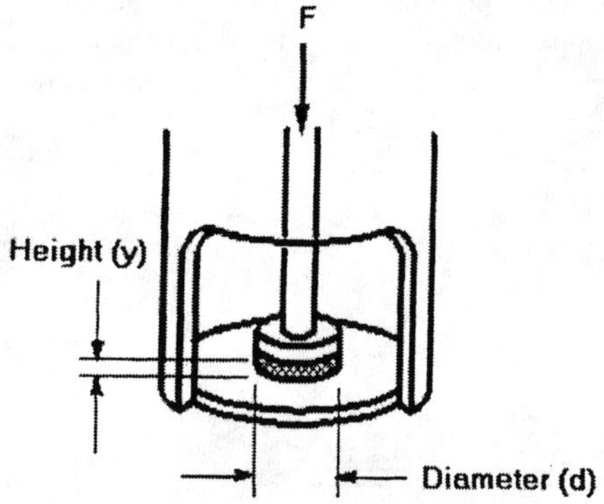

Figure 7 Quartz tube, probe, and specimen setup for TMA

The instrument consists of a platinum-wound furnace system that can be operated from -170 °C to 1000 °C. The scan rate is from 0.1°C to 100°C/min. The specimen is mounted between a quartz platform and probe, and then a static load (F), as shown in Figure 7, is applied to specimen while the dimensions of the specimen are monitored by the linear variable differential transducer (LVDT) throughout the analysis. The slope of the dimension change-versus-temperature curve obtains the TCE of the epoxy, and T_g of the material is obtained from the onset of the two different slopes of the curve.

Figure 8 shows the expansion curves of Epoxy A and Epoxy B. It can be seen that the TCE of Epoxy A ($61.1\times10^{-6}/^{o}C$) is larger than that ($48.3\times10^{-6}/^{o}C$) of Epoxy B. For fiber-optic transceiver application, epoxies with lower TCE are preferred because most of the transceiver materials are with TCE less than $30\times10^{-6}/^{o}C$.

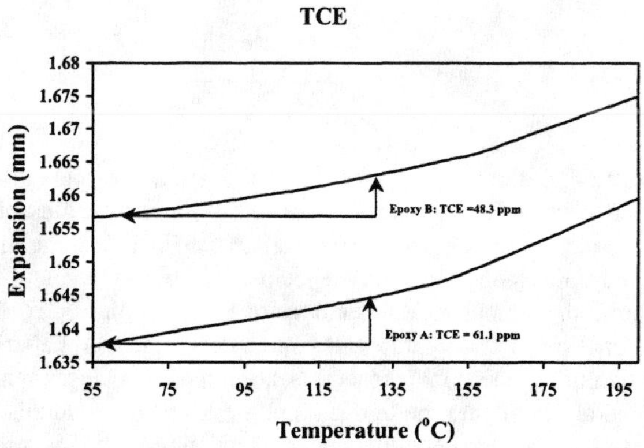

Figure 8 Typical expansion curves of epoxy A and B

(3b) Storage Modulus

The modulus of Epoxy A and Epoxy B is measured with a three-point bending specimen (3.0 ± 0.3 mm x 2.9 ± 0.3 mm x

19 ± 3 mm) in a DMA unit (50 to 200 °C) at a heating rate of 5 °C/min as shown in Figures 6 and 9. The objective of DMA is to measure mechanical properties such as modulus as a function of time, temperature, frequency, stress, or combinations of these parameters.

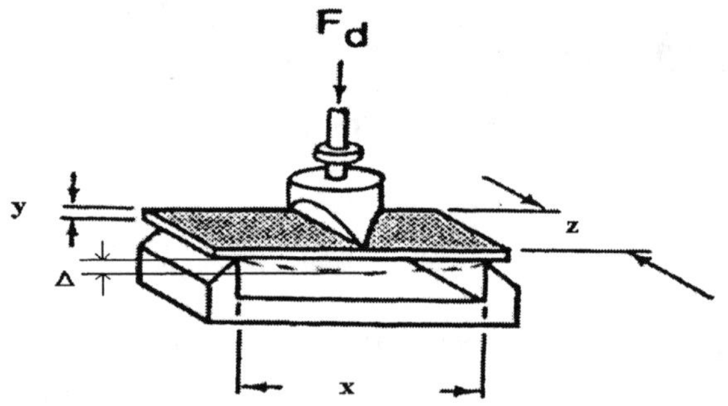

Figure 9 Three-point bending specimen setup for DMA

The instrument consists of a force motor that can be programmed to apply constant stress, dynamic stress, or combinations of both. The core rod applies stress to the sample and is held in place using an electromagnetic suspension. The ceramic furnace with platinum furnace element is capable of heating and cooling at a very high rate and can also reach temperatures to 1000 °C. For fiber-optic module materials, the flexural properties such as storage modulus, loss modulus, and tangent delta (tan δ) can be obtained with DMA.

In material sciences, storage modulus, E_s is a measure of the energy stored per cycle of deformation and can be expressed as:

$$E_s = \frac{F_d\, x^3 \cos\delta}{4y^3 z\, \Delta}$$

where F_d is the dynamic load, δ is the phase angle, x is the clear span between the supports, y and z are the dimensions of the specimen, and Δ is the maximum dynamic deflection of the three-point bending epoxy specimen (Figure 9). Figure 10 shows the storage modulus curves of Epoxy A and Epoxy B as a function of temperatures. It can be seen that: (1) the storage modulus of the epoxy materials is temperature dependent, in general, the higher the temperature the lower the modulus, and (2) for all the temperatures under consideration, the modulus of Epoxy A is larger than that of Epoxy B. For fiber-optic transceiver applications, epoxies with larger modulus are preferred.

The loss modulus, E_l is a measure of the energy lost per cycle of deformation and can be expressed as:

$$E_l = \frac{F_d\, x^3 \sin\delta}{4y^3 z\, \Delta}$$

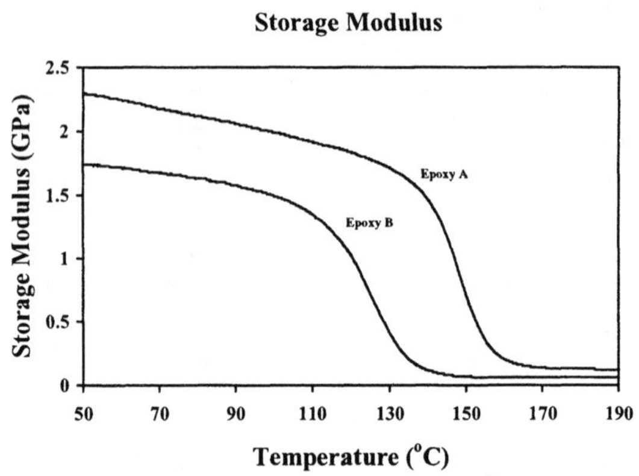

Figure 10 Typical storage modulus curves of epoxy A and B

(3c) Tan δ and T_g

Mechanical energy supplied at the temperature and frequency corresponding to the transition will be absorbed by the material, then partially stored elastically and partially dissipated in the form of heat. Tangent delta (tan δ) (which is a measure of material-related damping property) of Epoxy A and Epoxy B can be obtained by dividing E_l by E_s, or

$$\frac{E_l}{E_s} = \tan\delta$$

The temperature at the peak of a tan δ curve is often reported in literature as glass transition temperature. Figure 11 shows the tangent delta curves of Epoxy A and Epoxy B. It can be seen that T_g of Epoxy A (154.8 °C) is larger than that (135.2 °C) of Epoxy B. Usually, epoxies with higher T_g are preferred.

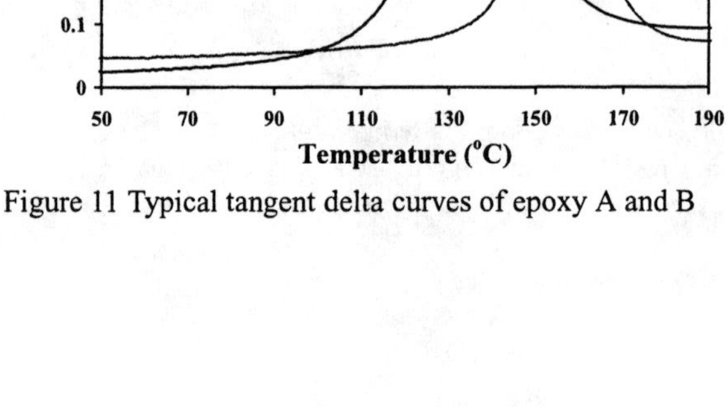

Figure 11 Typical tangent delta curves of epoxy A and B

The damping is the ratio of heat dissipation to energy storage, and is dependent on the temperature and frequency. Near the T_g, polymers display maximum values of loss modulus and tan δ, which are measures of energy dissipation - or, in other words, their "lossiness" is at a maximum. The wider range of tan δ may involve friction between fillers or between fillers and polymer and may provide additional means of dissipating energy. Contributions from the inhomogeneous distribution of fillers and from thermal stresses may also affect damping properties. In this study, it can be seen from Figure 11 that the tan δ curve of Epoxy A is sharper than that of Epoxy B, (that is, the transition temperature range is smaller), indicating that the stiffness of Epoxy A is better (stronger) than that of Epoxy B. The epoxies will become softer during a wider transition temperature range.

(3d) Moisture Absorption

Two sets of tests are carried out for determining the moisture content, one is with dry specimen and the other is with 85°C/85%RH (for 4 hours) specimen. All the specimen dimensions are: 6.4 ± 0.2 mm in diameter and 1.6 ± 0.1 mm in height. Weight change of Epoxy A and Epoxy B is measured with the TGA equipment as shown in Figure 12 under 105 °C for four hours.

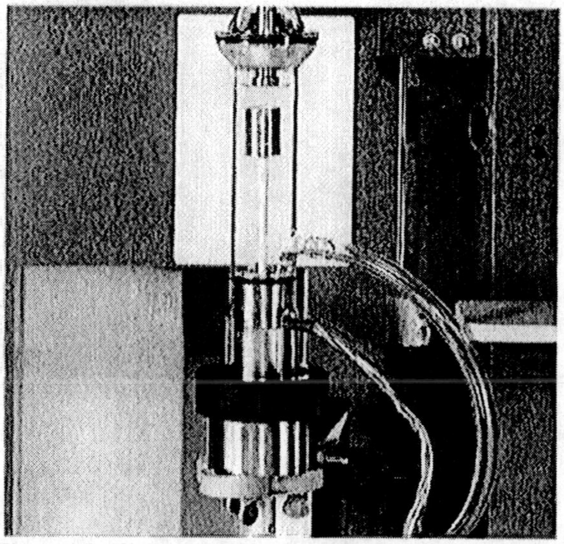

Figure 12 High-sensitivity weight measurements with TGA (thermal gravimetric analyzer)

The objective of TGA is to measure the change in mass of a sample as the sample is heated, cooled or held at a constant (isothermal) temperature. The instrument design consists of a microbalance that allows the sensitive measurement of weight changes as small as a few micrograms; a furnace; and a sample holder area. The vertical analytical design also serves to isolate the balance mechanism from the furnace, eliminating temperature fluctuations that cause drifts and nonlinearity. This system can measure Curie point, decomposition temperature, moisture uptake, and component separation.

The change in mass during thermal scan can be expressed as:

$$\frac{W_f - W_i}{W_i} \times 100\ \%$$

where W_f is the final weight after thermal scan and W_i is the initial weight before thermal scan. Figure 13 shows the percent weight change ratio (moisture content) of Epoxy A and Epoxy B after 15 hours and 1 hour, respectively of dry-air room conditions. It can be seen that the moisture absorption of Epoxy B (0.19%) is more than two times of that of Epoxy A (0.07%), even Epoxy A was exposed in dry-air room condition longer than Epoxy B. For fiber-optic transceiver applications, epoxies with lower moisture absorption are preferred, especially for protecting the moisture-sensitive part(s).

Figure 14 shows the percent weight change ratio of Epoxy A and Epoxy B after 4 hours of 85°C/85%RH. Even with this severe condition, the moisture absorption of Epoxy A is still slight better than Epoxy B.

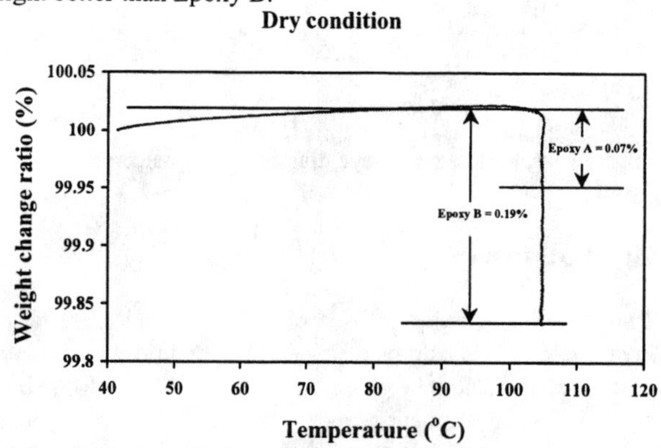

Figure 13 Moisture content of epoxy A and B at dry condition

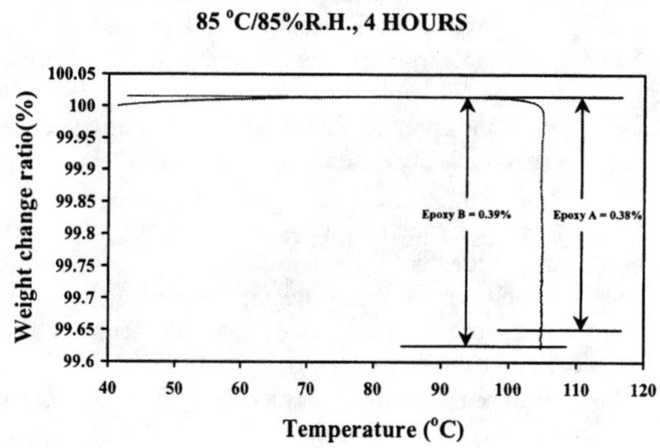

Figure 14 Moisture content of epoxy A and B after 4 hours 85°C/85%RH

(4) 85°C/85%RH TEST RESULTS

Transceivers with either epoxy A or epoxy B are subjected to the 85°C/85%RH test. Eye diagrams after 1000 hours of the test of most of the transceivers with epoxy A and with epoxy B are very similar. A typical set of eye diagrams is shown in Figure 15. It can be seen that the transceivers with either epoxy A or epoxy B work properly and that the mask margin is larger than 60%. Thus, both epoxies can be used for the present applications.

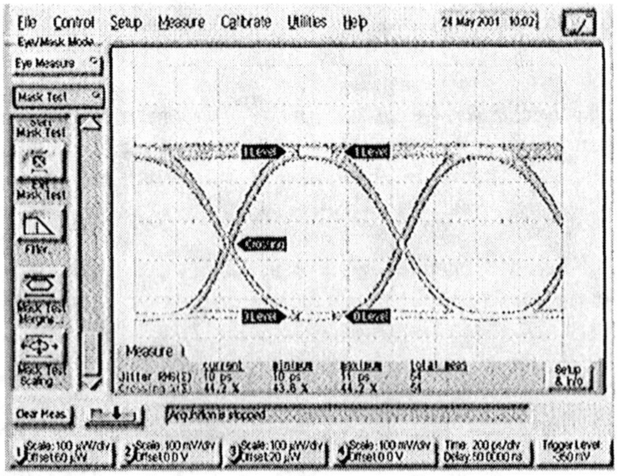

Figure 15 A typical set of eye diagrams of transceivers

(5) SUMMARY

Two different epoxies from two different vendors have been investigated in this study. Their curing conditions and material properties have also been measured using DSC, TMA, DMA, and TGA equipment. Furthermore, detailed specimen preparations, experimental procedures, and data analyses have been provided and discussed. Finally, transceivers with either epoxy A or epoxy B have been tested under 85°C/85%RH condition. Some important results are summarized as follows.

- For both epoxies considered, the curing time is temperature dependent, they cured faster at higher temperatures.
- Epoxy B can be fully cured faster at a lower temperature than Epoxy A.
- For both epoxies considered, the storage modulus is temperature dependent, the higher the temperature the lower the modulus.
- For all temperatures, the module of Epoxy A is larger (stronger) than that of Epoxy B.
- For both epoxies considered, they absorb moisture.
- The moisture absorption of Epoxy A is smaller (better) than that of Epoxy B, especially in dry-air room condition.
- The TCE of Epoxy B is smaller (better) than that of Epoxy A.
- With both epoxies, the transceivers have excellent responses (eye diagrams) even after 1000 hours of 85°C/85%RH test.

(6) REFERENCES

(1) Wong, C. P., *Polymers for Electronic and Photonic Applications*, Academic Press, San Diego, CA 1993.

(2) Lau, J. H., C. P. Wong, J. L. Prince, and W. Nakayama, *Electronic Packaging, Design, Materials, Processes, and Reliability*, McGraw-Hill, New Your, NY, 1998.

(3) Tummala, R. R., E. Rymaszewski, and A. Klopfenstein, *Microelectronics Packaging Handbook*, Chapman & Hall, New York, NY, 1997.

(4) Lau, J. H., *Low Cost Flip Chip Technologies for DCA, WLCSP, and PBGA Assemblies*, New York, NY, 2000.

(5) Seraphim, D. P., R. Lasky, and C. Y. Li, *Principles of Electronic Packaging*, McGraw-Hill, New York, NY

Proceedings of
2001 ASME International Mechanical Engineering Congress and Exposition
November 11–16, 2001, New York, NY
EPP-Vol. 1

IMECE2001/EPP-24721

Corrosion Analysis of a Fiber-Optic Transceiver Housing

J. H. Lau, C. L. Jiaa, and S. J. Erasmus
Agilent Technologies, Inc.
350 W. Trimble Road
San Jose, CA 95131

ABSTRACT

The corrosion responses of a fiber-optic transceiver's housing with zinc alloy die casting material are investigated in this study. Emphasis is placed on the determination of the weight change and corrosion rate of the housing when it is subjected to a 85% relative humidity and 85°C temperature (85%RH/85°C) test condition. Also, the average light optical power, receiver sensitivity, extinction ratio, and mask margin of transceiver modules subjected to 85%RH/85°C and 3.47V at 500, 1000, 1500, and 1600 hours are provided. Furthermore, metallography is performed on the surfaces and cross sections of the housing. Finally, measurements are made of the thickness of the oxide layer on the surfaces of the fiber-optic transceiver housing.

(1) INTRODUCTION

The small-form pluggable (SFP) transceivers [1, 2] are a family of DC low-power (3.3V) and low-cost small form factor (SFF) fiber optic modules. This fiber optic transceiver module allows the system designer to implement a range of solutions for Gigabit Ethernet and Fiber Channel applications. It can be inserted or removed from a fixed host chassis (e.g., computer networking equipment boxes) without removing power from the host system. This allows for convenient system configuration changes (e.g., field upgrades or port expansion) or maintenance without system down time. This benefit is critical to switch environments that cannot afford system downtime for online transaction processing applications.

Figure 1 shows a SFP transceiver. The dimensions of the SFP module are: 56.5 x 13.5 x (8.5 to 11.8) mm. It consists of several major parts, namely, (1) the receiver optical sub-assembly (ROSA) and its amplification and quantization circuitry; (2) the transmitter optical sub-assembly (TOSA) and its laser driver and safety circuitry; (3) the electronics sub-assembly (ESA) including the EEPROM on the printed circuit board; and (4) the housing with electromagnetic interference (EMI) shielding as shown in Figures 1 and 2.

The ROSA, containing a PIN photodiode and custom transimpedance preamplifier, is located at the optical interface and mates with the LC optical connector. The ROSA is mated to a custom IC that provides post-amplification and quantization. Also, it includes a loss of signal (LOS) detection circuit, [1, 2].

The TOSA, containing a 850 nm vertical cavity surface emitting laser (VCSEL) light source, is located at the optical interface and mates with the LC optical connector. The TOSA is driven by a custom IC, which converts differential logic signals into an analog laser diode drive current. This transmitter driver circuit regulates the optical power at a constant level provided the data pattern is DC balanced, e.g., 8B10B code, [1, 2]. These transceivers accept a transmit disable control signal input which shuts down the transmitter. A high signal implements this function while a low signal allows normal laser operation. In the event of a fault, e.g., eye safety circuit activated, cycling this control signal resets the module, [1, 2].

This transceiver [1, 2] features a serial identification module containing product data stored for retrieval by host equipment. This data is accessed via the 2-wire serial CMOS EEPROM protocol of the ATMEL AT24C01A or similar in compliance with the industry standard SFP multi-source agreement.

Housing is a very important part of transceivers. It not only protects all the delicate optical and electrical components but also provides shielding from EMI, which could disturb the system optical performance. The housing also serves as part of the receptacle for the fiber-optic connector. In this study, the effects of corrosion of the housing on the optical performance of transceivers [1, 2] are investigated.

The 85%RH/85°C test has been used for qualifying electronic and photonic components and products for a long time. The general consent is that, after 1000 hours of this test, if the characteristics such as the average light output power, extinction ratio, and mask margin of a photonic device meet the allowable specifications, then it is considered reliable for use.

In this study, two sets of 85%RH/85°C tests are performed. One is for determining the weight change and corrosion rate of the housing. The other is for determining the characteristics of the transceivers shown in [1, 2].

(2) CORROSION TESTS OF HOUSING

(2A) Specimen Preparation

The test coupons are prepared per typical corrosion testing procedures. Thirteen coupons are machined from 13 transceiver housings (3 of them are for controls). The die casting material of the housing is Zn-8Al-1Cu alloy. The dimensions (length, width, and thickness) of the 10 test coupons are measured and the corresponding total surface areas are determined, which are: 0.552, 0.558, 0.552, 0.560, 0.553, 0.552, 0.554, 0.552, 0.557, and 0.556 in^2. Per typical corrosion testing procedures, the coupon surfaces are scrubbed with a nylon brush to remove loose corrosion scale, rinsed with acetone, oven dried, desiccated, and then weighted. Figures 3 and 4 show some typical scanning electron microscopy (SEM) photos of the specimen before tests. It can be seen that the housing surface exhibits scattered holes. These holes are actually pores, inherent in the original cast part, that are exposed when the coupons are machined.

(2B) Test Set-Up

The test coupons are put into a temperature and humidity chamber at 85%RH/85°C. Two coupons are removed from the test chamber at time intervals of 48, 120, 216, 288, and 336 hours. The experimental set-up is attempts to minimize movement and coupon-to-coupon contact so as not to disturb the coupons. Per typical corrosion testing procedures, the coupon surfaces are scrubbed with a nylon brush to remove loose corrosion scale, rinsed with acetone, oven dried, desiccated, and then weighted.

(2C) Cumulative Weight Change and Corrosion Rate

The cumulative weight change and cumulative corrosion rate are calculated at each time interval. Figure 5 shows a typical SEM photo of coupon after the 85%RH/85°C test for 14 days. It can be seen that: (1) a tenacious oxide layer has grown on the surface, and (2) there is a pit or internal pore with some oxide scale visible inside it and on the surrounding surface. Figure 6 shows a typical SEM of the cross section of coupons after 14 days of 85%RH/85°C test. The corrosion type is uniform dissolution [4] and the average thickness of the oxide layer on the coupons is less than 1 mils.

The cumulative weight change of the coupons, computed as final weight minus initial weight, as a function of time is shown in Figure 7. It can be seen that oxides are forming on the surfaces of the coupons. Some loose oxide may have sloughed off, but the scale quickly became stable in this environment as shown in Figure 8. The data indicates that the oxide film on the coupon surfaces is formed rapidly and the corrosion rate is decreasing as the time is increasing. Even with the limited test time, a corrosion rate of 0.9 mpy (mils per year), as measured at Day 14, is minor by most standards. This is an acceptable rate if the electronic insulating property of the oxide layer is not of concern.

In this paper, the corrosion rate is defined as $[534 \cdot W/(D \cdot A \cdot t)]$, where "W" is the weight change of the coupon in mg, "D" is the material density in g/cc, "A" is the original surface area in in^2, and "t" is the time of exposure in hours. This equation is the most common method of expressing corrosion rates when evaluating or comparing the performance of materials in a corrosive environment.

It is noted that the net weight gain (as computed from the final weight minus the initial weight) experienced by all the coupons indicated that a tenacious oxide layer grew on the surfaces of the coupons, as expected, which could not be completely removed by a post-test cleaning procedure. The computed corrosion rates, therefore, do not reflect the true rate of metal loss, and may be somewhat understated (since the weight change of metal is also understated). However, this may not be significant over the long-term. The tenacious oxides that formed on the coupon surfaces during testing acted as a barrier layer to slow the corrosion (conversion) of metal over time (per diffusion controlled oxidation theory). Provided the layer remains intact, the loss or conversion of metal over the long-term should follow a decreasing trend.

(3) CORROSION TESTS OF TRANSCEIVERS

(3A) Transceivers Pre-Condition

Thirteen fiber-optic transceivers (3 are for controls), as shown in Figures 1 and 2, are used for this experiment. At time zero, the average light optical power (LOP), receiver sensitivity (RS), and extinction ratio (ER) of all these transceivers are measured and their results are tabulated in the second column of Table 1.

LOP is one of the most important characteristics of transceivers. It shows the optical output power of the laser and is often expressed as a milliwatt or dB value. The dB is a unit used to describe optical power with respect to one milliwatt. For the transceivers in [1, 2], the allowable light optical power is between –10 and 0 dB with a typical value of –6.3 dB.

One important characteristic of a transceiver is its receiver sensitivity (RS). This property defines how small a signal the receiver is able to detect. The receiver must be able to detect a signal over the top of any electromagnetic noise that may be interfering with the system. A careful design of the pre-amplification, post-amplification, and quantization circuitry, plus metal-house shielding around the receiver, can improve the sensitivity. For the transceivers in [1, 2], the allowable sensitivity is between –25 dB and -15 dB with a typical value of –20 dB.

ER is a measure of the amplitude of the digital modulation on the optical carrier and so affects the power penalty, or distance over which a fiber optic system can reliably transmit and receive a signal. ER is defined as the ratio of the average optical energy (output power) of a digital transmitter for a logic one bit (1 state) compared to a logic zero bit (0 state). This ratio is important in system design because it affects the recovered signal level at the receive end. Laser transmitters are usually designed with a compromise between switching speed and ER performance. ER is often expressed as a dB value. For the transceivers in [1, 2], the allowable ER is between 6 and 13 with a typical value of 9.

In viewing of all the values in the second column of Table 1, it can be seen that all the 13 transceivers meet the (LOP, RS, and ER) specifications. Ten of these transceivers are subjected to 20 thermal cycles at –40/+100°C with 5 minutes ramp and 15 minutes dwell (pre-condition). The measurement results of these transceivers after the pre-condition are shown in the third column of Table 1. Again, the LOP, RS, and ER of these transceivers meet all the specifications.

(3B) Transceivers Testing and Results

The 10 pre-conditioned transceivers are put into a temperature and humidity chamber at a 85%RH/85°C test condition. These transceivers are removed from the test chamber at time intervals of 500, 1000, and 1500 hours. The LOP, RS, and ER of these transceivers at each time interval are measured and their results are shown in the 4th, 5th, and 6th columns of Table 1. Again, it can be seen that they all meet the specifications.

The remaining columns of Table 1 show the drift of LOP, RS, and ER of all transceivers. Since the specifications for the drift of LOP, RS, and ER are, respectively, 1.5 dB, 1.5 dB, and 3 dB [1, 2], thus, the transceivers under consideration are within the specifications.

An oscilloscope display of a long stream of random bits (light pulses) is used to determine transmitter and receiver performance. Due to distortion of the waveform (by a variety of factors [6]), the displayed light pulses do not align, but instead overlay one another and appear blurred, as shown in Figure 9, which is called an eye diagram. Much information about system performance can be gathered from the eye-parameter (pattern), however, they are out of the scope of this paper. The important point is that the center of the pattern formed by these pulses must be open: the wider the opening, the less the distortion and jitter [6].

In production environments, as an alternative to eye-pattern analysis, mask tests are often used. A mask defines a set of keep-out regions (or polygons) within which a waveform must not intrude. By comparing an eye diagram against a predefined mask, the overall quality of the waveform can be assessed in one quick measurement.

Also, in a manufacturing environment it is often desirable to add a test-line margin to industry standard masks. At other times, it is useful to reduce the size of the mask to determine by how much a waveform fails the test. Some oscilloscopes (provided by Agilent Technologies, for example) allow this to be accomplished by adding a mask margin to the standard mask. The margin is defined as a percentage above or below the size of the standard mask. Since there are no industry standards that define mask margins this feature is more useful as a qualitative rather than quantitative measure of waveform performance.

Figure 9 shows a typical eye diagram of the transceivers after 1600 hours of the 85%RH/85°C test. It can be seen that the mask margin is 55%, which is much better than a typical mask-margin specification (30%) for these kind of transceivers [1, 2].

Transverse cross-sections of the controlled transceivers and tested transceivers over a period of 1600 hours at 85% relative humidity and 85°C temperature are metallographically prepared and thickness measurements of the surface layer are performed. The thickness of the controlled transceivers is about 0.5 mils as shown in Figure 10. The average thickness of the film on the transceivers' surface at the end of 1600 hours exposure is 2.2 mils as shown in Figures 11 and 12. Thus, the film concerning the exposure is the difference between the latter and the control, equal to 1.7 mils.

(4) SUMMARY

The effects of corrosion on the optical performance of a fiber-optic transceiver have been studied. Also, the cumulative weight change and cumulative corrosion rate of the housing specimen subjected to 85%RH/85°C have been measured and discussed. Furthermore, the characteristics of transceivers have been obtained after 20 thermal cycles at –40/+100oC with 5 minutes ramp and 15 minutes dwell (pre-condition), then 85%RH/85°C and 3.47V for 1600 hours. Finally, SEM photos are provided for a better understand of the zinc alloy die casting housing and the tenacious oxide layer on the housing surfaces. Some important results are summarized as following.

- The net weight gain experienced by all the housing specimens indicates that a tenacious oxide layer has grown on the surfaces of the specimen.
- The corrosion rate of the housing specimen is decreasing as the time is increasing. This is because the tenacious oxide layer will act as a barrier layer to slow the corrosion of housing metal over time.
- The characteristics such as the LOP, RS, and ER of transceivers after 1500 hours of 85%RH/85°C with 3.47V tests are not only meet the specifications but right around the typical values. Thus, normal operating corrosion conditions will not affect the optical performance of the transceivers.

(5) ACKNWLEDGEMENTS

The authors would like to thank Rolando Buted, Ken Pytlewski, Mia Caldwell, Fred Sporon-Fiedler, Matt Schwiebert, and Clinton Chao for their useful discussions and supports.

(6) REFERENCES

(1) HFBR-5710 Data Sheet by *Agilent Technologies*, Inc., 2001.

(2) HFBR-5720 Data Sheet by *Agilent Technologies*, Inc., 2001.

(3) J. Lau, S. Erasmus, F. Sporon-Fiedler, S. Murphy, R. Herms, and S. Pan, "Creep Analysis of a Fiber-Optic Transceiver Housing", *ASME Paper No. 15868, InterPACK'01*, Kauai, Hawaii, July 8-13, 2001.

(4) Porter, F. C., *Corrosion Resistance of Zinc and Zinc Alloys*, Marcel Dekker, Inc., New York, 1994.

(5) Goodwin, F. E., and A. L. Ponikvar, *Engineering Properties of Zinc Alloys*, International Lead Zinc Research Organization, 3rd edition, Research Triangle Pard, NC, 1989.

(6) Kazovsky, LL., S. Benedetto, and A. Willner, *Optical Fiber Communication Systems*, Artech House, Inc., Norwood, MA, 1966.

(a) Average Light Optical Power of Transceivers

S/N	0	20CYC	500HRS	1000HRS	1500HRS	D-20CYC	D-500HRS	D-1000HRS	D-1500HRS
Control 1	-6.46	-6.38	-6.27	-6.13	-6.18	0.08	0.19	0.33	0.28
Control 2	-6.02	-5.96	-5.96	-6.20	-5.88	0.06	0.06	-0.18	0.14
Control 3	-5.95	-5.89	-5.76	-6.15	-5.71	0.06	0.19	-0.20	0.24
1	-5.81	-5.98	-6.07	-6.08	-6.30	-0.17	-0.26	-0.27	-0.49
2	-5.97	-6.14	-6.12	-5.94	-5.97	-0.17	-0.15	0.03	0.00
3	-5.96	-6.09	-6.29	-6.61	-6.26	-0.13	-0.33	-0.65	-0.30
4	-5.74	-5.91	-5.92	-6.26	-5.80	-0.17	-0.18	-0.52	-0.06
5	-5.79	-5.56	-5.75	-6.26	-5.98	0.23	0.04	-0.47	-0.19
6	-5.97	-6.25	-6.41	-6.84	-6.72	-0.28	-0.44	-0.87	-0.75
7	-5.80	-5.98	-6.16	-6.77	-6.05	-0.18	-0.36	-0.97	-0.25
8	-6.07	-6.01	-6.31	-6.72	-6.18	0.06	-0.24	-0.65	-0.11
9	-5.95	-6.10	-6.02	-6.73	-6.72	-0.15	-0.07	-0.78	-0.77
10	-6.11	-5.88	-6.49	-6.41	-6.05	0.23	-0.38	-0.30	0.06

(b) Receiver Sensitivity of Transceivers

S/N	0	20CYC	500HRS	1000HRS	1500HRS	D-20CYC	D-500HRS	D-1000HRS	D-1500HRS
Control 1	-20.50	-21.69	-20.29	-19.90	-20.20	-1.19	0.21	0.60	0.30
Control 2	-20.80	-20.40	-20.36	-20.50	-21.71	0.40	0.44	0.30	-0.91
Control 3	-21.10	-20.67	-20.62	-20.80	-21.79	0.43	0.48	0.30	-0.69
1	-20.30	-20.40	-20.33	-20.30	-20.50	-0.10	-0.03	0.00	-0.20
2	-20.60	-20.95	-20.74	-21.10	-21.80	-0.35	-0.14	-0.50	-1.20
3	-21.00	-20.48	-20.29	-20.10	-21.80	0.52	0.71	0.90	-0.80
4	-19.40	-19.30	-18.81	-19.20	-18.77	0.10	0.59	0.20	0.63
5	-20.00	-20.05	-19.89	-19.70	-20.00	-0.05	0.11	0.30	0.00
6	-20.50	-20.19	-19.97	-19.90	-19.60	0.31	0.53	0.60	0.90
7	-20.80	-20.45	-20.61	-20.40	-20.60	0.35	0.19	0.40	0.20
8	-21.10	-20.86	-20.86	-20.70	-20.80	0.24	0.24	0.40	0.30
9	-20.40	-20.07	-19.97	-19.90	-20.10	0.33	0.43	0.50	0.30
10	-20.50	-20.72	-20.50	-20.50	-20.70	-0.22	0.00	0.00	-0.20

(c) Extinction Ratio of Transceivers

S/N	0	20CYC	500HRS	1000HRS	1500HRS	D-20CYC	D-500HRS	D-1000HRS	D-1500HRS
Control 1	9.32	9.72	9.82	9.84	9.95	0.40	0.50	0.52	0.63
Control 2	9.49	9.44	9.57	9.53	9.65	-0.05	0.08	0.04	0.16
Control 3	9.82	9.94	10.07	9.79	10.09	0.12	0.25	-0.03	0.27
1	9.93	9.92	9.94	10.01	9.69	-0.01	0.01	0.08	-0.24
2	9.83	9.65	9.57	9.79	9.63	-0.18	-0.26	-0.04	-0.20
3	10.03	10.07	9.88	9.88	9.66	0.04	-0.15	-0.15	-0.37
4	10.11	10.07	10.09	10.22	10.06	-0.04	-0.02	0.11	-0.05
5	9.32	9.83	9.79	9.68	9.44	0.51	0.47	0.36	0.12
6	9.69	9.53	9.41	9.59	9.67	-0.16	-0.28	-0.10	-0.02
7	10.33	10.29	10.28	10.20	10.13	-0.04	-0.05	-0.13	-0.20
8	9.88	10.04	9.80	10.05	10.12	0.16	-0.08	0.17	0.24
9	9.69	9.57	9.89	9.75	9.34	-0.12	0.20	0.06	-0.35
10	9.95	9.97	9.59	9.93	10.20	0.02	-0.36	-0.02	0.25

Table 1 Transceiver characteristics (85%RH/85C and 3.47V) at various hours
(Pre-condition: 20 cycles at -40/100C with 5 minutes ramp and 15 minutes dwell)

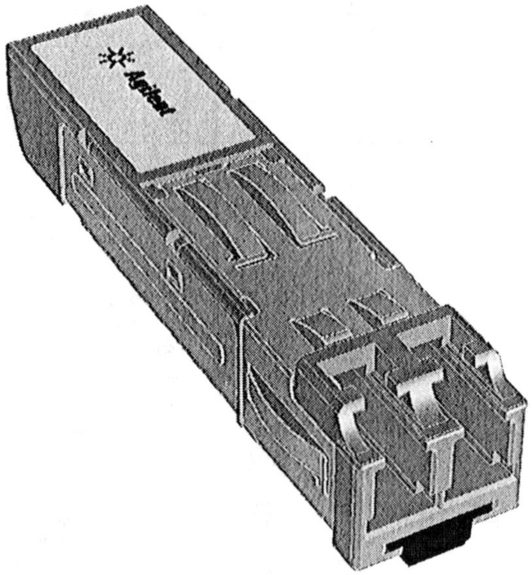

Fig. 1 A small form pluggable transceiver

Fig. 2 The housing is shown on top; the TOSA, ROSA and ESA are shown in the middle; and the cover is shown in the bottom

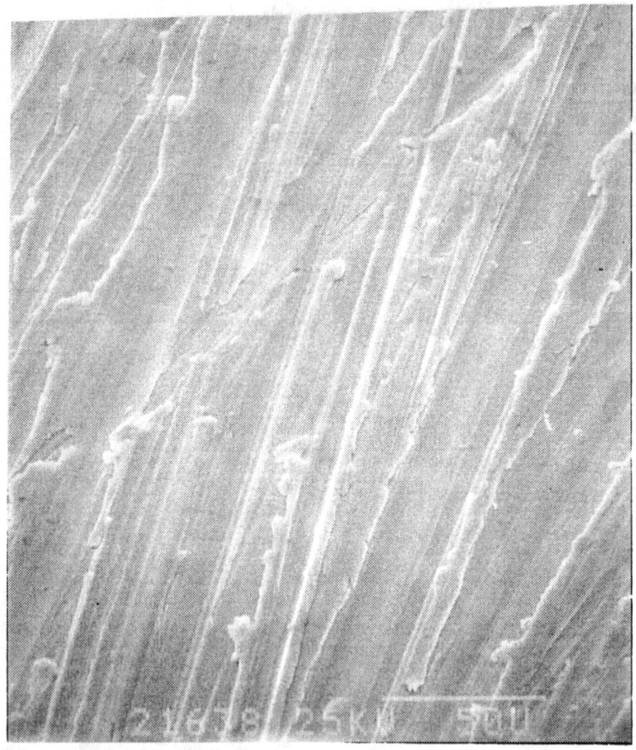

Fig. 3 A typical SEM photo of the control coupon (as-machined) at 500X

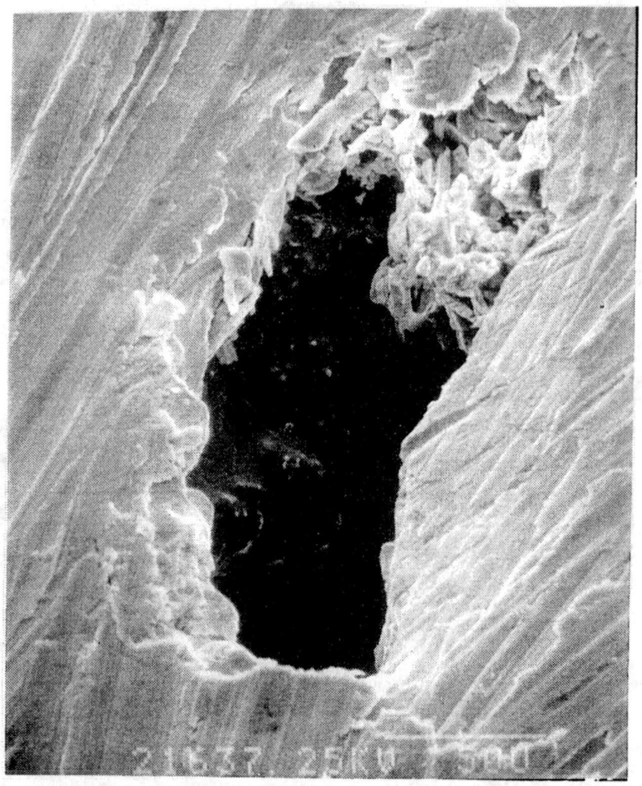

Fig. 4 In the (as-machined) control coupon, a pore beneath the surface that was opened to the surface by the grinding operation (500X)

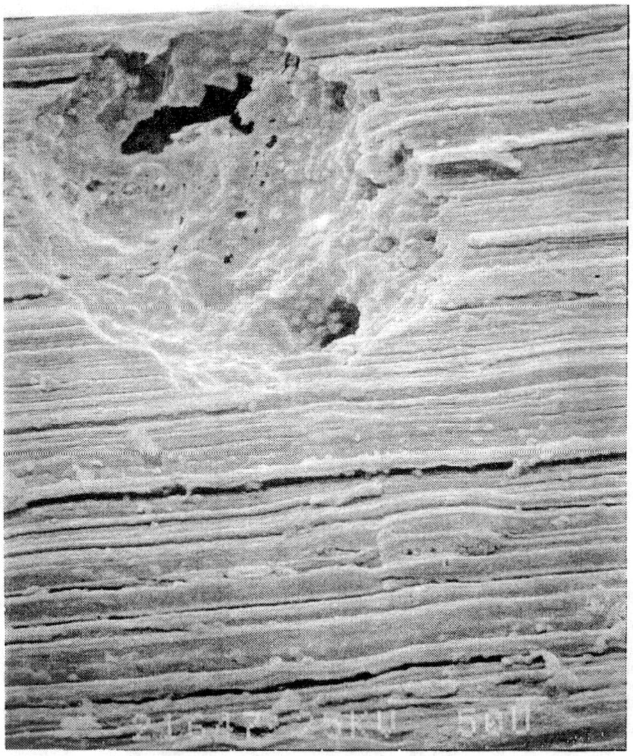

Fig. 5 A typical SEM photo of coupons after 85RH/85°C for 14 days (500X)

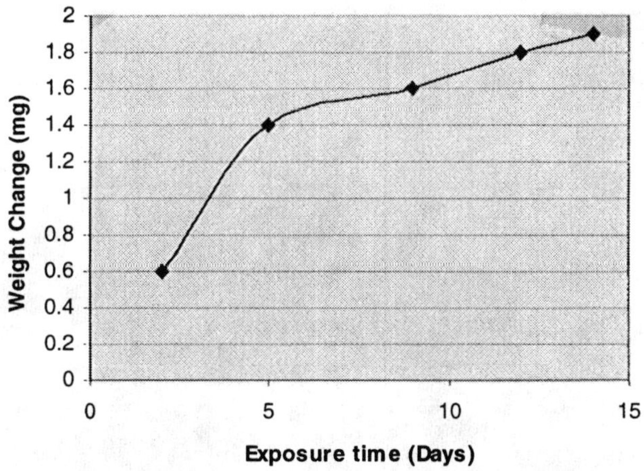

Fig. 7 Cumulative weight change (milligrams) vs. exposure time during the 85RH/85°C test

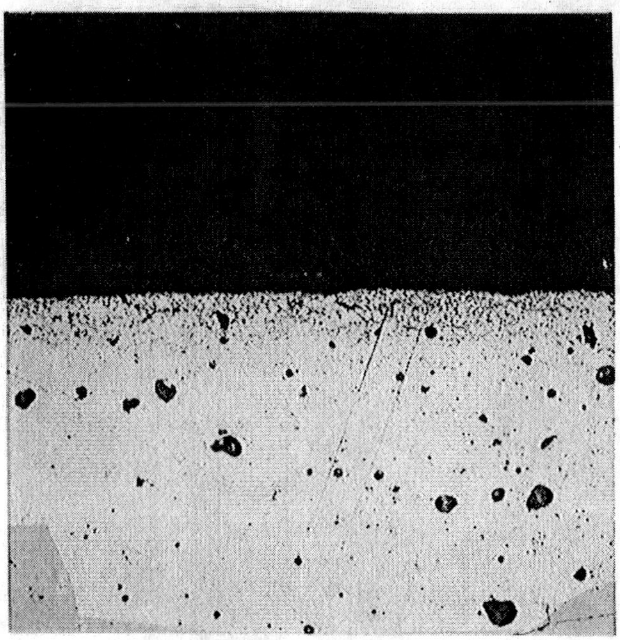

Fig. 6 A typical SEM photo of the cross-section of coupons after 14 days of 85/85 (160X)

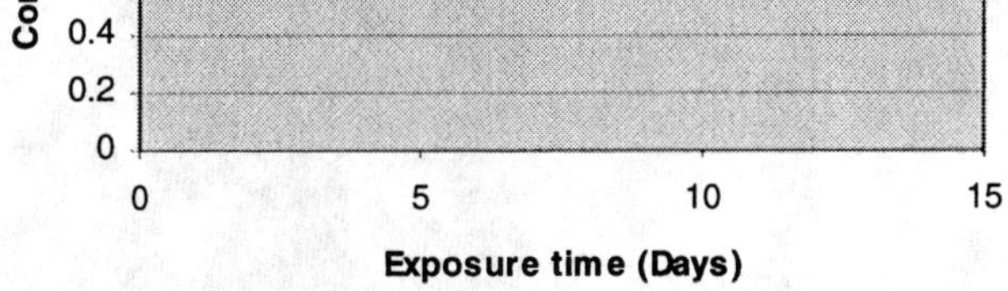

Fig. 8 Cumulative corrosion rate (mils/yr.) vs. exposure time under the 85RH/85°C test

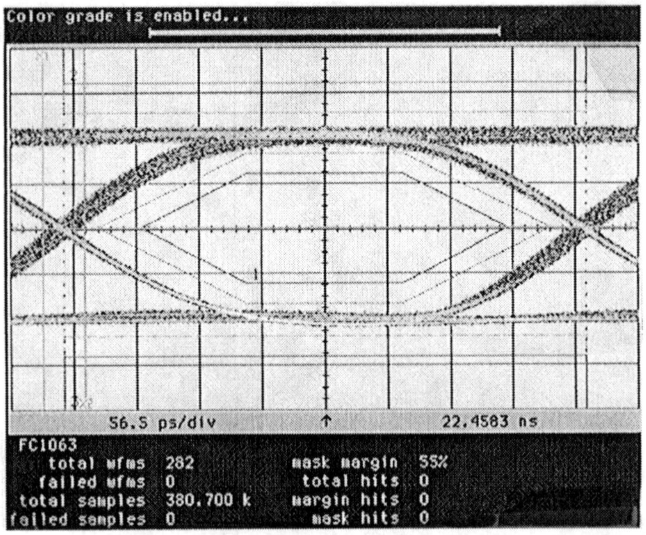

Fig. 9 A typical eye diagram showing the mask margin of transceivers subjected to 1600 hours of 85RH/85°C test

Fig. 10 A typical SEM photo of corrosion film thickness before the 85RH/85°C test

Fig. 11 A typical SEM photo of corrosion film thickness after 1600 hours of 85/85 test

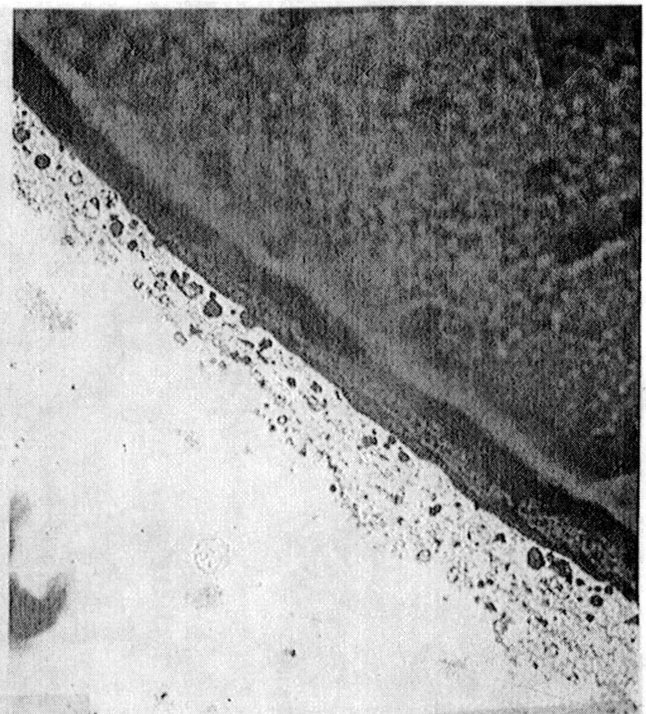

Fig. 12 A typical SEM photo of corrosion film thickness after 1600 hours of 85/85 test

Proceedings of
2001 ASME International Mechanical Engineering Congress and Exposition
November 11–16, 2001, New York, NY
EPP-Vol. 1

IMECE2001/EPP-24722

NEW HYBRID METHODOLOGY FOR THE DEVELOPMENT OF MEMS MULTIVARIABLE SENSORS

Emily J. Pryputniewicz*, **John P. Angelosanto****, **Gordon C. Brown*****,
Cosme Furlong, and **Ryszard J. Pryputniewicz**
NEST – NanoEngineering, Science, and Technology
CHSLT – Center for Holographic Studies and Laser micro-mechaTronics
Mechanical Engineering Department
Worcester Polytechnic Institute
Worcester, MA 01609

* Current address: IDA, Alexandria, VA 01752
** The Foxboro Company, Foxboro, MA, 02861
*** Current address: Corning Lasertron, Bedford, MA, 01730

ABSTRACT

Using recent advances in microelectromechanical systems (MEMS) technology, a new multivariable sensor was developed. This MEMS sensor, capable of measuring temperature, absolute pressure, and differential pressure on a single chip, is particularly suitable for applications in process control industry. However, functional operation of the sensor depends on validation of its performance under specific test conditions. We have developed a hybrid methodology, based on analysis and measurements, that allows such validation. In this paper, the MEMS multivariable sensor is described, the hybrid methodology is outlined, and its use is illustrated with representative results.

Keywords: hybrid methodology, MEMS, multivariable sensors, polysilicon piezoresistive technology, optoelectronic interferometry.

1. INTRODUCTION

Although standard piezoresistive pressure sensors, based on a single crystal technology, are widely and successfully used, they have a number of characteristics that limit their potential for future advancements [1]. That is, the conventional single crystal sensors employ p-n junction isolation and etched silicon diaphragm formation. In addition, junction and surface leakage plus parasitic components, such as unwanted capacitance, place limits on their long-term stability. Also, the etched diaphragm formation effectively limits the number of diaphragms that can be fabricated on a single chip, to just one. The need for both cost reductions and sensor performance improvements, e.g., collocation of two or more different sensors on a single chip, has led to development of polysilicon sensors that enable multivariable measurements.

2. MULTIVARIABLE POLYSILICON SENSORS

A MEMS multivariable sensor package is shown in Fig. 1. It was fabricated using the polysilicon piezoresistive sensor (PPS) technology that differs significantly from the single crystal technology and has a number of advantages. Namely, the PPS technology employs dielectric isolation as a result of a silicon-on-insulator (SOI) structure and, therefore, is not subject to junction leakage instability. In addition, the polysilicon sensing diaphragms are deposited to a specific, controllable thickness with consistency for enhanced fabrication.

Fig. 1. MEMS multivariable polysilicon piezoresistive sensor package.

Because of the deposited diaphragms used in polysilicon sensors, the amount of etching is reduced. One key advantage of the reduced etch volume is the ability to place multiple sensing diaphragms on a single sensing chip. Therefore, the technology is versatile because it can provide absolute pressure (AP), differential pressure (DP), as well as temperature (T) sensing elements on a single chip, resulting in a multivariable (MV) sensor, Fig. 2. In this paper, we concentrate on investigation of the DP capability of the MV sensor.

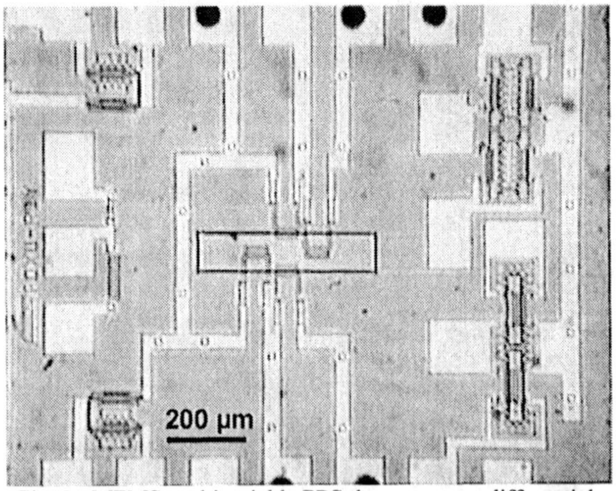

Fig. 2. MEMS multivariable PPS that measures differential pressure, absolute pressure, and temperature on a single sensing chip.

The diaphragm, which senses the applied DP, is shown in the center of the chip, Fig. 2. It is 160 μm wide, 650 μm long, its multilayer thickness is about 2 μm, and the cavity under the diaphragm is 0.9 μm deep.

Deformations of the diaphragm are sensed by the piezoresistive bridge elements (i.e., strain gauges) that are 0.4 μm thick. Determination of pressure from strains of the resistors is based on computations relying on a number of material specific and process dependent coefficients that certainly can vary, which may lead to uncertainties in displayed results. To establish an independent means for measurements of the PPS diaphragms and to validate the coefficients used, we have developed a hybrid methodology [2,3] for measurements and characterization of MEMS.

3. HYBRID METHODOLOGY

The hybrid methodology that we have developed for measurements and characterization of MEMS is based on optoelectronic laser interferometric microscope (OELIM) and finite element method (FEM) [3-6].

This methodology allows highly accurate and precise measurements of absolute and differential deformations of diaphragms as well as their computational modeling and simulation [6-8].

OELIM configuration used in this study is shown in Fig. 3. In this configuration, a beam of collimated coherent light is brought into the system and is directed into a spatial filter (SF) assembly consisting of a microscope objective and a pinhole filter. The resulting, expanded, light field is then collimated by lens L1, and redirected by the directional beam splitter (DBS) through the long distance microscope objective lens (MO) to illuminate the object. The proximal beam splitter (PBS) is placed close to the MEMS PPS. The reflected light is transmitted back through MO, DBS, and the relay lens to the CCD camera.

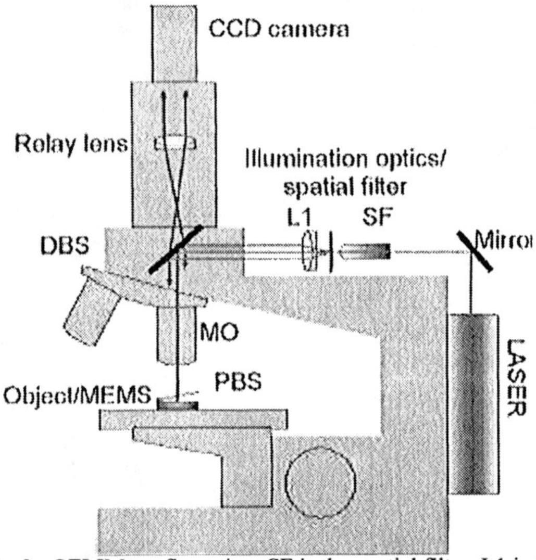

Fig. 3. OELIM configuration: SF is the spatial filter, L1 is the illumination lens, DBS is the directional beam splitter, MO is the long working distance microscope objective, and PBS is the proximal beam splitter.

The CCD camera captures n images, characterized by intensity distributions $I_n(x,y)$, which are defined by the equation

$$I_n(x,y)\quad I_o(x,y)\quad I_r(x,y)\\ \sqrt{I_o\ x,y\ I_r\ x,y}\cos\%M x,y\quad T_n\quad 8\ x,y\ ,\tag{1}$$

where $I_o(x,y)$ and $I_r(x,y)$ represent intensities of the two interfering, i.e., object and reference, respectively, beams, $\%M(x,y)$ represents the optical phase difference between the two beams, and T_n is the discrete phase step imposed during recording of the n-th image to facilitate determination of the fringe-locus function $8\,(x,y)$ which relates to displacements and deformations. If n=5, then from 5 relationships of the type of Eq. 1, we obtain

$$8\ x,y\quad \arctan\frac{2[I_4(x,y)\quad I_2(x,y)]}{2I_3\ x,y\quad I_1\ x,y\quad I_5\ x,y}\ .\tag{2}$$

The fringe-locus function determined from Eq. 2, in turn, relates to the unknown deformations $\mathbf{L}(x,y)$ via the equation [9]

$$\mathbf{K}(x,y)\varpi\mathbf{L}(x,y)\quad 8\,(x,y)\ ,\tag{3}$$

where $\mathbf{K}(x,y)$ is the sensitivity vector representing OELIM geometry. Therefore, $\mathbf{L}(x,y)$ computed from Eq. 3 provides quantitative measurements of deformations of the diaphragm of the PPS.

4. RESULTS AND DISCUSSION

The PPSs were subjected to positive and negative differential pressures with respect to the initial pressure. The initial pressure could be preset, in the test apparatus used in this study, to any value of interest, e.g., ambient atmospheric pressure. Typical OELIM fringe patterns, obtained due to such differential pressures, are shown in Figs 4 and 5.

Fig. 4. OELIM fringe patterns corresponding to: the diaphragm at rest, i.e., at zero psig (top), diaphragm subjected to a differential pressure of + 4 psig (middle), and a differential pressure of + 10 psig (bottom).

Fig. 5. OELIM fringe patterns corresponding to: the diaphragm at rest, i.e., at zero psig (top), diaphragm subjected to a differential pressure of - 4 psig (middle), and a differential pressure of - 10 psig (bottom).

It should be noted that the absolute shapes of the diaphragms at rest influence their response to the applied pressures. That is, deformed shapes of a specific diaphragm are different under positive and negative pressures of the same magnitude. Figure 6 shows a comparison of the OELIM fringe patterns, corresponding to the negative and positive differential pressures of the same magnitudes (10 psig) applied to a PPS. This figure vividly displays the effect that the absolute shape of the diaphragm at rest has on the deformed shape(s) due to the applied pressure.

Quantitative results corresponding to the OELIM fringe patterns of Fig. 6, determined along the longitudinal lines LL through the center of the diaphragm, indicate that the deformation at rest ranges from –20 nm to + 90 nm, for the PPS shown. Furthermore, at the positive and negative differential pressures of the magnitude of 10 psig, these deformations are + 813 nm and –792 nm, respectively [3]. These data are essential for proper interpretation of the signals produced by the PPS.

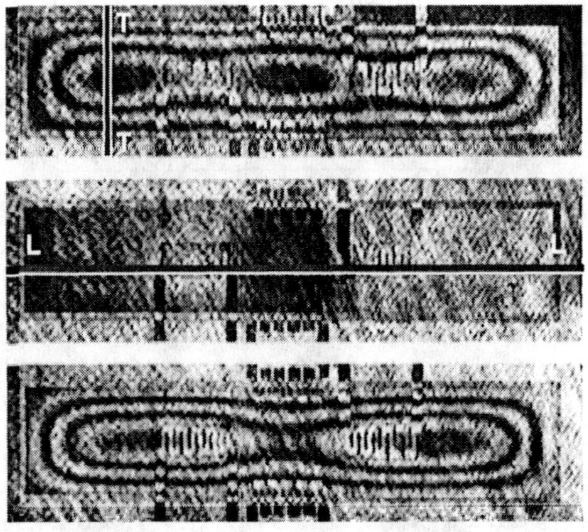

Fig. 6. OELIM fringe patterns corresponding to absolute shapes of the DP diaphragm at: rest, i.e., at zero psig (middle), +10 psig (top), and –10 psig (bottom).

Using OELIM methodology, deformations of the PPS diaphragm can be measured as a function of increments in pressure, i.e., differential pressure, both, along the longitudinal and transverse lines, LL and TT, respectively, Fig. 6. Representative results of such differential measurements are shown in Figs 7 and 8.

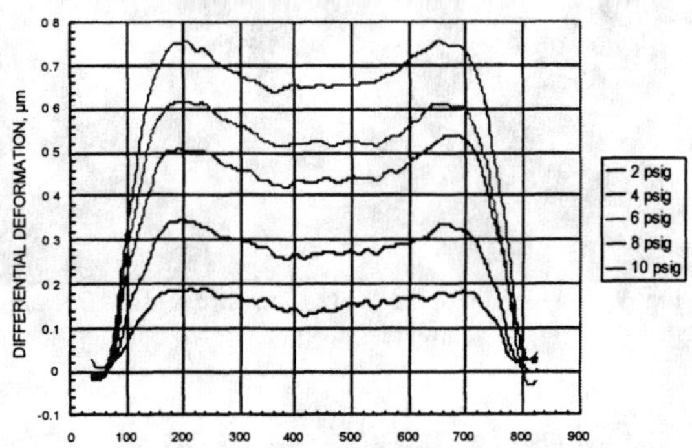

Fig. 7. Differential deformations of the DP diaphragm along the longitudinal line LL of Fig. 6, based on the OELIM measurements, for several specific positive DP levels. Similar results are obtained for the negative DP levels.

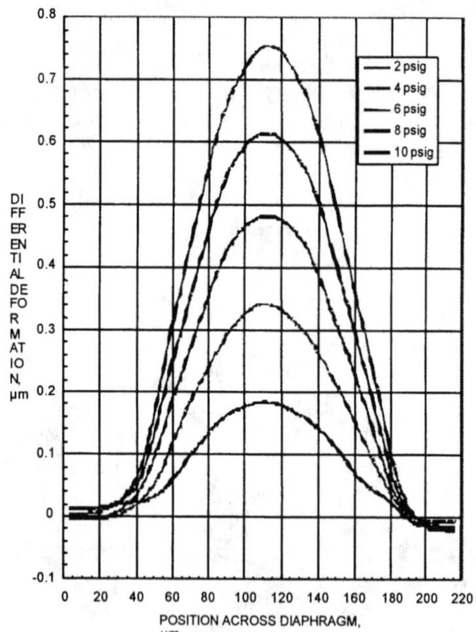

Fig. 8. Differential deformations of the DP diaphragm along transverse line TT of Fig. 6, based on the OELIM measurements, for several specific positive DP levels.

Finite Element Method (FEM) model of the PPS was developed in this study, Fig. 9, using solid elements. Because of the symmetry of the sensing element, only half of the element was modeled. This model incorporates the multilayer structure of the diaphragm and accounts for material properties of each layer. It also incorporates the strain gages and their operational characteristics.

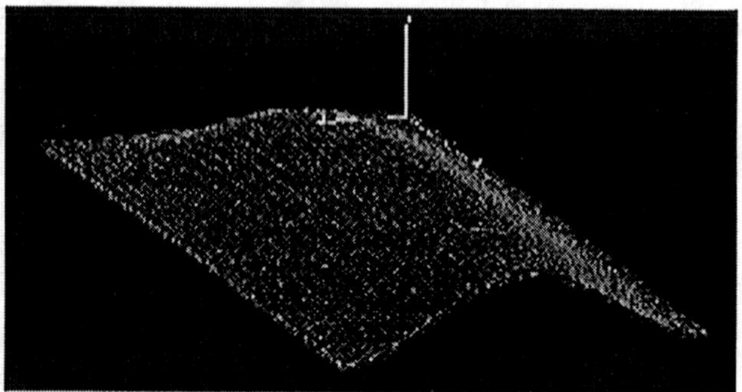

Fig. 9. FEM model of the MEMS pressure sensor. Because of symmetry of the diaphragm, half-model representation was used.

The FEM model was used to determine response of the PPS to a variety of the applied pressure conditions. It was also used to study the structure of the diaphragm and the effects that the strain gauges have on its response to the applied pressures. Representative FEM results are shown in Figs 10 to 12.

Figures 10 and 11 show deformations and von Mises stresses of the diaphragm, respectively, when the sensor was subjected to +10 psig load. Clearly, nonlinear deformations take place because of the design of the diaphragm and the strain gauges. Also higher stress concentrations are visible along the edges of the diaphragm than in its other parts.

Figure 12 shows typical results of investigations of the effects that the strain gauges have on the response of the diaphragm to the applied pressure. For example, Fig. 12a shows deformations corresponding to the conditions used to obtain the results shown in Figs 10 and 11, i.e., when all gauges are symmetrically located with respect to the center of the diaphragm, Fig. 12b represents the case when the lateral gauge is removed and all other gauges are present, while Fig. 12c represents the case of diaphragm deformations without any gauges. Therefore, as the gauge distribution changes from Fig. 12a to Fig. 12c, magnitude of deformations increases from 672 nm to 712 nm and the distribution of the deformations changes.

Comparison of the computational and experimental results shows good correlation.

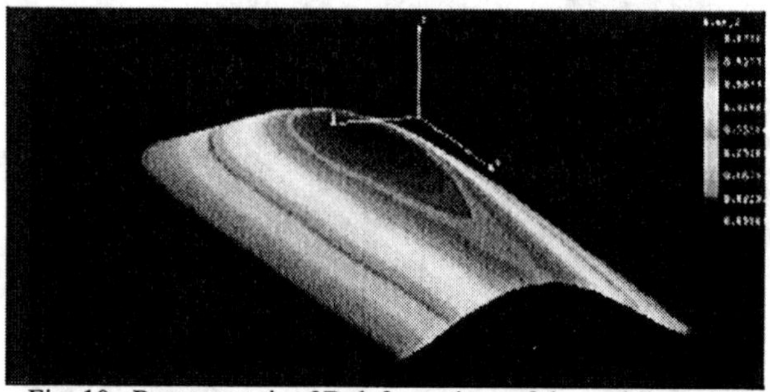

Fig. 10. Representative 3D deformations of the multilayer DP diaphragm subjected to the pressure of +10 psig, based on FEM model shown in Fig. 9.

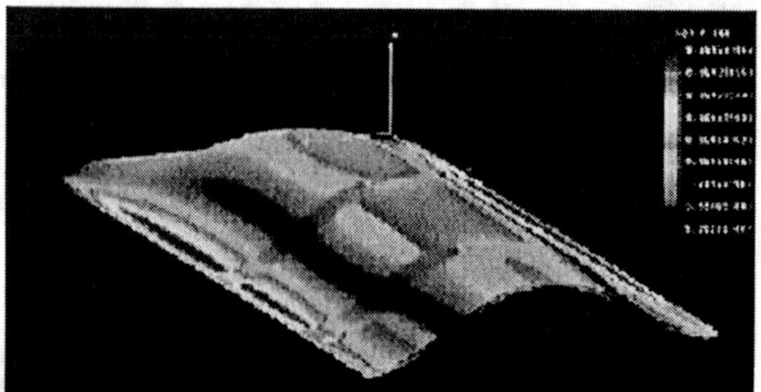

Fig. 11. Representative von Mises stresses of the multilayer DP diaphragm subjected to the pressure of +10 psig, based on FEM model shown in Fig. 9.

5. CONCLUSIONS

Using recent advances in MEMS technology, a new multivariable polysilicon piezoresistive sensor (PPS) for applications in process control industry has been developed. This sensor and a hybrid methodology used to validate its performance were described.

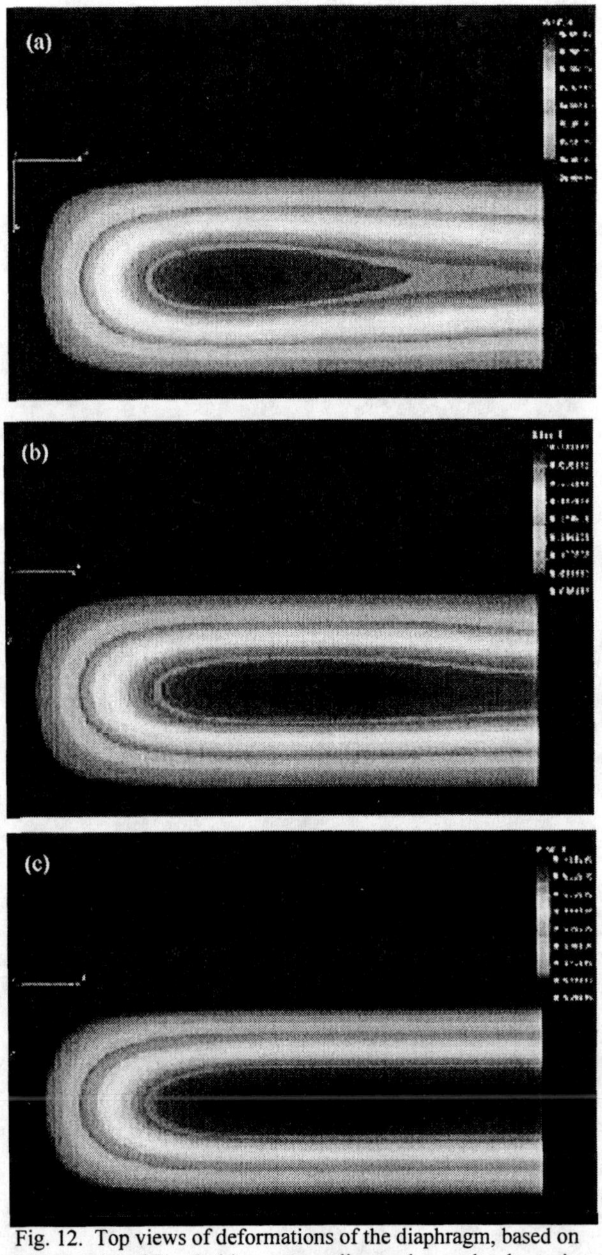

Fig. 12. Top views of deformations of the diaphragm, based on FEM model of Fig. 9: (a) corresponding to the results shown in Figs 10 and 11 with all strain gauges present, (b) lateral gauge removed, (c) no gauges at all.

This multivariable sensor collocates absolute pressure, differential pressure, and temperature measurement capabilities on a single chip, and requires only a single penetration of a process line for installation, which makes it particularly attractive.

The hybrid methodology, based on optoelectronic laser interferometry microscope (OELIM) and finite element method (FEM) has been outlined and its use for measurements of deformations of the PPS diaphragms, subjected to different pressure loads, has been illustrated with representative results. These measurements are used to validate performance of the PPS and in combination with computational results they help to optimize design of the sensor for specific applications.

Maximum differential deformations of the diaphragm subjected to 10 psig were measured to be on the order of 750 nm.

Comparison of the computational (i.e., FEM) and experimental (i.e., OELIM) results shows good correlation.

Validated correlations between computational and experimental results will lead to establishment of new tools for development of MEMS sensors.

6. REFERENCES

1. Rowe, J. R., 1999, "Advanced sensor technology key to new multivariable transmitter," *Proc. Instr. Soc. Am.*, pp. 124-132.
2. Pryputniewicz, R. J., 1994, "A hybrid approach to deformation analysis," *Proc. SPIE*, 2342:282-296.
3. Pryputniewicz, R. J., Champagne, R. P., Angelosanto, J. P., Brown, G. C., Furlong, C., and Pryputniewicz, E. J., 2000, "Multivariable MEMS polysilicon sensor: analysis and measurements," *Proc. Internat. Symp. on Microscale Systems*, Orlando, FL, pp. 76-79.
4. Furlong, C., 1999, *Hybrid, experimental and computational, approach for the efficient study and optimization of mechanical and electro-mechanical components*, Ph.D. Dissertation, Worcester Polytechnic Institute, Worcester, MA.
5. Brown, G. C., 1999, *Laser interferometric methodologies for characterizing static and dynamic behavior of MEMS*, Ph.D. Dissertation, Worcester Polytechnic Institute, Worcester, MA.
6. Przekwas, A. J., Turowski, M., Furmanczyk, M., Hieke, A., and Pryputniewicz, R. J., 2001, "Multiphysics design and simulation environment for microelectromechanical systems," *Proc. Internat. Symp. on MEMS: Mechanics and Materials*, Portland, OR, pp. 85-89.
7. Brown, G. C., and Pryputniewicz, R. J., 2000, "New test methodology for static and dynamic shape measurements of microelectromechanical systems," *Opt. Eng.*, 39:127-136.
8. Furlong, C., and Pryputniewicz, R. J., 2000, "Absolute shape measurements using high-resolution optoelectronic holography methods," *Opt. Eng.*, 39:216-223.
9. Pryputniewicz, R. J., 1994, "Quantitative determination of displacements and strains from holograms," Ch. 3 in Springer Series on Optical Sciences, Vol. 68, *Holographic interferometry*, Rastogi, P. K., Ed., Springer-Verlag, Berlin, pp. 33-72.

Proceedings of
2001 ASME International Mechanical Engineering Congress and Exposition
November 11–16, 2001, New York, NY
EPP-Vol. 1

IMECE2001/EPP-24723

FINITE ELEMENT MODELING OF PROCESSES IN OPTOELECTRONIC ALIGNMENT

Ben Ting

Vincent P. Manno
Corresponding Author

Department of Mechanical Engineering
Tufts University
Medford, Massachusetts 02155

ABSTRACT

For semiconductor lasers, fiber and optical source alignment is crucial for maintaining high optical transfer efficiency. Traditional optoelectronic manufacturing, production of butterfly packages for example, involves laser welding of fiber mountings followed by a tedious realignment procedure to reverse thermally-induced distortions. An alternate technique, laser hammering, entails manipulation of the fiber to light alignment through deformation of the fiber housing with high precision laser beams. A detailed understanding of the material and mechanical behavior, characteristics, and dynamic response is vital to successfully apply an efficient controller that can choose an optimal weld pattern based on a light to fiber misalignment. Modeling provides an effective means to determine an optimal fiber alignment control technique. Modeling is difficult due to the dynamic thermal-mechanical coupling of these processes. This paper presents the preliminary results of a series of parametric studies regarding thermal-mechanical coupling models employed in finite element analysis in order to assess the behavior and dynamic response of representative materials and geometries under various boundary conditions. Fiber ferrule and ferrule housing dimensions affect resistance to bending and torsion, which in turn governs the magnitude of the displacement field. The models are then applied to geometries typical of alignment fixtures used in laser diode packages. The effects of laser energy deposition location and resolution as well as assumed boundary and initial conditions are also discussed. Convection and the small variations in ferule geometry do not have a strong effect on the overall response.

INTRODUCTION

Both laser hammering and traditional alignment techniques involve aligning a semiconductor laser with a fiber. An analogy would be shinning a flashlight through a pipe. A misalignment of the flashlight with the pipe prevents full illumination of the pipe resulting in an inefficient use of electrical power. Likewise, a misalignment with the semiconductor laser and the fiber results in an inefficient use of the laser and poor network transmission performance.

During the manufacturing process, an alignment feedback control is necessary to maximize the light to fiber efficiency by regulating the position between the fiber and laser (Yamauchi, 2000). Sub-micron positioning is required for the optoelectronic devices to operate efficiently. Plots of coupling power vs. maximum distance from maximum coupling are used to depict the light to fiber efficiency. A typical light to fiber efficiency plot, shown in Figure 1, illustrates that precise positioning is vital (Jablonski, 2000).

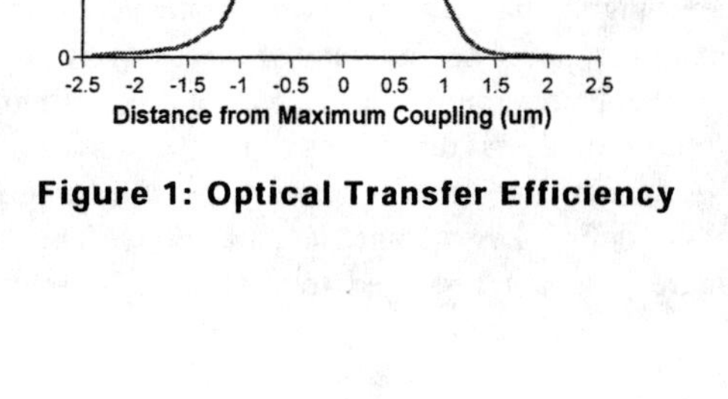

Figure 1: Optical Transfer Efficiency

The Laser Hammering (LH) technique involves regulating the fiber to light alignment through deformation of the fiber housing with high precision laser beams (see Fig. 2), (Song, 1996). The centroid of the fiber is assumed to be located at the centroid of the bay. The legs deform when irradiated by laser beam resulting in a thermal induced alignment or misalignment. Geometric parameter changes directly affect the thermal impedance, which in turn governs the magnitude and topology of the displacement field.

form. A solid rectangular shaped bay is used as a baseline for geometric configuration studies. Although the fiber is supposed to rest in the side the bay, the area of the fiber is small in comparison to the fiber ferrule, so the bay is assumed to be solid for most of the analyses in the paper. The sensitivity to hollow shapes within the ferrule is also reported. It is typical for multiple lasers simultaneously excite the system. The goal of this paper however, is to understand the response to one laser excitation before looking at responses due to multiple beams.

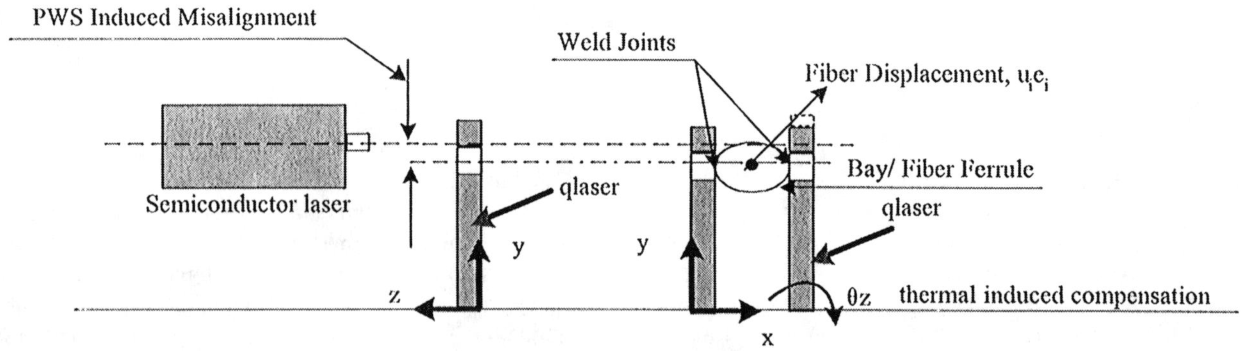

Figure 2: Laser Hammering Concept

Although, the LH technique can be used for multi-channel optoelectronic devices, only single channels devices are considered in the analyses presented. This paper describes parametric studies conducted to identify the system's behavior and dynamic response. The sensitivity studies are divided into two parts, steady state sensitivity and transient sensitivity. It is important to first assess the system's steady state characteristics before a full dynamic parametric study, because the results from steady state will help define the transient studies. Geometric sensitivity, material sensitivity, heat flux sensitivity, focused spot diameter sensitivity, and convective sensitivity studies, are conducted at steady state. Transient cases are simulated to investigate the sensitivity of the magnitude of the imposed energy deposition (heat flux) and pulse duration.

APPROACH

A simple rectangular based geometry is employed to model the laser hammering process. The goal of the study is not to model the laser hammering technique for a particular application, but to address the systems characteristics and sensitivities. A rectangular based geometry provides a good starting point to define key geometric parameters. The bay shown in Figure 3 is not restricted to a specific geometric

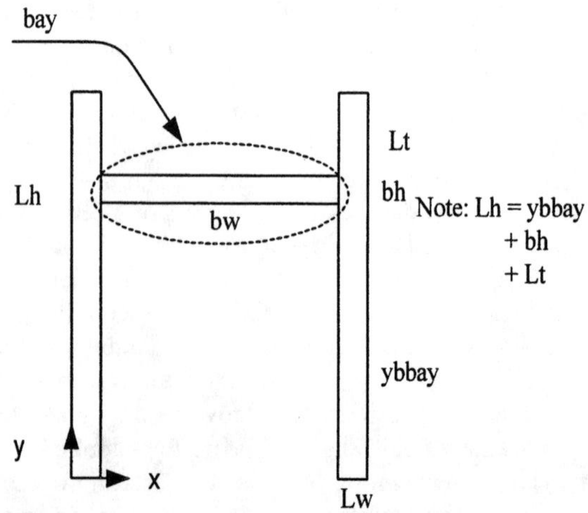

Figure 3: Fiber Housing for Laser Hammering

Geometric sensitivity studies are conducted first to find a baseline geometry to use for the other cases. Likewise, material sensitivity studies are performed to find a baseline material to use for other cases. The determined baseline material and geometry are used for the transient studies. Preliminary studies have shown that the expected stress fields are low compared to the yield and shear strength, thus high stresses are not anticipated in the LH process.

OPTICAL TRANFER EFFICIENCY

The displacement field of the fiber housing is used to get a sense of the laser cavity to fiber transfer efficiency. The analyses assume that the fiber housing is initially positioned such that the fiber is perfectly aligned with the semiconductor laser. Laser beams are used to induce a thermal strain on the fiber housing, resulting in misalignment. The optical transfer efficiency change is directly proportional to the induced strain field. No attempt is made to back out the fiber transfer efficiency from thermal induced displacement. However, for each considered geometry, local displacements and rotations are monitored at the nodal point that mimics the centroid of the fiber. A maximum allowable deflection for a particular alignment technique is dependent upon its use in application.

MATERIAL PROPERTIES

Three different materials are considered to assess the sensitivity of the thermal induced displacement to material properties. Aluminum 6061-T6 and stainless steel 304L where chosen, because they are commonly used for welding materials. A copper tungsten alloy, 25% Cu and 75% W, was also chosen for its high thermal conductivity. Difficulties in obtaining material properties for the CuW alloy, led to assuming that the tungsten properties would have a greater contribution than Cu.

assumed. Future work will consider material behavior and temperature dependent properties.

MODEL OF THE LASER

The laser is modeled as an imposed heat flux on a defined surface area (i.e. the focused spot diameter). No attempt is made to model the optical dynamics of the photon discharge from the laser. Typical focused spot diameters and laser heat fluxes range from 0.05 to 1 mm and 10^5 to 10^7 W/cm^2 respectively (Grigoryants, 1994).

DISCRETIZATION

The Finite Element Method (FEM) was chosen for discretization and simulation. ANSYS® 5.5 is used as the finite element solver. The 2-D, four-node quadrilateral coupled field element is used through out the analyses. Although the coupled field element supports thermal mechanical coupling, the coupled thermo mechanical equations are not solved simultaneously. The thermal mechanical solver activated by the coupled field element solves the Uncoupled Quasi-Static Thermoelastic equations.

$$\frac{\partial \sigma_{ij}}{\partial x_j} + \rho \cdot Fb_i = 0 \tag{1}$$

$$\rho \cdot C_p \cdot \frac{\partial T}{\partial t} = \frac{\partial}{\partial x_i} \cdot (k \cdot \frac{\partial T}{\partial x_i}) + q_{gen} \tag{2}$$

Material Property	AL 6061-T6	SS 304L	Cu W (0.25-cu, 0.75 W)
Density (kg/m3), ρ	2700	7900	14,840
Elastic Modulus (GPa), E	73.1	193	400
Shear Modulus (GPa), G	27.48	75	175
Poisson's Ratio, ν	0.33	0.27	0.28
Shear Strength (MPa), Sxy	205		400
Specific Heat (J/kg K), Cp	963	477	134
Thermal Conductivity (W/mK), k	155.8	14.9	220
Coefficient of Thermal Expansion (1/K), α	24.3e-6	17e-6	4.4e-6
Melting Temperature (K)	854.15	1698.15	1358.15

Table 1 - Material Properties

The values for the elastic modulus, shear modulus, Poisson's ratio, shear strength, and specific heat, and coefficient of thermal expansion are pure tungsten values. Constant thermal and mechanical properties are assumed at standard conditions (see Table 1). Isotropic, temperature independent properties are

Uncoupled quasi-static thermoelastic theory assumes that the strain rate and inertial terms are small. Thus, the assumption results in the absence of a strain rate term from the heat equation and an inertial term from the equation of motion, resulting in the equilibrium equation (Boley, 1960). High strain rates are not expected in the laser alignment process.

The following procedure is used in determining the quasi-static case. The temperature distribution is determined first and used to initialize the strain tensor. The determination of the thermoelastic response starts with calculating the temperature distribution. The temperature distribution is then used to initialize the strain tensor.

$$\varepsilon_{ij} = (\frac{\partial u_i}{\partial x_j} + \frac{\partial u_j}{\partial x_i}) - \alpha \cdot \Delta T \quad (3)$$

With the use of Hooke's Law, substitution of the strain tensor into the equilibrium equation yields an equation that can be solved for the displacement field. The use of the coupled field element tells the solver, ANSYS®, to determine the coupled response by converging to the L2 norm of the applied force and heat flow load vectors (ANSYS®, 1999). For all simulations, an automatic mesh generator was used to create the mesh. To track the assumed fiber displacement, nodes were placed at and near the centroid of the bay. Three nodes were used to define the focused spot diameter. Nodes were also placed at the center and on the ends of the focused spot region. The mesh generator allows for a non-uniform mesh and element size. The average number of elements used through out the analyses is 1172 (468 for the right leg, 280 for the bay, and 424 for the left leg). The mesh in the bay for the hollow bay sensitivity studies is on the order of 1086 elements.

BOUNDARY CONDITIONS

Thermal

An isothermal boundary condition, set to an ambient temperature of 298 K, is imposed on the bottom of the legs (see Fig. 4). The centroid of the laser beam heat flux is applied at half of ybbay (see Fig. 3) on surface of the right leg for all LH cases, except for the laser location sensitivity study. The baseline Focused Spot Diameter (see FSD), is 0.05 mm for all cases except for the FSD parametric study. The remaining surfaces are assumed to be adiabatic. Convection is not expected to have an important role in the thermoelastic response; a sensitivity is conducted to investigate its dependency. Convection is modeled as a constant film coefficient for the convective parametric study. Radiation effects are thought to have a more dominant role than the convection, but radiation is not currently modeled. The fiber housing is assumed to be initially in thermal equilibrium with the ambient.

Mechanical

A zero displacement boundary condition is imposed at the leg bottoms, preventing motion both in the x and y directions. The mechanical model is excited through a thermally induced strain field (Fig. 4). Plane stresses is assumed because excitations act in the x-y plane.

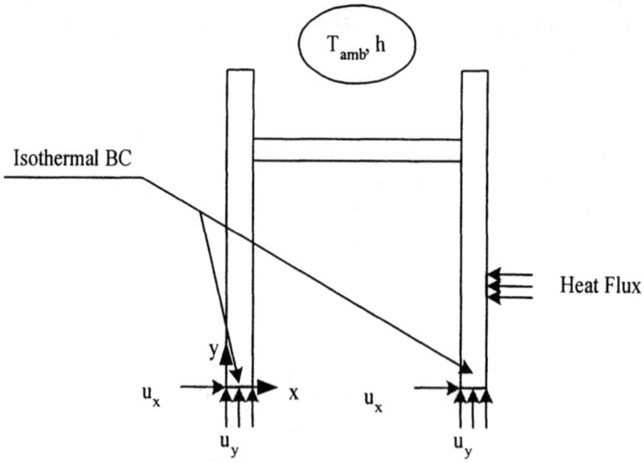

Figure 4: Thermal and Mechanical BC's

STEADY STATE SENSITIVITY STUDIES

MATERIAL SENSITIVITY

Large melt regions occurred for stainless steel, because it's low thermal conductivity offers little resistance to the energy discharge from the laser (see Table 2). Ratios of the fiber displacements (centroidal bay displacements) over the energy deposition show that the CuW alloy is a better material than aluminum as it offers a smaller displacements for the same heat flux (see Fig. 5).

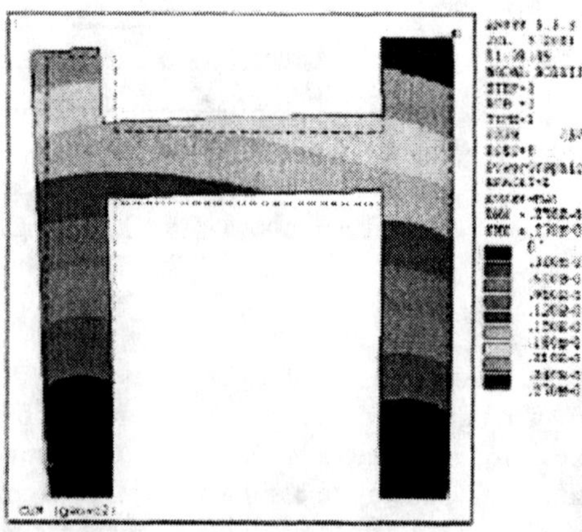

Figure 5: Displacement Magnitude Contours for CuW

Material	ux/q (mm/W/cm2)	uy/q (mm/W/cm2)
Al	-6.20E-05	1.15E-04
CuW	-7.97E-06	1.47E-05
SS304L	n/a - materila melted	

Table 2: Material Table

GEOMETRIC SENSITIVITY

The rectangular based geometry shown in Figure 3 with governing geometric parameters, bw, bh, Lw, ybbay, and Lt, was chosen as the baseline geometry. The initial geometric values used in the geometric sensitivity studies were physically inadequate, resulting in melt regions for energy depositions of in the range of 10^5 to 10^7 W/cm^2. The short leg width offered little resistance to the laser source, which in turn induces melting. Although certain applications are likely to have strict requirements on size, volume, and area, no geometric bounds have been defined in this study. The geometric values used as datum for the geometric sensitivity cases can be found in Table 3.

bh (mm)	bw (mm)	Lw (mm)	ybbay (mm)	Lt (mm)	Lh (mm)
0.5	2	0.5	2	0.5	3

Table 3: Baseline Geometry

Leg Width

A 50% increase in the leg width (Lw) produced a 9.5% temperature drop at the centroid of the FSA (TFSC) and an 11.3% temperature drop at the fiber (centroid of the bay, Tbayc). The thermal resistance is directly proportional to the leg thickness, thus the decrease in temperature distribution with a 50% increase of Lw. The resistance to bending is proportional to Lw. Increasing Lw 50% yielded a 31% decrease of the horizontal fiber displacement and 26% decrease for the vertical fiber displacement (see Table 4). It is currently uncertain as to why the bay centroidal position experienced higher displacements with a thicker leg width. A first thought would suggest the displacement of the bay should be lower with an increase in the leg thickness, because the leg thickness is proportional to the resistance to bending. Both the thermal and mechanical responses are inversely proportional to the leg thickness. More work needs to be done to determine the behavior of the thermoelastic response with respect to the leg width.

Lw (mm)	TFSC (K)	Tbayc (K)	ubay_cx (μm)	ubay_cy (μm)
0.25	n/a material melted			
0.5	674	456	-0.797	1.473
0.75	610	404	-0.547	1.08

Table 4: Leg Width Sensitivity Summary

Bay Height

Initial bay height parametric studies conducted at the baseline geometry (see Table 3) show that the thermoelastic response is nonlinear. The temperatures and fiber displacements do not behave in a monotonic fashion. Although the thermal mechanical response is nonlinear with respect to variations with the bay height, the thermoelastic sensitivities are relatively low (see Table 5). The vertical fiber displacement parameter experienced the highest deviation from the baseline.

bh (mm)	TFSC (K)	Tbayc (K)	ubay_cx (μm)	ubay_cy (μm)
0.25	690	457	-0.704	1.658
0.5	674	457	-0.797	1.47
0.75	669	457	-0.8176	1.684
1	670	459	-0.808	1.717

Table 5: Bay Height Sensitivity

Ybbay

The geometric parameter ybbay, the vertical height of the portion of the leg below the bay (see Fig. 3), exhibited the highest degree of sensitivity amongst all the geometric parameters. A 100% reduction of ybbay produced fiber displacements that dropped about 60 % from the baseline ybbay parameter, 2 mm (see Table 6). Small displacements can be achieved when the distance between the bay and bottom of the legs are shortened. Temperature variations appear to be fairly linearly with ybbay, as TFSC and Tbayc changed approximately 20% when ybbay was both increased and decreased a 100%. The magnitude of the temperature distribution is proportional to ybbay but the response appears to be non-linear.

ybbay (mm)	TFSC (K)	Tbayc (K)	ubay_cx (μm)	ubay_cy (μm)
1	556	378	-0.255	0.597
2	674	457	-0.797	1.47
3	802	537	-1.87	2.86

Table 6: Ybbay Sensitivity

Bay Width

Bay width is a dominant factor governing the magnitude and deformation of the bay. The axial stiffness of the bay, EI, is proportional to its moment of area, which in turn is proportional to the bay length. As the ratio bw to ybbay increases, a distinct bow becomes present in the deformed bay geometry (see Fig. 6). Increasing the bay length can reduce the horizontal fiber displacement, but the trade off is that increasing the length too much will result huge displacements in the vertical direction. The total temperature variation at the centroid of the bay is less than 1% and less than less than a 3% temperature variation occurred at the FSC during the parametric study (see Table 7). Thus, the bay width does not have a substantial effect on the temperature distribution.

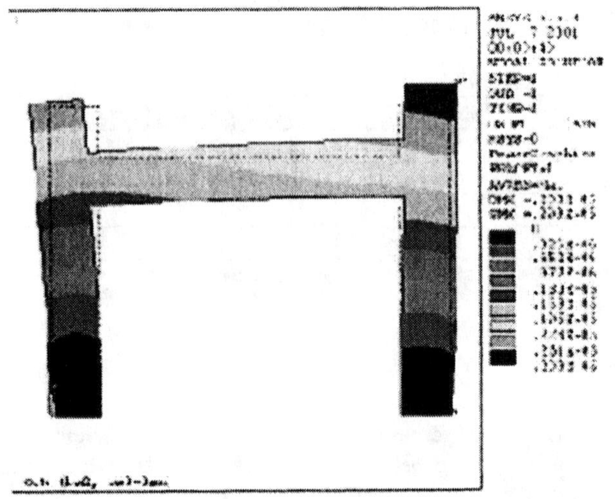

Figure 6: Magnitude Displacement Contour

bw (mm)	TFSC (K)	Tbayc (K)	ubay_cx (μm)	ubay_cy (μm)
0.5	665	457	-1.189	1.25
1	670	459	-0.933	1.32
2	674	457	-0.797	1.47
3	684	459	-0.758	1.78

Table 7: Bay Width Sensitivity

Lt

Lt, the vertical portion of the leg above the bay has a small effect on the overall thermoelastic response. As Lt increases, the vertical fiber displacement increases while the horizontal fiber displacement decreases (see Table 8). The leg width, Lw, and ybbay, are the two dominating geometric parameters. The thermal mechanical response is extremely sensitive to the magnitude of Lw and ybbay. Ironically, this is works out nicely, because Lw and ybbay are the only two parameters were both the horizontal and vertical parameters increased/decreased as the geometric parameter changed. A decrease in the both ubay_cx and ubay_cy can be achieved by either increasing the leg width or by reducing the leg height between the bottom of the bay and the bottom of the legs. For the remaining geometric parameters, the vertical and horizontal fiber displacements do not increase or decrease at the same time while one of the remaining parameters is varied.

Lt (mm)	TFSC (K)	Tbayc (K)	ubay_cx (μm)	ubay_cy (μm)
0.25	679	459	-0.767	1.5
0.5	674	457	-0.797	1.47
1	678	460	-0.748	1.52

Table 8: Lt Sensitivity

LASER EFFECTS

Focused Spot Diameter

Typical focused spot diameters in laser material processing can range from 0.05 to 1.0 mm (Grigoryants, 1994). A small focused spot area is likely to allow for finer tuning. The focused spot area sensitivity study can help determine how critical of a tolerance is necessary. A supplier may rate a laser with an associated nominal focused spot diameter and it is important to know how much variation to expect from a given tolerance. The later portion of the study is aimed in determining focused spot diameter tolerance levels that lead to low sensitivities.

FSD defines the area of heating on the work piece exposed by the laser. Melting began to occur when the FSD was increased to 0.15 mm (see Table 9). A larger FSD allows a greater energy flux to cross the surface causing a rise in with the temperature distribution.

FSD (mm)	TFSC (K)	Tbayc (K)	ubay_cx (μm)	ubay_cy (μm)
0.05	674	456	-0.797	1.47
0.075	842	537	-1.17	2.21
0.1	1019	617	-1.561	2.95
0.15	1358	n/a, melting occurred		

Table 9: FSD Sensitivity

Laser Discharge Energy

For the baseline geometry, the focused spot temperature is near the melting temperature when the laser applies a heat flux of 2 x 10^5 W/cm2 of an area with a 0.05 mm diameter. Low heat fluxes are required for so sub-micron alignment can be achieved through sub-micron displacements (see Table 10). Displacements and temperatures quickly become large as the heat flux varies from 10^4 to $2x10^5$ W/cm^2. Small displacements can also be achieved at high power by increasing the leg width allowing the energy to diffuse through the fiber housing.

qlaser (W/cm^2)	TFSC (K)	Tbayc (K)	ubay_cx (μm)	ubay_cy (μm)
10^4	336	319	-0.078	0.147
$5x10^4$	488	378	-0.39	0.737
10^5	674	457	-0.797	1.47
$2x10^5$	1056	617	-1.562	0.737
$2.75x10^5$	1344	736	-2.148	4.054
$10^{5.5}$	n/a - melting occurred			
10^6	n/a - melting occurred			

Table 10: Laser Discharge Energy Sensitivity

Focused Spot Location

The studies mentioned above all kept the location of the applied energy deposition constant, on the surface of the right leg at y = ybbay/2. It is important to understand the differences in the magnitude of the displacement field and the temperature distribution due to the location of the applied heat flux in order to efficient apply to a controller to the LH process. In an automated application, a controller will need to assess the misalignment of the fiber and determine the optimal location of applied laser heat flux that minimizes the time required for alignment.

FSL	FSD (mm)	FSCX (mm)	FSCY (mm)	ydflo (mm)	ydfhi (mm)	TFSC (K)	Tbayc (K)	ubay_cx (μm)	ubay_cy (μm)
ybbay/4	0.05	3	0.5	0.475	0.525	561	378	-0.562	0.841
ybbay/2	0.05	3	1	0.975	1.025	674	457	-0.797	1.47
3ybbay/4	0.05	3	1.5	1.475	1.525	778	536	-0.851	1.93
bh/2	0.05	3	2.25	2.225	2.275	906	632	-0.764	2.28
Lt/2	0.05	3	2.75	2.725	2.775	1029	649	-0.733	2.33

Table 11: FS Location Sensitivity

h (W/m^2 K)	TFSC (K)	Tbayc (K)	ubay_cx (μm)	ubay_cy (μm)
0	679	456.5	-0.797	1.47
20	684	456.9	-0.78	1.5

Table 12: Convective File Coefficient Sensitivity

The other issue that needs to be addressed is the sensitivity of moving the focused spot area (FSA) by a small tolerance. This is crucial for an automated process. The controller may think it's moving the FSA to a specified location (by moving the laser), but it is likely that the desired FSA location could be off by some tolerance. This misalignment of the desired FSA location could cause problems with the control.

Focusing the laser 25% from the bottom resulted in temperatures and displacements that were smaller compared to the other cases were the FS location was varied (see Table 11). Higher temperatures arose when the heat flux was applied at the beam. The horizontal displacement became smaller as the vertical displacement got bigger. Applying the energy deposition at the surface of the bay centroid produced a 4% decrease in the horizontal displacement and a 31% vertical displacement increase. Laser beams focused on ybbay are extremely sensitive to where the to location of the focused spot area.

SEPARATE EFFECTS

Convective Film Coefficient

Two extreme cases were considered, h=20 W/m^{2}K and h=0, the adiabatic case. Adding convection had little effect on the overall thermal mechanical response. The temperature as the centroid of the bay and focused spot diameter rose less than 1% with the addition of the convective impedance (see Table 12). Latitude and longitudinal displacements increased about 2% with the addition of convection.

Hollow Bay

The model shown in Figure 2 is most likely too simplistic to adequately represent an actual laser hammering process. The bay is solid and does not account for any hollow space to allow where the fiber can rest. Stress concentrations are likely to arise that would not appear in the baseline model with the solid bay. The hollow bay sensitivity study is the first attempt at the investigation of more complicated bay geometries. Initial hollow bay parametric studies will address the ratio of the bay area to the hollow area that produces high sensitivities. These preliminary hollow baseline studies assume a rectangular bay with a hollow circle centroid at the centroid of the bay. Thermal boundary condition on the hollow surface is treated as adiabatic. The iAbay parameter defined as a scaling parameter for the hollow area.

$$Acirc = \frac{Abay}{iAbay} \tag{4}$$

Adding small circular hollow holes centered at the centroid of the bay had little effect on the response. A temperature rise less than 1% occurred from the solid bay to the hollow bay with the bigger circle. Changes in displacement are relatively small. The biggest displacement deviation from the solid baseline geometry, was a 12% increase in the vertical component for the case with the bigger circle (see Table 13).

heat fluxes to diffuse through the geometry. Likewise, melting can occur if the leg width becomes too thin. The leg width and ybbay are geometric parameters that the system is tremendously sensitive to. The system is exceedingly sensitive to the geometric parameters, Lw, and ybbay.

A high sensitivity to the geometric parameter Lt, was not found for the baseline geometry. It is likely that if Lt is large in comparison to ybbay, or close to the ybbay, then the thermoelastic response could be sensitive to the Lt. Future work will include revisiting the sensitivity to Lt over broader ranges. Variations in the bay width had a dominating effect on the vertical centriodal displacement. The deformed shape for the bay bowed upwards when the bay width became large. Bay width variations had a small effect on temperature and the horizontal centroidal bay displacement.

Variations in the bay height didn't have a dominating effect on the thermoelastic response. The thermal mechanical response however, was not a monatomic function with respect to the bay height. Future studies will require the investigation of the non-monatomic phenomena.

case	iAbay	Acirc (mm2)	ID (mm)	TFSC (K)	Tbayc (K)	ubay_cx (μm)	ubay_cy (μm)
baseline	1	0	0	679	457	-0.797	1.47
small circ.	1000	0.001	0.036	678	458	-0.763	1.48
bigger circ.	16	0.0625	0.282	681	469	-0.822	1.64

Table 13: Hollow Bay Sensitivity

STEADY STATE SENSITIVITY SUMMARY

The paper has presented the initial studies taken to identify the system's behavior and dynamic response. The thermoelastic response increased in magnitude with the replacement of the adiabatic boundary condition with the convective one. Changes in the thermoelastic response due to the replacement of the adiabatic boundary condition with the convective one were small. The study confirmed that the sensitivity to the convective impedance is low and thus, had no real dominating effect in the thermal mechanical response. The focused spot diameter sensitivity study confirmed that the magnitude of the thermal mechanical response increases with as more energy is passed through the surface by increasing the focused spot area. For CuW, melting began to occur for a focused spot diameter of 0.15 mm. However, melting can be avoided with focused spot diameters higher than 0.15 mm, through expansion of the leg width (Lw). The expansion of the leg width increases the thermal resistance, allowing for higher

TRANSIENT SENSIVITY STUDEIS

The baseline geometry, chosen from the steady state geometry sensitivity results, is used for the transient analysis. The dynamic response of the system is a function of the magnitude of the energy deposition, pulse length, pulse period, the number of pulses, and the number of lasers used. Laser pulse lengths are typically around 10^{-3} to 10^{-15} seconds (Hecht, 1994). Only one laser is considered in these initial transient studies to determine how the dynamic thermoelastic response varies with the pulse length, pulse magnitude, pulse spacing, and the number of pulses. The transient studies presented in this paper involve a constant energy deposition boundary condition and stepping the energy deposition off after some specified pulse length. Future studies will entail a full sensitivity study on the pulse length, pulse magnitude, pulse spacing, and number of lasers. A single laser with an imposed heat flux with 10^5 W/cm2, acting on the surface on the right leg at half of ybbay.

Work on the transient sensitivity studies is ongoing and only a single case of a constant power deposition is reported here. The transient response is reported in terms of the base centroid displacements (see Fig. 7). Figure 8 shows the transient displacements at the center node. A smooth displacement transient, with a response time on the order of 0.5 milliseconds. is observed. The asymptotic displacement values show a net displacement up and to the left on the order of 1 micron.

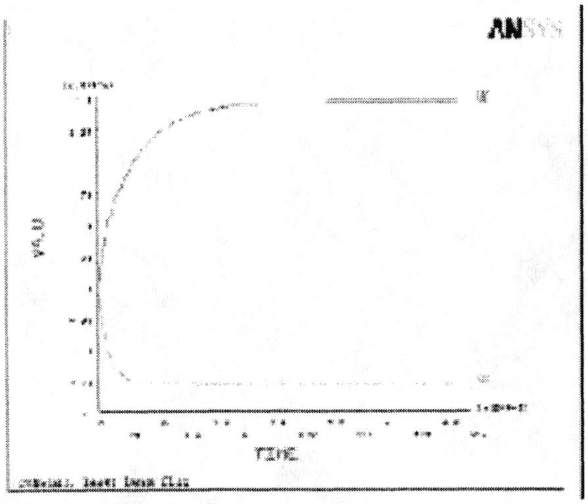

Figure 7: Centroidal Bay Displacements vs.Time, for Constant Laser Heat Flux

CONCLUSIONS AND FUTURE WORK

These initial studies indicate that finite element analysis can be used to predict, at least qualitatively, the thermo-mechanical dynamics of a laser hammering alignment process. Such sensitivity studies can be used to develop a basic understanding of material, geometry and process parameter effects. Much work remains to be done. From a modeling perspective, four outstanding critical issues are: the appropriateness of the quasi-static approach employed in this study to describe the spectrum of LH processes, the allowable geometry simplifications that can be employed to obtain reasonable results, the proper boundary conditions to be used in such simulations, and the appropriate meshing strategy. Once these questions are addressed to a reasonable level of certainty, more realistic cases, representative of actual manufacturing processes, can be simulated. For example, Fig. 8 is a schematic of a two beam laser hammering process that might be employed to achieve temporal and spatial responses unobtainable with a single beam process.

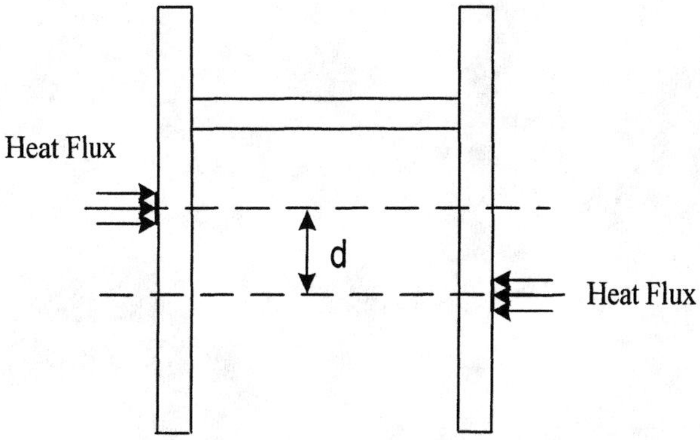

Figure 8: Thermal Moment

REFERENCES

Boley, B., and Weiner, J., 1960, "Theory of Thermal Stresses", John Wiley & Sons Inc., pp 41-61

Song, M., et al., 1996, "Laser Weldability Analysis of High-Speed Optical Transmission Device Packaging", IEEE Transactions on Components, Packaging and Manufacturing Technology – Part B, Vol. 19, No. 4, pp. 758-763.

Yamauchi, K., et al., 2000, "Automated Mass Production Line for Optical Module Using Passive Alignment Technique", Proceedings of the 2000 Electronic Components and Technology Conference, IEEE, 2000, pp. 15-20.

Grigoryants, Alexander G., 1994, "Basics of Laser Material Processing", CRC Press, pp 197-203

Hecht, J., 1994, " Understanding Lasers, An Entry-Level Guide", 2nd Edition, IEEE Press Understanding Science & Technology Series, The IEEE Inc., New York

Jablonski, D., 2000, "Laser Welding Familiarization Module", Tufts Thermal Manufacturing Workshop, Tufts University, Department of Mechanical Engineering, http://ase.tufts.edu/mechanical/thermal/infopages/Program2000.htm

ANSYS®, 1999, "ANSYS Theory Manual", SAS IP Inc., pp 11.1 – 11.12

Proceedings of
2001 ASME International Mechanical Engineering Congress and Exposition
November 11–16, 2001, New York, NY
EPP-Vol. 1

IMECE2001/EPP-24724

FAILURE MECHANISMS AND MODELING IN HIGH POWER OPTICAL DEVICE PACKAGING: SEMICONDUCTOR LASER DIODES - A CASE STUDY

Ajit R. Dhamdhere
Department of Mechanical Engineering
High Density Electronics Center (HiDEC)
University of Arkansas, Fayetteville, AR 72701

Ajay P. Malshe*
Department of Mechanical Engineering
Department of Electrical Engineering
High Density Electronics Center (HiDEC)
University of Arkansas, Fayetteville, AR 72701

S.N. Yedave
Department of Mechanical Engineering
University of Arkansas, Fayetteville, AR 72701

W.F. Schmidt
Department of Mechanical Engineering
High Density Electronics Center (HiDEC)
University of Arkansas, Fayetteville, AR 72701

W.D. Brown
Department of Electrical Engineering
High Density Electronics Center (HiDEC)
University of Arkansas, Fayetteville, AR 72701

* Contact Information: Tel (501) 575-6561 / E-mail apm2@engr.uark.edu

Keywords: high power semiconductor laser diode, die attach, solder, reliability, thermomechanical stress

ABSTRACT

High power semiconductor laser diodes are finding increasing applications in diverse areas such as the biomedical field, materials processing, cosmetics, etc. Dissimilar materials integration and thermal management are some of the primary challenges in the fabrication and packaging of reliable laser diode systems. For example, in a specific case, a high power diode must be mounted with its epitaxy-side down, so as to achieve efficient heat transfer. The device and package junction, particularly the die attach interface, is desired to be void-free (i.e., micro-defect free), stress-managed, and contamination-free for uniform optical delivery and reliable optical alignment. Soft solders used in optoelectronic applications are typically subjected to extreme service conditions that may exceed two-thirds of the melting point. As a result, it is expected that the solder microstructure will be metastable and subjected to nearly constant modification during processing and device operation. In this study, we investigated high power laser diode packages aged under various conditions. Die attach interface, device stress, and microscopic defect analyses were carried out using primarily metallography, scanning electron microscopy (SEM), scanning acoustic microscopy, and micro-Raman spectroscopy. It was observed that the intermetallic compounds and microscopic physical defects at the die attach interface are detrimental to transient heat transfer, and thus, overall package reliability. Further, accurate device modeling can be used to assist in the overall understanding of the effect of packaging-induced stress on the performance and reliability of semiconductor lasers. The purpose of this modeling work was to better understand the interplay between electrical, optical, mechanical, and thermal effects in the operation of laser diodes. Micro-Raman spectroscopy was employed to explore the degradation of high power lasers due to mechanical stresses that appear in an active region of the diode during its growth, packaging, and operation. Using micro-Raman spectroscopy, we found that, although stress increases with aging, the shape of the stress profile and the strength changes randomly from one active region of the diode to another.

INTRODUCTION

Increasing optical efficiency, new package designs, optical coupling methods, a growing number of applications, and steadily declining prices have accelerated the transition of high power laser diodes from research and development into mainstream applications. As applications continue to demand increasingly higher optical output power and longer lifetime, mechanical stresses on dissimilar materials interfaced for packaging pose an ever-growing challenge to the realization of a durable system. Thus, it has become increasingly important to analyze the real root causes of specific degradation modes at optimum laser operating conditions so as to manufacture reliable systems.

A packaged high power laser diode system is a combination of delicate multilayered quantum structures, along with integrated packaging schemes that combine dissimilar materials, all functioning under harsh temperature conditions. Typically, the failure of a packaged system is due to interrelated reasons. For example, temperature cycling of a packaged laser bar attached using soft solder results in creep and stress relaxation at the die attach interface causing mechanical deformation of the laser diode array. Such deformation causes variations in optical emission across the bar. During the product development and manufacturing, understanding the influence of packaging parameters and operating conditions on optical device performance, thermo-mechanical stresses on the device and at the die attach interface, and microstructural changes in die attachment material under continuously evolving/degrading conditions is crucial.

Over the years, many analytical and experimental studies have been used for state-of-the-art assessment of solder joint integrity from a physics-of-failure perspective [1-21]. However, a detailed analysis, correlating device performance and failure, packaging and operating conditions, and thermo-mechanical stresses and material behavior is necessary and timely for application specific systems. Many inconsistencies result from a lack of knowledge of the unique properties of each solder, such as age and cycle softening, hardening due to intermetallic grain-growth, dynamic recrystallization, strain-rate hardening, superplasticity, etc. Many analytical inconsistencies are traced to differing interpretations of the effects of temperature, current, cycle frequency, and period, and to results obtained from different cycle life prediction algorithms. [1]

Hence, the objective of this research was to explore and understand various failure mechanisms under different testing/operating conditions through theory and experimentation. We have observed various macro and microscopic defects at various interfaces in packaged high power laser diode bars using complementary analytical techniques.

THERMO-MECHANICAL ANALYSIS

In the packaged high power laser diode system, electrical input, thermally managed and optical output parameters are very closely related in deciding the ultimate performance of the system. The purpose of this model is to seek a better understanding of the interplay between the electrical, optical and thermal effects, which is important in designing and manufacturing optical system with optimum performance.

One of the known modes of degradation of quantum-well lasers is due to the formation of dark-line-defects corresponding to a dislocation network.[22] Such a network is usually created as a result of mechanical stresses, which appear in the active region of device during its growth, packaging and also during its operation.[22] All the primary properties of interest namely the wavelength of peak gain, the wavelength of the individual lasing mode, the threshold current, and the external differential quantum efficiency, the internal loss, the internal quantum efficiency; are related intimately to the structure of energy bands. The band structure is influenced by both temperature and stresses in the device region. Effect of mechanical stresses on high power laser diode performance can be incorporated in the analysis by introducing the change in optical gain and band width as a function of mechanical stresses.[23] For this thermo-mechanical analysis, a two dimensional model is assumed.

The lumped model describing the isothermal behavior of semiconductor lasers in terms of two ordinary differential equations for electron and photon concentrations (the single mode equations) is established. Thermal effects can be incorporated into the single-mode rate equation by modifying the recombination mechanisms to include Arrhenius law for temperature dependence of the coefficients. This is represented by equations (1) and (2). [24]

$$\frac{dN}{dt}=\frac{J}{ed}-(N-N_t)f(S)S+\alpha_{\text{int}}C\frac{S}{\mu_g}+\frac{1}{V_3}\int_{V_3}\gamma\frac{(1-\Gamma\beta)}{(1+\gamma)}BN^2e^{\frac{E_2}{k}(\frac{1}{T_a}-\frac{1}{T})}dV$$
$$-\frac{1}{V_3}\int_{V_3}N[A_{nr}e^{\frac{E_1}{k}(\frac{1}{T_a}-\frac{1}{T})}+BNe^{\frac{E_2}{k}(\frac{1}{T_a}-\frac{1}{T})}+CN^2e^{\frac{E_3}{k}(\frac{1}{T_a}-\frac{1}{T})}]dv$$

.......(1)

$$\frac{ds}{dt}=\Gamma a(N-N_t)f(S)S-\frac{S}{\tau_p}+\frac{1}{V_3}\int_{3}\tau\beta BN^2e^{\frac{E_2}{k}(\frac{1}{T_a}-\frac{1}{T})}dv$$

$$f(S)=(1+\frac{S}{S^*})^{-1}$$

.......(2)

The symbol represents:
N = concentration of electrons in the conduction band

S = photon density in the lasing mode; t = time; J = driving current density; e = charge on the electron; d = thickness of the active layer; V_i = volume the ith layer; a = linear gain rate; N_t = electron gain at transparency; S^* = saturation coefficient Γ = confinement factor; T_a = ambient temperature = temperature of the heat sink (assumed to be constant); k = Boltzmann constant; T = temperature; α_{int} = absorption of stimulated emission; μ_g = group refractive index.

In the above equation (1), on the right of the equality sign, the first term represents uniform supply of electrons resulting from the applied current, second term represents the optical gain, third term the electrons produced by absorption of photons from the lasing mode, fourth term the electrons produced by absorption of spontaneous emission and the last term carrier recombination.

In the equation (2), on right side of the equality sign, the first term represents confined proportion of the optical gain, second term represents the photon loss and the last term the fraction of the spontaneous emission that enters the lasing mode.

Since optical gain is not confined to the active layer, the confinement factor Γ indicates the extent to which the lasing mode is confined to the active region.

Optical gain, $$G(N,S) = \frac{a(N-N_t)}{(1+\frac{S}{S^*})}$$

........(3)

$\Gamma\beta$ proportion of spontaneous radiation enters lasing mode. A proportion $\gamma(1-\Gamma\beta)/(1+\gamma)$ produces electrons, while a proportion $\gamma(1-\Gamma\beta)/(1+\gamma)$ is absorbed in the surrounding as heat.

The four main sources of heating are:

1. non-radiative recombination in active region
2. re-absorption of stimulated and spontaneous radiation
3. scattering loss at the heterojunction
4. Joule heating of the substrate (n-GaAs).

The following equation represents these four contributions.

$$Q(x,t) = [A_{nr}e^{\frac{E_1}{k}(\frac{1}{T_a}-\frac{1}{T})} + CN^2e^{\frac{E_3}{k}(\frac{1}{T_a}-\frac{1}{T})}]NE_p + a(N-N_t)f(S)S(1-\Gamma)E_p\frac{V_5}{V_2+V_4}$$
$$+\frac{1}{V_5+V_5}\int_{V_5}\alpha_{ss}BN^2E_pe^{\frac{E_2}{k}(\frac{1}{T_a}-\frac{1}{T})}dv$$

.......(4)

The band structure of the device is influenced by both temperature and stress, which are influenced by packaging process parameters and further the aging of the bar and thus, the system during operation life, which is typically about $\geq$10,000 hrs. The stress induced changes in the heavy hole and light hole valance-bands for a direct-gap cubic semiconductor are given as (at the center of the Brillouin zone, k=0)

$\Delta E_{1,2} = H \pm S$

Where, $H = a(\varepsilon_1+\varepsilon_2+\varepsilon_3)$ (shift term)

$S = \{b^2/2*[(\varepsilon_1-\varepsilon_2)^2 + (\varepsilon_2-\varepsilon_3)^2 + (\varepsilon_3-\varepsilon_1)^2 + d^2(\varepsilon_4^2+\varepsilon_5^2+\varepsilon_6^2)]^{0.5}\}$ (splitting term)

$\Delta E_{1,2}$ are the shifts of the light and heavy hole bands, relative to conduction band edge, ε_i are strains, while a, b, d are measured deformation potentials. [23]

The following are three cases for the applied stress, during fabrication, packaging and operation, on the device bar:

(1) Under hydrostatic pressure:

Normal strains = $\varepsilon_1 = \varepsilon_2 = \varepsilon_3 = \varepsilon$

Shear strains = $\varepsilon_4 = \varepsilon_5 = \varepsilon_6 = 0$ ($S = 0$)

The widening of the band gap,

$\Delta E_g = a(\varepsilon_1+\varepsilon_2+\varepsilon_3) = a3\varepsilon$ and $S = 0$

Hydrostatic deformation potential is negative, hence compression increases the bandgap.

(2) In-plane biaxial strain:

This is basically introduced by lattice-mismatched epitaxy.

$|S/H| = b(S_{11} - S_{12}) / 2a(S_{11} + 2S_{12})$

S_{ij} = elements of the compliance tensor. For cubic III-IV semiconductors (eg. GaAs), the ratio is less than one. Hence biaxial strain widens the bandgap.

(3) In-plane uniaxial stress:

This is applied in the [110] direction. The cleaved mirrors are usually [110] faces.

This has both normal and shear components.

$|S/H| = [b^2(S_{11} - S_{12})^2 + d^2S^2_{44}]^{0.5} / 2a(S_{11} + 2S_{12})$

Since $|S/H| \rangle 1$, therefore for in-plane compression the band-gap narrows.

In the presence of compressive uniaxial stress, the optical gain decreases for the TE mode and increases for TM mode. [23] From equation 1 and 2, the above described stresses will affect the device bandgap and hence, the ultimate device's optical performance. For example, for an aged bar, as the optical gain decreases higher pumping current is required, which increases the temperature of the bar resulting into the change in the carrier density and thus, change in the emission wavelength.

The following experimental analysis builds the synergistic database for various failure mechanisms, in light of the above-described analysis and understanding.

EXPERIMENTATION

In this paper, we describe an investigation of packaged linear array continuous wave (CW) laser bars. A schematic diagram of a packaged bar is shown in Figure 1. The GaAs quantum well bars (19 emitters; 1mm x 1cm x 100μm) were mounted on a polished copper heatsink in an *epitaxy side down* configuration using soft indium solder and Ti-Pt-Au interface metallurgy. A copper heat sink provided anode contact to the device. A metal foil mounted on top of the bar provided cathode contact to the device. During die attachment solder reflow, a load of about 50 grams was applied in the direction normal to the plane of the bar. The reflow process was carried out in a vacuum furnace. The bars were then aged for 0, 96, 744, 1000 and 7,000 hours at 40 A input current for an output power density of 200 W/cm^2. It was observed that some bars succumbed to infant mortality, while most of the bars survived and operated satisfactorily for over 7,000 hours.

In order to identify and understand the defects and related failure modes, samples were characterized using the following complementary analytical techniques: optical microscopy and scanning electron microscopy (SEM, Hitachi) to observe interface microstructure, scanning acoustic microscopy (SAM, Sonoscan) to explore void distribution at the device bar and solder interface, energy dispersive spectroscopy (EDS, Kevex) for chemical analysis of the interface, and micro-hardness (Buehler) to test the mechanical response as a function of thermal degradation. The purpose of this analysis was to determine the relationship between material microstructure, micro-hardness, and chemical changes at the interface as a function of aging of the package.

In an innovative approach, thermo-mechanical stresses present in the active region of the device were calculated using micro-Raman spectroscopy by observing the shift of the characteristic GaAs peak [25]. Positive and negative peak shifts correspond to compressive and tensile stress in the GaAs material [26,27]. Micro-Raman measurements were performed using a 488 nm Ar^+ laser. Observations were made at different locations along and across both active regions of the laser bars. Observations were made on the front [110] facet of the device. In this configuration, scattering from the transverse optical (TO) phonons is allowed in accordance with the symmetry selection rules. The shift in the TO peaks was studied as a function of aging. Since the peak shifts are small, peak fitting was employed systematically. The Gaussian peak fit for the laser peak and mixed Gaussian and Lorenztian peak fits for the Raman peak were used.

RESULTS AND DISCUSSION

I. Physical defect and morphological observations:

The following is a detailed discussion of the various micro-defect structures observed in the aged laser bars.

Non-uniform physical contacts between the top cathode metal foil contact and the laser diode bar, the bar and the multilayer metallization, the metallization and the die attach solder, and the solder and the copper heat-sink were observed at various locations across the length of the device. These non-uniform contacts were found to be different in their morphology, such as delamination (Figure 2), crack propagation in metallization (Figure 3), and cavitaion at GaAs ridges and crack propagation near GaAs ridge structures.

Coefficient of thermal expansion (CTE) mismatches between the metallization, die attach, and copper heatsink is likely to cause delamination and cracks that run along the interface. Such non-uniform physical contact between a high power laser diode bar and a copper heatsink causes excessive heating at the device junction, resulting in occasional burning at the bar emitter surface.

Figure 4 shows the output of the laser emission analyzer for a mounted bar aged for 1000 hrs. Physical bending/displacement of the bar can be clearly seen. This is the signature of the various observations on different samples, for which random bending of the bars across their lengths was observed. Further, it is observed that, during the operation of thes high power laser diode bars, a high steady-state temperature of ~ 70°C initiated thermo-mechanical stress relief of the die-attach metallization, which allowed for transfer displacement of the GaAs bar, commonly known as the "smile problem".

Figure 5 demonstrates that non-uniform physical contact, resulting from non-uniform microloading on the GaAs bar along the metallization and die-attach interface during the die-attachment process, particularly at the ridges, can cause cavitations and stress gradients. This results in cracking of the bar. Furthermore, dynamic recrystallization of the solder is observed at the interface, particularly in the 7000 hour aged samples. Since the heat generated is concentrated in a few regions, thermal gradients contribute to the nucleation and grain boundary motion for dynamic crystallization. Figure 5 shows crystal growth in the direction of the shear stress. The crack seen in the bar is due to excessive stress.

Other physical defects were observed more frequently at the ends of the laser diode bars. After rigorous SEM and optical analysis, we believe that the stress was concentrated near the ends of the bars, causing frequently observed delamination at the ends of the bars.

Further, it was observed that, after aging the packaged GaAs laser bar samples, the solder layer and the heat sink front surface were misaligned up to 15 microns. This misalignment varied from sample-to-sample, and along the width of a bar. This is a second order reliability problem.

Figures 6 and 7 show typical distributions of voids in a good and a defected bar, respectively. Although it was observed that voids present in the solder region were distributed randomly, it was observed that the void density is relatively higher at the center of most of the bars. After an important synchronized set of observations, we conclude that both void density and delamination at the bar and die attachment interface increases with aging. This is detrimental since these are major barriers for efficient heat transfer and can cause accelerated failure of the packaged device.

II. Chemical and hardness observations

Figure 8 shows the metal microstructure and large grain growth for a sample aged for 7000 hours. The grain size at the interface of the copper (Cu) and indium (In) increased with aging time. Based on the earlier morphological observations, we suggest that excessive heating due to insufficient physical contact between the device and heat sink gave rise to grain growth at the interface as a function of aging time.

Figure 9 provides energy dispersive x-ray analysis (EDS) data for aged samples. The measurements were performed in the solder region and the points 1-10 on the x-axis represent sampling points from the solder near the GaAs laser diode bar to copper heat sink. The measurements were performed at various emitter regions across the bar to collect better statistics. We observe copper diffusion from the copper heatsink into the indium solder. Further, we observed that the diffused copper concentration varies along the width of the bar (data not shown). Based on the Cu-In phase diagram and earlier observed grain growth, we conclude that, at the interface, there is intermetallic formation, which is known to be a brittle material and can result in crack development under stress conditions [28]. We also observed that copper diffusion occurs at higher rates near the metallization to die attach interface during the initial hours of aging, and that the change is relatively small, though gradual, over the remaining aging period. Thus, to confirm the increase in hardness caused by the intermetallic formation, we performed microhardness analysis on the interface metallurgy. Table 1 gives Knoop micro-hardness data as a function of aging. The hardness of the solder region increases with aging time, which agrees with earlier predicted intermetallic phase formation.

III. Micro-Raman observations

Figure 10 shows micro-Raman shifts as a function of aging time. Micro-Raman measurements were performed on the optically sensitive GaAs device, across the width of the bar, at about 1μm spacings. Sampling points 1-6 on the x-axis of the plot represent measurement locations across the width of the front face of the active region of the GaAs bar. Point 1 is the first point toward the device-heatsink interface and 6 is the fartherest one. For the purpose of the analysis, we used the GaAs signature Raman peak at 269.3 cm^{-1} [26]. The graph clearly shows a shift in the peak, particularly toward lower wavenumbers for regions of the device near the die-heatsink interface, and high wavenumbers for the region away from the interface. The lower and higher shifts are responsible for tensile and compressive stresses within the GaAs lattice [25,26]. Using the known pressure dependence of the TO-mode shift of 0.5 $cm^{-1}/1$ kbar, these experimental mode shift profiles can be converted to stress profiles [25], and this is shown in the plot for various regions of the device. Two stress zones, one under compressive and the other under tensile stress, can be seen. The zone near the GaAs bar-heatsink interface is under tensile stress, while the zone near the quantum well active layer is under compressive stress. In addition, from the plot where we compare a 0 hr sample to a 7,000 hr aged sample, we believe that the tensile stress, which is introduced during packaging near the interface, is being relieved and, at the same time, intentionally-introduced stress in the quantum well region is almost the same. The authors note that further *insitu* micro-Raman analysis during aging is essential to understanding the gradual changes in the stresses as a function of aging time.

CONCLUSIONS

In summary, we have demonstrated that, in the high power laser diode systems, microstructural, optical, mechanical, and electrical parameters are inter-related and form the basis of the synergistic analysis of the failure modes of the system. We have applied complementary analytical techniques to evaluate micro-cracks and related physical defects at the die attach-device interface and aging-induced grain growth and change in the chemistry of the metallurgy. We observed significant copper diffusion into the die attach layer, causing formation of brittle intermetallics (confirmed by microhardness) which adds to the brittle failure at the interface. Further, we conclude that stress relief during aging and excess heating caused by inadequate physical contacts between the active regions and the heat sink results in uneven bending of the bar causing non-uniform optical emission. Furthermore, changes in the stress across the bar and changes due to aging can be effectively characterized by micro-Raman spectroscopy.

Hence, physical defects, optical response, and chemical composition analysis, along with micro-hardness measurements, provide valuable insight into packaging and operation-induced defects affecting the reliability of high power laser diodes. In combination with a clear understanding of parametric correlations, enabled by modeling, the work described here should ultimately provide assistance in predicting packaging failure modes that will lead to better processing and packaging schemes in the future.

ACKNOWLEDGEMENTS

The authors acknowledge Semiconductor Group of Coherent, Inc. for financial support of this research. Also, we acknowledge Dr. John Nightingale, Mr. Robert Miller and Mr. John Morales, of Coherent, Inc. for financial support and valuable technical discussions. We sincerely acknowledge Dr. Richard Bormett and Ms. Diane Allen of Renishaw Inc. and Dr. John Shultz of Arkansas Analytical Laboratory for micro-Raman and related analytical work support.

REFERENCES

1. L. Wen, G. R. Mon, R. G. Ross, Design and Reliability of Solders and Solder Interconnections: Symposium Proceedings of TMS Annual Meeting (1997) 219.
2. S. A. Merritt, P. J. S. Heim, S. H. Cho, M. Dagenais, 1997 IEEE Transactions on Components Packaging and Manufacturing Technology - Part B 20 (2) (1997) 141.
3. M. Plotner, H. Klett, K. Melzer, G. Blasek, Proced. of the International Conference on Micro Electro Opto Mechanical Systems Technologies (1991) 358.
4. P. T. Vianco, P. F. Hlava, A. C. Kilgo, Journal of Electronic Materials 23 (7) (1994) 583.
5. Y. M. Liu, T. H. Chuang, Journal of Electronic Materials 29 (4) (2000) 405.
6. W. Nakwaski, Electron Technology 23 (1-4)(1990) 33.
7. C. C. Lee, D. H. Chien, Proceedings of the Technical Program NEPCON (1997) 1395.
8. J. W. Morris, H. L. Reynolds, Conference Proceedings of International Intersociety Electronic and Photonic Packaging Conference INTERpack, Advances in Electronic Packaging 2 (1997) 1529.
9. J. L. FreerGoldstein, J. W. Morris, Journal of Electronic Materials 23 (5) (1994) 477.
10. R. B. Cinque, J. W. Morris, Journal of Electronic Materials 23 (6) (1994) 533.
11. M. Antler, Plating 6 (1970) 615.
12. J. M. Parsey, S. Valocchi, W. Cronin, J. Mohr, B.L. Scrivner, K. Kyler, Journal of Materials 31 (3) (1999) 28.
105. P. W. Epperlein, SPIE proceedings 3001 (1997) 13.
13. W. L. Phillipson, D. J. Diaz, Materials Developments in Microelectronics Packaging Conference Proceedings, (1991) 289.
14. A. Zubelewicz, R. Berriche, L. M. Keer, M. E. Fine, Journal of Electronic Packaging 111 (1989) 179.
15. N. Strifas, A. Christou, Mat. Res. Soc. Proc. 356 (1995) 869.
16. N. Zhu, IEEE 15th SEMI-THERM Symposium, (1999) 22.
17. W. Nakwaski, Electron Technology 23 (1-4) (1990) 33.
18. V. D. Bo, F. Bartels, A. Scandurra, Ch. Luchinger, S. Radeck, 11th European Microelectronics Conference (1997) 308. 19. W. W. So, C. C. Lee, 1998 IEEE Electronic Components and Technology Conference Proceedings (1998) 284.
20. Q. Tan, Y. C. Lee, 1996 IEEE Electronic Components and Technology Conference (1996) 26.
21. A. Y. Kuo, K. L. Chen, Conference proceedings of ASME Winter Annual Meeting (1991) 1.
22. M. E. Polyakov, Sov. J. Quantum Electronics 19 (1) (1989) 26.
23. D. A. Cohen, L. A. Coldren, 1997 IEEE Electronic Components and Technology Conference (1997) 768.
24. W. R. Smith, J. R. King, B. Tuck, IEE Proc.-Optoelectron. 144 (6) (1997) 389.
25. P. W. Epperlein, SPIE proceedings 3001 (1997) 13.
26. P. Puech, G. Landa, R. Carles, C. Fontaine, J. Appl. Phys. 82 (9) (1997) 4493.
27. J. P. Landesman, A. Flore, J. Nagle, V. Berger, E. Rosencher, P. Puech, Appl. Phys. Lett. 71 (17) (1997) 2520.
28. C. C. Lee, Y. C. Chen, Mat. Res. Soc. Symp. 390 (1995) 225.

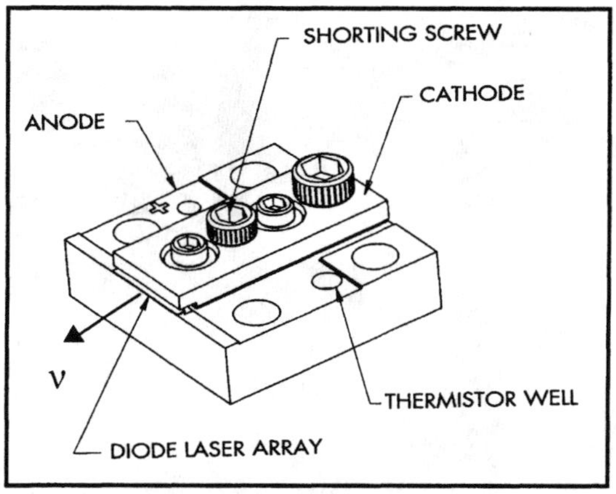

Figure 1: Schematic of the package

Figure 2: Delamination of metallization

Figure 3: Crack running along the die-attach

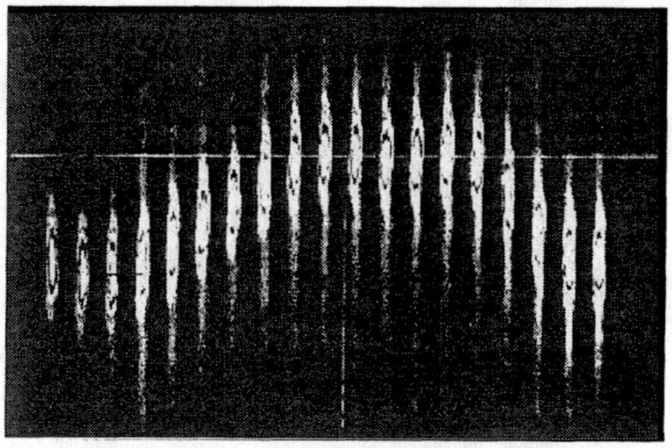

Figure 4: Beamview analyzer picture of the "Smile Problem"

Figure 5: Higher physical defects and stresses at ridges. A cracked bar.

Figure 6: SAM picture of randomly distributed (light regions) lesser voids in a good bar.

Figure 7: SAM picture of randomly distributed (light regions) larger voids in a defected bar.

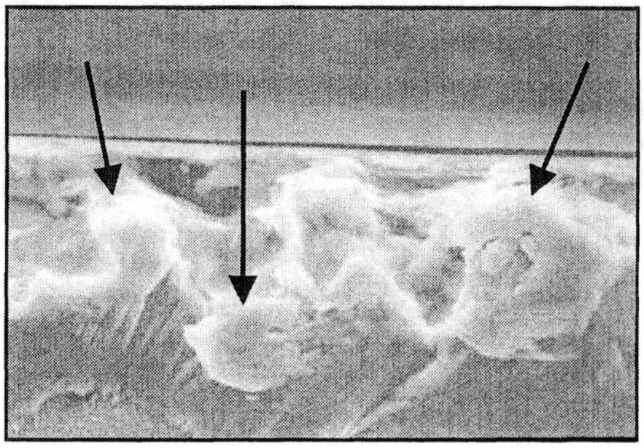

Figure 8: Large size grain growth in 7000 hour aged sample

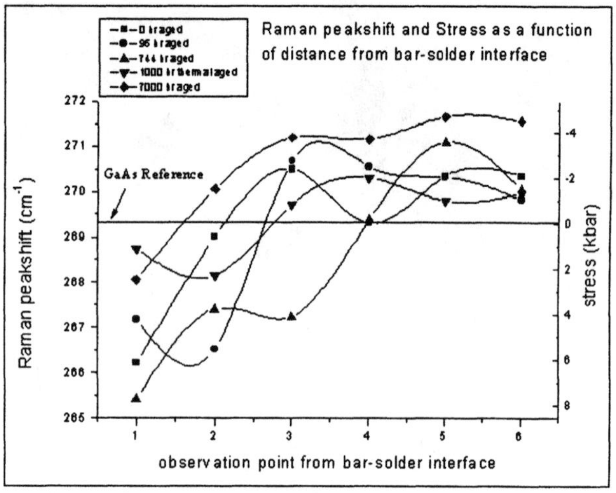

Figure 10: Typical graph of Raman peak-shift as a function of aging.

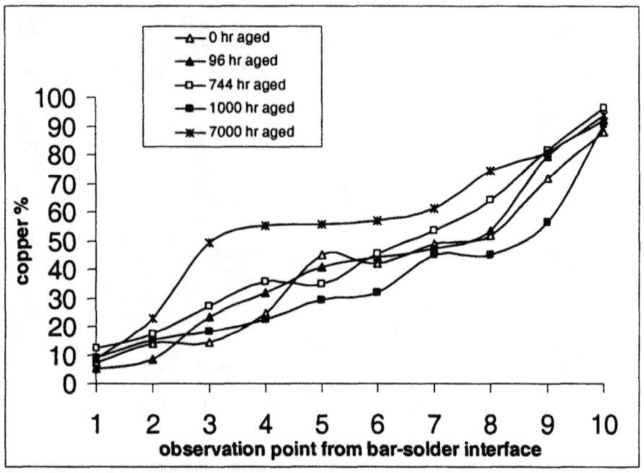

Figure 9: Copper concentration is highest near die-attach interface for 7000 hour aged sample

Sample description	Average Knoop Micro-hardness
0 hr aged	157
96 hr aged	160
744 hr aged	196
1000 hr thermal aged	203
7000 hr aged	214

Table 1: Knoop micro-hardness as a function of aging

Proceedings of
2001 ASME International Mechanical Engineering Congress and Exposition
November 11–16, 2001, New York, NY
EPP-Vol. 1

IMECE2001/EPP-24725

COMPUTATIONAL AND EXPERIMENTAL STUDY OF LASER MICROWELDING PROCESSES FOR ELECTRONIC PACKAGING

Wei Han and **Ryszard J. Pryputniewicz**
NEST – NanoEngineering, Science, and Technology
CHSLT – Center for Holographic Studies and Laser micro-mechaTronics
Mechanical Engineering Department
Worcester Polytechnic Institute
Worcester, MA 01609-2280

ABSTRACT

Advances in microelectronics depend on reliable packaging. As sizes of components decrease and number of components per package increases, demands on interconnections, or joining, of two parts, e.g., wire to tab, or tab to tab, increase rapidly. Since traditional joining techniques, such as resistance welding, may no longer produce satisfactory microinterconnections, alternative methods are being developed. One of such methods is based on laser microwelding. In this paper, we discuss fundamental processes involved in laser microwelding, outline a computational methodology to simulate thermal effects produced, describe facilities used, present results, and summarize correlations between the experimental and computational data for applications in electronic packaging.

Keywords: laser microwelding, laser parameters, heat affected zone, material properties, nanoindentation, computational modeling, experimental validation, electronic packaging.

1. INTRODUCTION

Competitive trends in microelectronics and packaging require features that cannot be readily produced using existing manufacturing capabilities. One of such features is a connection, or joining of two parts, e.g., wire to tab, or tab to tab [1]. Normally, these connections are made by resistance welding. However, as the size of the parts to be joined decreases, and/or interconnection densities increase, resistance welding cannot meet requirements of the process [2]. An alternative is to use laser microwelding, or more specifically, pulsed laser microwelding. Quality of the laser microwelding, however, depends on a number of parameters characterizing parts to be joined, laser output, environmental conditions, and specific requirements of the processes to be performed, such as spot diameter, center to center distance, penetration depth, and microweld strength [3,4].

To gain better understanding of the fundamentals governing laser microwelding processes, we characterize "raw" materials before joining them, monitor laser microwelding process parameters, and characterize materials after they have been laser joined, by measuring mechanical properties in the heat affected zones (HAZ). In this way, we are able to determine changes, if any, in the material properties due to the laser microwelding. In addition, we also develop computational procedures for modeling and simulation of the laser microwelding, which we validate experimentally.

The objective of our study is to develop a fundamental understanding of the processes governing interaction of a laser beam with metals during microwelding. This development concentrates on understanding equations governing laser microwelding processes (LMP), conducting computational and experimental study of the LMP, and developing correlations between computational and experimental results.

In the following sections, we define the geometry used and present analytical considerations, describe the facilities, and discuss preliminary results of our study.

2. ANALYTICAL CONSIDERATIONS

Typical geometry used in our study is shown in Fig. 1. It represents laser microwelding of two parts [1]. The laser beam irradiates the top part, which is to be joined to the part underneath.

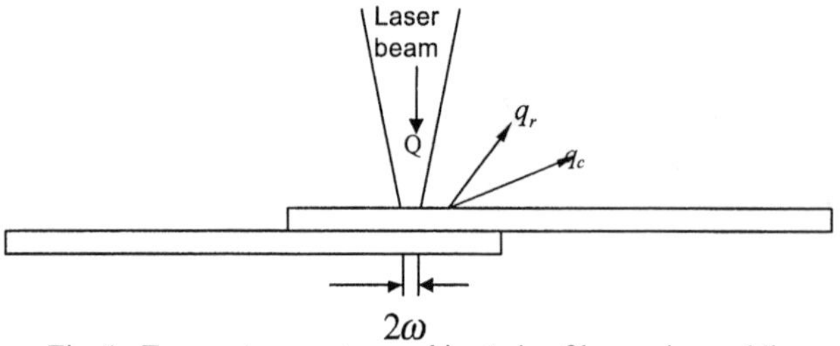

Fig. 1. Two-part geometry used in study of laser microwelding.

In the configuration shown in Fig. 1, laser beam energy Q is coupled into the parts to be joined. Because of the thermal processes involved, part of this energy is lost to the surroundings by convection and radiation, q_c and q_r, respectively. These boundary conditions can be described by the following equations:

$$q_c = h(T - T_a) \tag{1}$$

and

$$q_r = \varepsilon\sigma\left(T^4 - T_a^4\right) \ , \tag{2}$$

where h is the convective heat transfer coefficient, ε is the emissivity, σ is the Stefan-Boltzmann constant, T_a is the ambient temperature, and T is the instantaneous temperature of the parts being welded. In fact, the temperature of the parts is a function of three dimensional space and time, i.e.,

$$T = T(x, y, z, t) \ , \tag{3}$$

where x, y, and z are the spatial coordinates, while t represents time.

The equation governing temperature distribution within the HAZ may be expressed as [1,5-8]

$$\rho \frac{\partial}{\partial t} \int_{T_o}^{T} c(T)dT = \frac{\partial}{\partial x}\left[k(T)\frac{\partial T}{\partial x}\right] + \frac{\partial}{\partial y}\left[k(T)\frac{\partial T}{\partial y}\right] + \frac{\partial}{\partial z}\left[k(T)\frac{\partial T}{\partial z}\right] + Q(x, y, z, t) \ , \tag{4}$$

where ρ, c, and k are the density, specific heat, and thermal conductivity, respectively, of the material, T_o is the initial temperature, and other symbols are as previously defined. Also, for simplification, the arguments (x,y,z,t) were omitted in all but one term in Eq. 4.

In addition to analytical modeling, finite element method (FEM) is used to determine thermal field distributions in the HAZ during LMP [1,9]; the computed temperatures are correlated with measured temperatures. Understanding of correlation between computational and experimental results facilitates development of robust laser processes [10].

3. FACILITIES USED

In this study, two special facilities were used. One, laser facility, was used to produce microwelds. The other, nanoindentation facility, was used to measure material properties.

3.1. Laser facility

Microwelding was performed using the facility shown in Fig. 2. This facility utilizes pulsed Nd:YAG laser operating at 1.06 μm, capable of delivering 130 J per pulse, duration of which can be controlled form 0.1 msec to 10 msec.

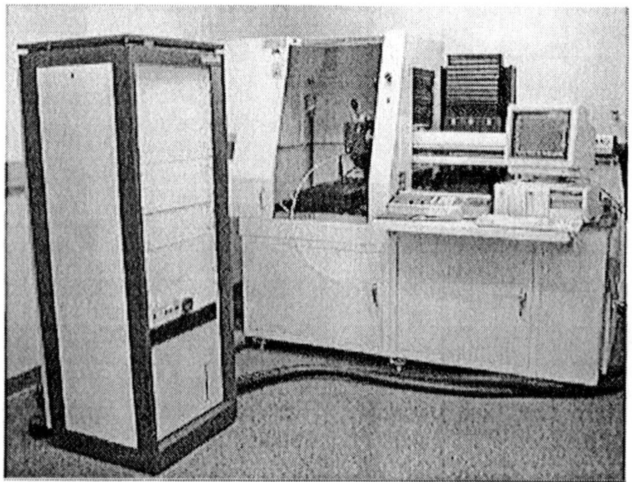

Fig. 2. Pulsed Nd:YAG laser facility used in this study.

Temperature measurements, made during the laser microwelding processes, were facilitated using computer based data acquisition system shown in Fig. 3.

Fig. 3. Computer based temperature measurement system.

3.2. Nanoindentation facility

Nanoindentation is a process of pressing a hard, rigid indenter, usually an accurately ground diamond, into the surface of the sample of interest, Fig. 4. The shape of the indenter is selected based on the specific material properties to be investigated [11]. Nanoindentation provides a record of both force and penetration depth during the pressing process, which consists of two parts: loading and unloading. During the unloading part of the process, elastic restoration occurs as the load is reduced from the maximum value to either a partial or full unloading value, Fig. 5. Initial slope of the unloading portion of the cycle allows determination of the elastic material properties of the sample.

Fig. 4. Close-up view of the nanoindenter and a sample.

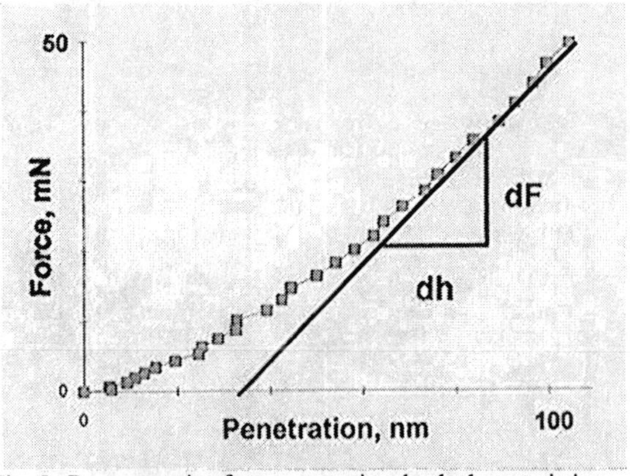

Fig. 5. Representative force penetration depth characteristics at a single nanoindentation site.

More specifically, the slope *dF/dh* of the unloading curve at the maximum penetration depth, Fig. 5, is related to material properties of the sample by the following equation:

$$\frac{dF}{dh} = \beta\sqrt{A}E^{*} \quad , \qquad (5)$$

where β is the constant characterizing the indenter used, A is the area of the specific indentation site, and E^{*} is the quantity that relates to the modulus of elasticity of the sample. This relationship is given by the equation

$$\frac{1}{E^{*}} = \frac{1-\nu_s^2}{E_s} + \frac{1-\nu_n^2}{E_n} \quad , \qquad (6)$$

where E is the modulus of elasticity, ν is the Poisson's ratio, and subscripts s and n refer to the sample and nanoindenter, respectively.

Using Eq. 6, modulus of elasticity of the materials that are laser microwelded is determined and then used in the computations of the quality of the microweld.

4. RESULTS AND DISCUSSION

Results presented in this section are based on our study of laser microwelding of parts made out of copper-based alloys. First, an example of computational results is given. Then, typical laser microweld is shown and correlation between measured and computed temperatures is discussed. Finally, nanoindentation results in the HAZ are presented.

Typical FEM computed temperature distribution during laser microwelding is shown in Figs 6 and 7. To obtain these results, quarter-model of the laser microwelded parts was used, because of the geometrical symmetry of the parts. Clearly, maximum temperatures are obtained along the centerline in the direction of the propagation of the laser beam.

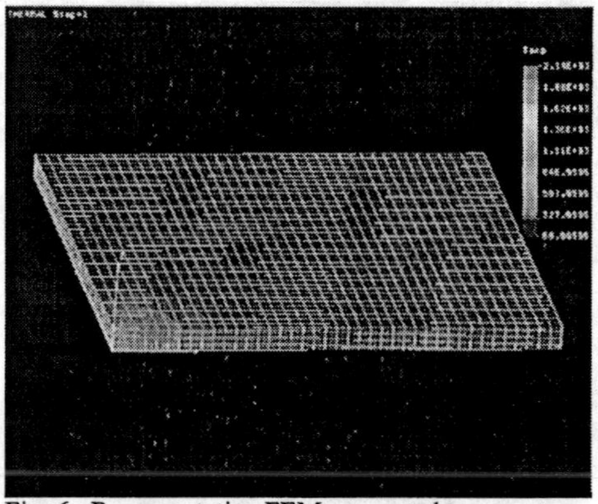

Fig. 6. Representative FEM computed temperature distribution during the laser microwelding process.

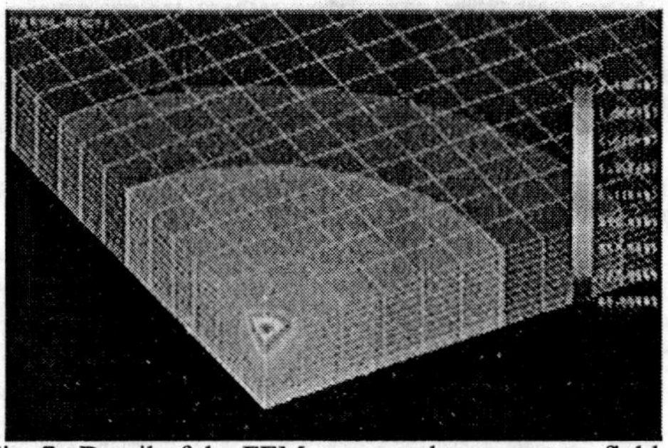

Fig. 7. Detail of the FEM computed temperature field.

Correlation between computed and measured temperatures during the laser microwelding process is shown in Figs 8 and 9. The laser pulse duration, for this specific case, was 3 msec. Clearly, because of the thermal processes governing energy interactions during the laser microwelding processes, properties of the thermocouples used, as well as the geometrical relationships between the point of laser incidence and location of thermocouples, time at which maximum temperature is observed is about 60 msec, which is much longer that duration of the laser pulse. However, agreement between computed and measured temperature profiles is good.

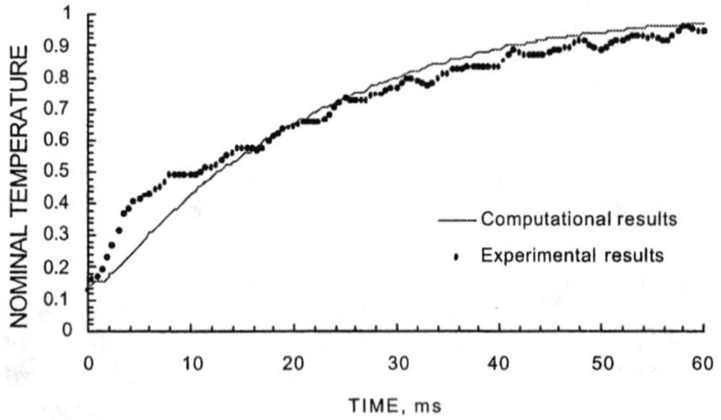

Fig. 8. Comparison of computational and experimental temperature profiles during the first 60 msec after initiation of the laser pulse.

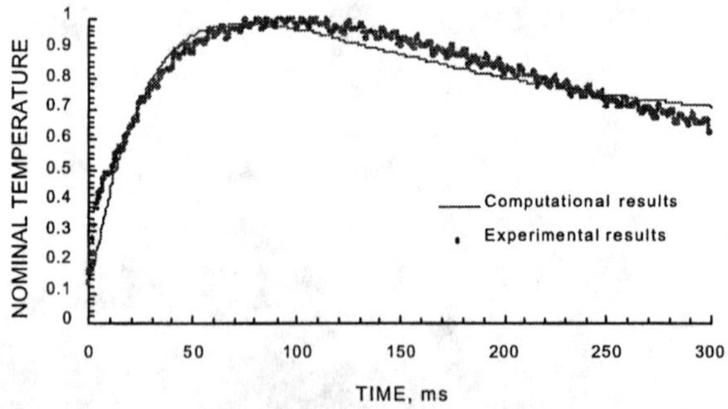

Fig. 9. Comparison of computational and experimental temperature profiles during the time period of 300 msec, following initiation of the laser pulse.

Figure 10 shows typical laser microwelds obtained during this study. The figure exhibits good interconnection by a well-formed microweld.

Cross-section of another laser microweld, shown in Fig. 11, indicates substantial mixing of the materials during the process. Series of dark-dots in Fig. 11 show nanoindentation sites, at which measurements were made. A close-up view of a few nanoindentation sites at the edge of the HAZ is displayed in Fig. 12. Clearly, since size of a specific nanoindentation relates to the material properties at that site, variations in material properties across the HAZ are discernible. Quantitative interpretation of the force penetration depth characteristics at individual nanoindentation sites resulted in a spatial distribution of the material properties which were used in computational modeling and simulation of reliability of the laser microwelds.

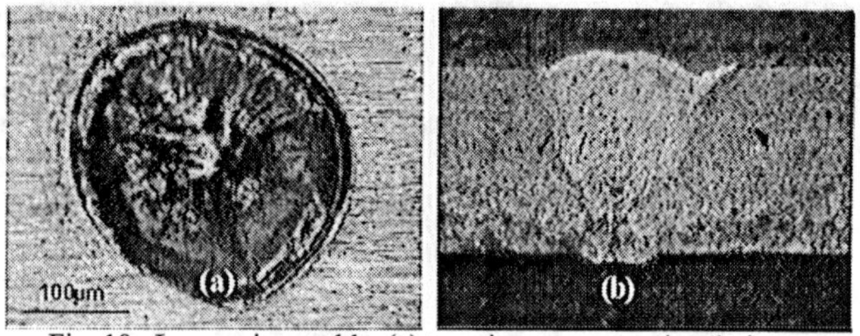

Fig. 10. Laser microwelds: (a) top view, cross-sectional view.

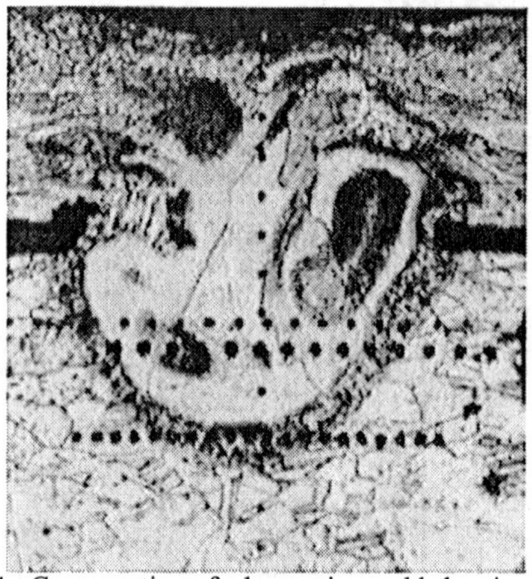

Fig. 11. Cross-section of a laser microweld showing HAZ, dark-dots indicate nanoindentation sites.

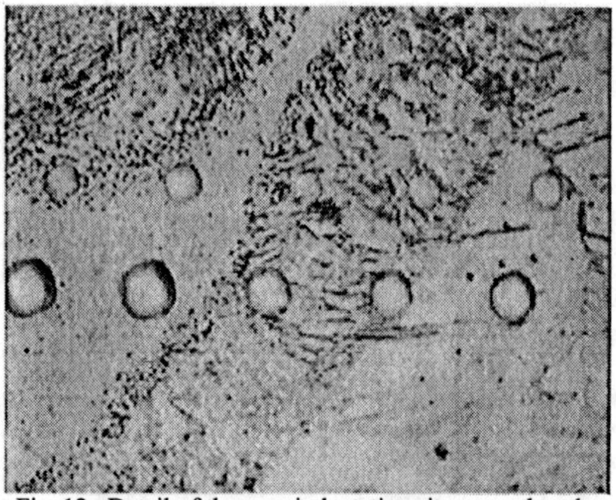

Fig. 12. Detail of the nanoindentation sites near the edge of the HAZ produced during laser microwelding.

5. CONCLUSIONS

Thermal processes governing laser interaction with metals during microwelding were investigated. This investigation consisted of computational and experimental studies. The computational studies were conducted using FEM. The experimental studies were conducted using pulsed Nd:YAG facility for making microwelds and nanoindentation facility for measuring material properties within the HAZ formed during microwelding.

Temperatures generated during the LMP were measured using computer based data acquisition system. Good correlation between measured temperatures and computed temperatures was obtained, during the preliminary investigations described in this paper.

Our work on the laser microwelding will continue and will concentrate on development of processes to produce reliable microwelds for applications in electronic packaging.

6. REFERENCES

1. Han, W., 1999, *Computational and experimental investigation of laser drilling of small holes in metals*, MS Thesis, Worcester Polytechnic Institute, Worcester, MA.
2. Wojcicki, M. A., and Pryputniewicz, R. J., 1997, "Feasibility study of laser microwelding of high density cable assemblies with applications to portable electronics," *Proc. 47th Electronics Components and Technology Conf.*, San Jose, CA.
3. Nowak, T., 1990, *Theoretical and experimental investigation of laser drilling in partially transparent medium*, MS Thesis, Worcester Polytechnic Institute, Worcester, MA.
4. Nowakowski, K. A., 1990, *Computer simulation and experimental investigation of laser beam interaction with metals*, MS Thesis, Worcester Polytechnic Institute, Worcester, MA.
5. Steen, W. M., 1991, *Laser material processing*, Springer-Verlag, New York.
6. Bruggemann, G. T., 1996, "Prediction of weld data using process control based on surface temperature measurement for high power energy flow processes," *Proc. SPIE*, 2888:168-177.
7. Chamberland, R. P., 1988, *Experimental and numerical study of laser interaction with a semi-infinite solid*, MS Thesis, Worcester Polytechnic Institute, Worcester, MA.
8. Dowden, J. M., 2001, *The mathematics of thermal modeling*, Chapman & Hall, Boca Raton, FL.
9. Fan, G., Guan, Y., Zhang, J., and Wang, H., 1998, "Finite element method analysis for temperature field influence by applied stress during laser processing," *Proc. SPIE*, 3550:357-365.
10. Nowak, T., and Pryputniewicz, R. J., 1992, "Theoretical and experimental investigation of laser drilling in a partially transparent medium," *J. Electronic Packaging*, 14:71-80.
11. Furlong, C., and Pryputniewicz, R. J., 2001, "Measurements and modeling of a nanoindentation process for MEMS," *Proc. Internat. Symp. on MEMS: Mechanics and Measurements*, Portland, OR, pp. 28-33.

Proceedings of
2001 ASME International Mechanical Engineering Congress and Exposition
November 11-16, 2001, New York, NY
EPP-Vol. 1

IMECE2001/EPP-24726

RELIABLE LOW-COST FLIP CHIP INTERCONNECTIONS

Z. W. Zhong
School of Mechanical & Production Engineering
Nanyang Technological University
Singapore 639798

ABSTRACT

A few reliable low-cost flip chip assembly processes using gold bumps with NCA, ACF or ACP that involved daily mass production activities in the industry are reported in this paper. Some key issues of material selection and assembly for reliable low-cost flip chip interconnections are discussed. This paper also discusses reliable low-cost processes of flip chip on FR-4 using eutectic solder bumps with possible fewer process steps compared to the full assembly process. Some interesting results with respect to the reliability performance of the processes have been obtained.

Introduction

The chip on board technology meets many current industrial needs because of its proven reliability and ease of use. However, demands are increasing for better electrical performance systems. Flip chip on board and flip chip in package have been considered as the next generation of the interconnect technology (Zhong and Goh, 1999). With the push towards miniaturization in electronic products, flip chip technology is emerging as one of the alternatives to complement the surface mount technology in board assembly and electronics packaging. Also, there is increased availability of various bonding, underfill and encapsulation materials, bumped bare dies, substrate and equipment. Many simple, low cost and low-temperature processes for flip chip assemblies have been developed (Lau, 1995; Kloeser *et al.*, 1997; Zhong *et al.*, 1997; Zhong, 1999), since the C4 (controlled collapse chip connection) process was invented over 35 years ago.

Adhesive materials can be used in flip chip assemblies as the electrical and/or mechanical bonding materials. These materials include isotropic conductive adhesive (ICA), non-conductive adhesive (NCA) and anisotropic conductive adhesive (ACA). ACA materials have two forms: film form (namely, ACF) and paste form (namely, ACP).

Flip chip attachment using gold bumps with adhesive is a solderless and low temperature flip-chip-mounting method. For some applications where high temperature is not desirable and reflow of a solder alloy is not allowed, flip chip using gold bumps with NCA, ACP or ACF can be a solution. The demand for elimination of alpha emission, environmental requirements, requirements for gold bumps with fine pitch, applications of flip chip on glass (liquid crystal displays), etc. can be other reasons for choosing flip chip assembly processes using gold bumps with NCA, ACP and ACF.

NCA is low cost. ACF and ACP are still relatively expensive because they are still under research and development and the market is yet relatively small. However, the cost of ACF and ACP may drop gradually in the future when their applications increase. Also, ACF and ACP processes are reworkable. Flip chip attachment with ACP, ACF or NCA has fewer process steps compared with other flip chip processes. Its reliability may need to be improved compared to that of flip chip attachment using solder bumps with underfill epoxy in order to meet stringent reliability requirements for many applications (Zhong, 1999).

The first half of this paper discusses reliable low-cost processes for flip chip on board or in package using gold bumps with NCA, ACF, or ACP. The bonding parameters and flip chip assemblies using different adhesive materials were investigated by design of experiments (DOE) and yield runs. The assembly yield and reliability performance of the processes was compared.

Meanwhile, there are many parameters affecting flip chip reliability (Määttänen *et al.*, 2000). In flip chip assemblies using solder joints, CTE (coefficient of thermal expansion) mismatch between the silicon chip and the substrate results in stress created in the solder joints especially near the corners of the chip. Underfill epoxy is used to reduce the effect of the global thermal expansion mismatch. It reduces the stresses and strains in the solder joints and redistributes them over the entire chip area. This is done with the underfill epoxy rigidly holding the chip to the substrate to reduce the magnitude of the force on the solder during operation and temperature cycling. Underfill also helps to protect the chip from ionic contaminants and the hostile environments such as mechanical shock and vibration (Goh, 1999). Underfilling can significantly improve the reliability of soldered flip-chip interconnections (Roesner, 1998).

The underfill epoxy extends its use to a variety of organic and inorganic substrate materials resulting in ten to hundred fold improvement in fatigue life compared to a package without underfill. Therefore, underfill has been the key to the development of flip chip technology (Lau *et al.*, 1998).

Bare chips need protection from the atmosphere, moisture, and various chemicals used in the manufacturing process and to which the end product may be exposed to during its functional lifetime. The means for doing this fall basically into two categories: glob-top coatings and lids. Epoxy coatings are applicable for chip protection or glob-top applications (Harper, 1997).

However, it requires some time, process steps and equipment to place underfill epoxy under the chip and an encapsulation material on the chip. A curing oven is needed and it takes long time to cure the underfill and encapsulation epoxies separately. The curing time of the epoxies in a curing oven is typically several hours, excluding cooling time.

Can we omit one or two of these process steps to reduce the total assembly time and cost? Does encapsulation also improve fatigue life of the solder joints? Should we use the same epoxy or different epoxies for underfill and encapsulation? The second half of this paper tries to answer some of these questions. It discusses reliable low-cost processes of flip chip on FR-4 using eutectic solder bumps with possible fewer process steps compared to the full assembly process. Flip chip on FR-4 assemblies using eutectic solder bumps with and without underfill and encapsulation have been investigated by DOE runs and a temperature cycling test. Some interesting results with respect to the reliability performance of the processes have been obtained.

Flip Chip on Board Using NCA or ACF

The test die used for the evaluation trials had a size of 3.5 mm × 6.0 mm, a minimum terminal pitch of 310 μm, and 33 gold bumps. The gold bump had a size of 100 μm × 100 μm. The substrate boards were multi-layer rigid flex with gold pads. The reliability tests included temperature cycle tests, temperature humidity test, stringent temperature cycle test, artificial sweat test, drop test, and temperature behavior test.

Flip Chip Attachment Using NCA. A typical flip chip attachment using NCA is very simple, and involves only two process steps: NCA placement, and IC (integrated circuit) placement and bonding. This process can give reasonable reliability and high assembly yield at low cost, when the parameters for epoxy placement (dispensing volume and pattern, shelf life of the epoxy) and bonding conditions (bonding force, temperature and time, cooling temperature) are optimized.

An L_{18} orthogonal array was selected for the DOE. Seven parameters (dispensing volume and pattern, shelf life of the epoxy, bonding force, temperature and time, cooling temperature) with three levels and one parameter (surface cleanliness) with two levels were investigated in the experiments. Eighteen experiments with four runs for each experiment were conducted, and the in-circuit voltage and functional tests were performed after the bonding experiments.

The analysis of variance (ANOVA) technique was used to establish the relative significance of the individual parameters. The ANOVA analysis revealed with 99% confidence that dispensing volume and pattern of the NCA, and bonding force and temperature were the important parameters significantly affecting the assembly yield.

Based on the DOE results, the settings for five parameters were determined. However, the other three parameters needed further investigation. Therefore, another DOE was carried out. After the two DOE analyses, the settings for all the eight parameters were determined. Confirmation and yield testing experiments were then conducted. The assembly yield was found to be over 97%. Failure analysis helped to improve the assembly yield even more.

All of the samples tested passed the temperature cycle test, the temperature behavior test, the temperature humidity test and the drop test. 80% of the tested samples passed the artificial sweat test. 95%, 90%, 85% and 70% of the samples passed 84 cycles (1 week), 168 cycles (2 weeks), 336 cycles (4 weeks) and 588 cycles (7 weeks) of the stringent temperature cycle test, respectively.

Flip Chip Attachment Using ACF. Six ACF materials were evaluated. The ACFs have different types of conductive particles and densities. Preliminary studies were carried out to select ACF materials. Bonding tests, functional tests, temperature cycling tests and artificial sweat tests were conducted to evaluate the ACF materials and the process. The dies and substrate used were the same as those used for the NCA process.

In order to optimize the bonding parameter settings, DOE was performed. Three ACF materials were selected for the DOE runs because they did not produce any failures in the preliminary studies. Four factors were selected and three levels for each factor were determined. There were a total of 9 experiments and 4 runs for each experiment in this DOE using an L_9 orthogonal array (Ross, 1988). Table 1 shows the details of the DOE. The maximum differences of the three levels of bonding temperature (T1 and T3) and force (F1 and F3) were 20°C and 17 N respectively.

Table 1 Details of the DOE using an L_9 orthogonal array

Experi-ment	Bonding temperature	Bonding force	Second bonding	Adhesive material
1	T1	F1	With	ACF A
2	T1	F2	Without	ACF B
3	T1	F3	Rework	ACF C
4	T2	F1	Without	ACF C
5	T2	F2	Rework	ACF A
6	T2	F2	With	ACF B
7	T3	F1	Rework	ACF B
8	T3	F2	With	ACF C
9	T3	F2	Without	ACF A

The attribute frequency for each level or each interaction was calculated based on the in-circuit voltage and functional test results. The best combination of parameter settings and material was determined. Confirmation and yield testing experiments were then conducted. The assembly yields were 100%.

All of the samples assembled during the yield runs passed the temperature cycle test, the temperature behavior test, the temperature humidity test and the drop test. 90% of the tested samples passed the artificial sweat test. 95% of the samples passed 336 cycles (4 weeks) of the stringent temperature cycle test. The assembly yield and the reliability performance of the ACF process were better than those of the NCA process.

Flip Chip in Package Using ACP

The test die used for the evaluation trials had a size of 3 mm × 4.6 mm, a minimum terminal pitch of 220 μm, and 22 gold bumps. The gold bump had a size of 100 μm × 100 μm. The substrate boards were multi-layer ceramics with gold pads.

DOE Runs. DOE was conducted using the fractional factorial technique. A modified L_8 orthogonal array (Ross, 1988) was selected for the DOE. There were a total of eight experiments and five runs for each experiment in this DOE. Table 2 shows the details of the DOE.

Table 2 Details of the DOE using a modified L_8 array

Experi-ment	Adhesive material	Bonding time	Bonding temperature	Bonding force	Laser marking
1	ACP 1	t1	T1	F1	P1
2	ACP 1	t2	T2	F2	P2
3	ACP 2	t2	T1	F1	P2
4	ACP 2	t1	T2	F2	P1
5	ACF 1	t1	T2	F1	P2
6	ACF 1	t2	T1	F2	P1
7	NCA 1	t2	T2	F1	P1
8	NCA 1	t1	T1	F2	P2

The one factor with four levels was the adhesive material. Two ACP, one ACF and one NCA materials were the four levels. The important bonding parameters and one material processing parameter of the application were the other four factors. The differences of the two levels of bonding time (t1 and t2), temperature (T1 and T2) and force (F1 and F2) were 30 seconds, 10°C and 10 N respectively. P1 and P2 refer to the processes with and without laser marking, respectively.

Inspection tests were conducted after packaging experiments. Only one failure was produced using the condition of Experiment 1. The packaging yield using the conditions of Experiments 2-8 was 100%. This means that the flip chip attachment using any of the NCA, ACF and ACP materials investigated can produce 100% assembly yield.

A temperature cycling test (–30°C/+60°C) was conducted using all the samples that passed the inspection. This test was one of the several reliability tests required for an application. A further temperature cycling test (–40°C/+125°C) was conducted using 24 samples (three samples from each DOE experiment) that passed the first temperature cycling test.

Analysis of DOE Runs. Analysis of the DOE runs was performed to determine parameter settings and select the adhesive material. Pareto analysis (Ott, 1990) was carried out. The estimated attribute frequencies of failure and the estimated effects (Cobb, 1998; Montgomery, 1991) of the five factors on the estimated failure frequencies were calculated. The results showed that the most significant factor is the adhesive material selection and the most insignificant factor is laser marking.

After the analysis, ACP 2, T2, F2 and t1 were determined to be the adhesive material, bonding temperature, force and time, respectively. The final combination of the important factors was the same parameter setting of Experiment 4. Therefore, confirmation runs could be omitted in this study.

Yield Runs and Reliability Tests. Yield runs were carried out using the parameter settings determined after the DOE analysis. The packaging yield of the process was 100%. Various reliability tests required for the application were conducted. The reliability tests included burn in, artificial sweat and humidity tests, and several temperature-cycling tests, etc. There was no failure occurred. The flip chip packages attained 100% reliability required for the application.

Again, a further temperature cycling test (–40°C/+125°C) was conducted using the samples that passed a required temperature cycling test (–30°C/+70°C, 112 cycles). Figure 1 shows the result of this reliability test. The reliability performance of the packages was quite satisfactory. 100% of the tested samples survived 836 cycles of the temperature cycling test. 95%, 85% and 80% of the samples survived 1,401, 2,266 and 4,329 cycles of the test, respectively.

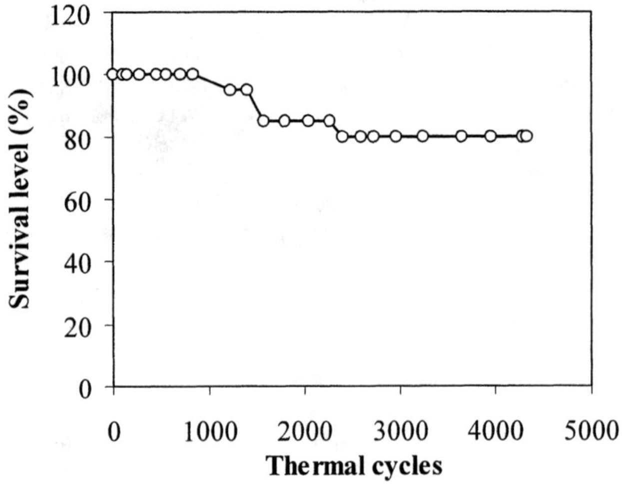

Fig. 1 Survival level of the packages during a further temperature cycling test (–40°C/+125°C) after the samples passed a required temperature cycling test (–30°C/+70°C, 112 cycles)

Flip Chip on FR-4 Using Eutectic Solder

Test Die and Equipment. The test die used for the experiments had a size of 6.3 mm × 6.3 mm, bump pitch of 457 μm and 48 eutectic solder bumps. The solder bump had a size of ϕ190 μm. An FR-4 test vehicle with daisy chained circuitry matching the die's daisy chained pattern was designed and fabricated. A pick and place station was used to conduct the mounting experiments. The station could be used to reflow the solder alloy. An eight-zone forced air convection reflow oven was also used for reflow. A dispensing machine was then used for dispensing underfill and encapsulation epoxies. Curing of the underfill epoxy and/or encapsulant was carried out in a curing oven.

Design of Experiments. DOE was conducted using the fractional factorial design techniques. An L_9 orthogonal array was selected for the DOE. Three factors with three levels were investigated. There were a total of nine experiments in this DOE. Table 3 shows the details of the DOE.

Epoxy 1 and Epoxy 2 from an epoxy manufacturer were selected as the underfill and encapsulation materials to be investigated. The curing time of them excluding cooling time was 3.5 hours and 4.5 hours respectively. Epoxy 1 was recommended by the epoxy manufacturer as an underfill material, while Epoxy 2 was recommended as an encapsulation material. The pick and place station and the reflow oven were used to study the effect of reflow methods. In Table 3, S1 and S2 refer to the two speeds of the conveyer used in the refolw oven: S1 = 0.68 m/min and S2 = 0.9 m/min.

Table 3 Details of the DOE using an L_9 orthogonal array

Experiment	Underfill	Encapsulant	Reflow
1	Without	Without	Station
2	Without*	Epoxy 1	Oven (S1)
3	Without*	Epoxy 2	Oven (S2)
4	Epoxy 1	Without	Oven (S1)
5	Epoxy 1	Epoxy 1	Oven (S2)
6	Epoxy 1	Epoxy 2	Station
7	Epoxy 2	Without	Oven (S2)
8	Epoxy 2	Epoxy 1	Station
9	Epoxy 2	Epoxy 2	Oven (S1)

* See the explanation in the text of this paper

When Experiments 2 and 3 were carried out, underfilling and cure of underfill were omitted. However, because the viscosity of the epoxy (Epoxy 1) used for Experiment 2 was low, the epoxy could seep, although not completely, into the gap between a flip chip and a substrate during encapsulation. When the encapsulation epoxy (Epoxy 2 for Experiment 3) with high viscosity was used for encapsulation without carrying out underfilling and cure of underfill, the epoxy could not seep into the gap easily.

The assembly yield of all the nine experiments was 100%. The flip chips were assembled on the FR-4 boards for a temperature cycling test. The temperature cycling test condition was –40°C (15min) / +125°C (15min), one hour per cycle.

Results of the Reliability Test. Figure 2 shows Weibull plots of the results of the temperature cycling test. The Weibull distribution curve of Experiment 1 is not shown in Fig. 2 because all the samples assembled for Experiment 1 had open interconnections when they were inspected after 128 cycles of the thermal cycling test. Due to the CTE mismatch, the solder joints of the flip chip on FR-4 assemblies without underfill and encapsulation would develop cracks and open interconnections quickly during temperature cycling.

Experiment 2 resulted in the best reliability out of the nine experiments based on the fractional factorial design. This was a very interesting result because underfilling and cure of underfill were not carried out for Experiment 2 as explained above. When Epoxy 1 was used for both underfill and encapsulation, the reliability of the flip chip on FR-4 samples was also quite good as shown by the curve of Experiment 5. It was the second best out of the nine experiments. However, when the encapsulation epoxy (Epoxy 2) with high viscosity was used for encapsulation without carrying out underfilling and cure of underfill, the reliability of the flip chip on FR-4 assemblies was very bad as shown by the curve of Experiment 3. It was the worst curve shown in Fig. 2.

This suggests that underfilling and curing of underfill can be omitted when a suitable underfill epoxy with relatively low viscosity is used for encapsulation. Even if both underfill and encapsulation are conducted, instead of the four separate process steps for underfill and encapsulation, the following two process steps can be performed: placing underfill and encapsulation epoxy (or epoxies) and curing of the epoxy (or epoxies). Thus, the whole assembly time and cost can be reduced.

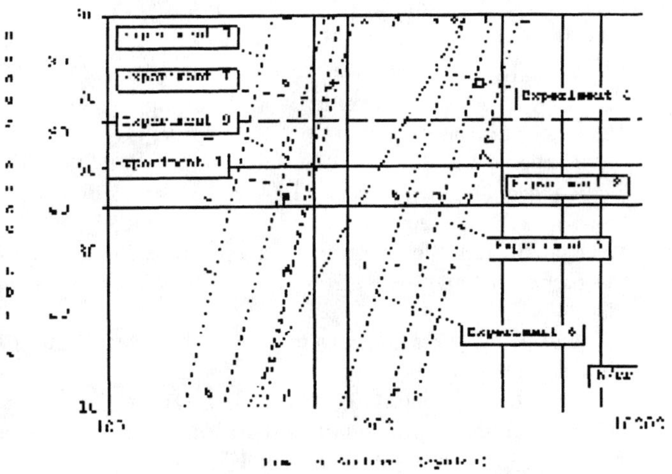

Fig. 2 Weibull distribution curves of Experiments 2 to 9 for flip chip on FR-4 with eutectic solder joints

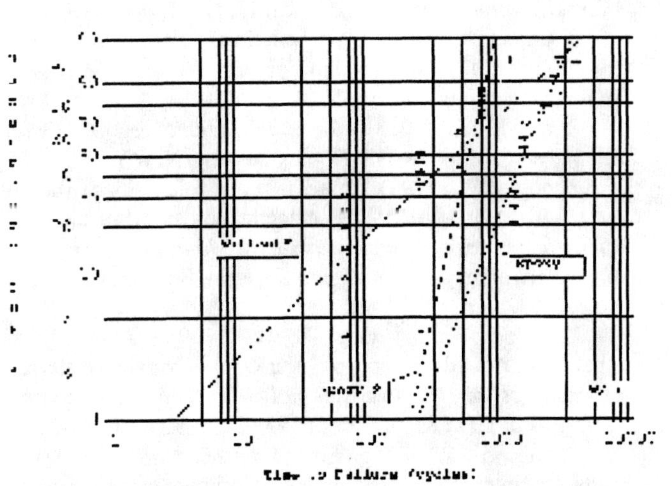

Fig. 3 Weibull distribution curves of flip chip on FR-4 assemblies with eutectic solder joints showing effect of underfill. The curve with the label of "Without*" shows the plots of the reliability testing results of the flip chip on FR-4 samples assembled for Experiments 1, 2 and 3

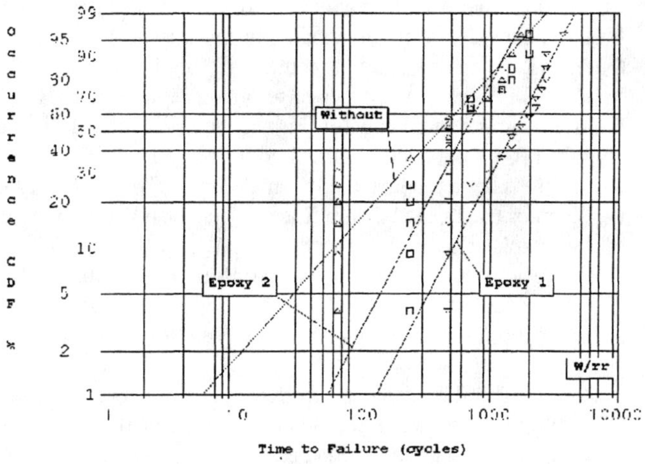

Fig. 4 Weibull distribution curves of flip chip on FR-4 assemblies with eutectic solder joints showing effect of encapsulation

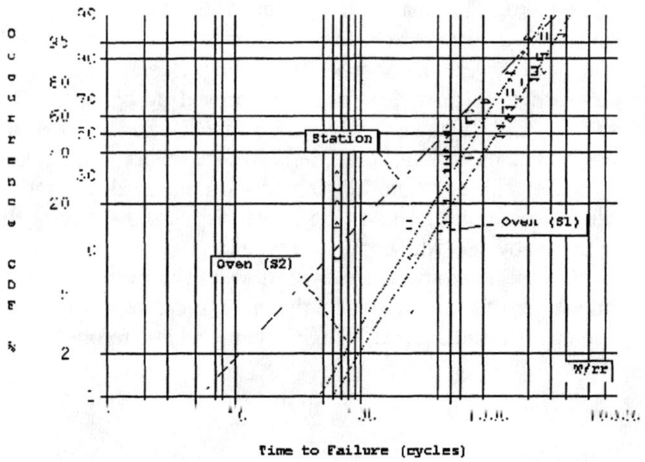

Fig. 5 Weibull distribution curves of flip chip on FR-4 assemblies with eutectic solder joints showing effect of different reflow operations

The effects of underfill, encapsulation and reflow methods on the reliability performance of the flip chip on FR-4 assemblies are shown in Figs. 3, 4 and 5 respectively. It should be noted that the DOE was conducted using the fractional factorial design technique. Using fractional factorial techniques can significantly reduce the number of experiments needed but one of the drawbacks of the techniques is that interactions of the individual factors investigated cannot be distinguished completely from the effects of the individual factors. The analysis using Figs. 3, 4 and 5 was conducted under an assumption that the interaction effects of the three factors investigated were relatively small compared with the effects of the individual factors and thus could be ignored.

Material selection is important to obtain reliable flip chip on FR-4 assemblies. As shown in Fig. 3, the reliability of the flip chip on FR-4 samples underfilled with Epoxy 1 was better than that of the samples underfilled with Epoxy 2. The reliability of the samples encapsulated with Epoxy 1 was also better than that of the samples encapsulated with Epoxy 2, as shown in Fig. 4.

The reflow time and methods also affected the reliability of flip chip on FR-4 samples. The longest reflow time of the three reflow operations investigated resulted in the best result as shown in Fig. 5.

Reflow profiles can be optimized to achieve the best soldering performance based on defect mechanism analysis. In general, a slow ramp-up rate is desired in order to minimize hot slump, bridging, skewing, opens, solder beading, solder balling, and so on. A minimized soaking zone reduces voiding, poor wetting, solder balling, and opens (Lee, 1999).

When the pick and place station was used, reflow of the solder joints could be carried out for only one chip at the time. The whole time (about 1.5 min) for preheat and reflow was much shorter than that (about 4 min for Speed 1, about 3 min for Speed 2) when the reflow oven was used. Thus, reliable solder joints were not obtained when the pick and place station was used for reflow. This can be improved by optimizing the reflow profile with a slow ramp-up rate. However, when the pick and place station is used, reflow of the solder joints can be carried out for only one flip chip at the time. A slow ramp-up rate therefore means more time is needed to reflow several flip chips one by one, which is not economical.

On the contrary, the oven reflow might result in longer fatigue life of solder joints. Furthermore, it can reflow flip chip assemblies manufactured in high volume, and therefore is more economical.

Conclusion

Flip chip on board using gold bumps with NCA or ACF was studied. The assembly yields of the NCA and the ACF process were 97% and 100%, respectively. A number of reliability tests were also carried out. The reliability performance of the ACF process was better than that of the NCA process.

A reliable flip-chip packaging process using gold bumps and ACP was developed. The process parameter settings and adhesive selection were determined by DOE. The packaging yield was 100%. All of the packages assembled during the yield runs passed various reliability tests. The reliability of the packages was 100% meeting the requirements for an application.

Flip chip on FR-4 assemblies using eutectic solder bumps with and without underfill and encapsulation have been evaluated by DOE runs and a temperature cycling test. The process steps of underfilling and curing of underfill can be omitted when a suitable epoxy with relatively low viscosity is used for encapsulation. On the other hand, when underfill is conducted, encapsulation may not be necessarily needed from reliability point of view. The reliability of flip chip on FR-4 assemblies with encapsulation might be worse than that of the assemblies without encapsulation.

References

Goh, K. S. (1999), *Chip on flex and flip chip on flex technology*, MSc Dissertation of Nanyang Technological University, Singapore.

Harper, C. A. (1997), *Electronic Packaging & Interconnection Handbook*, 2nd Edition, McGraw Hill, NY.

Kloeser, J., Kutzner, K. and Gross, K. (1997), "A new production line for low cost flip chip assembly," *Proceedings of the Technical Program, Surface Mount International*, USA, pp. 257-266.

Lau, J. H., (1995), *Flip Chip Technologies*, MacGraw-Hill, New York, NY.

Lau, J. H., Wong, C.P., Prince, J. L. and Nakayama, W. (1998), *Electronic Packaging: Design, Materials, Process, and Reliability*, McGraw-Hill, New York, NY.

Lee, N.-C. (1999), "Optimizing the reflow profile via defect mechanism analysis", *Soldering & Surface Mount Technology*, Vol. 11, No. 1, pp. 13-20.

Määttänen, J., Palm, P. and Tuominen, A. (2000), "Reliability of flip chip components soldered on FR5 board", *Microelectronics International*, Vol. 17, No. 1, pp. 27-30.

Roesner, B. (1998), "The beneficial effect of underfilling on the reliability of flip-chip joints", *Soldering & Surface Mount Technology*, Vol. 10, No. 2, pp. 14-18.

Ross, P. J. (1988), *Taguchi Techniques for Quality Engineering*, McGraw-Hill Book Company, New York, NY.

Zhong, Z.W. (1999), "Assembly and reliability of flip chip on boards using ACAs or eutectic solder with underfill", *Microelectronics International, Journal of The Microelectronics and Packaging Society - Europe*, Vol. 16, No. 3, pp. 6-14.

Zhong, Z.W. and Goh, K. S. (1999), "Assembly and reliability of flip chip on various substrate materials", *Proceedings of 1st International Workshop on Electronics Materials and Packaging (EMAP99)*, Singapore, pp. 12-20.

Zhong, Z. W., Wong, S. and Nishikawa, T. (1997), "A study of flip chip on board mounting process using anisotropic conductive films", *Proceedings of SEMICON Test, Assembly and Packaging 97*, Singapore, pp. 157-164.

Proceedings of
2001 ASME International Mechanical Engineering Congress and Exposition
November 11–16, 2001, New York, NY
EPP-Vol. 1

IMECE2001/EPP-24727

DYNAMIC CHARACTERIZATION OF SHAPE MEMORY ALLOYS FOR INTEGRATION WITH ELECTRONIC PACKAGING

Shivananda P. Mizar and **Ryszard J. Pryputniewicz**
NEST – NanoEngineering, Science, and Technology
CHSLT – Center for Holographic Studies and Laser micro-mechaTronics
Mechanical Engineering Department
Worcester Polytechnic Institute
Worcester, MA 01609-2280

ABSTRACT

Integration of Shape Memory Alloys (SMAs) with electronic packaging depends on detailed knowledge of thermomechanical behavior of SMAs. In this paper, thermomechanical behavior of SMAs is studied using a new approach based on Analytical, Computational, and Experimental Solutions (ACES) methodology. More specifically, variation in modulus of elasticity of equiatomic NiTi alloy is studied, as a function of temperature. The results show that the modulus of elasticity of the NiTi alloy studied herein varies from 38 GPa to 72 GPa as the temperature changes from -15°C to 190°C, corresponding to phase transition from martensite to austenite. Comparison of analytical, computational, and experimental results shows good correlation.

Keywords: shape memory alloys, shape memory effect, electronic packaging, ACES methodology, optoelectronic holography, time-average optoelectronic holography.

1. INTRODUCTION

Shape memory alloys exhibit complex thermomechanical behavior and have characteristics unique in comparison to commonly used materials. One of these characteristics is related to Shape Memory Effect (SME) [1-3], which is attributed to a phase transformation from martensite to austenite resulting in a variation in the modulus of elasticity as a function of temperature. Effective integration of SMAs with electronic packaging, e.g., in designs of microsensors and microactuators, depends on an accurate knowledge of this variation and hence the modulus of elasticity should be well characterized as a function of temperature.

SME is a nonlinear process, and requires statistical and computational methodologies to model its unique characteristics. The transformation can be achieved either by stressing the SMA or by thermal treatment.

Equiatomic NiTi is the particular SMA studied in this paper. The study concentrates on detailed measurements and characterization of the modulus of elasticity of the NiTi as a function of temperature.

More specifically, this characterization is made using Analytical, Computational, and Experimental Solutions (ACES) methodology [4].

In this paper, analytical investigations are based on the Euler-Bernoulli theory, finite element method (FEM) is used for computational investigations, and time-average optoelectronic holography (TAOEH) is used for experimental investigations. The experimental results obtained from TAOEH are complemented by such other experimental methods as X-ray diffraction (XRD), scanning electron microscopy (SEM), and energy dispersive spectroscopy (EDS).

2. METHODOLOGY USED

In this paper, ACES methodology was used to evaluate the modulus of elasticity of SMAs as a function of temperature.

ACES methodology depends on coordinated use of analytical, computational, and experimental solutions to obtain an answer to a problem that might not otherwise be obtainable [4]. Using this methodology a hybrid optimum solution is sought that provides reliable design information. Information obtained from experimentation can be used in computational analysis and for validation of analytical results. Also, uncertainty analysis is an integral part of ACES methodology and facilitates investigation of critical variables affecting SME of NiTi.

2.1. Analytical considerations

A linear approach has been used for analysis considered in this study [5]. Such an approach is valid when only material properties are sought and the microstructural changes as a result of the phase transition are not taken into consideration. Also, in the linear approach, the SMA is analyzed by itself because the approach will be different when the SMA is a part of a composite structure.

In this study, determination of material properties as function of temperature was based on dynamic characterization of the SMA. This characterization required excitation of the test samples at their natural frequencies.

Euler-Bernoulli theory was used to formulate the equations of motion to model dynamic behavior of the SMA [5]. Solving the equations of motion, relationship for the modulus of elasticity, Y, was obtained to be

$$Y = \frac{4\pi^2 f_n^2 L^4 \rho A}{\beta_n^4 I} \quad , \tag{1}$$

where f_n is the frequency of vibration of the nth mode characterized by the coefficient β_n, L is the effective length of the sample used, ρ is the density, A is the cross-sectional area of the sample, and I is the moment of inertia of this cross-section.

Based on Eq. 1, the fundamental equation for the modulus of elasticity can be written as

$$Y = Y(f_n, L, \rho, A, \beta_n, I) \quad , \tag{2}$$

from which the RSS-type uncertainty in Y can be shown [6] to be

$$\delta Y = \left[\left(\frac{\partial Y}{\partial f_n} \delta f_n \right)^2 + \left(\frac{\partial Y}{\partial L} \delta L \right)^2 + \left(\frac{\partial Y}{\partial \rho} \delta \rho \right)^2 + \left(\frac{\partial Y}{\partial A} \delta A \right)^2 + \left(\frac{\partial Y}{\partial \beta_n} \delta \beta_n \right)^2 + \left(\frac{\partial Y}{\partial I} \delta I \right)^2 \right]^{1/2} , \tag{3}$$

where $\delta()$ indicates the uncertainty in the parameter () and $\partial Y / \partial$ () is the partial derivative of Y with respect to the parameter ().

Equations 1 and 3 allow determination of the modulus of elasticity and the corresponding uncertainty in this modulus, respectively, based on the TAOEH measurements.

2.2. Computational considerations

Dynamic FEM analysis deals with the prediction of natural frequencies of vibrating structures and corresponding mode shapes of structures undergoing free undamped vibrations. The modal analysis is a eigenvalue problem for which the governing equation is

$$[M]\{\ddot{u}\} + [K]\{u\} = 0 \quad , \tag{4}$$

where $[M]$ is the total mass matrix, $[K]$ is the stiffness matrix, while $\{\ddot{u}\}$ and $\{u\}$ are the nodal accelerations and displacements, respectively.

In this study, material properties, in particular the modulus of elasticity, needed to run FEM, were determined using Time-Average Opto-Electronic Holography (TAOEH).

2.3. Experimental considerations

The TAOEH setup shown in Fig. 1 was used for the determination of the modulus of elasticity of NiTi from the measured resonant frequencies. Using the TAOEH method, the mode shapes corresponding to the first four bending frequencies were visualized and the corresponding frequencies recorded [7,8]. The NiTi samples were excited in a cantilever beam configuration. Care was taken to use a fixture that consisted of flat surfaces, to ensure "fixed" boundary conditions. The resonance frequencies were monitored over the temperature range from −15°C to 190°C, for all of the lengths of the samples. In TAOEH method the deformation of the object can be obtained by solving for Ω_t defined [9,10] as

$$\Omega_t = \frac{1}{2} \tan^{-1} \left\{ \left[\frac{1 - \cos 2B}{\sin 2B} \right] \frac{I_3 - I_2}{2I_1 - I_2 - I_3} \right\} \quad , \tag{5}$$

where I represents the laser irradiance fields, Ω_t is the time varying fringe-locus function relating to displacements of the object, and B is the bias modulation.

Fig. 1. TAOEH setup for measurements of resonance frequencies of NiTi: EC is the environmental chamber, LS is the laser subsystem, and OEH is the optoelectronic holography subsystem.

For XRD and EDS analyses, the NiTi samples were prepared using standard hot press technique and were polished to ensure flatness. Samples were mounted on the X-ray diffractometer and the diffraction patterns were recorded from 10° to 165°. The radiation used to generate the X-rays was Cu$K\alpha$. From the patterns obtained, the composition of the microstructure was determined. The spectrum obtained was analyzed to determine the distribution of martensite and austenite. This was necessary as the microstructure of the NiTi sample has a direct impact on the range of modulus of elasticity as determined from TAOEH. The Hanawalt search technique was used to determine the phases present in the specimen. The percentage compositions by weight were determined for the NiTi samples using X-ray analysis on the SEM, which houses the Kevex system. Quantitative analysis on the NiTi samples provided results pertaining to the amounts of Nickel and Titanium and also traces of other elements.

3. TEST SAMPLES

The SMA used in this study was in the form of a ribbon with cross-sectional dimensions of 500 μm and 2250 μm for the thickness and width, respectively. The samples were of three different lengths: 39.85 mm, 28.82 mm, and 25.0 mm. They were subjected to vibration excitation and the first four resonant frequencies were recorded using the state-of-the-art TAOEH methodology [5]. The experiments were performed over a temperature range from -15°C to 190°C. Scanning electron microscopy (SEM), X-ray diffraction (XRD), and energy dispersive spectroscopy (EDS) techniques were used to determine the phase structure of the alloy.

Composition of the alloy strongly affects material properties being measured. Measurements made using SEM, XRD, and EDS provided quantitative information and allowed determination of composition of the SMA used in this study.

4. RESULTS

Determination of modulus of elasticity of SMA was facilitated by correlation between composition of the test samples used and the resonance frequencies measured as a function of temperature.

4.1. Phase and composition

SMAs have only two phases present at any given temperature, i.e., martensite and austenite. The NiTi samples used in this study were heat treated to eliminate work hardening from the manufacturing process. Heat treatment (annealing) was followed by natural cooling (absence of quenching) to retain both the martensitic and austenitic phases. Heat treatment, however, eliminates most of the martensite in the sample and an XRD analysis should result in a spectrum where most of the peaks should correspond to austenite. Also, based on X-ray diffraction results, it was possible to determine the crystallographic structure by comparing the *d*-spacing with the typical powder metallurgy file standards 18-199 (austenite) and 35-1281 (martensite). The austenite was found to have a cubic structure and martensite was found to be monoclinic. Figure 2 shows the XRD spectrum for a heat-treated NiTi ribbon measured at room temperature.

Quantitative SEM, XRD, and EDS analyses on the NiTi samples provided results pertaining to the amounts of Nickel and Titanium and also traces of other elements. The composition was found to be close to equiatomic (47.7% Ni, 51.42% Ti). Traces of aluminum and silicon were detected in the analyses. The presence of aluminum and silicon can be attributed to metallographic polishing (Al) and the detector on the SEM (Si).

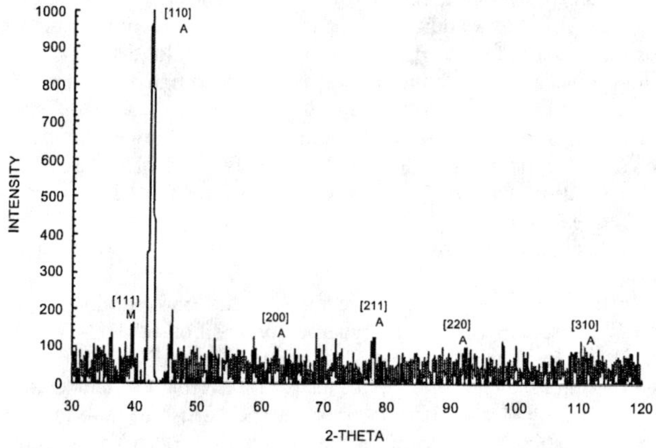

Fig. 2. XRD spectrum, at room temperature, of a heat treated NiTi ribbon.

4.2. Modulus of elasticity

The modulus of elasticity was determined using Eq. 1 and the resonance frequencies measured by TAOEH. The modulus of elasticity of SMAs is a strongly nonlinear function of temperature. Therefore, accurate knowledge of the temperature of the sample at the instant that measurements are made is necessary. The time constant for these measurements was 60 min, i.e., the measurements of parameters used in Eq. 1 were made 60 min after the temperature change was imposed on the samples located within the environmental chamber shown in Fig. 1.

The data were obtained for three different lengths of the samples and for the temperature range from –15°C to 190°C. All samples were placed in a specially designed environmental chamber, Fig. 1, during these experiments.

Figure 3 shows a representative dependence of the modulus of elasticity of NiTi on temperature, based on the measurements made during this study. These results indicate a hysteritic behavior due to heating and cooling cycles, which is typical of SMAs.

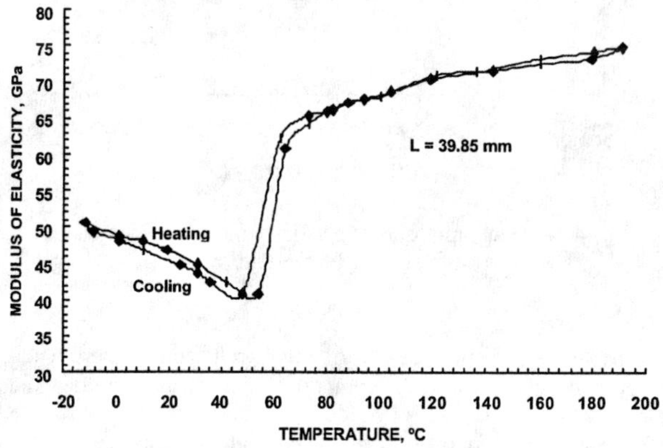

Fig. 3. Modulus of elasticity of NiTi as a function of temperature.

According to Fig. 3, modulus of elasticity initially decreases at the beginning of heating and cooling stages of the process. Then, it rapidly increases, or decreases, during the phase transformation stages, depending whether heating, or cooling, respectively, takes place. Finally, it increases during the last stages of the heating, or cooling, processes. The results of Fig. 3 clearly show that, for the NiTi alloy of the composition studied in this paper, the phase transformation takes place at 60°C. That is, the phase transformation from martensite to austenite is at 60°C. Martensitic transformations, however, are never complete in nature and there will always be residual martensite [5]. The results shown in Fig. 3 were obtained for the NiTi sample of effective length of L = 39.85 mm, and the modulus of elasticity was calculated from the frequencies corresponding to the first bending mode.

It should be noted, however, that the phase transformation temperature depends on the specific composition of the NiTi alloy. That is, for the NiTi alloys of compositions different than the composition studied herein, the phase transformation temperatures other than 60°C would be obtained. It should also be noted that the transformation temperature for SMAs is known to shift as a result of thermomechanical fatigue of the microstructure.

The modulus of elasticity based on the second, the third, and the fourth bending modes was also determined. Then, the calculations were repeated for effective lengths of 28.82 mm and 25.0 mm. The average modulus of elasticity as a function of temperature was finally calculated as a function of temperature for each of these lengths, Fig. 4.

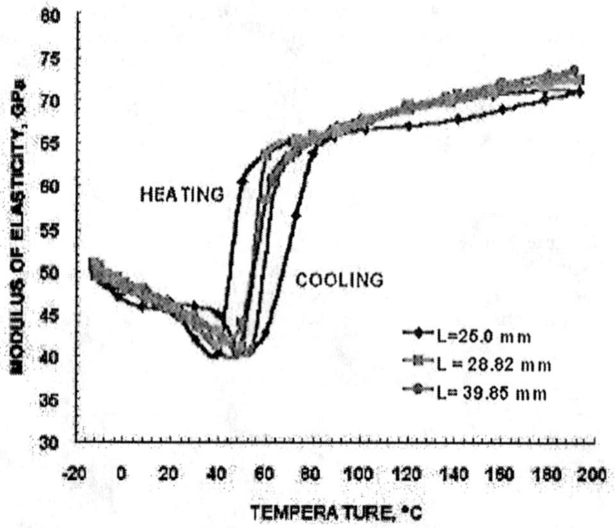

Fig. 4. Average modulus of elasticity for a NiTi sample as a function of temperature.

The results from FEM correlated well with experimentally measured frequencies. A comparison between analytical and experimental methods [5] yielded a difference of –9.2% and comparisons between experimental and computational results have provided a difference of 0.8%. Figure 5 shows representative comparison of ACES results.

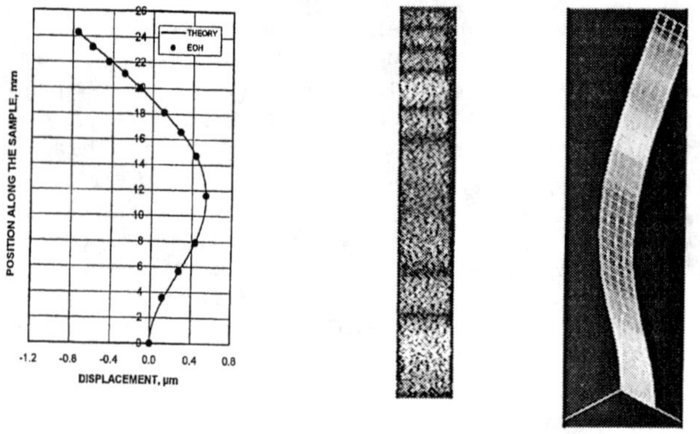

Analytical f_a = 2352 Hz | Experimental f_e = 2330 Hz | Computational f_c = 2347 Hz

Fig. 5. Comparison of ACES results for NiTi with effective length of 28.82 mm.

5. CONCLUSIONS

The results presented in this paper show validity of using a hybrid approach, based on the ACES methodology, in characterizing SMAs, more specifically NiTi alloy. Material property values determination was achieved by TAOEH, a noninvasive holographic methodology.

Many times, analyses are performed based on "book values", i.e., published material properties. This usually results in inaccuracy in evaluation and analysis of a design. The errors resulting from poorly defined material property values are even more critical for SMAs, which exhibit a strong dependence of the material property values on stress, strain, and temperature. The dependence of modulus of elasticity on temperature was measured by TAOEH and the values obtained reflect actual behavior of the material.

In this paper, a comparison between analytical and experimental methods yielded a difference of –9.2%, while comparison between experimental and computational results provided a difference of 0.8%. These differences are in good agreement with the uncertainties determined from Eq. 3.

For the 500 μm thick NiTi samples used in this study, the XRD/EDS revealed material composition to be 57% Ni and 43% Ti. For these samples, the TAOEH results indicate a highly nonlinear change in modulus of elasticity ranging from 38 GPa to 72 GPa as the temperature ranged from –15°C to 190°C. The changes in the modulus of elasticity, as a result of the changes in temperature to which the samples were subjected, correspond to the phase transformation from martensite to austenite.

Comparison of the analytical, computational, and experimental results shows good correlation and provides basis for use of the measured characteristics of SMAs for integration with electronic packaging, e.g., in development of microsensors and microactuators to be made of these materials.

Overall, this study has shown that there exists good correlation between analytical, computational, and experimental methodologies in terms of material property determination and prediction of resonance frequencies. Furthermore, following sufficient convergence analysis as well as experimental validation, the FEM can be used as a primary source of analysis in place of expensive and time-consuming experimental work. Finally, a hybrid approach, such as ACES methodology, proves to be an important step in the characterization of material behavior as it provides results with a higher level of confidence.

Continued developments of the methodology presented in this paper will provide new tools and information for effective and efficient integration of SMAs with electronic packaging.

5. REFERENCES

1. Hornbogen, E., 1989, "Shape memory materials," *Practical Metallurgy*, 26:279-294.
2. Ford, D. S., and White, S. R., 1995, "Thermomechanical behavior of 55Ni45Ti NiTiNOL," *Acta Metallurgica*, 44:2295-2307.
3. Baz, A., and Ro, J., 1992, "Thermodynamic characteristics of NiTiNOL-reinforced composite beams," *Composite Engineering*, 2:527-542.
4. Pryputniewicz, D. R., 1997, *ACES approach to the development of microcomponents*, MS Thesis, Worcester Polytechnic Institute, Worcester, MA.
5. Mizar, S. P., 1999, *Thermomechanical characterization of Shape Memory Alloys by ACES methodology*, MS Thesis, Worcester Polytechnic Institute, Worcester, MA.
6. Pryputniewicz, R. J., 1992, *Engineering experimentation*, Worcester Polytechnic Institute, Worcester, MA.
7. Pryputniewicz, R. J., 1985, "Holographic and finite element studies of vibrating beams," *Proc. SPIE*, 599:54-62.
8. Pryputniewicz, R. J., 1985, "Time-average holography in vibration analysis," *Optical Engineering*, 24:843-848.
9. Pryputniewicz, R. J., 1994, "Quantitative determination of displacements and strains from holograms," Ch. 3 in *Holographic interferometry*, Vol. 68, Springer Series in Optical Sciences, Springer-Verlag, Berlin.
10. Furlong, C., and Pryputniewicz, R. J., 2000, "Absolute shape measurements using high-resolution optoelectronic holography methods," *Optical Engineering*, 39:216-223.

Proceedings of
2001 ASME International Mechanical Engineering Congress and Exposition
November 11–16, 2001, New York, NY
EPP-Vol. 1

IMECE2001/EPP-24728

CREEP BEHAVIOR OF PB-FREE SOLDERS

F. Hua and C. M. Garner
Materials Technology Operation
Intel Corporation
3065 Bowers Ave.
Santa Clara, CA 95054

H. G. Song and J. W. Morris
Center for Advanced Materials
University of California, Berkeley CA 94720s

ABSTRACT

This study reports results of shear creep behavior of four Pb-free solders, Sn-3Ag-0.5Cu, Sn-3.5Ag, Sn-0.7Cu and Sn-10In-3.1Ag at 95θC and 130θC. At the stress levels tested, all the four solders showed the stress components close or larger than 5, typical for matrix creep. The calculated activation energies for Sn-0.7Cu, Sn-3.5Ag and Sn-3Ag-0.5Cu are from 103kJ/mol to 117kJ/mol, which are very close to the pure Sn self-diffusion activation energy (107kJ/mol). It suggested that the creep process is controlled by Sn bulk self-diffusion rate. The creep activation energy for Sn-10In-3.1Ag is higher in the range of 173-193kJ/mol.

The Sn-0.7Cu, Sn-3Ag-0.5Cu and Sn-10In-3.1Ag solder joints were also prepared with two different cooling rates, 3.5θC/min. (furnace-cooling) and 2.7θC/s (air-cooling) and tested at 130θC. It was observed that faster cooled solder joints have faster creep strain rates than slower cooled solder joints at the stress levels tested for all three solders, due to the fine and even distribution of intermetallic particles.

INTRODUCTION

European legislation pressure and Japan marketing pressure have led intensive research of Pb-free solders in microelectronics industry. There are at least two trends in the design of solder joints in microelectronics, that makes understanding of relationship of solder composition – microstructure – mechanical behavior more critical. The first trend is the increasing use of new Pb-free solder compositions to meet environmental concerns. The new solders do not have the empirical bases as Sn-Pb solders, and cannot automatically be assumed to be able to use same life prediction model as Sn-Pb solders. The second trend is the miniaturization of solder joints as miniaturization of micro-electronic packaging. The microstructure of the mini-joints is different from bulk solder joints due to rapid cooling at assembly process, therefore leads to different mechanical behaviors.

Creep behavior of solder joints is critical to the reliability of the solder joints in application because the service temperatures of solder joints are usually high homologous temperatures (T/T_{melt}) for the solder joints and the thermal strain rate is ordinarily slow. The impetus for study of creep is the need to understand and improve the fatigue life of solder joints. Much work [1-4] has been done in the Sn-Pb system and suggested good correlations in creep behavior and fatigue life.

Work by Shine *et al.* [1-3] demonstrated a good correlation between the isothermal fatigue life of eutectic Sn-Pb solder and the amount of matrix creep in the fatigue cycle. This work suggests that exceptional fatigue life could be obtained by refining the microstructure of the solder joints to achieve super-plastic creep. Wild [4] has also shown that solder joints with large grains have a shorter fatigue life than those with smaller grains.

Eutectic Sn-Ag-Cu has been recommended by the industry as a surface mount assembly solder. Most of the works in Pb-free solder creep behavior were done in Sn-3.5Ag system and reported by several investigators [5-7, 13-16]. Relatively, very limited research data reported on eutectic Sn-Cu and Sn-10In-3.1Ag solders.

While the influence of cooling rate on the mechanical behavior of eutectic Sn-Pb solder have been discussed extensively[8-10] by investigators, the cooling rate effects on creep behavior of Sn-0.7Cu, Sn-3Ag-0.5Cu and Sn-10In-3Ag solder joints have not reported.

EXPERIMENTS

Making solder alloys and foil. All raw materials used for alloying had a purity of 99.999%. Sn-0.7wt.%Cu, Sn-10wt.%In-3.1wt.%Ag, and Sn-3wt.%Ag-0.5wt.%Cu alloy ingots were made by a vacuum arc melting technique. The 96.5Sn-3.5 wt.% Ag ingot was made by heating the material in a vacuum-sealed quartz tube. After melting, all alloys were homogenized for 55hrs at 180°C before cold rolling. The compositional analysis was not performed for ingots, but the mass loss during melting was below 0.05g for the initial weight of 20g. Rolled foil thickness of all alloys was 155-15 Πm.

Making creep test samples. Figure 1 shows the geometry of the single shear specimen. The substrate pad size is about 48mil x 88mil(1.2192mm x 2.2352mm). The sandwich specimen samples were assembled with one side as Cu and the other side as electroless Ni and immersion Au. The thickness of the electroless Ni and Au is about 160 micro inches and 5 micro inches, respectively. Solder joint thickness was controlled by spacer thickness. The solder joint thickness made for the experiment is about 6 mil. The shear specimens were reflowed at N_2 atomsphere with peak temperature of 235θC for Sn-3.5Ag, Sn-10In-3.1Ag, and Sn-3.0Ag-0.5Cu and 245θC for Sn-0.7Cu. Samples were cooled at two different cooling rate, 3.5θC/min.(furnace cooling) and 2.7θC/s (air cooling). After reflow, samples were aged at 160θC for about 4 hours.

Creep Testing. Creep tests were done in a dead-load creep machine. The sample is loaded with a fixed weight while its temperature is controlled by an oil bath. The creep specimen is held by frictional grips. The top grip is attached to a pull-rod that transmits a load to the specimen. The displacement of the top grip is monitored with a Linear Voltage Displacement Transmitter (LVDT). The output voltage signal is amplified, recorded and converted into a displacement measurement. The specific apparatus used has a detection limit below 1 μm displacement. The shear stress is measured as load divided by the solder-substrate contact area (total area of the wetted pads). Shear strain is the relative displacement of the sample plates divided by solder joint thickness. Creep tests were conducted until steady state was reached, and the reported creep rates were the steady-state creep rates. Samples were tested at 2 different temperatures of 130θC and 95θC. Data was taken at least 5 different load levels at one temperature. Displacement was recorded every 30 sec. Table 1 shows homologous temperature at testing conditions for different solders.

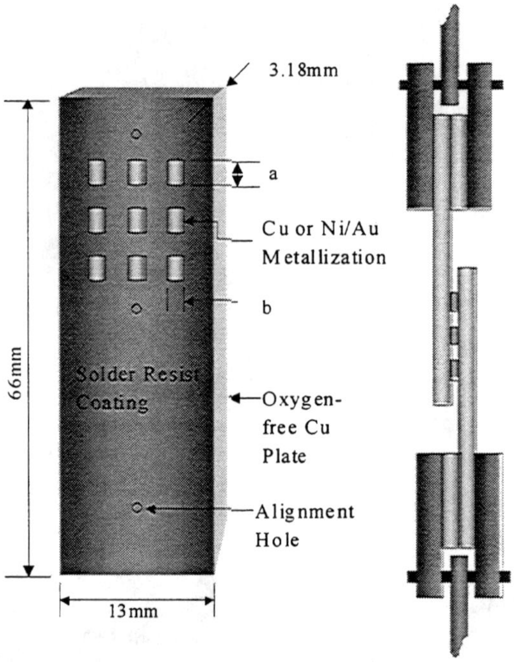

Fig. 1 Geometry of the single shear specimen

Table 1. Homologous temperatures for creep tests of solders.

Solders	M.P.(θC)	130θC	95θC
Sn-37Pb	183	0.884	0.807
Sn-3.5Ag	221	0.816	0.745
Sn-0.7Cu	227	0.806	0.736
Sn-3.0Ag-0.5Cu	217	0.821	0.749
Sn-10In-3.1Ag	201-204	0.845	0.771

RESULTS

Shear Creep. Figure 2(a-h) contain creep curves of Sn-0.7Cu, Sn-3.5Ag, Sn-3.0Ag-0.5Cu and Sn-10In-3.1Ag solders measured at 130θC and 95θC under various stress levels. It is seen that only very brief or non-primary creep was observed and followed by steady state creep, and then tertiary creep. A steady state strain rate was determined by the slope of the linear portion of the creep curve (minimum strain rate) through curve fitting.

Figure 3(a and b) are plots of creep curves of Sn-0.5Cu, Sn-3.5Ag, Sn-3.0Ag-0.5Cu and Sn-10In-3.1Ag solder at 130θC under stress level of 5.44 MPa and 95θC at 9.07MPa respectively. Under 5.44MPa stress, Sn-0.5Cu shows the largest creep strain rate, followed by Sn-10In-3.1Ag, then Sn-3.5Ag and Sn-3.0Ag-0.5Cu at both temperatures. Rupture time for Sn-0.5Cu solder is about 22 minutes while Sn-10In-3.1Ag is about 3.5 hours and Sn-3.0Ag-0.5Cu and Sn-3.5Ag are about 8.5 hours at 130θC. The same trend is also seen in testing at 95θC.

The constitutive equation that governs steady state creep by Dorn:

$$\dot{\gamma} = A\tau^{n}\exp(-Q/RT) \qquad (1)$$

where ϑ is the shear strain rate, Ωis the shear stress, n is stress components, Q is an activation energy and A is a pre-exponential factor, which is governed by microstructures of solder joints. For eutectic Sn-Pb [8-10], solder materials exhibit more than one creep mechanisms as a function of stress at given temperature, indicated by changes in the stress exponent, n, and in the activation energy, Q.

Data in Figure 4 (a and b) shows the relationship between the steady state strain rate and stress for Sn-0.5Cu, Sn-3.5Ag, Sn-3.0Ag-0.5Cu at 95θC and 130θC in steady state creep regime. Sn-0.5Cu solder joints clearly have the fastest strain rates under stress tested among all the solders at 130θC and 95θC. At 95θC, the strain rates of Sn-3.5Ag, Sn-3.0Ag-0.5Cu and Sn-10In-3.1Ag solders are very close in steady state creep while that of Sn-10In-3.1Ag solder is significantly faster than Sn-3.0Ag-0.5Cu solder at 130θC. Creep rate of Sn-3.5Ag solder joints at 130θC is close to Sn-10In-3.1Ag at lower strain rates, and close to Sn-3.0Ag-0.5Cu at higher strain rates.

A stress component (n) is the slope of a Log strain rate vs. Log applied stress plot. In the strain rate range tested, one stress component fits well on for all the data points at one temperature. The numbers are slightly different for different temperatures. They are in the range of 5.4 to 7.8 for all cases.

The activation energies (Q) under different stress levels at temperature from 95θC and 130θC are calculated by equation 2:

$$Q = -R\,(\mathrm{Ln}\ ϑ_1 / \mathrm{Ln}\ ϑ_2)\,/(\,1/T_1 - 1/T_2) \qquad (2)$$

Where $ϑ_i$ is the strain rate at temperature T_i, and R is the gas constant. The activation energies obtained from the creep tests of Sn-0.7Cu, Sn-3.5Ag, and Sn-3.0Ag-0.5Cu solders are close to the activation energy of the pure Sn bulk self-diffusion (107 K J/mol). The activation energy for creep in Sn-10In-3.1Ag solder ranges from 173 to 194 kJ/mol, higher than the Sn self-diffusion activation energy, Table 2.

Fig. 5 (a-c) compares creep stress vs. strain plots of Sn-0.7Cu, Sn-3.0Ag-0.5Cu and Sn-10In-3.1Cu solder joints with different cooling rates. For all three solders, joints made at faster cooling rate shows lower creep resistance than those made at slower cooling rate at the strain rates tested (> 10^{-5}/s).

Microstructure. Microstructure of solder joints has direct effect on solder joints creep deformation behavior. Figure 6 shows the microstructures of the solder joints at cooling rate of 2.7θC/s and 3.5θC/s/min.. It was noticed that all the solders have similar microstructures - intermetallics particles distributed in Sn matrix. Although for different solders, the types of intermetallics, and their size and distribution are different. The intermetallics in Sn-3.5Ag, Sn-0.7Cu and Sn-3.0Ag-0.5Cu solder joints are Ag_3Sn, Cu_6Sn_5 and a combination of Cu_6Sn_5 and Ag_3Sn, respectively. In Sn-10In-3.1Ag solder joints, Ag_2In and Ag_3Sn Intermetallics are expected [11,12]. A more detailed microstructure analysis for Sn-10In-3.1Ag solder joint is needed. Attempts were made to get clear view of grain size of the Sn matrix, but were not successful. It is possible that Sn grows as dendrites not equal-axis grains.

Table 2 Parameters in Dorn's equation for the Sn-0.5Cu, Sn-3.5Ag, Sn-3.0Ag-0.5Cu and Sn-10In-3.1Ag solder joints

Sn-0.5Cu			
Stress (MPa)	n		Q (KJ/mol)
	95θC	130θC	
2.72	6.6	5.7	103
3.63	6.6	5.7	106
5.44	6.6	5.7	104
Sn-3.5Ag			
Stress (MPa)	n		Q (KJ/mol)
	95θC	130θC	
5.44	6.2	5.4	104.5
7.25	6.2	5.4	104.3
9.07	6.2	5.4	109
Sn-3.0Ag-0.5Cu			
Stress (MPa)	n		Q(KJ/mol)
	95θC	130θC	
5.44	7.8	6.6	117
7.25	7.8	6.6	115
9.07	7.8	6.6	108
Sn-10In-3.1Ag			
Stress (MPa)	n		Q(KJ/mol)
	95θC	130θC	
5.44	7.4	7.0	173
6.35	7.4	7.0	184
7.25	7.4	7.0	194

DISCUSSION

All of the solders here are tested above 0.5 of their melting points, so creep is expected to dominate the deformation process. In general, creep response of a material reflects a continually changing interaction between strain hardening and softening and typically showing three regions in strain-time space, Figure 7. Strain hardening at elevated temperature involves subgrain formation associated with the rearrangement of dislocations, while thermally activated cross slip and dislocation climb represent recovery processes. Stage I – primary creep, is related to substructure changes to increase overall resistance to dislocation motion; Stage II – steady state, indicates a stable substructure and a dynamic balance of hardening and softening processes; Stage III – tertiary creep, is dominated by a number of weakening metallurgical instabilities, such as localized necking, intercrystalline facture, micro-voiding formation, recrystallization of hardened grains, etc.

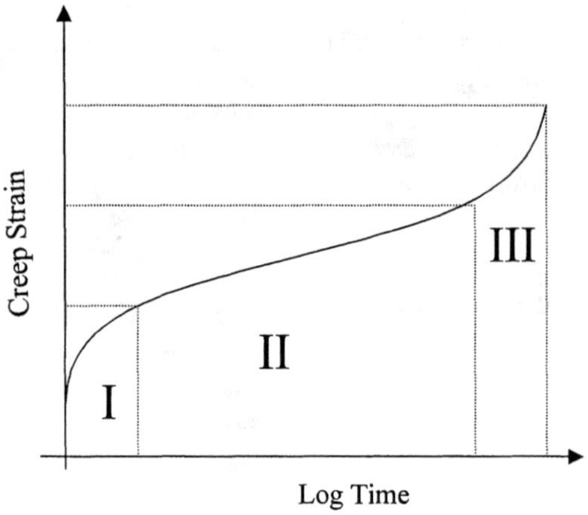

Figure 7. Typical creep curve for alloys

Creep behavior of Sn-3.5Ag, Sn-1Cu and Sn-3.8Ag0.7Cu solders has been investigated by a few researchers [13-16], Primary creep was observed. In this work, a creep displacement was not recorded until the reading is stabilized. Therefore the short primary creep reported here may be due to incomplete data recording, but it is also possible that at higher temperature, the rearrangement of dislocation was achieved very quickly.

Under the same stress, the Sn-0.7Cu solder joints reaches tertiary creep stage in the fastest rate, followed by Sn-10In-3.1Ag, then Sn3.5Ag and Sn-3Ag-0.5Cu in all different stresses at both 130θC and 95θC, Fig. 3. Comparing microstructures of all the solders, there is much fewer dispersed intermetallic particles in Sn-0.7Cu solder than all the other solders. The resistance to dislocation motion is much less than the other solders.

Table 2 is a summary of the creep parameters in Dorn's equation generated from this work. The stress components range from 5.4 to 7.8 for all cases, indicating dislocation climb-controlled deformation. The determined activation energies for Sn-0.7Cu, Sn-3.5Ag, Sn-3.0Ag-0.5Cu solders are from 103kJ/mol to 117kJ/mol, is very close to the activation energy of pure Sn self diffusion. It confirms matrix creep dominated creep mechanism for all the three solders. However, the activation energy for Sn-10In-3.1Ag solder is somewhat higher, about 2x of that for all the other three solders. Similar appearance in the microstructures for all the solders makes it difficult to explain the higher activation energy for Sn-10In-3.1Ag solder based on available data.

Table 3 Literature comparison of creep parameters in Dorn's equation.

Sn-xAg		
	n	Q (kJ/mol)
Daeveaux [14] Sn-3.5Ag	5.5	72.675
Schubert [13], Sn-3.5Ag (20θC-180θC)	4.75	57.2
Mavoori [16] (20θC - 80θC) Sn-3.5Ag	11.3	82.3
Liang [15], Sn-5Ag (23θC - 150θC)	7.83	69.5
This work, Sn-3.5Ag (95θC - 130θC)	5.8	105.7
Sn-xCu		
	n	Q(kJ/mol)
Liang [15], Sn-1Cu (23θC - 150θC)	8.1	79.4
This work, Sn-0.7Cu (95θC - 130θC)	6.15	104
Sn-xAg-yCu		
	n	Q(kJ/mol)
Schubert [13], Sn-3.8Ag-0.7Cu (20θC-150θC)	3.02	38.76
This work , Sn-3.0Ag-0.5Cu (95θC - 130θC)	7.2	113

Table 3 compares the parameters in Dorn's equation obtained in the literature and in this work. It shows that the value of stress components are very similar to those by other investigators. However, the activation energies obtained from this work are higher than others and closer to activation energy of pure Sn self-diffusion. It is believed that at higher temperature, the dislocation climb controlled matrix creep becomes a more dominated mechanism than that at lower temperature.

Cooling rate effects on microstructure are shown in Figure 6. Due to limited intermetallics precipitation in Sn-0.7Cu solder joints, it is hard to see the differences from cross section of optical microstructure of solder joints made from cooling rate of 2.7θC/s and 3.5θC/min. However, there are clear differences in microstructures of Sn-3.0Ag-0.5Cu and Sn-10In-3.1Ag solder joints cooled at the two different cooling rates. Slower cooled solder joints show large primary Sn phase and segregated intermetallics particles while faster cooled joints show fine, uniformly dispersed intemetallics. The differences in the microstructures are responsible for the difference in creep behaviors of solder joints made at different cooling rate. Smaller the particles size and larger interparticle spacing results in higher creep rate.

CONCLUSIONS

Shear creep behavior of Sn-0.7Cu, Sn-3.5Ag, Sn-3.0Ag-0.5Cu and Sn-10In-3.1Ag solders was investigated. Sn-0.7Cu showed lowest creep resistance in steady state creep while Sn-3.0Ag-0.5Cu shows strongest. At the stresses and temperatures tested, dislocation matrix creep dominates the creep mechanism for all the solders. Cooling rate has significant effects solder joints microstructure, that ultimate effects solder joints creep behavior. Faster cooling rate results in fine and uniformly distributed intermetallics, that is responsible for the faster creep rate for fast cooled solder joints.

ACKNOWLEDGMENTS

Authors would like to thank supports from management of Intel Corporation.

REFERENCES

[1] M.C.Shine, L.R.Fox and J.W. Sofia, Brazing & Soldering, No.9, 11 (1985).

[2] M.C. Shine and L.R.Fox, Low Cycle Fatigue, ASTM STP 942, 588 (1987).

[3] M.C. Shine, U.S. Patent 4947341, Aug. 7 (1990).

[4] R.N. Wild, IBM report 73z000421, RC14858, Owego, NY(1973).

[5] H. Mavooris, J. Chin, S. Vaynman, B. Moran, L. Keer and M. Fine, "Creep, stress relaxation, and Plastic Deformation in Sn-Ag and Sn-Zn Eutectic Solders," J. Electronic Materials, 26 (7), 78(1997).

[6] H. Mavoori, Semyon, Jason Chin, Brian Moran, Leon M. Keer and Morris E. Fine, "Mechanical Behavior of Eutectic Sn-Ag and Sn-Zn solders", Mat, Res. Soc. Symp. Proc. Vol. 390, p 161 (MRS 1995).

[7] Alan Woosley, Geoff Swan, T.S. Chong, Linda Matsushita and Kenon Simmons, "Development of Lead and Halogen Free Peripheral Leaded and PBGA Components to meet MSL3 260C reflow peak Reflow" from NEMI 2001.

[8] J.W. Morris, Jr. and H.L. Reynolds, Design and Reliability of Solders and Solder Interconnections, pp.49-57 (1997).

[9] Z. Mei, J.W. Morris, Jr., M.C. Shine and T.S.E.Summers, Journal of Electronic Materials, Vol. 20, No.8, pp.599-608 (1991).

[10] Z. Mei and J.W. Morris, Jr., ASME Journal of Electronic Packaging, 114, 104 (1992).

[11] T. M. Korhonen and J. K. Kivilahti, Journal of Electronic Materials, Vol. 27, No. 3, 1998, pp149

[12] I. Ohnuma, Y. Cui, X. J. Liu, Y. Inohana, S. Ishihara, H. Ohtani, R. Kainuma, and . Ishida, Journal of Electronic Materials, Vol. 29, No. 10, 2000, pp1113-1121.

[13]A. Schubert, H. Walter, R. Dudek, B. Michel, G. Legranc, j. otto, G. Mitic, 2001 International Symposium on Advanced Packaging Materials, pp129-134.

[14] R. Darveaux, R. banerji, K. Mawer, A. Dody, G. "Reliability of Pastics Ball Grid Array Assembly", in Ball Grid Array Technology (ed. J. Lau), 1995, pp.379-442.

[15] J. Liang, P.S. Lee, N. Gollhardt, S. Schroeder and w. Morris, Mat. Res. Soc. Symp. Proc. Vol. 445, 1997 Materials Research Society, pp 307-312.

[16] H. havoori, J. Chin, S. Vaynman, B.Moran, L. Keer, and M. Fine, Journal of Electrical Materials, Vol. 26, No. 7, 1997, pp783-790.

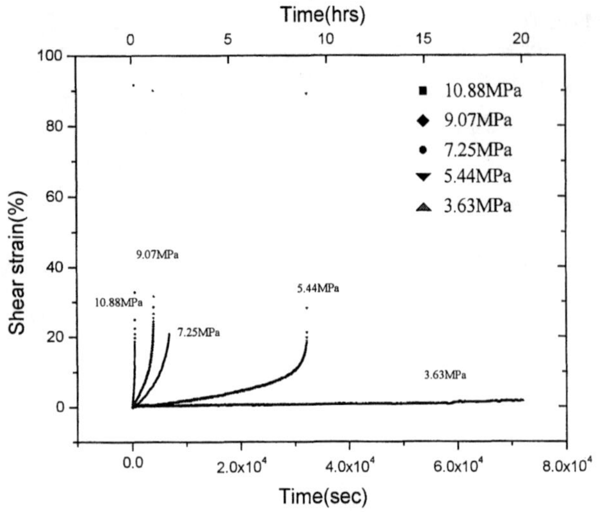

(e) Sn-3.0Ag-0.5Cu at 130θC

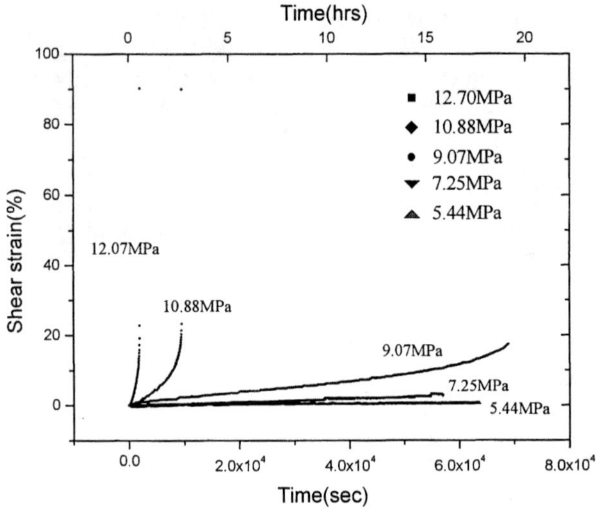

(f) Sn-3.0Ag-0.5Cu at 95θC

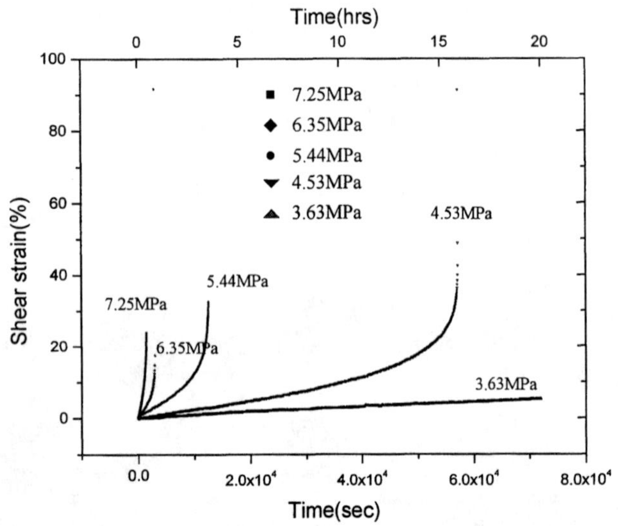

(g) Sn-10In-3.1Ag at 130θC

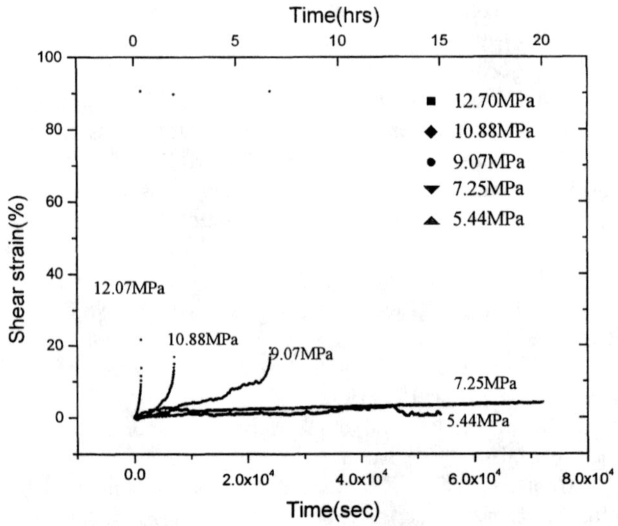

(h) Sn-10In-3.1Ag at 95θC

Fig. 2 (e-h) Creep curves (Shear strain vs. Time) of Sn-3.0Ag-0.5Cu and Sn-10In-3.1Ag at 130θC and 95θC

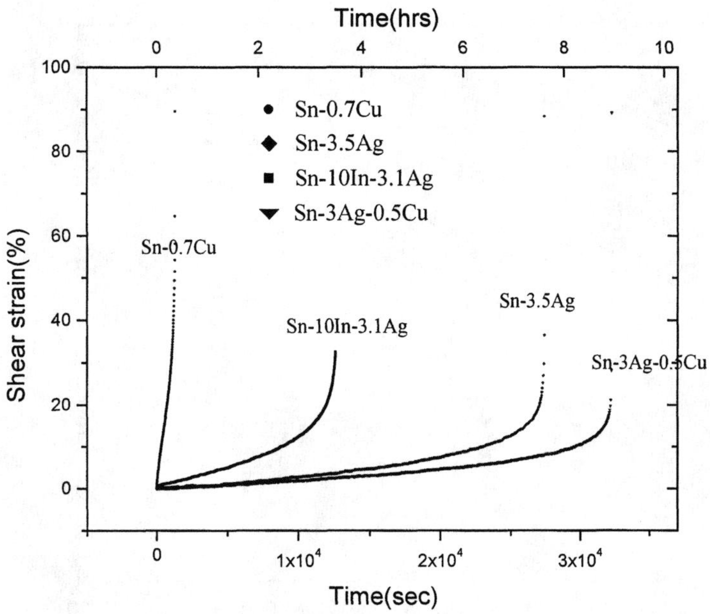

(a) Creep Curves at 5.44MPa, 130°C

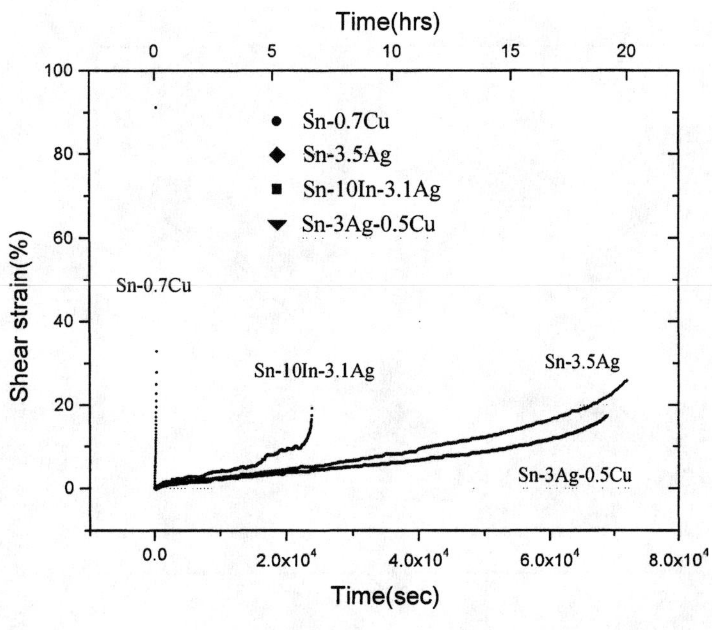

(b) Creep Curves at 9.07MPa, 95°C

Fig. 3 Creep Shear strain (%) vs. Time for Sn-0.7Cu, Sn-3.5Ag, Sn-3.0Ag-0.5Cu and Sn-10In-3.1Ag under 5.44 MPa, 130°C and 9.07MPa, 95130°C

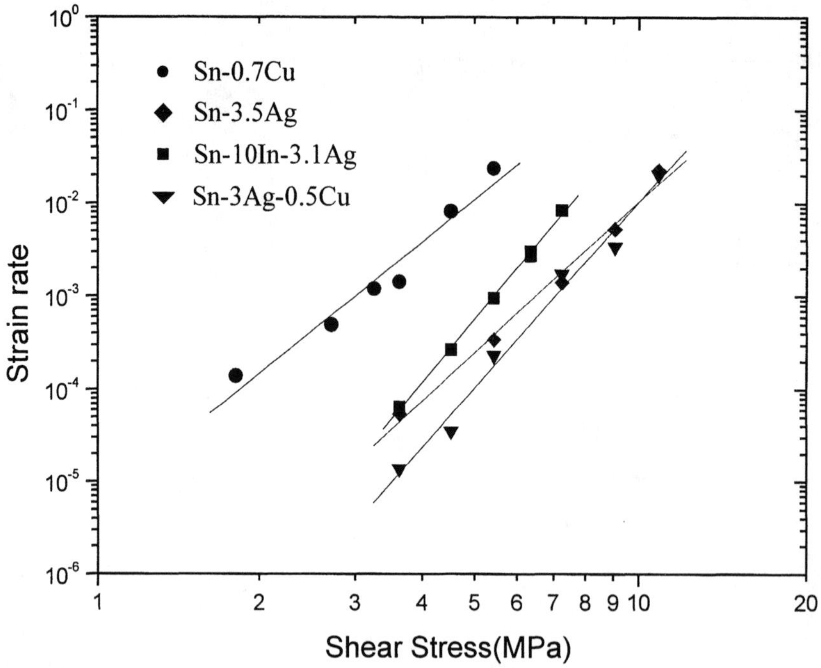

(a) at 130θC

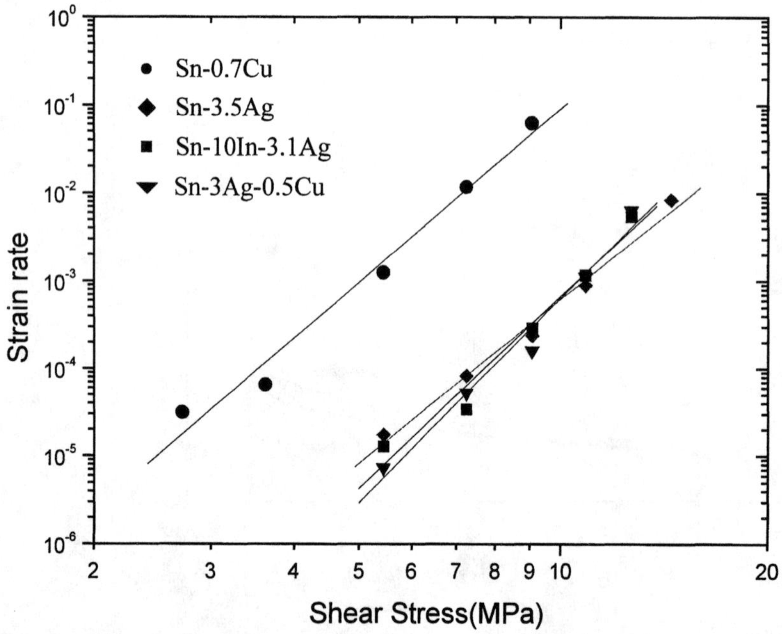

(b) at 95θC

Fig. 4 Stress vs. steady state Strain rate plots for Sn-0.7Cu, Sn-3.5Ag, Sn-3.0Ag-0.5Cu, Sn-10In-3.1Ag solders at 130θC and 95θC.

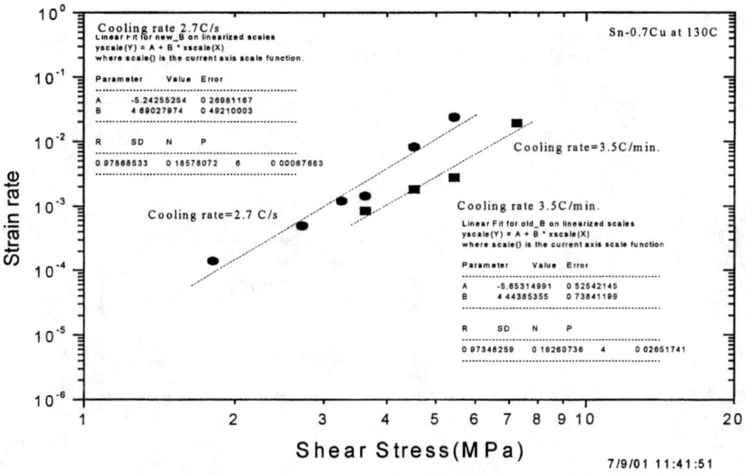

(a) Sn-0.7Cu solder joints, cooling rate of 3.5θC/min and 2.7θC/s, tested at 130θC

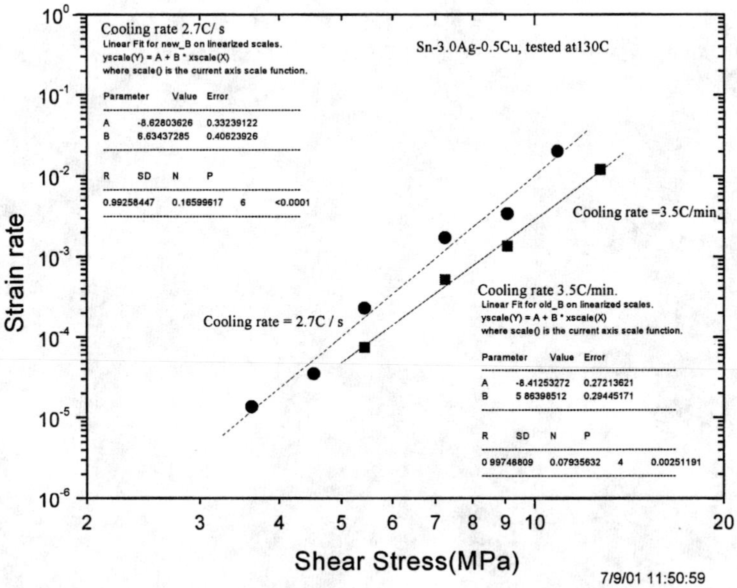

(b) Sn-3.5Ag-0.5Cu solder joints, cooling rate of 3.5θC/min and 2.7θC/s, tested at 130θC

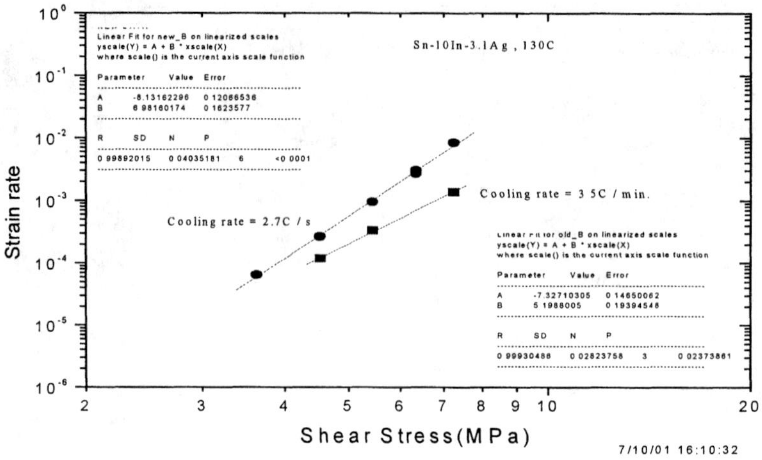

(c) Sn-10In-3.1Ag solder joints, cooling rate of 3.5θC/min and 2.7θC/s, tested at 130θC

Fig. 5 Creep stain rate vs. stress plots of Sn-3.0Ag-0.5Cu, Sn-0.7Cu and Sn-10In-3.1Ag solder joints with cooling rate of 3.5θC/min and 2.7θC/s, tested at 130θC

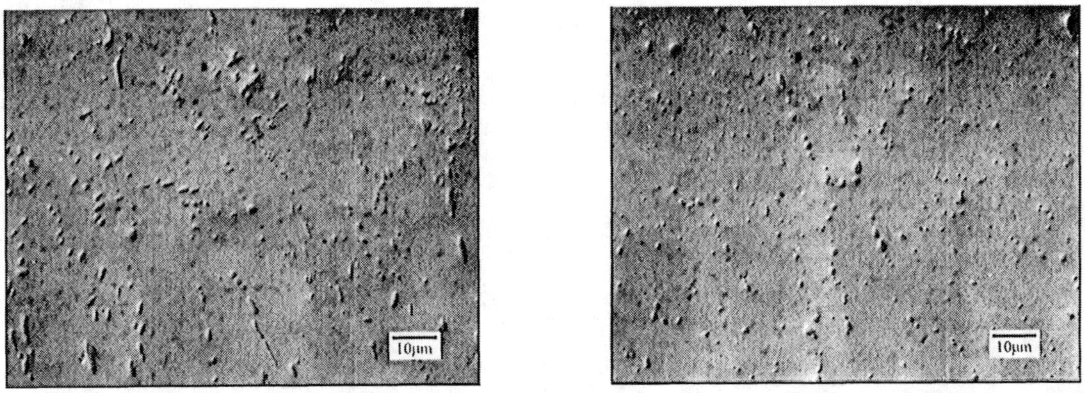

(a) Faster cooled (b) Slower cooled

Figure 6 (a) and (b) Optical microstructure of Sn-0.7Cu solder joints cooled at 2.7θC/s and 3.5θC/min.

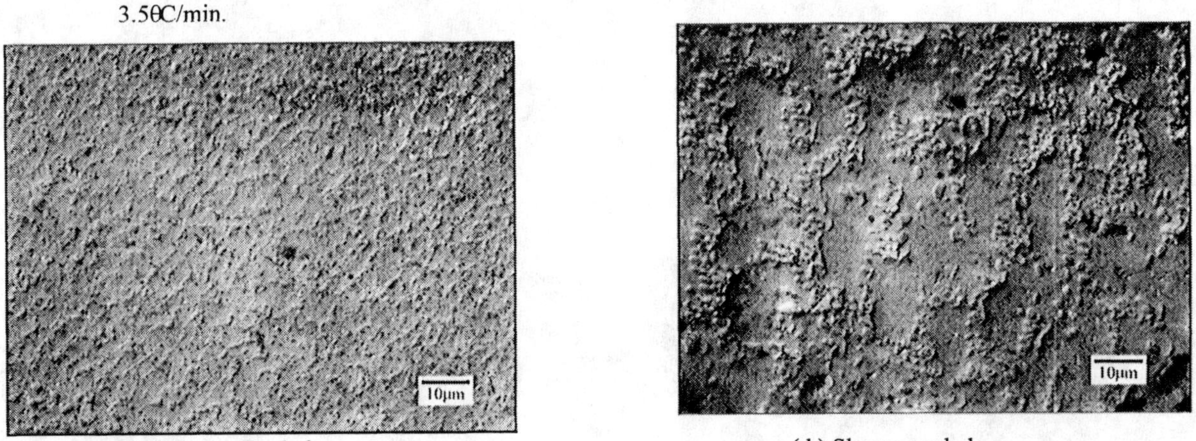

(c) Faster Cooled (d) Slower cooled

Figure 6 (c) and (d)) Optical microstructure of Sn-3.0Ag-0.5Cu solder joints cooled at 2.7θC/s and 3.5θC/min.

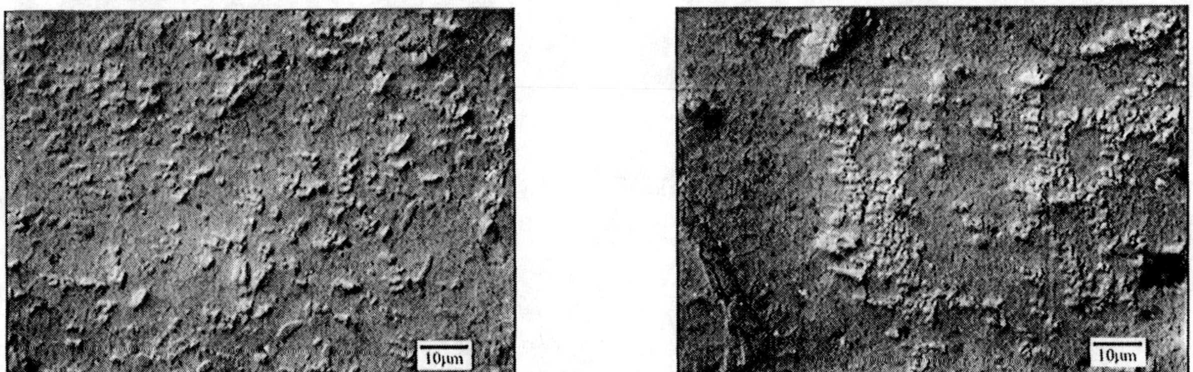

(e) Faster cooled (f) Slower Cooled

Figure 6 (e) and (f)) Optical microstructure of Sn-10In-3.1Ag solder joints cooled at 2.7θC/s and 3.5θC/min.

Proceedings of
2001 ASME International Mechanical Engineering Congress and Exposition
November 11–16, 2001, New York, NY
EPP-Vol. 1

IMECE2001/EPP-24729

Multi-axial Cyclic and Monotonic Behavior of a 63Sn-37Pb Solder Alloy

T. Jesse Lim and Wei-Yang Lu
Sandia National Laboratories
Material Mechanics Department
7011 East Avenue, MS 9721
Livermore, CA 94550
e-mail: tjlim@sandia.gov

Abstract

In this work, the cyclic and monotonic loading of both pure torsion and uniaxial behavior of 63Sn-37Pb solder alloy are compared. By comparing the monotonic loading, it is shown that the ultimate equivalent stress of both torsion and uniaxial tensile behavior is comparable; and the failure strain in uniaxial tension is considerably less than that of pure torsion. The fatigue life of this solder alloy under the same equivalent strain amplitude for both uniaxial tension-compression and pure torsion are also comparable. These data provide a baseline for investigating the behavior of the solder alloy in the multi-axial stress-strain space.

Introduction

Surface mount technology (SMT) become prevalent in packaging printed wiring board(PWB), because of ease of manufacturing and a more compact board size. However, in this case, solder joints become load bearing structures holding the electronic component and the PWB together. Because of the mismatch in thermal expansion coefficient between various components and the PWB, the solder joint goes through load cycling during on-off cycles or if the component physically go through different temperature zones. (e.g. airplane parking on runway on a hot day and fly up to several km high; Electronic components near/inside the hood of a car) In order to understand the long term reliability of the SMT, it is necessary to study the long term behavior of the solder joint such as fatigue life and creep fracture. The study of these properties under multiaxial stress state is necessary since the stress in the solder joint can be multiaxial.

Uniaxial Testing of the Solder Alloy

Uniaxial dog-bone type specimens of the PbSn alloy were prepared in three steps: 1) Molten PbSn alloy was solidified between two brass rods, 3.18mm diameter and about 25mm length each, inside two half round section mold as shown in Figure 1. The resulting solder alloy section is about 6.35mm in length. 2) The solder section was then machined down to nominal 1mm in diameter and 3mm in gage length. The schematic drawing of the specimen is illustrated in Figure 2. 3) The specimen was aged in 100°C, atmospheric air for 16 hours.

Figure 1 Solidified PbSn Alloy in the mold

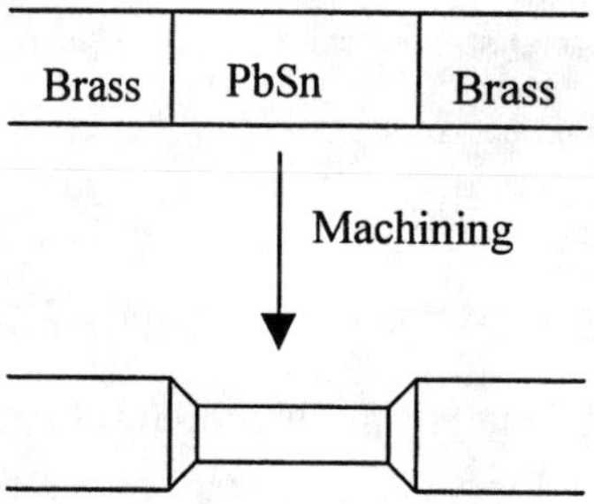

Figure 2 Machining operation to make the specimen gage section.

The specimens were then attached to the MTS Tytron loading frame as shown in Figure 3. (in the figure, indicate extensometer, load cell, actuator, specimen, ...) A monotonic ramp function was applied to the actuator so the axial strain rate at the gage section was $10^{-4}s^{-1}$. A typical stress-strain curve is shown in Figure 4. The ultimate tensile stress is about 30Mpa. After yielding, the alloy goes through strain softening and final failure occurs at about 0.25 axial strain.

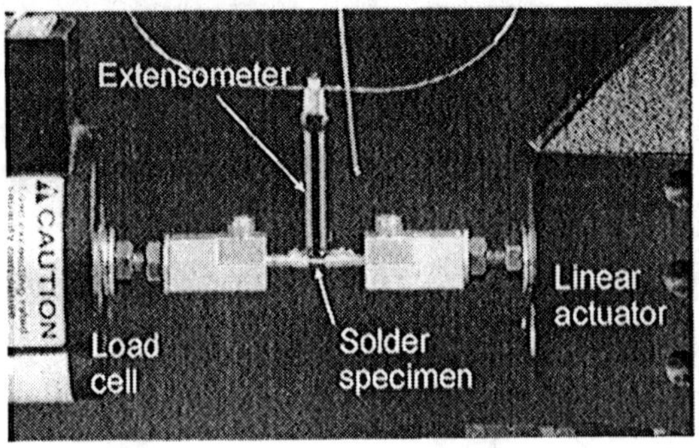

Figure 3 Specimen as mounted on the Tytron loading frame.

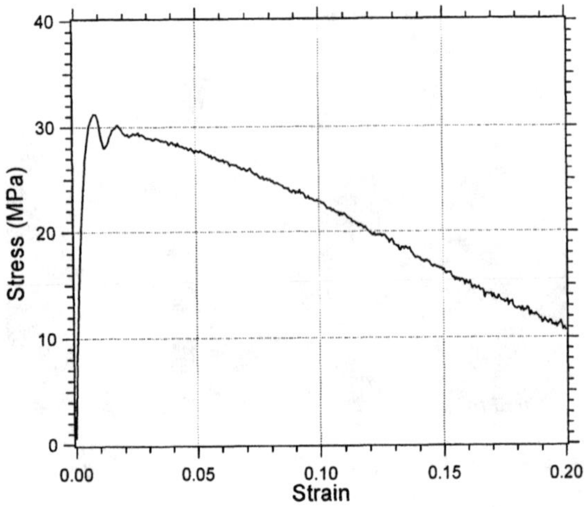

Figure 4 Uniaxial stress-strain response of a typical Pb-Sn specimen.

Uniaxial Fatigue Testing of the Solder Alloy

For uniaxial fatigue testing, two nominal strain amplitude was chosen for R = -1, ±0.5% and ±1% axial strain with axial strain rate = $10^{-4}s^{-1}$. Two specimens were cycled at each strain level. The load cycling would stop once the stress amplitude drops 50% from the average of the first 20 cycles. In this case, fatigue crack was apparent in the specimen gage section and the cycle number was considered the fatigue life of the specimen. Table 1 summarizes the fatigue life, N_f, with respect to the strain amplitude. The stress-strain response of specimens 2 and 4 are shown in Figures 5 and 6 respectively.

Table 1 Fatigue life of various specimens

Specimen #	Strain, ε	Fatigue life, N_f
1	0.5% < ε < 0.5%	1600
2	0.5% < ε < 0.5%	1349
3	1.0% < ε < 1.0%	390
4	1.0% < ε < 1.0%	376

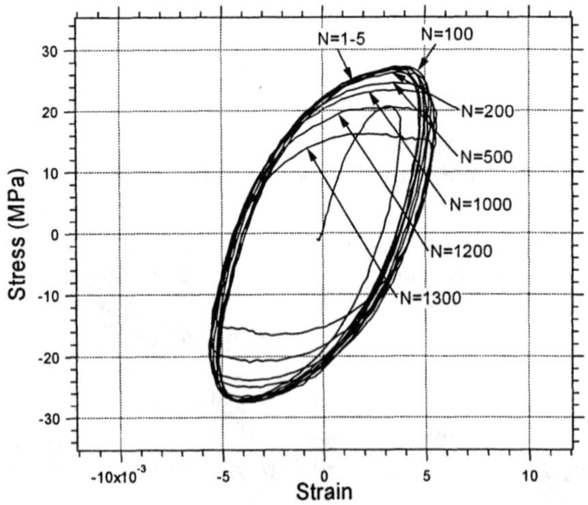

Figure 5 Stress-strain response for specimen cycling at 0.5% < ε < 0.5%, N indicates cycle number.

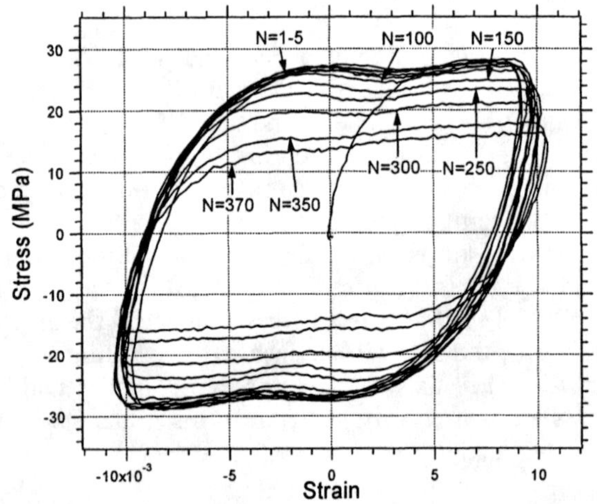

Figure 6 Stress-strain response for specimen cycling at 1.0% < ε < 1.0%, N indicates cycle number.

Considering Figure 5, cyclic stress-strain response for 0.5% < ε < 0.5% was stabilized and consistent throughout cycle 1 to about cycle 200, after cycle 200, damage started to accumulate and the stress-strain response is degrading as cycling proceeds. Same conclusion can be drawn for

Figure 6, where 1.0% < ε < 1.0%, the stress-strain response started to degrade after cycle 100. Comparing Figure 5 and 6, it is interesting to see that the ultimate stress for both strain amplitude is at about 27Mpa, and for the 1.0% < ε < 1.0% cycling, about 0.5% of strain is pure plastic strain. In Figure 5, there was a slight dip in the plastic strain range because there is a slight decrease of strain rate after the elastic range.

Monotonic Torsional Testing of the Solder Alloy

Torsional specimens were prepared using the method as outlined in previous sections. Besides solid dog-bone type specimens, thin-walled tubular specimens were also prepared by drilling through the solder alloy section through one brass end. The specimens were prepared in two sizes, 2.54mm outside diameter (o.d.) and 5.08 mm o.d.. Each size includes both solid and thin walled tube specimens. For thin walled tube specimens, the nominal o.d. to wall thickness ratio is 10.0. All specimens have a nominal gage length equal to the o.d.

Torsional experiments were performed on an MTS TorsionMaster System, Figure 7. During loading, the specimen was twisted at a constant ramp rate such that the equivalent strain rate at the surface of the specimen was $10^{-4}s^{-1}$. Where the equivalent strain, ε_{eq} and equivalent stress, σ_{eq}, are respectively given by

$$\varepsilon_{eq} = \gamma / \sqrt{3} \quad \text{and} \quad \sigma_{eq} = \sqrt{3}\tau$$

Where γ and τ are shear strain and shear stress, respectively.

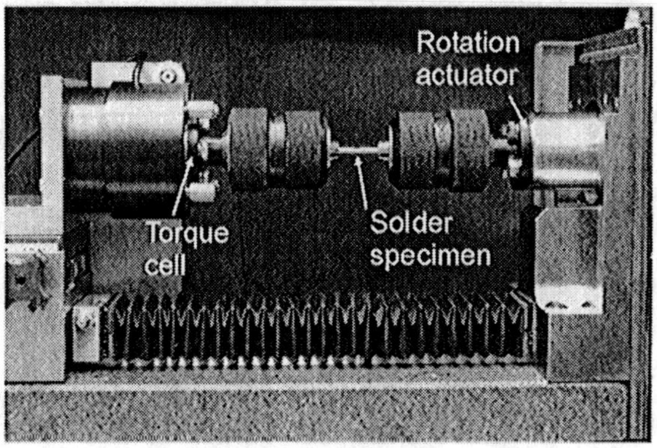

Figure 7 A torsional specimen as attached to the MTS Torsion Master torsional loading frame.

Figures 8 and 9 show the equivalent stress-strain response of each specimen type of 2.54mm o.d. and 5.08mm 0.d., respectively. The equivalent stress and strain were evaluated at the outer surface of the specimen. All specimens have an ultimate stress at about 33Mpa, suggesting the shear stress-strain behavior is quite size and geometry independent for the specimens tested.

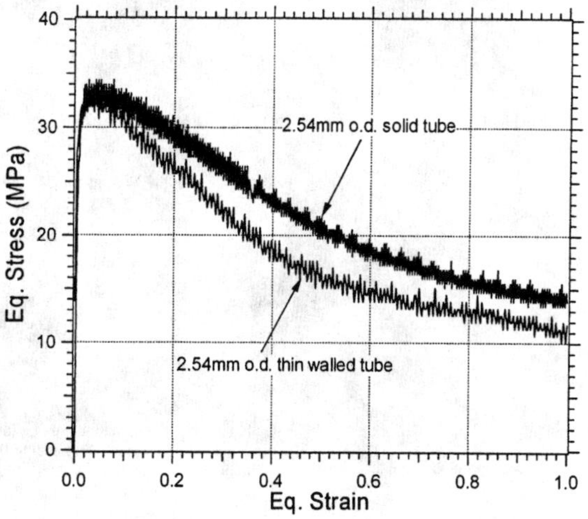

Figure 8 Equivalent shear stress-strain response of 2.54mm o.d. torsional specimens.

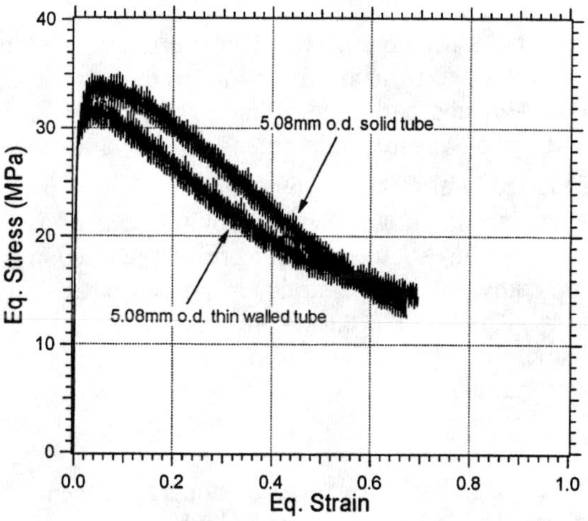

Figure 9 Equivalent shear stress-strain response of 5.08mm o.d. torsional specimens.

Since the solder alloy is viscoplastic in nature, they exhibits extremely large shear strain without failure (>100% equivalent shear strain). In most cases, weak shear bands will form at certain point of torsion and from that point on, viscoplastic (?) slips will concentrate on that particular shear band, as seen in Figure 10. Comparing Figures 4, 8 and 9, shear deformation can go for a relatively large distortion than that of uniaxial loading without failing, since there is no stress concentration in the shear band. In the case of uniaxial loading, once necking is formed in the gage section, stress concentration occurred in the necking area and fails earlier than that of monotonic torsion.

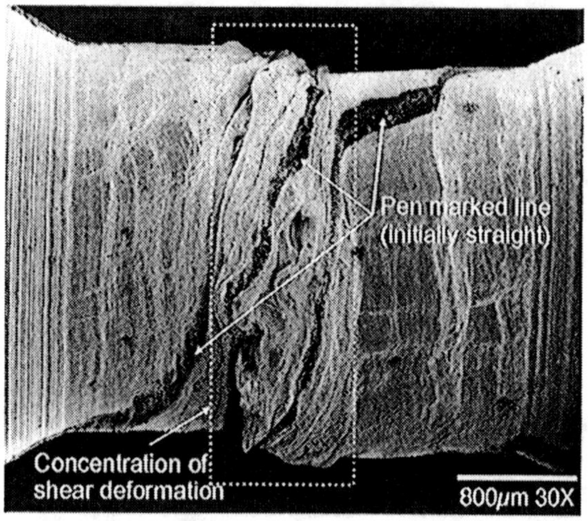

Figure 10 Torsional specimen after loading, showing localized shear band.

Torsional Fatigue of the Solder Alloy

In addition to monotonic torsional tests, torsional fatigue test was also performed on 2.54mm o.d. specimens, both thin walled tube and solid cylinder. In torsional fatigue tests, the equivalent strain rate = $10^{-4}s^{-1}$ and equivalent strain range = ±1.0% or ±0.5%, and R = -1, which were the same as uniaxial fatigue for easy comparisons of data. The fatigue lives of various specimen are summarized in table 2. Their equivalent stress-strain responses are shown in Figures 11 to 14. The curves are wavy in the plastic range since there is slight variation of rotation rate on the Torsion Master frame, and the solder alloy is quite rate-dependent.

Table 2 Fatigue life of various torsional specimens. (TW specimen type indicates thin walled tube.)

Specimen #	Specimen type	Eq. Strain, ε_{eq}	Fatigue life, N_f
1	TW	$0.5\% < \varepsilon_{eq} < 0.5\%$	1663
2	TW	$1.0\% < \varepsilon_{eq} < 1.0\%$	206
3	Solid	$0.5\% < \varepsilon_{eq} < 0.5\%$	1166
4	Solid	$1.0\% < \varepsilon_{eq} < 1.0\%$	459

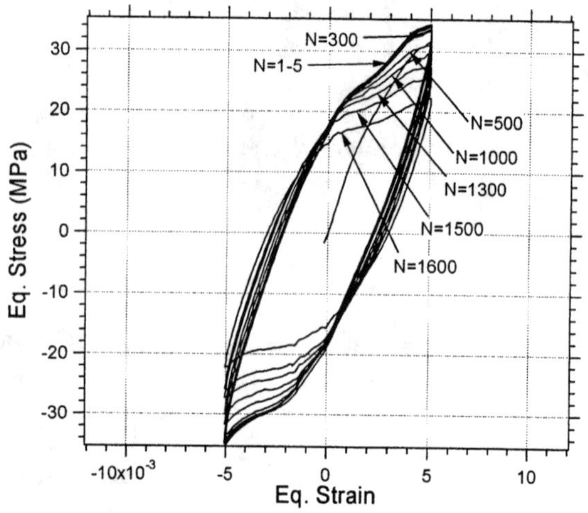

Figure 11 Stress-strain response of torsional fatigue specimen 1, N indicates cycle number.

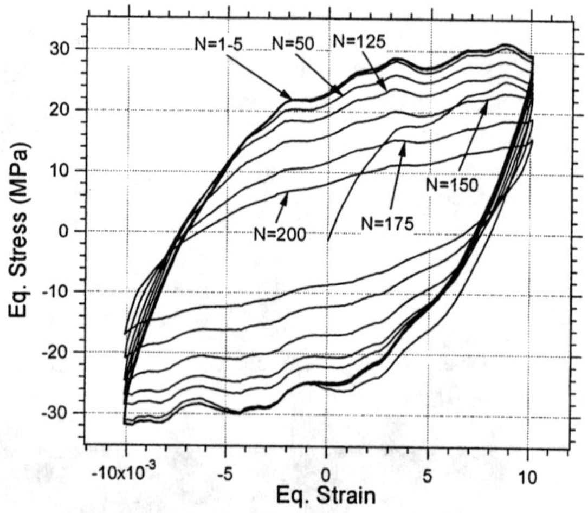

Figure 12 Stress-strain response of torsional fatigue specimen 2, N indicates cycle number.

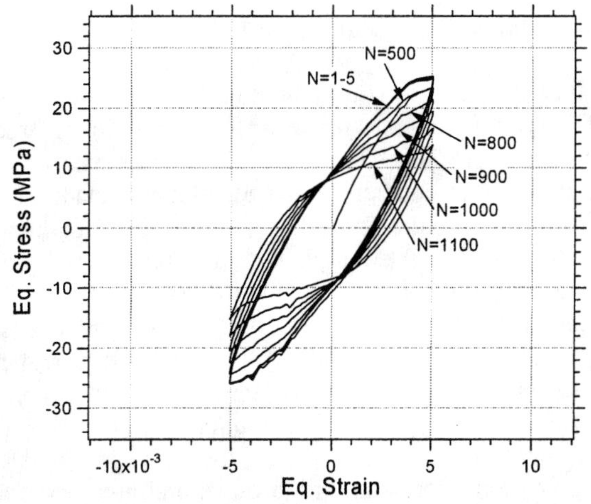

Figure 13 Stress-strain response of torsional fatigue specimen 3, N indicates cycle number.

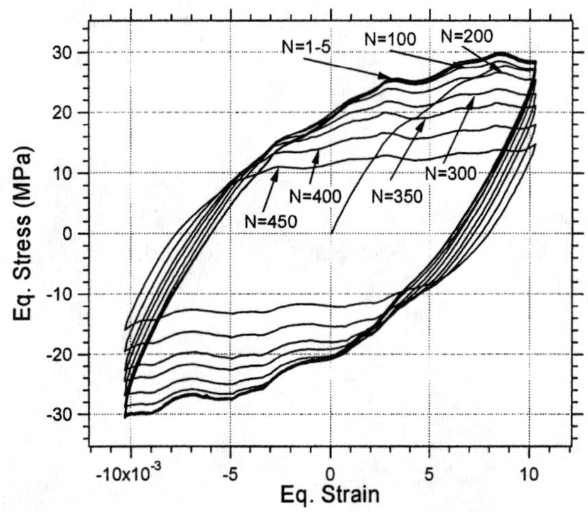

Figure 14 Stress-strain response of torsional fatigue specimen 4, N indicates cycle number.

Comparing the data in Table 2, there is no general trend of whether thin walled tube or solid specimen has a longer fatigue life. In addition, comparing the fatigue data in Table 2 and Table 1, and comparing the stress-strain response of torsional fatigue specimens (Figures 11 to 14) and uniaxial fatigue specimens, (Figures 5 and 6), The fatigue life for the two strain amplitudes for the two mode of loadings are in general comparable. There are variations in the peak stresses because of statistical variations. More experiments have to be carried out to establish a statistical average.

The fatigue data for Pb-Sn alloy from various sources are quite varied. However, the results reported in this work are comparable to that reported by Jiang et.al.[1] for uniaxial fatigue testing on bulk specimens of 63Sn-37Pb alloy.

Micro-structural Damage Accumulation

Three solid torsion specimens were put under Scanning Electron Microscope (SEM) to study the micro-structural damage evolution: 1) as-received specimen; 2) fatigue specimen after failure for $\varepsilon_{eq}= \pm 0.5\%$; and 3) Fatigue specimen after failure for $\varepsilon_{eq}= \pm 1.0\%$. Figures 15-17 are the SEM images of the surfaces of the three respective specimens. The two different phases, the Pb-rich phases (light area) and Sn-rich phases (dark area) can be clearly seen in Figure 15. As cycling proceeds, damage accumulates. As evidenced in Figures 16 and 17, most of the fatigue damage occurred at the interface between Pb-rich phase and Sn-rich phase. And for a higher magnitude of cycling strain, damage is more severe.

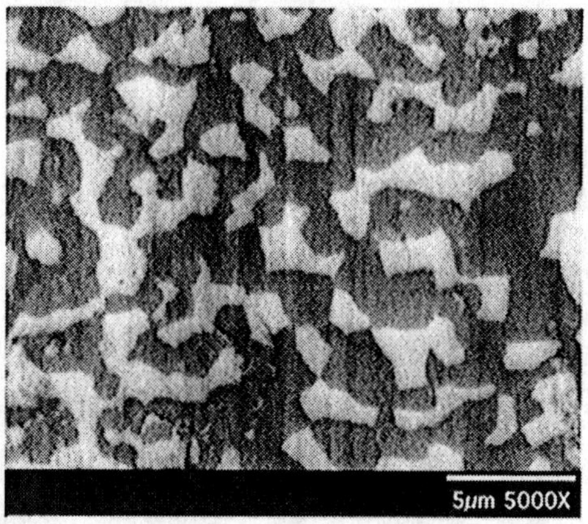

Figure 15 SEM image of the surface of as-received specimen.

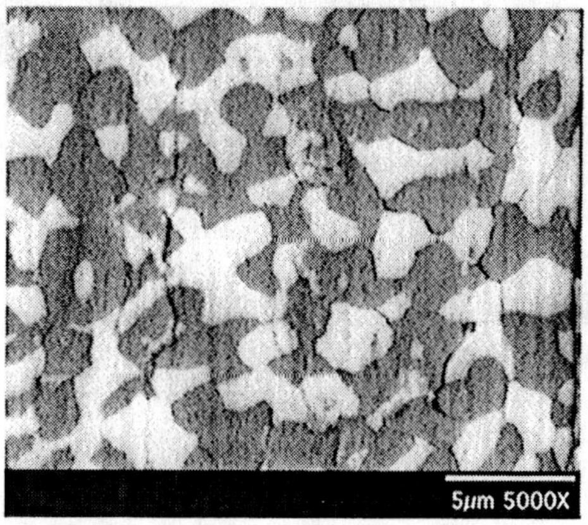

Figure 16 SEM image of the surface of fatigued specimen with cycling $\varepsilon_{eq}=\pm 0.5\%$.

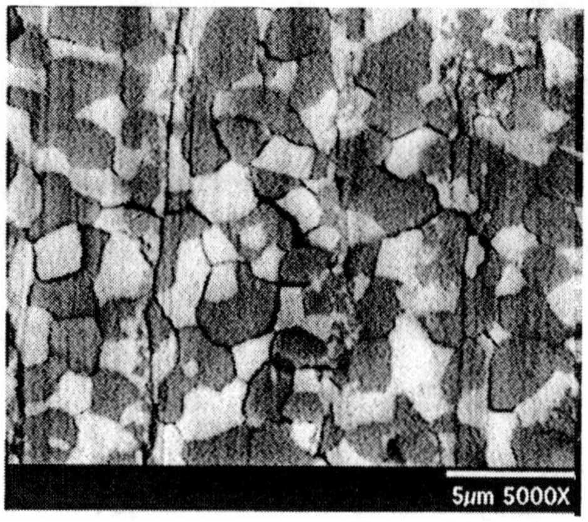

Figure 17 SEM image of the surface of fatigued specimen with cycling $\varepsilon_{eq}=\pm1.0\%$.

The fatigue damage as appeared in Figures 16 and 17 occurred all over the specimen surface. And as cycling damage proceeds, some damages coalesce to become a dominant crack, as appeared in Figure 18, for the $\varepsilon_{eq}=\pm0.5\%$ fatigue specimen. And from that point onwards, the dominant crack grows deeper as cycling proceeds until the specimen fails.

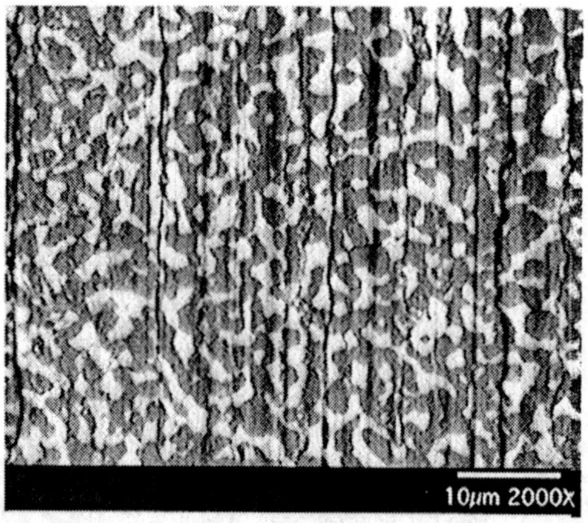

Figure 18 Coalescence of fatigue damage into a dominant crack.

Conclusion and Recommendation

In this work, both the monotonic loading and fatigue cycling of uniaxial and torsional loading are compared. In monotonic loading, the equivalent ultimate stress is comparable for both uniaxial and torsional loading, while the strain to failure is higher for torsional loading than uniaxial loading. The fatigue life and damage accumulation during uniaxial and torsional cyclic loading are also comparable. The work done here serves as a baseline understanding for PbSn solder alloy under different mode of loading. In order to understand the complete deformation and fatigue behaviours of this alloy under multi-axial state of loading, more experiment with combined axial-torsional proportional and non-proportional loading need to be carried out.

Acknowledgement

Sandia is a multiprogram laboratory operated by Sandia Corporation, a Lockheed Martin Company, for the United States Department of Energy under contract DE-AC04-94-AL85000.

References

1. Jiang, H., Herman, R. and Plumbridge, W.J., "High-strain fatigue of Pb-Sn eutectic solder alloy," *Journal of Material Science,* **31**, pp. 6455-6461, 1991.

Proceedings of
2001 ASME International Mechanical Engineering Congress and Exposition
November 11–16, 2001, New York, NY
EPP-Vol. 1

IMECE2001/EPP-24730

Characterization of Strain Rate-Dependent Behavior of 63Sn-37Pb Solder Using Split Hopkinson Torsional Bars (SHTB)

Shi-Wei Ricky Lee
Department of Mechanical Engineering
Hong Kong University of Science and Technology, Kowloon, Hong Kong, PRC

Lan Hong Dai
State Key Laboratory of Nonlinear Mechanics (LNM)
Institute of Mechanics, Chinese Academy of Sciences, Beijing, PRC

ABSTRACT

The present study is aimed at the experimental characterization of strain-rate dependent behaviour of solder materials under impulsive shear loading. In order to achieve this objective, a unique testing technique, namely, split Hopkinson torsion bar (SHTB) is employed. The solder material under investigation is 63Sn-37Pb. The experimental results indicate that the shear behavior of the solder joint is very sensitive to the strain rate and the dynamic shear strength of the solder joint is much higher than the static one.

INTRODUCTION

Surface mount technology (SMT) was developed in the 80s and has become the main stream for board level assembly in the electronics manufacturing industry [1]. For surface mounted components (SMC), the solder joints are not only the passage of electrical signal and power/ground, but also the mechanical fixture to hold the components in position on the printed circuit board (PCB). Since the solder joints are rather small and are subjected to various thermal mismatch during service, they are relatively susceptible to temperature cycling (TC) [2]. Because the thermal fatigue life of solder joints is a critical issue for SMCs, the investigation on solder joint reliability under TC has been a popular research topic in the past decade [3-5]. Besides thermal loading, in practice, the electronic components may be subjected to mechanical vibrations and shocks during shipping and service [6]. For instance, automotive electronics are constantly under vibrations and hand-held devices may easily drop on the ground. As a consequence of these undesired loading conditions, severe solder joint reliability issues may be induced. However, the investigations on the aforementioned topics are relatively scarce, especially in the area of modeling [7]. One of the major reasons for this situation is that the dynamic mechanical properties of solder materials are not yet available [8]. Before any meaningful modeling can be reached, the mechanical behavior of solders under impulsive loading must be characterized [9].

In the present study, the strain-rate dependent mechanical behavior of 63Sn-37Pb solder under impulsive shear loading is characterized using the split Hopkinson torsion bar (SHTB) technique [10-12]. The solder specimen is reflowed to joint two split torsion bars, which are made of copper (Cu). Two dynamic shear strain gauges are installed on the input bar and the output bar, respectively, for data acquisition. Part of the input bar is clamped and subjected to a twisting torque. Once the clamping force is suddenly removed, the pre-stored energy will turn into a sharp torsional pulse to load the solder specimen. With the signals from the two strain gauges on the torsion bars, the dynamic shear modulus and strength of solder specimen can be characterized. The outcome of this study should be very valuable to the researchers in the area of solder joint reliability and will provide a foundation for modeling solder joints under dynamic loading in the future.

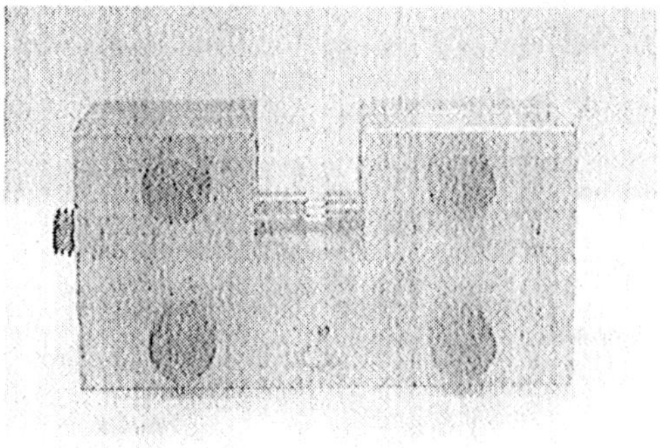

Fig. 1: Fixture for Making SHTB Specimen

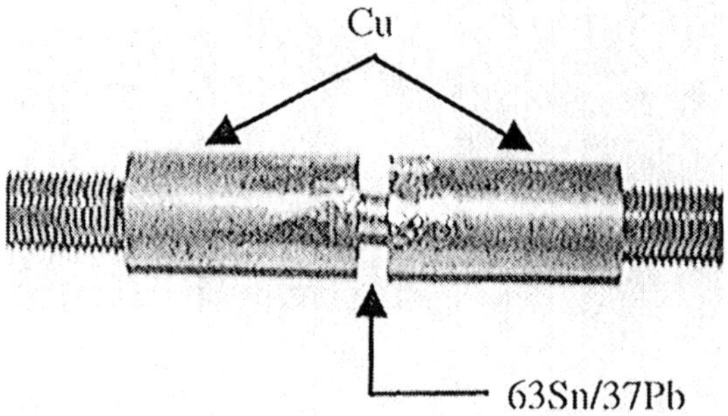

Fig. 2: SHTB Specimen with Solder and Copper

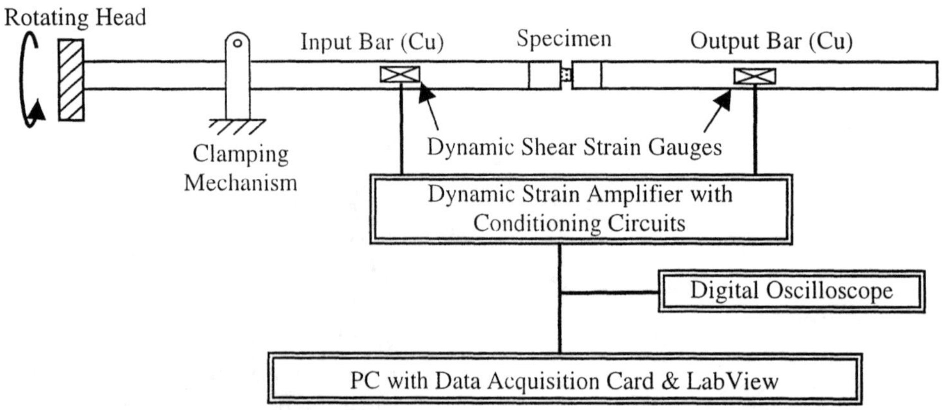

Fig. 3: Schematic Diagram for SHTB Testing System

(a) Overview of SHTB Device

(b) Testing Specimen Section

(c) Torsional Energy Storage Section

(d) Clamping Mechanism

Fig. 4: Slip Hopkinson Torsional Bar (SHTB) Testing System

EXPERIMENTAL PROCEDURES

In the present study, the SHTB specimen is made by jointing a pair of Cu cylinders with 63Sn-37Pb eutectic solder. For the fabrication of specimens, an aluminium (Al) fixture is designed as shown in Fig. 1. The two Cu cylinders (with a diameter of 8 mm) are aligned and clamped by the Al fixture with a fixed gap apart. The gap is filled with solder paste at room temperature. Then the whole fixture is placed in a bench-top forced convection reflow oven and a standard reflow temperature profile is applied to form the solder joint. After the reflow process, the specimen is removed from the Al fixture. Certain surface polishing work may be required in order to maintain a uniform diameter for the whole specimen. In addition, non-destructive inspection with X-ray imaging system is performed to verify the integrity of solder joint. The last step for specimen fabrication is to trim the solder to form a "dog-bone" shape as shown in Fig. 2. A lathe machine is employed for the trimming. The purpose of such arrangement is to ensure that, under the applied uniaxial torsion, the failure would occur in the solder, not at the interface between the Cu cylinder and the solder. Note that the gauge length is 2 mm and the diameter is about 2 mm.

Fig. 3 illustrates the schematic diagram of SHTB testing system. The actual device used in the present study is shown in Fig. 4. This equipment consists of two members, namely, input bar (1000 mm long) and output bar (700 mm long), which are made of Cu and have a diameter of 8 mm (see Fig. 4b). The afore-mentioned SHTB specimen is attached to the two split torsion bars by screw fastening (refer to Figs. 2 and 4b). Two dynamic shear strain gauges are installed on the input bar and the output bar, respectively, for data acquisition. One end of the input bar is attached to a mechanical loading device and subjected to a twisting torque (see Fig. 4c). The rotating head for providing torque pre-load is controlled by a servo-motor so that a precise initial twisting angle can be specified.

At a distance (400 mm) from the loading end, the input bar is clamped by a mechanism with special design (see Fig. 4d). This mechanism mainly consists of a U-shape clamp, a fastening bolt, and a fast-response hydraulic cylinder. The clamping force is provided by tightening the bolt. The torsional energy is stored in the front end of the input bar by twisting with the rotating head. Once the test is ready to launch, the hydraulic cylinder can provide a pulling force (up to 40 kN) instantaneously to break the fastening bolt. Consequently, the stored elastic energy will be released and generate a sharp torsional pulse to load the solder specimen. With the signals from the strain gauges on the torsion bars, the dynamic constitutive relation between shear stress and shear strain of the solder specimen can be characterized.

It should be noted that the direct output from the SHTB device is the voltage from the circuits of strain gauges. Before the actual tests, it is necessary to perform a calibration in order to turn the voltage reading into twisting torque for the subsequent calculation of shear stress and strain. In the present analysis, an initial twisting angle of 7.2° is applied to one end of the input bar. The corresponding twisting torque can be calculated as

$$T_o = GJ\frac{\phi_o}{L_o} = 5.18\ (\text{N}\cdot\ \text{m}) \tag{1}$$

where G is the shear modulus of Cu (41 GPa), J is the polar moment of inertia of the input bar (402 mm^4), ϕ_o is the initial twisting angle in radian, and L_o is length of input bar between the loading end and the clamping point (400 mm). Once the clamping force is removed, a torsional pulse would be released and propagate along the Hopkinson bars. It should be noted that, for the calibration test, the input and output bars are jointed together by a single piece of Cu cylinder with screws at both ends. Therefore, the two bars may be considered as a continuous bar and the two strain gauges on input and output bars, respectively, would generate similar signals but with a time delay as shown in Fig. 5. From the peak voltage values read from the two strain gauges (~1.65 V), it is determined that the sensitivity of the present SHTB device is 0.318 (V/N m). This value will be used in the subsequent evaluation of shear stress and strain.

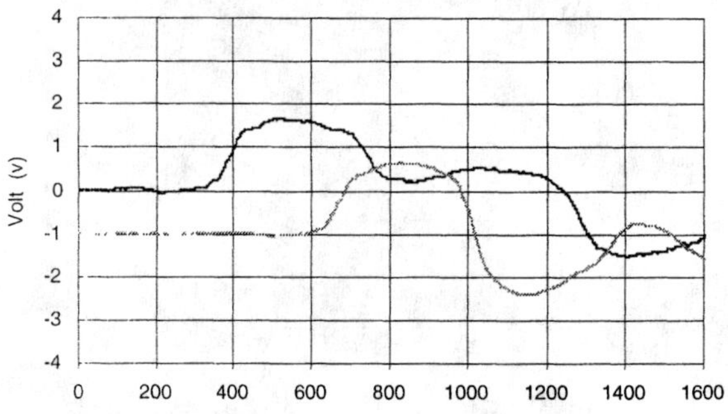

Fig. 5: Typical Wave Signal from SHTB

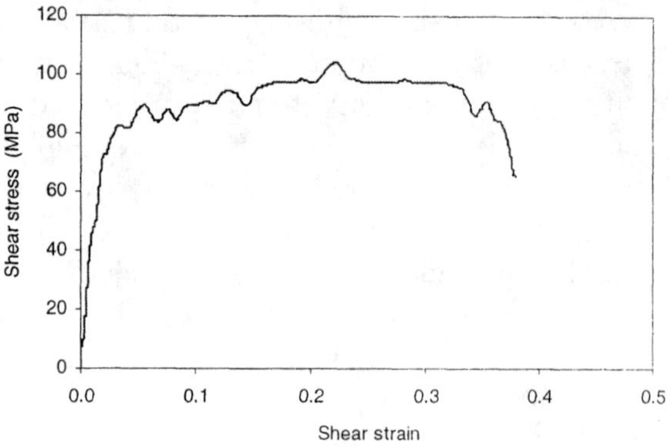

(a) Specimen #1

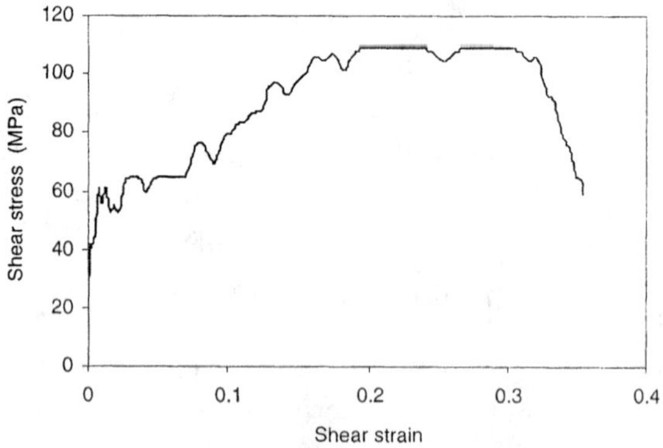

(b) Specimen #2

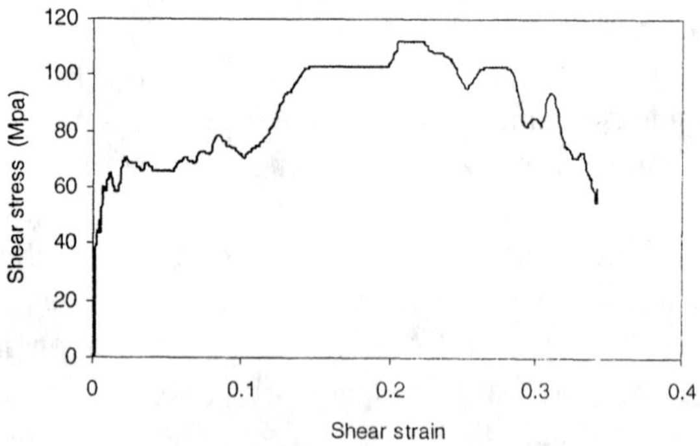

(c) Specimen #3

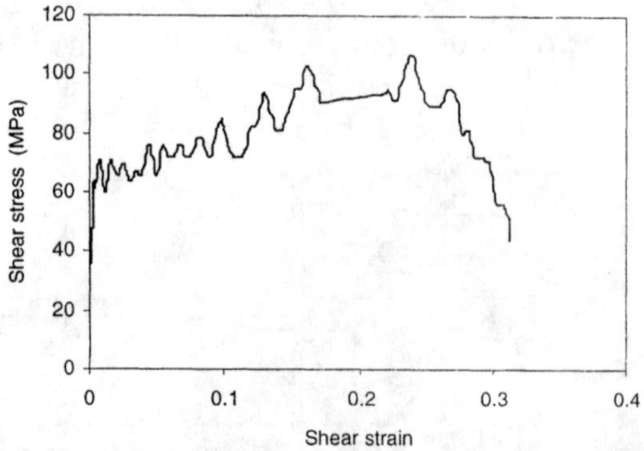

(d) Specimen #4

Fig. 6: Shear Stress-Strain Relation
($\dot{\gamma} \approx 1.5\times10^{3}\ s^{-1}$)

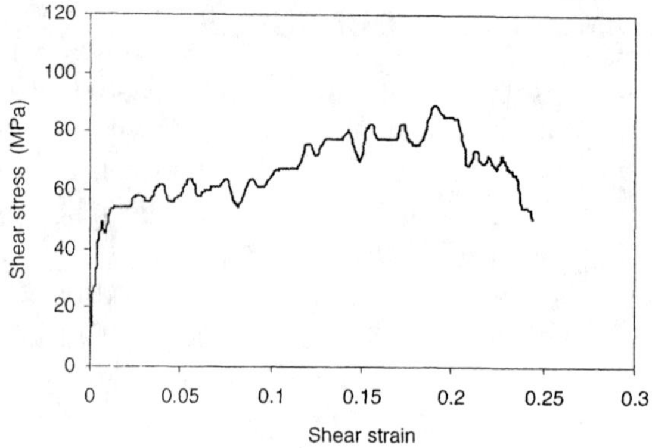

(a) Specimen #5

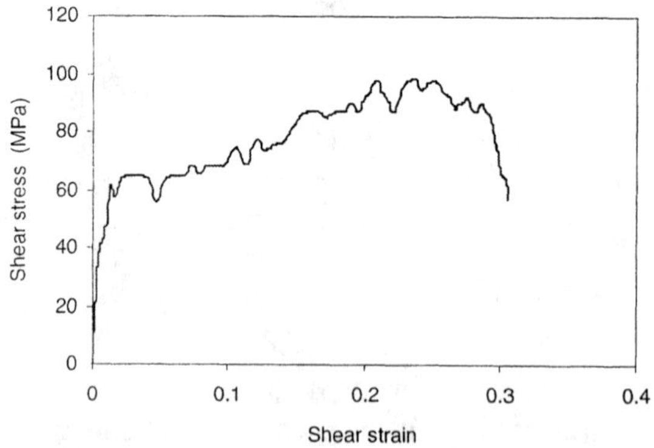

(b) Specimen #6

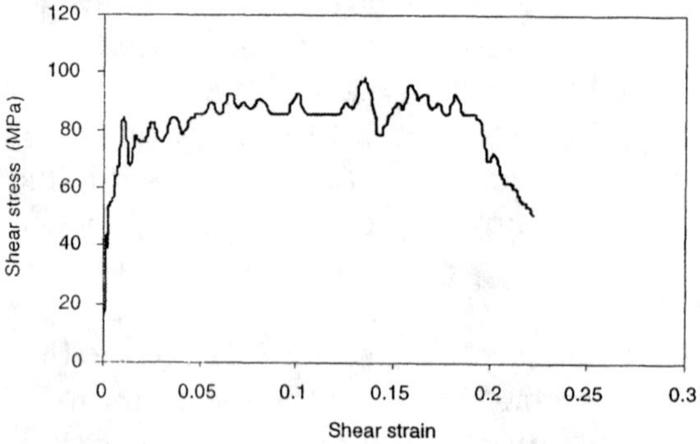

(c) Specimen #7

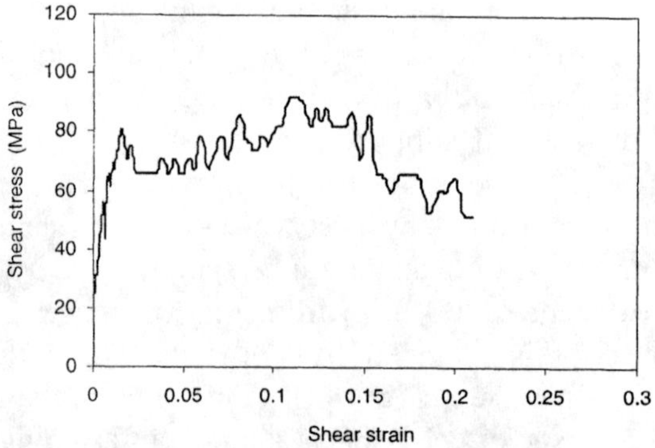

(d) Specimen #8

Fig. 7: Shear Stress-Strain Relation
($\dot{\gamma} \approx 5\times10^{2}\ s^{-1}$)

EVALUATION OF STRESS AND STRAIN

After the calibration, the twisting torques can be obtained from the strain gauge outputs. However, to deduce the constitutive relation, it is necessary to calculate the stress and the strain. Therefore, further analysis is required. From a previous study [13], the maximum shear stress in a cylindrical specimen under uniaxial torsion can be expressed as

$$\tau_S = \frac{1}{2\pi R_S^3}(3T + \phi\frac{dT}{d\phi}) \quad (2)$$

where T is the torque applied on the specimen, ϕ is the twisting angle, and R_S is the radius of the specimen. On the other hand, the average shear strain rate in the specimen can be determined as

$$\dot{\gamma}_S = R_S(\dot{\phi}_i - \dot{\phi}_r - \dot{\phi}_t)/L_S \quad (3)$$

where L_S is the gauge length of the specimen and $\dot{\phi}_i, \dot{\phi}_r, \dot{\phi}_t$ are the incident, reflective, transmitted angular velocities at the input and output ends of the specimen, respectively. According to one-dimension torsion elastic stress wave theory,

$$\dot{\phi} = C\frac{\phi}{L_S} \quad (4)$$

where $C = \sqrt{G/\rho}$ (ρ is the density of Hopkinson bars) is the shear wave propagation speed of Hopkinson bars (C = 2100 m/s for Cu). Besides, the relationship between the torque and the twisting angle can be written as

$$T = GJ\frac{\phi}{L_S} \quad (5)$$

From Eqs. (4) and (5),

$$\dot{\phi} = \frac{T}{CJ\rho} \quad (6)$$

Consequently, with Eq. (3), the average shear strain rate in the specimen can be expressed as

$$\dot{\gamma}_S = \frac{R_S}{L_S}\frac{(T_i - T_r - T_t)}{CJ\rho} \quad (7)$$

From the torque equilibrium of SHTB,

$$T_i + T_r = T_t \quad (8)$$

Combining Eqs. (7) and (8) leads to

$$\dot{\gamma}_S = \frac{2(T_i - T_t)}{CJ\rho}\frac{R_S}{L_S} \quad (9)$$

and, subsequently, the shear strain in the specimen can be evaluated as

$$\gamma_S = \int_0^t \frac{2(T_i - T_t)}{CJ\rho}\frac{R_S}{L_S}dt \quad (10)$$

where T_i and T_t are incident and transmitted torques, respectively, and can be calculated from the strain gauge outputs and the SHTB sensitivity.

Table 1: Comparison of Shear Strength (MPa)

$\dot{\gamma}$ (s^{-1})	5 x 10^{-4}	5 x 10^{-3}	5 x 10^2	1.5 x 10^3
1	24.8	38.8	89.3	101.6
2	23.4	37.1	96.8	108.8
3	29.2	37.6	98.3	111.9
4	29.4	38.5	91.7	106.6
Ave.	26.7	38.0	94.1	107.2

RESULTS AND DISCUSSION

A series of SHTB tests was performed to characterize the dynamic mechanical behaviours of 63Sn-37Pb solder. The experimental constitutive relations between the shear stress and the shear strain under two different shear strain rates are measured and presented in Figs. 6 and 7. In general, the experimental data are quite consistent. Although the difference in loading speed is only three fold, the effect of strain rate still can be clearly observed.

Table 1 summarizes the comparison of shear strengths with various shear strain rates. The data with two rather slow loading rates are obtained in a previous study with quasi-static torsional loading [13]. From this comparison, it is obvious that 63Sn-37Pb eutectic solder has substantial strain rate-dependent mechanical properties and the shear strength under dynamic loading is much higher than that subjected to quasi-static loading.

CONCLUSIONS

In the present study, an experimental work was performed to investigate the strain rate-dependent mechanical behaviour of 63Sn-37Pb eutectic solder under impulsive shear loading. SHTB tests were conducted to characterize the dynamic constitutive relation between the shear stress and the shear strain. Comparison was made among various cases with different shear strain rates. The results indicate that the mechanical behavior of 63Sn-37Pb solder is very sensitive to the strain rate and the dynamic shear strength is much higher than the static one.

ACKNOWLEDGMENTS

The Research Grant Council of Hong Kong sponsored this study through the grant of HKUST6047/99E to the Hong Kong University of Science and Technology (HKUST). The authors wish to acknowledge this support.

REFERENCES

[1] Hwang, J.S., *Modern Solder Technology for Competitive Electronics Manufacturing*, McGraw-Hill, New York, 1996.

[2] Lau, J.H., and Y.H. Pao, *Solder Joint Reliability of BGA, CSP, Flip Chip, and Fine Pitch SMT Assemblies*, McGraw-Hill, NY, 1997.

[3] Darveaux, R. and Banerji, K., 1992, "Constitutive Relations for Tin-Based Solder Joints," *IEEE Trans. Comp., Hybr. and Manuf. Tech.*, Vol. 6, pp. 1013-1024.

[4] Lee, S-W.R., and J.H. Lau, "Design for Plastic Ball Grid Array Solder Joint Reliability," *Circuit World*, Vol. 23, No. 2, 1997, pp. 11-14.

[5] Sandstrom, R. *et al.*, 1993, "Deformation Behavior During Low Cycle Fatigue Testing of 60Sn-40Pb Solders," *Mater. Sci. Tech.*, Vol. 9, pp. 811-819.

[6] Steinberg, D.S., *Vibration Analysis for Electronic Equipment*, John Wiley & Sons, New York, 1988.

[7] Leung, M.S., *Experimental Characterization of Constitutive Behaviours for Pb/Sn Solders under High Strain-Rate Loading*, Master Thesis, National Taiwan University, Taipei, Taiwan, 1998.

[8] Enke, N.F. *et al.*, 1989, "Mechanical Behavior of 60/40 Tin-Lead Solder Lap Joints," *IEEE Trans. Comp., Hybr. and Manuf. Tech.*, Vol. 12, pp. 459-468.

[9] Meyers, M.A., *Dynamic Behavior of Materials*, John Wiley & Sons, New York, 1994.

[10] Johnson, W., *Impact Strength of Materials*, Crane Russak, New York, 1972.

[11] Zukas, J.A. et al., *Impact Dynamics*, John Wiley & Sons, New York, 1982.

[12] Dai, L.H., Bai, Y.L. and Lee, S-W.R., "Experimental Investigation on the Shear Strength of Unidirectional Carbon/Aluminum Composite under Dynamic Torsional Loading," *Composites Science & Technology*, Vol. 58, No. 10, 1998, pp. 1667-1673.

[13] Dai, L.H., Lee, S.-W.R. and Huang, X., 2000, "Characterization of Strain Rate–Dependent Shear Response of 63Sn/37Pb Solder under Uniaxial Torsion," *Proc. 5th Asia-Pacific Symposium on Advances in Engineering Plasticity and Its Applications*, Hong Kong, pp. 267-272.

Proceedings of
2001 ASME International Mechanical Engineering Congress and Exposition
November 11–16, 2001, New York, NY
EPP-Vol. 1

IMECE2001/EPP-24731

FAILURE ANALYSIS OF MINIATURE SOLDER SPECIMEN

Y. Wei and C. L. Chow
Department of Mechanical Engineering
University of Michigan-Dearborn
Dearborn, Michigan 48128

H. E. Fang and M. K. Neilsen
Sandia National Laboratories
Albuquerque, NM 87185
T.J. Lim and W. Lu
Sandia National Laboratories
Livermore, CA 94550

ABSTRACT

The effects of specimen geometry size on the behavior of 63Sn-37Pb solder are investigated both experimentally in the laboratory and analytically with finite-element simulations. The simulations are achieved by developing a constitutive model for solder which couples viscoplasticity with a unified damage theory. The unified damage theory is characterized by a damage surface in strain space which separates fatigue damage from inelastic damage. The damage evolution equations are derived within the framework of irreversible thermodynamics. A series of uniaxial tension, tensile creep, and strain-controlled fatigue experiments are performed to obtain material parameters for the solder damage model. The solder damage model is then implemented into a finite element code and used to simulate a uniaxial tension test on a miniature specimen and on a standard ASTM specimen (ASTM standards, 1999). Predictions from these simulations are then compared with each other and with experimental results in order to examine specimen size effects.

INTRODUCTION

Solder alloys have been extensively used in the electronic industry. The conventional approaches for electronic package reliability are based on empirical relations from laboratory testing. However, testing for solder performance under service condition is very expensive and a time consuming process. Recently, several researchers have presented their work based on the theory of damage mechanics (Basaran, et al., 1998, Wei, et al., 1999, 2000a). Wei, et al. (1999) proposed a damage-coupled viscoplastic constitutive model taking into account the effects of change in grain or phase size and damage for solder material. In this paper, the model is extended by introducing the concept of a damage surface to distinguish between the types of damage that may accumulate during a complex loading history. The damage model is implemented in a finite element package, ABAQUS™, through its user defined material subroutine, UMAT. The damage coupled finite element analysis is applied to characterize monotonic tensile behavior of specimens with different sizes. The predicted results are compared with test data.

THEORY OF DAMAGE MECHANICS

Following the theory of damage mechanics, material degradation due to nucleation and growth of micro-defects can be characterized using an internal state variable **D** known as the damage variable. Then the effective stress tensor $\bar{\sigma}$ is introduced to include the effects of damage on material behavior, and can be used to substitute the Cauchy stress tensor σ in the development of constitutive equations of materials. In general, the effective stress tensor is defined as (Lemaitre and Chaboche, 1990)

$$\bar{\sigma} = \mathbf{M} : \sigma \tag{1}$$

where **D** is the damage variable, **M(D)** is the damage effect tensor. The success of a damage model to a large extent depends on the form of the damage effect tensor to effectively characterize material degradation. One conventional form was introduced with a scalar damage D variable by Kachanov (1958) as:

$$\mathbf{M(D)} = \frac{1}{1-D}\mathbf{I} \tag{2}$$

where **I** is the fourth-order symmetric identity tensor. However, this simple model fails to agree with the experimental observation that the values of Poisson's ratio of damaged materials vary with the increase in applied stress under tension loading (Chow and Wang, 1987). Accordingly, a two-scalar damage model was proposed to characterize changes in both

Young's modulus and Poisson's ratio for isotropic materials [Chow and Wei, 1991a, 1999]. The damage effect tensor is expressed in a matrix form as

$$\mathbf{M(D)} = \frac{1}{1-D}\begin{bmatrix} 1 & \mu & \mu & 0 & 0 & 0 \\ \mu & 1 & \mu & 0 & 0 & 0 \\ \mu & \mu & 1 & 0 & 0 & 0 \\ 0 & 0 & 0 & 1-\mu & 0 & 0 \\ 0 & 0 & 0 & 0 & 1-\mu & 0 \\ 0 & 0 & 0 & 0 & 0 & 1-\mu \end{bmatrix} \tag{3}$$

where D and μ are two scalar damage variables. The model has been successfully applied to predict the S-N curves for AL2024-T3, failure analyses under both monotonic and cyclic loading for notched specimens, and fatigue crack propagation (Chow, et al., 1991b, 1996, Wei, et al., 2000b).

DAMAGE EVOLUTION EQUATIONS

Based on the free energy equivalence principle: *the free energy for a damaged material may be expressed in a form similar to that for a material without damage except that all stresses (true stress and back stress) are replaced by their corresponding effective stresses* (Wei, et al., 1999), the free energy for damaged solder materials is expressed as

$$\Psi = \frac{1}{2}\sigma^T : \mathbf{C}^{-1} : \sigma + \frac{3}{4}C_k^{-1}\mathbf{X}^T : \mathbf{X} \tag{4}$$

where **C** is the effective isotropic, elastic tensor taking into account the effects of damage for a damaged material,

$$\mathbf{C}^{-1} = \frac{1}{E}\begin{bmatrix} 1 & -\nu & -\nu & 0 & 0 & 0 \\ -\nu & 1 & -\nu & 0 & 0 & 0 \\ -\nu & -\nu & 1 & 0 & 0 & 0 \\ 0 & 0 & 0 & 2(1+\nu) & 0 & 0 \\ 0 & 0 & 0 & 0 & 2(1+\nu) & 0 \\ 0 & 0 & 0 & 0 & 0 & 2(1+\nu) \end{bmatrix} \tag{5}$$

E and ν are the Young's modulus and the Poisson's ratio of a damaged material, and can be expressed in terms of the damage variables D and μ as (Chow and Wei 1991a)

$$E = \frac{E_0(1-D)^2}{1-4\nu_0\mu+2(1-\nu_0)\mu^2}$$

$$\nu = \frac{\nu_0-2(1-\nu_0)\mu-(1-3\nu_0)\mu^2}{1-4\nu_0\mu+2(1-\nu_0)\mu^2} \tag{6}$$

where E_0 and ν_0 are the Young's modulus and the Poisson's ratio for undamaged or as-received material. **X** is the back stress tensor. C_k in Eqn. (4), known as a kinematic hardening variable for a damaged material, is defined as:

$$C_k^{-1} = \frac{(1-\mu)^2}{(1-D)^2}C_{k0}^{-1} \tag{7}$$

where C_{k0} is the as-received material parameter for kinematic hardening. The damaged-coupled elastic equation is derived as

$$\varepsilon^e = \frac{\partial\Psi}{\partial\sigma} = \mathbf{C}^{-1} : \sigma \tag{8}$$

and the kinematic hardening equation is deduced as

$$\alpha_k = \frac{\partial\Psi}{\partial\mathbf{X}} = \frac{3}{2}C_k^{-1}\mathbf{X} \tag{9}$$

where ε^e is the elastic strain and α_k is the back strain. The damage energy release rates corresponding to the damage variables D and μ are derived using the theory of irreversible thermodynamics as

$$Y_D = -\frac{\partial\Psi}{\partial D} = -\frac{1}{1-D}\sigma^T : \mathbf{C}^{-1} : \sigma - \frac{3}{2}\frac{C_k^{-1}}{1-D}\mathbf{X}^T : \mathbf{X} \tag{10}$$

$$Y_\mu = -\frac{\partial\Psi}{\partial\mu} = -\frac{1}{1-D}\sigma^T : \mathbf{Z} : \sigma + \frac{3}{2}\frac{C_k^{-1}}{1-\mu}\mathbf{X}^T : \mathbf{X} \tag{11}$$

Accordingly, the equivalent inelastic damage energy release rate, Y_d, is established as

$$Y_d = \left[\frac{1}{2}(Y_D^2 + \gamma Y_\mu^2)\right]^{1/2} \tag{12}$$

where γ is a material constant. The form of the fourth-order tensor **Z** in equation (11) is derived as

$$\mathbf{Z} = \frac{1}{E_0(1-D)} \begin{bmatrix} z_1 & z_2 & z_2 & 0 & 0 & 0 \\ z_2 & z_1 & z_2 & 0 & 0 & 0 \\ z_2 & z_2 & z_1 & 0 & 0 & 0 \\ 0 & 0 & 0 & 2(z_1 - z_2) & 0 & 0 \\ 0 & 0 & 0 & 0 & 2(z_1 - z_2) & 0 \\ 0 & 0 & 0 & 0 & 0 & 2(z_1 - z_2) \end{bmatrix}$$

$$z_1 = 2\mu(1-\nu_0) - 2\nu_0 \qquad z_2 = (1+\mu)(1-\nu_0) - 2\mu\nu_0 \tag{13}$$

In order to establish damage evolution equations, a damage dissipation potential function ϕ^d is introduced as

$$\phi^d = \frac{Y_h}{B_1+1}\left(\frac{Y_d}{Y_h}\right)^{B_1+1} \tag{14}$$

where B_1 is the damage-related material constant, and Y_h is the damage hardening variable defined in equation (20). Accordingly, the damage evolution equations can be derived as (Wei, et al., 1999, 2000a)

$$\dot{D} = -\beta_{in}\frac{\partial \phi^d}{\partial Y_D} = -\dot{w}\frac{Y_D}{2Y_d}$$
$$\dot{\mu} = -\beta_{in}\frac{\partial \phi^d}{\partial Y_\mu} = -\dot{w}\frac{\gamma Y_\mu}{2Y_d} \tag{15}$$

where β_{in} is a multiplier, w is the equivalent damage variable.

A damage surface in strain space is introduced to characterize the type of damage that is being generated. The damage surface is mathematically described by

$$F_d = p^{in} - p^{in}_{max} = 0 \tag{16}$$

where p^{in} is current equivalent inelastic strain, p^{in}_{max} is maximum equivalent inelastic strain in previous loading history. The equivalent inelastic strain is defined as

$$p^{in} = \sqrt{\frac{2}{3}\varepsilon^{in} : \varepsilon^{in}} \tag{17}$$

ε^{in} is the inelastic strain tensor. The damage surface bounds the regime of fatigue damage. This is analogous to a yield surface in plasticity theory which bounds the regime of elastic behavior. When the inelastic strain is inside the damage surface, fatigue damage is accumulated. When the inelastic strain is on the damage surface and trying to enlarge the damage surface, inelastic damage is generated.

Accordingly, the equivalent damage evolution in equation (15) can be written in a unified form as

$$\dot{w} = \begin{cases} \dot{p}^{in}\left(\dfrac{Y_d}{Y_h}\right)^{B_1}, & \text{if } F_d = 0 \text{ and } \dfrac{\partial F_d}{\partial \varepsilon^{in}} : \dot{\varepsilon}^{in} > 0 \\ \dot{p}^{in}\dfrac{Y_d}{Y_{hf}}, & \text{if } F_d < 0 \text{ or } F_d = 0 \text{ and } \dfrac{\partial F_d}{\partial \varepsilon^{in}} : \dot{\varepsilon}^{in} \le 0 \end{cases} \tag{18}$$

The different damage hardening variables are given as

$$Y_h = Y_0\, e^{B_2 w + \frac{B_3}{T}}$$
$$Y_{hf} = Y_{0f}\, e^{\frac{B_3}{T}} \tag{19}$$

where Y_0, Y_{0f}, B_1, B_2 and B_3 are damage-related material constants, $\dot{p}^{in}$ is the equivalent inelastic strain rate, T is the absolute temperature.

A failure criterion is proposed using the concept of equivalent damage accumulation: *a material element is said to have ruptured when the equivalent damage w in the element reaches a critical value w_c, a material constant.*

For inelastic deformation, a constitutive model has been established to include the effects of grain/phase coarsening and damage accumulation (Wei, et al., 1999). The inelastic strain rate can be expressed as

$$\dot{\varepsilon}^{in} = \dot{p}^{in}\frac{3}{2}\frac{\mathbf{S}-\mathbf{X}}{J_2} \tag{20}$$

where J_2 is a second invariant of the stress difference and defined as

$$J_2 = \left\{\frac{3}{2}(\mathbf{S}-\mathbf{X})^T : (\mathbf{S}-\mathbf{X})\right\}^{\frac{1}{2}} \tag{21}$$

S is the deviatoric stress. The equivalent inelastic strain rate $\dot{p}^{in}$ is expressed with the consideration of damage accumulation and grain/phase size coarsening as (Frear, et al., 1997, Wei, et al., 2000a)

$$\dot{p}^{in} = \frac{1-\mu}{1-D} f \exp\left(\frac{-Q}{RT}\right)\left(\frac{\lambda_0}{\lambda}\right)^p \sinh^m\left(\frac{1-\mu}{1-D}\frac{J_2}{\alpha(c+\hat{c})}\right) \tag{22}$$

where f, p, m and Q are material parameters, R is the gas constant, T is the absolute temperature, λ is the current grain diameter, λ_0 is the initial grain diameter, α is a scalar function of the absolute temperature, c and $\widehat{c}$ are state variables. The scalar state variable *c* is given by (Frear, et al., 1997)

$$\dot{c} = A_1\dot{p}^{in} - (A_2\dot{p}^{in} + A_3)(c - c_0)^2 \tag{23}$$

The back strain tensor is expressed as

$$\dot{\alpha}_k = \dot{\varepsilon}^{in} - C_{k0}(A_5\dot{p}^{in} + A_6)\alpha_k\sqrt{\frac{2}{3}\alpha_k{}^T : \alpha_k} \tag{24}$$

The state variable $\hat{c}$ is related to the grain/phase size λ by

$$\hat{c} = A_7\left(\frac{\lambda_0}{\lambda}\right)^{A_8} \tag{25}$$

where c_0, A_1, A_2, A_3, C_{k0}, A_5, A_6, A_7 and A_8 are material parameters. The evolution of the grain/phase size for 63Sn-37Pb is given by Vianco et al.'s (1999) experiments which generated the following equation

$$\dot{\lambda} = \frac{1.05\times10^{-5}e^{-11023/T} + 4.00\times10^{-8}e^{-3123/T}\dot{p}^{in}}{(\lambda - \lambda_0)^{2.9}} \tag{26}$$

NUMERICAL SIMULATION

Although the constitutive equations of damaged material are more complex than those of a conventional model without damage consideration, the existing finite element method can be conveniently modified to include the effects of damage accumulation. The modification is accomplished by introducing an iteration procedure of the damage-coupled viscoplastic constitutive equations through a new material stiffness matrix incorporating the effects of damage accumulation and other internal state variables. Accordingly, the proposed constitutive model is implemented in ABAQUS™ (version 5.8) through its user-defined material subroutine UMAT. A semi-implicit time-integration scheme has been adopted for finite element analyses (Fu, et al., 1998).

Several uniaxial tests under different loading conditions, i.e. monotonic tension, tensile creep and strain-controlled fatigue, were carried out to determine the material constants for the 63Sn-37Pb solder material. The measurement for Young's modulus and Poisson's ratio were made at certain strain intervals until final rupture. The measured data were used to evaluate the damage parameters required in the model. The material parameters determined for the 63Sn-37Pb solder material are summarized in Table 1 and 2.

TABLE 1 63Sn-37Pb solder material parameters (temperature independent)

Poisson's ratio	0.4
Young's modulus (GPa)	33.26
A_1 (MPa)	0.00E+00
A_2 (1/MPa)	0.00E+00
A_3 (1/MPa-sec)	0.00E+00
A_7 (MPa)	5.66
A_8	0.5
Flow rate f (1/sec)	1.802E+06
Sinh exponent m	3.04
Growth exponent p	3.00
Flow stress C_0 (MPa)	2.83
Phase size λ_0 (mm)	2.257E-03
Activation energy Q (cal/mol)	1.376E+4
Gas constant R (cal/mol-K)	1.987
Scalar function α	1.0
Damage constant B_1	0.4
Damage constant B_2	5.0
Damage constant B_3 (K)	3.34E+03
Damage constant γ	-0.2
Damage constant w_c	0.58
Damage constant Y_0 (MPa)	6.0E-08
Damage constant Y_{0f} (MPa)	5.6E-06

TABLE 2. 63Sn-37Pb solder material parameters (temperature dependent)

Temperature (^{0}C)	25	75	100
Temperature (^{0}K)	298	348	373
C_{k0} (GPa)	28.15	22.08	19.32
A_5 (1/MPa)	44.95	103.0	153.7
A_6 (1/MPa-sec)	8.42E-03	5.95E-02	0.2

APPLICATIONS

The proposed model was used to predict monotonic tensile behavior of 63Sn-37Pn solder geometry manufactured in accordance with the ASTM standard. The cross section of the specimen is a rectangular shape of 5mm×10mm and the gauge length is 30mm. The loading is applied with a true strain rate of 10^{-4}/s at 25^{0}C and 75^{0}C. The predicted true stress-true strain curves are compared with the test results shown in Fig.1. The softening behavior of 63Sn-37Pb solder material observed from the tests can be successfully characterized with the damage-coupled constitutive equations. It is evident from the figure that the damage accumulation contributes to the reduction in loading capability.

The size effect of the miniature specimen and microstructure were next examined. The dimensions of the specimens chosen for the investigation are depicted in Fig.2. Two testing specimens with different aging treatments, namely ps_a7_03_16h (16 hours in 100^0C) and ps_a7_03_25d (25 days in 100^0C), were employed. The gage diameter chosen is 0.354 mm and the gage length is 2.58 mm. The curves describing the testing and average loading rate for the whole process are shown in Fig.3. For ease of computation, an average loading curve was applied for the numerical simulations.

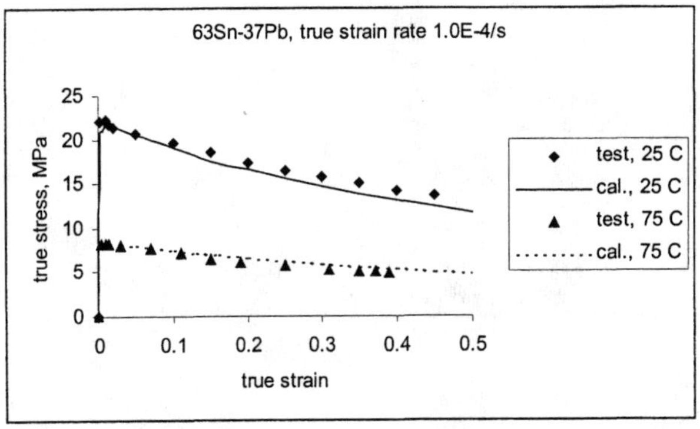

Fig.1 Softening behavior under monotonic tension at strain rate 10^{-4}/s

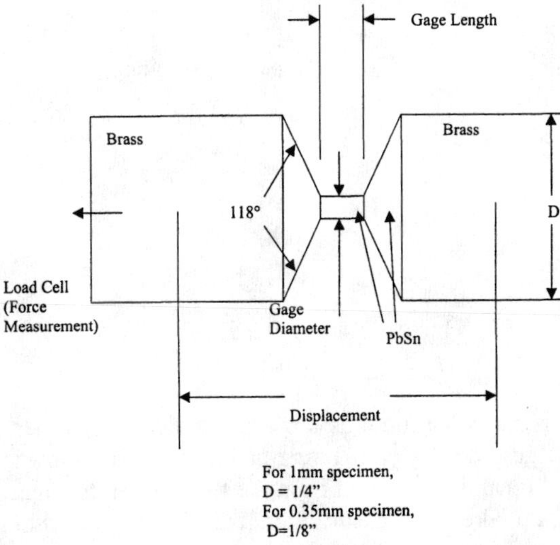

Fig. 2 Small-size specimen (gage diameter: 0.354 mm, gage length: 2.58 mm)

Since both specimen geometry and applied load depicted in Fig. 2 are axisymmetric, a two-dimensional, axisymmetric finite element model was used for the FEA analysis. Numerical simulations were carried out with ABAQUS^{TM} (version 5.8). The loading curve was simulated with a displacement boundary condition. Two axisymmetric solid elements CGAX4R (4-node bilinear, reduced integration with hourglass control) and CGAX4 (4-node bilinear, full integration) were applied. Three different element sizes, namely 0.01mm, 0.02mm and 0.036mm, were used for the straight portion of the specimen. The numerical results are found to be almost the same for different combinations. The load-displacement curve in axial direction was calculated and compared with test results shown in Fig.4. It can be observed from the figure that the predicted results are found to be 10% lower than the test data at the peak load, but are closer at larger displacements. The discrepancy in the prediction may be attributed in part to the size effects of the different test specimens and of the microstructures between the miniature and the conventional ASTM specimens. Additional experimentation will be carried out to examine more closely the size effects of the miniature specimen and real solder joints.

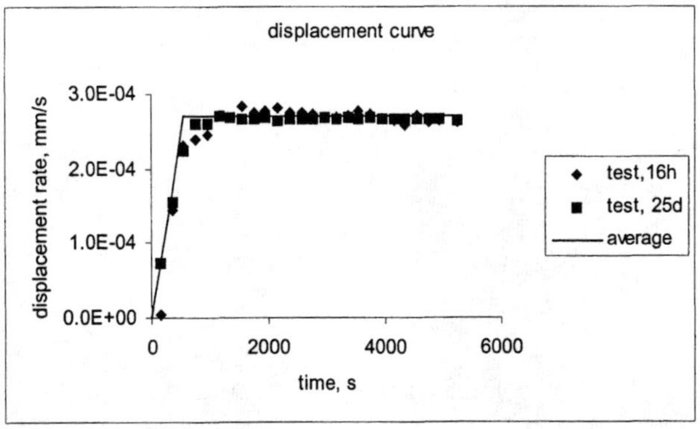

Fig.3 Loading rate curves for miniature specimen

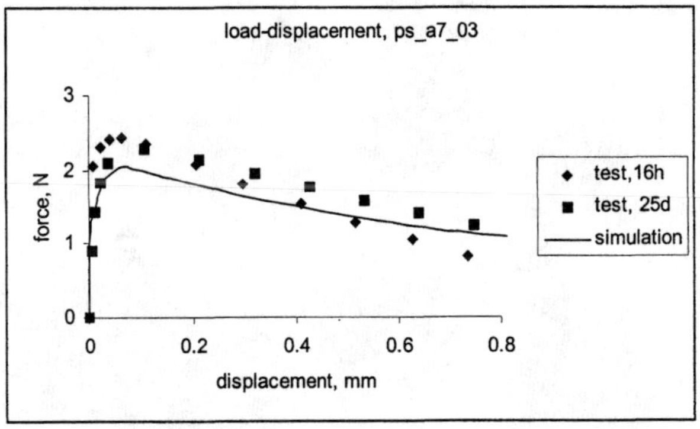

Fig.4 Load-displacement curves for miniature specimen

Fig. 5 depicts the undeformed elements of the specimen. Slight necking can be observed at the region shown in Fig.6 with deformed elements at the displacement of 1.3 mm.

The inelastic strain rate distributions were calculated at different applied displacements, namely 0.283 mm and 1.3 mm, as shown in Figs.7-8. It can be observed from the figures that there is a slight difference in the inelastic strain rates at the straight portion of the specimen (about 1.0E-4 /s) upon the initial load application. The largest increase in the inelastic strain rate is found to be confined within a small region, despite continued increase in applied displacements. In the straight region outside the confined one, the inelastic strain rates decrease from the beginning 1.0E-4 /s to the final about 7.0E-5 /s. Finally, the necking is developed at the region where the highest inelastic strain rate occurs.

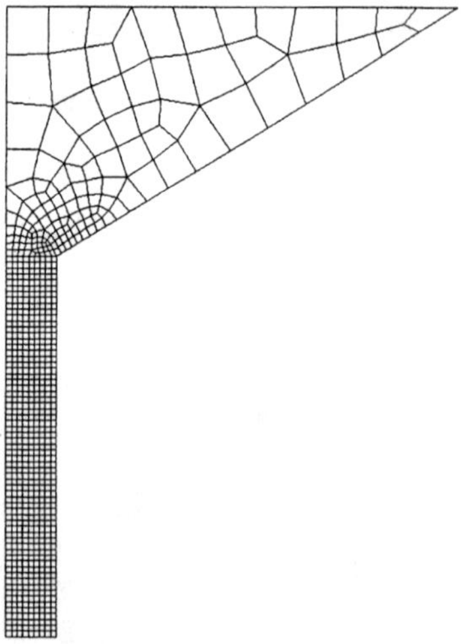

Fig.5 Undeformed elements

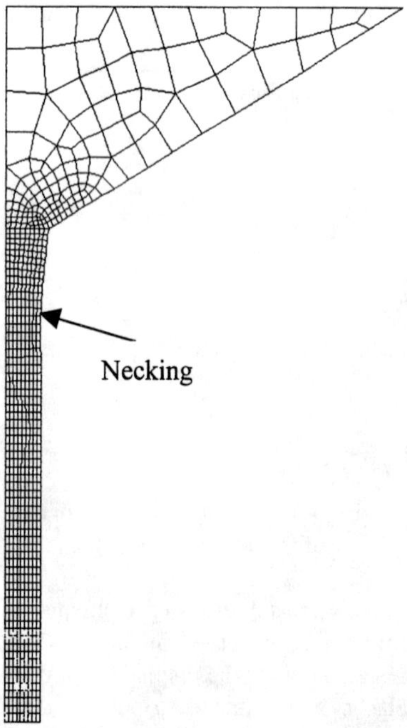

Fig.6 Deformed elements at displacement 1.3 mm

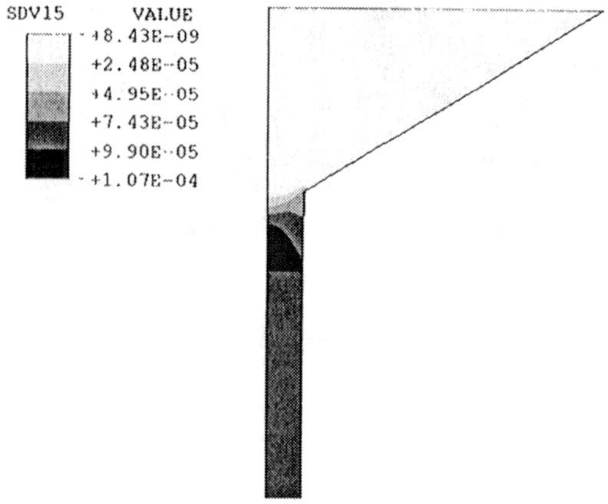

Fig.7 Inelastic strain rate distribution at displacement 0.283 mm

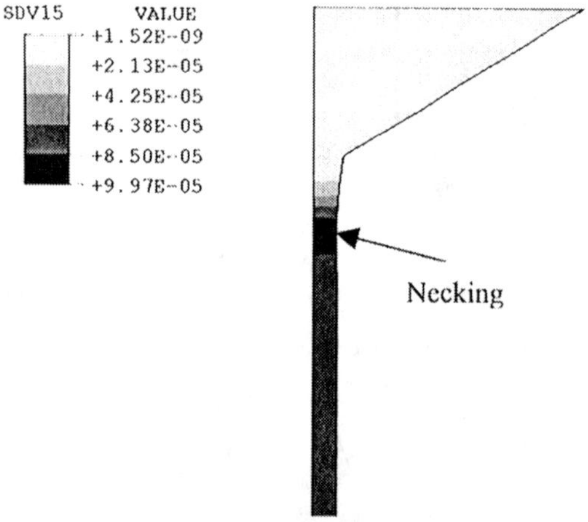

Fig.8 Inelastic strain rate distribution at displacement 1.3 mm

The damage distributions are also predicted for different loading stages, namely 0.283 mm and 1.3 mm, as shown in Figs.9-10. It can be observed from the figures that the damage accumulation, like the inelastic strain rates, is confined within a small region where the accumulated damage is found to reach its critical value. Consequently, the specimen is expected to fail at this location.

CONCLUSIONS

The damage-coupled constitutive model has been developed and shown to successfully characterize the behavior of 63Sn-37Pb solder specimen manufactured with the conventional ASTM standard size under monotonic tension conditions, especially for softening characterization. The model is also capable of quantifying damage accumulation and predicting failure pattern/location in a miniature component.

Further research is in progress to examine the applicability of the proposed model to include the effects of specimen size and material microstructure under different loading conditions.

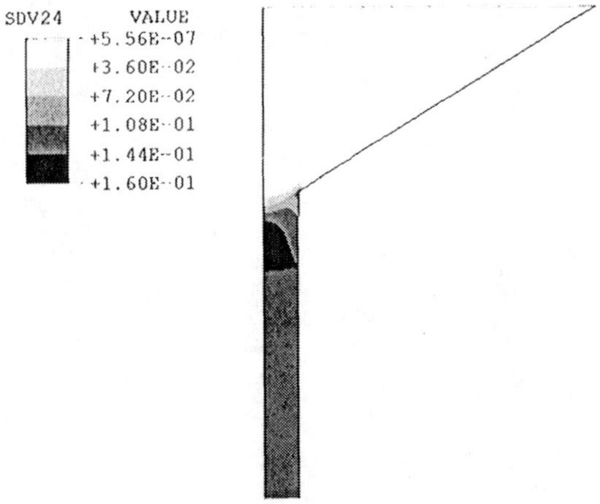

Fig.9 Equivalent damage distribution at displacement 0.283 mm

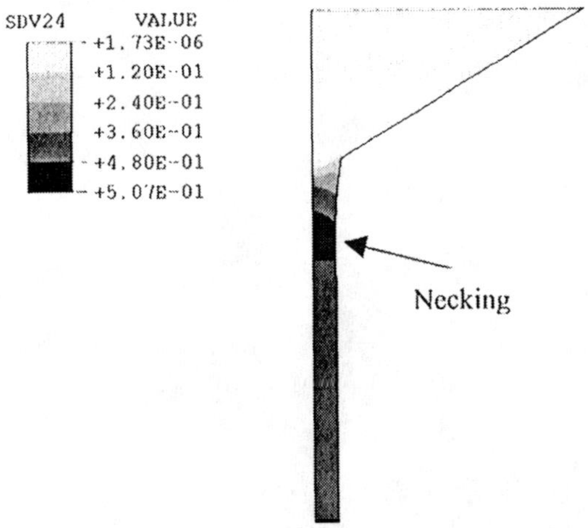

Fig.10 Equivalent damage distribution at displacement 1.3 mm

ACKNOWLEDGMENTS

This work was supported by the United States Department of Energy under Contract DE-AC04-94AL85000. Sandia is a multiprogram laboratory operated by Sandia Corporation, a Lockheed Martin Company, for the United States Department of Energy.

REFERENCES

ASTM standards, 1999, "Tension testing of metallic materials", ASTM E8-78.

Basaran, C. and Yan, C.Y., 1998, "A thermodynamic framework for damage mechanics of solder joints", Journal of Electronic Packaging, Vol.120, pp.379-384.

Chow, C. L. and Wang, J. 1987. "An anisotropic theory of continuum damage mechanics for ductile fracture", Engng. Fract. Mech., Vol.27, pp.547-558.

Chow, C. L. and Wei, Y., 1991a. "A model of continuum damage mechanics for fatigue fracture", Int. J. of Fracture, Vol.50, pp.301-316.

Chow, C.L. and Wei, Y., 1991b, "A damage mechanics model of fatigue crack initiation in notched plates", Theoretical and Applied Fracture Mechanics, Vol.16, pp.123-133.

Chow, C.L. and Wei, Y., 1996, "A fatigue damage model for crack propagation", Advances in Fatigue Lifetime Predictive Techniques, ASTM STP 1292, pp.86-99.

Chow, C.L. and Wei, Y., 1999, "Constitutive modeling of material damage for fatigue failure prediction", Int. J. of Damage Mechanics, Vol.8, pp.355-375.

Frear, D. R., Burchett, S. N., Neilsen, M. K. and Stephens, J. J., 1997, Microstructurally based finite element simulation of solder joint behaviour, Soldering & Surface Mount Technology Vol.2, pp.39-42.

Fu, C., McDowell, D.L. and Ume, I.C., 1998, "A finite element procedure of a cyclic thermoviscoplasticity model for solder and copper interconnects", Journal of Electronic Packaging, Vol.120, pp.24-34.

Kachanov, L. M., 1958. "On creep rupture time", Izv. Acad. Nauk SSSR, Otd. Techn. Nauk, Vol.8, pp.26-31.

Lemaitre, J. and Chaboche, J.L., 1990, Mechanics of Solid Materials, Cambridge University Press.

Vianco, P.T., Burchett, S.N., Neilsen, M.K., Rejent, J.A., and Frear, D.R., 1999, "Coarsening of the Sn-Pb Solder Microstructure in Constitutive Model Based Prediction of Solder Joint Thermal Mechanical Fatigue," Journal of Electronic Materials, Vol. 28, No. 11, pp. 1290-1298.

Wei, Y., Chow, C.L., Fang, H.E. and Neilsen, M.K., 1999, "Characteristics of creep damage for 60Sn-40Pb solder material", 99-IMECE/EEP-15.

Wei, Y., Chow, C.L., Fang, H.E. and Neilsen, M.K., 2000a, "Constitutive modeling of viscoplastic damage in solder material", IUTAM Symposium on Creep in Structures, KLUWER ACADEMIC PUBLISHERS, pp.131-140.

Wei, Y., Chow, C.L. and Liu, C.T., 2000b, "Damage analysis for mixed mode crack initiation", Computer Modeling in Engineering & Sciences, Vol.1, pp.115-121.

Proceedings of
2001 ASME International Mechanical Engineering Congress and Exposition
November 11–16, 2001, New York, NY
EPP-Vol. 1

IMECE2001/EPP-24732

NOVEL EXPERIMENTAL-COMPUTATIONAL METHOD FOR QUANTITATIVE APPLICATIONS IN ELECTRONIC PACKAGING

Cosme Furlong and **Ryszard J. Pryputniewicz**
NEST – NanoEngineering, Science, and Technology
CHSLT – Center for Holographic Studies and Laser micro-mechaTronics
Mechanical Engineering Department
Worcester Polytechnic Institute
Worcester, MA 01609

ABSTRACT

Rapid advances in microelectronics require design and optimization of components and packages, for new and ever more demanding applications, in relatively short periods of time while satisfying electrical, thermal, and mechanical specifications, as well as cost and manufacturability expectations. In addition, reliability and durability have to be taken into consideration. As a consequence, some of the most sophisticated analytical, computational, and experimental methods are being used for development, optimization, and quantitative characterization of electronic packages. In this paper, a novel experimental-computational method, based on combined use of recent advances in laser-based optics and computational modeling, is described and its application is demonstrated by case studies of microelectronic components subjected to electro-thermo-mechanical loads. Results of these studies show that this methodology provides an effective engineering tool for nondestructive testing (NDT) applications in electronic packaging and provides indispensable quantitative data for development, optimization, and applications in electronic packaging.

Keywords: experimental-computational method, nondestructive testing, electro-thermo-mechanical deformations, electronic packaging, optoelectronic holography.

1. INTRODUCTION

Continued advances in packaging of electronic circuits require miniaturization of components and their denser placements on electronic printed circuit boards (PCBs) [1-3]. A PCB (categorized as the second level of the electronic packaging hierarchy) may contain a large number of electronic components (which are classified as the first level of electronic packaging) made of materials with different characteristics and properties and, therefore, each of the components behaving differently under specific loading conditions.

The PCB, as well as the individual electronic components assembled on it, is subjected to electromagnetic, electrical, thermal, and/or mechanical loads, which may generate failure of an electronic package. This failure can be due to any individual or a combination of the following mechanisms: fatigue, creep, stress relaxation, stress concentration, and/or bonding fractures [4]. Reliability of an electronic package depends on the correct functionality, under operating conditions, of the components at each electronic packaging level, starting with the first level. Therefore, due to the quantity of components assembled in an electronic package and the complexity of possible failure mechanisms, reliability assessment requires the application of effective quantitative methodologies [5]. These quantitative methodologies comprise the computational and experimental techniques [6-8].

An effective study and characterization of electro-mechanical components should include the use of both computational and experimental methodologies. Computational investigations enable parametric studies and the determination of critical engineering conditions, while experimental investigations, especially optical, provide quantitative information on the actual response of the component of interest to the applied load conditions [8].

Although the new microelectronic packages offer reduced size and increased functionality, thermal management issues do not diminish, but may become an unsurpassable challenge, unless concurrently considered at the beginning of the product development process.

In this paper, we describe a novel approach, based on computational and experimental methods, to study thermomechanical effects in microelectronics and illustrate its use by representative applications.

2. EXPERIMENTAL-COMPUTATIONAL APPROACH

The experimental-computational approach used in this study consists of a computational method and experimental method [9].

2.1. Computational method

The computational methodology used in this study is based on finite element method (FEM). This section reviews some of the fundamental relationships used in FEM modeling and simulation.

An initial investigation consists of studying a linear, steady state loading condition. Therefore, computationally, it is necessary to solve the electric potential equation

$$\nabla \cdot (\boldsymbol{\kappa} \nabla V) = 0 \ , \tag{1}$$

where $\boldsymbol{\kappa}$ is the electrical conductivity of the material and V is the spatially varying electric potential. Solution of Eq. 1 allows evaluation of heat generated Q,

$$Q = \kappa |\nabla V|^2 \ , \tag{2}$$

and determination of spatially dependent temperature field T by solving the thermal equilibrium equation

$$\nabla \cdot (k \nabla T) + Q = 0 \tag{3}$$

in which k is the thermal conductivity of the material. Solution of Eq. 3 can be applied to the strain-displacement elasticity equations and to the solution of the stress-strain equilibrium equation

$$\sigma = E(\varepsilon - \alpha \Delta T) \ , \tag{4}$$

where σ is the thermo-mechanical stress tensor, E is the modulus of elasticity of the material, ε is the mechanical strain tensor, α is the coefficient of thermal expansion of the material, and ΔT is the temperature difference between the temperature obtained by solving Eq. 3 and an ambient temperature T_a.

Solutions of Eqs 1 to 4 can be obtained by FEM [8] as illustrated by representative results shown in Section 4.

2.2. Experimental method

The experimental method used in this study is based on recent developments in optoelectronic holography (OEH) methodology for nondestructive (NDT) measurements of deformations and shapes of objects [8]. Figure 1 shows a typical fiber optic based OEH system used for quantitative NDT and shape measurement investigations based on single-illumination and single-observation geometry.

In the OEH system shown in Fig. 1, laser light is coupled into a single fiber optic by means of a microscope objective (MO). Then, the single fiber is divided into two parts by means of a directional coupler (DC). One of the resulting parts is used to illuminate the object, while the other part provides reference against which the signals from the object are recorded. Both, the object and reference beams are combined by the interferometer (IT) and recorded by the system camera (CCD).

The images recorded by the CCD are processed to determine the fringe-locus function, Ω, constant values of which define fringe loci on the surface of object under investigation [8-10]. The values of Ω relate to the system geometry and the unknown deformations via the relationship

$$\Omega = (\mathbf{K}_2 - \mathbf{K}_1) \cdot \mathbf{L} = \mathbf{K} \cdot \mathbf{L} \ , \tag{5}$$

where $\mathbf{K}$ is the sensitivity vector defined in terms of vectors $\mathbf{K}_1$ and $\mathbf{K}_2$ identifying directions of illumination and observation in the OEH system, Fig. 1, and $\mathbf{L}$ is the displacement vector.

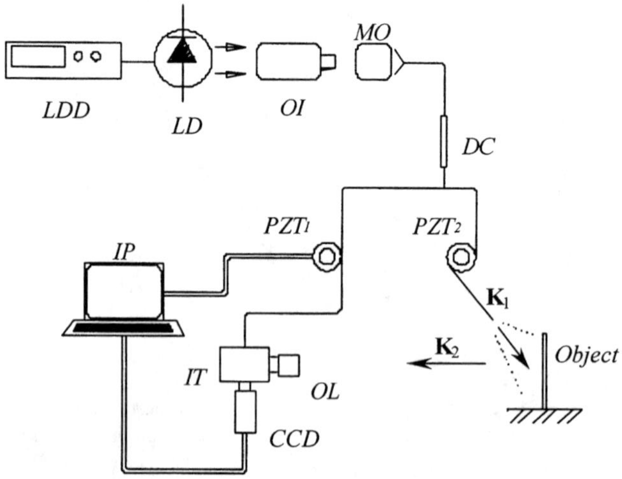

Fig. 1. Single-illumination and single-observation geometry of a fiber optic based OEH system for applications in electronic packaging: *LDD* is the laser diode driver, *LD* is the laser diode, *OI* is the optical isolator, *MO* is the microscope objective, *DC* is the fiber optic directional coupler, PZT_1 and PZT_2 are the piezoelectric fiber optic modulators, *IP* is the image processing computer, *IT* is the interferometer, *OL* is the objective lens, *CCD* is the camera, while $\mathbf{K}_1$ and $\mathbf{K}_2$ are the directions of illumination and observation, respectively.

Quantitative determination of structural deformations due to thermal fields induced by electric loads can be obtained, by solving a system of equations similar to Eq. 5, to yield [10-12]

$$\mathbf{L} = \mathbf{L}(\mathbf{K}, \Omega) \ . \tag{6}$$

Representative results obtained by solution of Eq. 6 are discussed in Section 4.

3. TEST SAMPLES

The second level of electronic packaging depends on attachment of components to a board, Fig. 2. These attachments can be of either of the following two types: (1) through the hole, or (2) surface mount. In this study, components mounted using both types of attachments were considered.

To illustrate applicability of the novel experimental-computational method, two tests samples are used in this paper: (1) a simple through the hole mounted cylindrical carbon resistor, Fig. 3, and a more complex collision avoidance radar (CAR) system, Fig. 4.

More specifically, in the cylindrical 3-Ohm ± 3 % carbon resistors, Fig. 3, the carbon/silica core is encapsulated in a silica shell package and two copper/lead leads are molded-in in order to provide connections/attachments to the board. The resistor has nominal dimensions of 3.48 mm for the diameter and 10 mm for the length.

Fig. 2. The second level electronic packaging configurations.

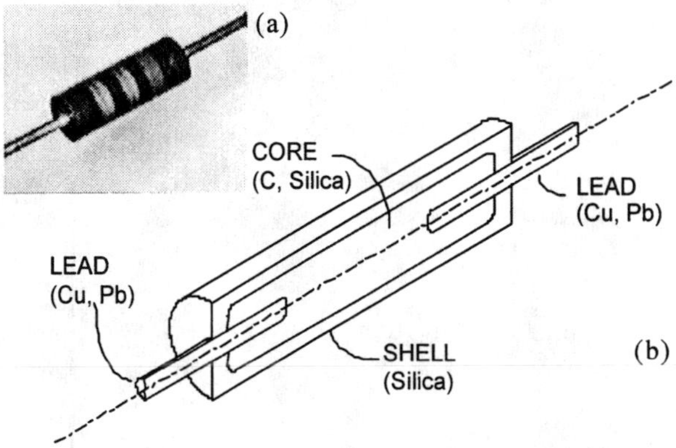

Fig. 3. Electronic carbon resistor of interest: (a) actual package, and (b) cross sectional view.

An electrical resistor is used in an electronic circuit in order to provide a specific voltage difference. This voltage difference induces heat in the resistor and, therefore, temperature changes that produce mechanical strains and thus mechanical stresses, which may produce undesirable mechanical failure. The behavior of the resistor can be analyzed by considering electromagnetic, thermal, and mechanical stress-strain effects, as discussed in Section 2.1. Representative results showing experimentally and computationally obtained deformations of the resistor are presented in Section 4.

The CAR system shown in Fig. 4 was used to demonstrate capability of the OEH method to measure temporal changes in spatial deformation fields of complex electronic packages. Typical OEH results obtained on the CAR system are shown in Section 4.

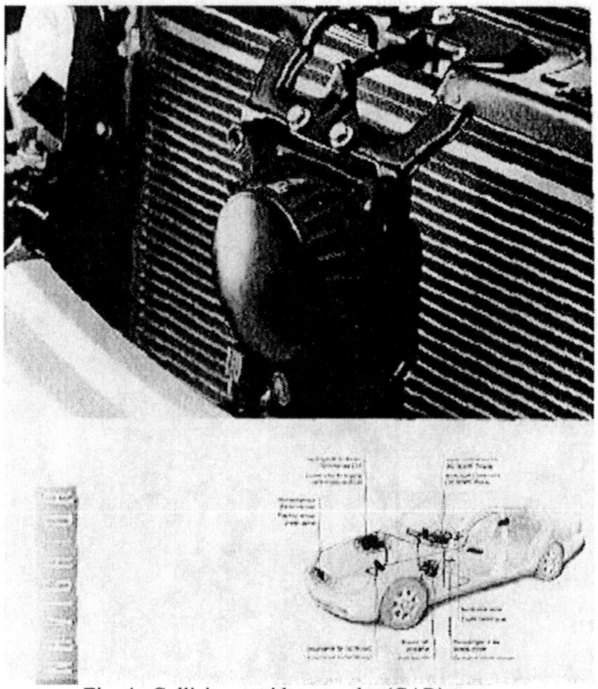

Fig. 4. Collision avoidance radar (CAR) system.

4. RESULTS AND DISCUSSION

Figure 5 shows an image of a cylindrical resistor without any load. Representative interferograms recorded during loading of the resistors in real-time are shown in Fig. 6. The fewer number of interference fringes observed in Fig. 6a than in Fig. 6b clearly indicates that the resistors behave differently under the same loading conditions. Rapid visualizations of deformations, such as those displayed in Fig. 6, are one of particular features of OEH methodology as it applies to studies of thermomechanical deformations of electronic components.

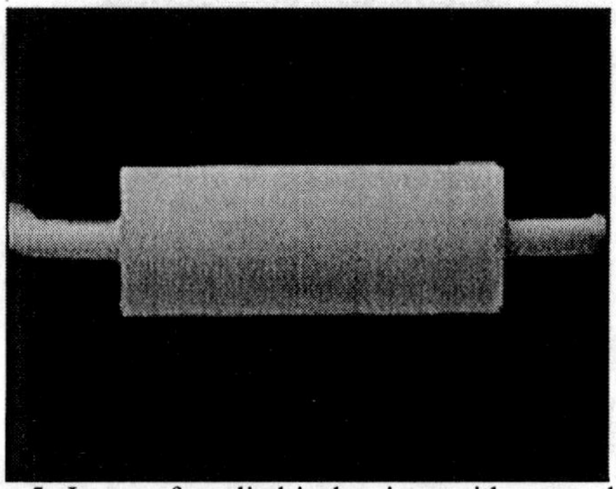

Fig. 5. Image of a cylindrical resistor without any load.

Assuming symmetry with boundary and load conditions similar to operating conditions used during experiments and using published properties of the materials of the resistors studied herein, FEM was used to compute deformations of the resistors.

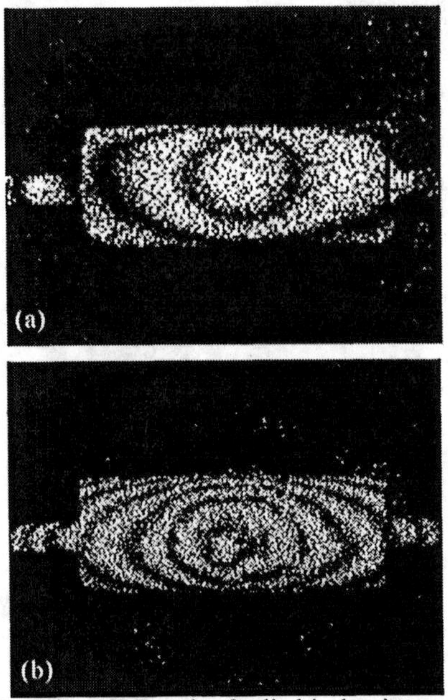

Fig. 6. Real-time NDT study of cylindrical resistors subjected to the same loads of 2 V_{pp} at 1 kHz, using the OEH method: (a) interferogram of a resistor in good condition, (b) interferogram of a resistor in damaged condition.

There are a number of ways that FEM computed results can be displayed. In this study, to facilitate direct comparison of the FEM results with the OEH results, computational algorithms for simulation of interference fringes based on FEM results were used [8]. Representative fringe pattern based on FEM computations, using the published material properties as well as the same boundary and loading conditions as those under which the OEH measurements were made, is shown in Fig. 7.

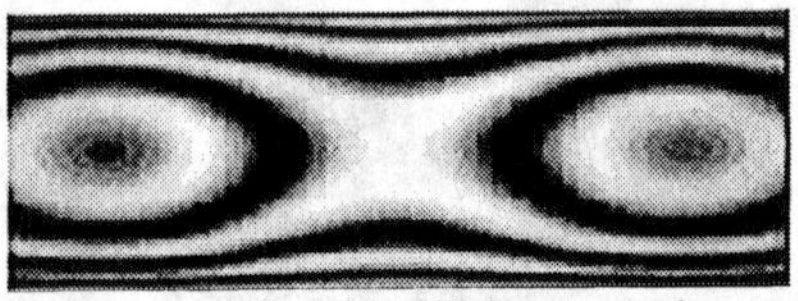

Fig. 7. Fringe pattern based in FEM computations, using *published* material properties.

By comparing the measured and computed, Figs 6a and 7, respectively, fringe patterns, differences between them are noticed. Clearly, input parameters that FEM computations were based on must be revised. Since the boundary and loading conditions represented actual experimental conditions, the only parameters that were questionable were the material properties, because they were based on published (general) data.

To overcome the above difficulty, material properties used in the FEM solution were measured in our laboratory. Then, the FEM computations were performed again with the representative results shown in Fig. 8. Now, these results are drastically different than the results of Fig. 7. Moreover, the new FEM results, based on the measured material properties, compare well with the measured deformations, shown in Fig. 6a. In fact, based on Fig. 6a, actual electro-thermo-mechanical deformations, as a function of position on the outside surface of the resistor, are shown in Fig. 9.

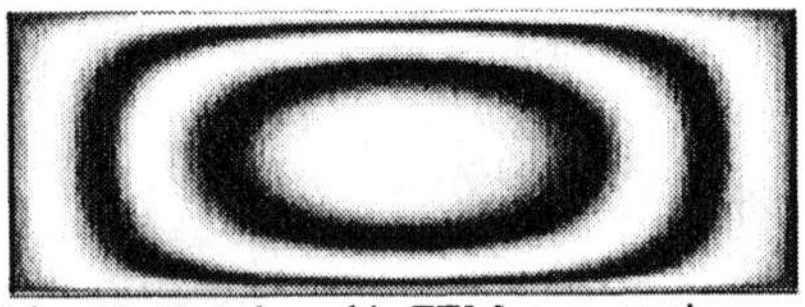

Fig. 8. Fringe pattern based in FEM computations, using *measured* material properties.

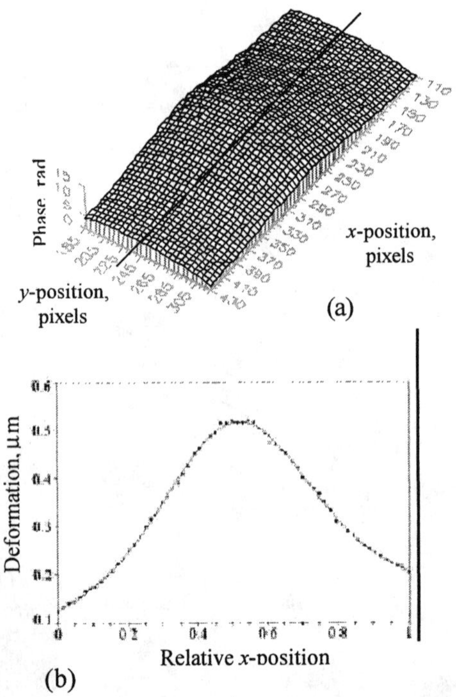

Fig. 9. OEH measured deformations of the resistor, based on the fringe pattern of Fig. 6a: (a) out-of-plane spatial distribution, (b) deformation along the line shown in (a) indicating maximum deformation of 520 nm measured with the accuracy better than 5 nm.

Representative results on the CAR system are shown in Figs 10 to 12. More specifically, Fig. 10a shows the OEH fringe pattern representing absolute shape of the transceiver of the CAR system at ambient temperature, while Fig. 10b shows the differential fringe pattern shortly after the system was powered on.

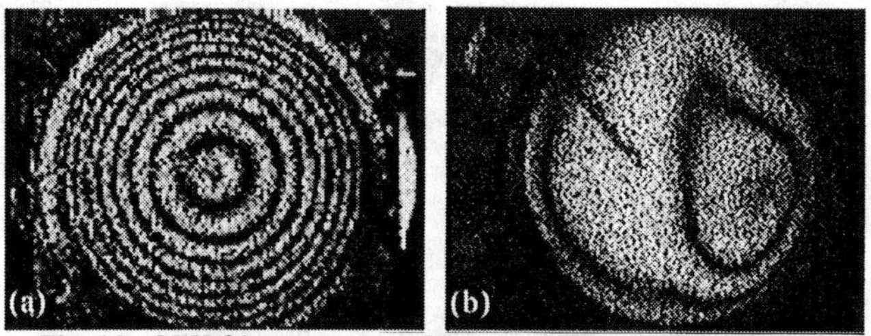

Fig. 10. OEH fringe patterns of the transceiver: (a) representing absolute shape at ambient temperature, (b) representing differential changes caused by an increase in temperature due to heating during the power-on condition.

Quantitative analysis of the OEH fringe patterns leads to the results such as those shown in Fig. 11. These results are particularly useful in measuring deformations of the transceiver due to heating (i.e., during power on) and cooling (i.e., during power off) of various components in the control unit of the CAR system. These changes cause deformations to vary in space and in time as the system undergoes various cycles during its functional operation.

(a)

(b)

(c)

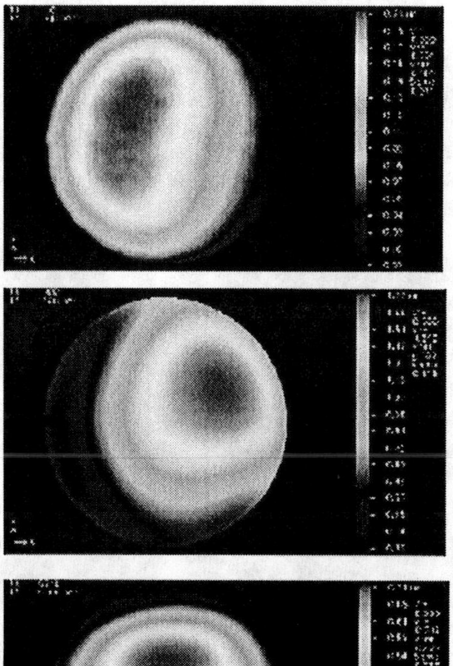

Fig. 11. Spatio-temporal changes in the deformation patterns of the transceiver as it undergoes various stages of its functional operation, time increases from (a) to (c).

Observation of the results shown in Fig. 11 shows that the extent of heat affected zone changes as the time into the operation of the CAR system goes on. Animation of the experimental data, such as those shown in Fig. 11, vividly indicates dynamics of thermomechanical effects in electronic packaging, Fig. 12.

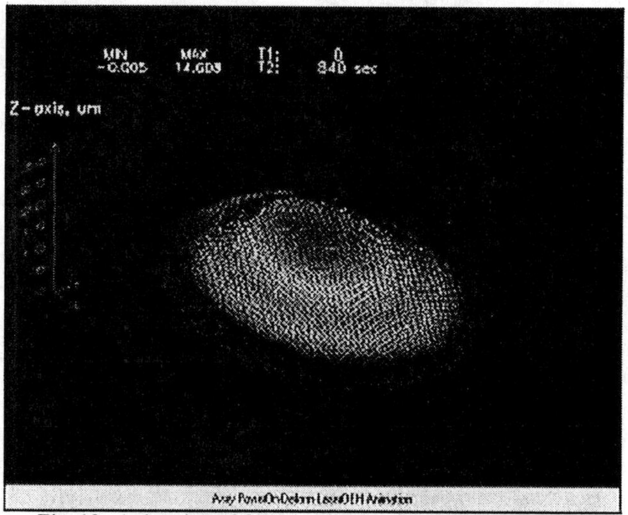

Fig. 12. Animation of OEH results showing spatio-temporal changes in the transceiver during the power-on cycle of the CAR system.

5. CONCLUSIONS

A novel computational-experimental method for noninvasive quantitative studies of thermomechanical deformations in electronic packaging has been presented and illustrated with representative applications. The applications demonstrate that neither only computational method nor only experimental method should be used to solve for the thermomechanical effects, but rather the methods should complement each other.

The results demonstrate that in addition to the accurate knowledge of boundary, initial, and loading (BIL) conditions, material properties must also be well known. Otherwise, erroneous information may be obtained. For example, using the published data, FEM computed deformations exhibit two maxima along the cylindrical resistor while in actuality there is only one maximum, as demonstrated by experimental results. This maximum has the magnitude of about 520 nm, determined with the accuracy better than 5 nm, based on the fringe patterns, determined using the OEH configuration used in this study.

Continued development of the experimental-computational method presented in this paper will provide a new tool for quantitative applications in electronic packaging.

6. REFERENCES

1. Tummala, R. R., and Rymaszewski, E. J., 1989, *Microelectronic packaging handbook*, Van Nostrand, New York.
2. Dally, J. W., 1990, *Packaging of electronic systems*, McGraw-Hill, New York.

3. Engel, P., 1995, "Experimental structural analysis topics in printed circuit board systems," *Proc. SEM Spring Conf.*, Grand Rapids, MI, pp. 215-217.
4. Dasgupta, A., and Pecht, M., 1991,"Material failure mechanisms and damage models," *IEEE Transactions on Reliability*, 40(5):531-536.
5. Pryputniewicz, R. J., ed., 1993, *Unification of analytical, computational, and experimental solution methodologies in micromechanics and microsystems*, Proc. 11th Internat. Invitational UACEM Symp., SEM, Bethel, CT.
6. Pryputniewicz, R. J., 1985, "Heterodyne holography applications in studies of small components," Opt. *Eng.*, 24:849-854.
7. Pryputniewicz, R. J., 1994, "A hybrid approach to deformation analysis," *Proc. SPIE*, 2342:282-296.
8. Furlong, C., 1999, *Hybrid, experimental and computational, approach for the efficient study and optimization of mechanical and electro-mechanical components*, Ph.D. Dissertation, Worcester Polytechnic Institute, Worcester, MA.
9. Pryputniewicz, R. J., and Furlong, C., 2000, "Novel optical-computational approach for NDT applications in microelectronics," *Proc. IX Internat. Congress on Experimental Mechanics*, SEM, Bethel, CT.
10. R. J. Pryputniewicz, 1992, *Holographic numerical analysis*, Worcester Polytechnic Institute, Worcester, MA.
11. R. J. Pryputniewicz, 1981, "High precision hologrammetry," *Internat. Arch. Photogramm.*, 24:377-386.
12. C. Furlong and R. J. Pryputniewicz, 2000, "Absolute shape measurements using ultra-high resolution optoelectronic holography methods," *Opt. Eng.*, 39:216-223.

Proceedings of
2001 ASME International Mechanical Engineering Congress and Exposition
November 11–16, 2001, New York, NY
EPP-Vol. 1

IMECE2001/EPP-24733

Reliability Analysis On the Chaos induced by the Effect of the Transient Temperature Variation of a Printed Wiring Board

Xiaoling He[1], Robert Fulton[2]
xiaoling@agere.com, robert.fulton@me.gatech.edu

[1]Bell Laboratory
Agere Systems/Lucent Technology
9333 South John Young Parkway,
Orlando FL 32819

[2]Mechanical Engineering,
Georgia Institute of Technology
801 Ferst Drive N.W.
Atlanta, Georgia 30332-0405

Abstract

The transient response of a printed wiring board (PWB) under mechanical and thermal loading is analyzed. The thermal load is caused by the time-dependent temperature variation of the PWB. Analytical solution is obtained for the equation of motion in nonlinear dynamics analysis. Numerical results showed that the response is in chaotic behavior under harmonic excitation due to the transient thermal effect. PWB response is characterized with respect to the thermal and mechanical load effect in both phase diagram and temporal oscillation.

Index: Nonlinear Dynamics, Printed wiring board, Laminated structure, Chaos.

Nomenclature

A : Stiffness matrix of the laminate
A^* : inverse matrix of A
A^*_{ij} : Elements of A^*
$(E)^{(k)}$: kth lamina Young's Modulus
$F(x,y,t)$: Airy's stress function
$F_1, F_2, F_{11}, F_{22}, F_{66}$: Coefficients of Failure function
N : In-plane force vector
$\mathrm{N}(w)$: Function of in-plane force and deflection
PWB: Printed wiring board
$R(x,y)$: Failure Criteria function
r^* : Stiffness coefficient of the Duffing equation
r_{mn} : Nonlinear spring coefficient
$T_0(x,y)$: In-plane spatial temperature variation;
$T_1(x,y)$: Through the thickness temperature variation;
$u_0(x,y)$: Mid-plane deformation of PWB in x direction
$u(x,y)$: Deformation of PWB in x direction
$v_0(x,y)$: Mid-plane deformation of PWB in y direction
$v(x,y)$: Deformation of PWB in y direction
$|W(t)|$: Out-of-plane deflection magnitude of PWB
$(\sigma)^{(k)}$: kth lamina plane stress vector
σ^Y : Yield strength of the lamina
τ^F : Shear failure strength of the lamina
ν : Thermal oscillation frequency
ω : Excitation frequency
ω_{mn} : Linear system frequency

Introduction

The application condition of the printed wiring board (PWB) implies that it undergoes non-uniform and time-dependent temperature variation due to the components on board as well as the conduction and dielectric pattern distribution in the lamina. The difference in the coefficient of thermal expansion between different lamina materials also leads to the through thickness temperature variation, other than the deflection of the laminate. Transient temperature variation introduces deflection and stress oscillation. For a thin laminated structure, the deflection could be nonlinear, hence the induced large deflection tends to lead to mechanical and electrical failure of the PWB.

The nonlinear vibratory behavior of the PWB is proven to be in the range of large deflection by Suhir in 1992[1] and He in 2000[2] in single layer plate and laminated structure, respectively. The equation of motion is governed by the Duffing equation. The response under constant load has been obtained analytically in Suhir's work[1].

The nonlinear dynamics of laminated composites and plates has been developed based on the Von-Karman strain field in elastic deformation. Significant work has been accomplished by Whitney in 1969[3], Lisa in 1969 and 1980[3,4], and by Chia in 1980[5]. Work in the past has been mainly focused on the response induced by either mechanical load[3-5] or by thermal load alone, e.g. by Pal in 1970[6]. Joint thermal and mechanical effect has not been addressed in the nonlinear dynamics analysis. Nor the response induced by the transient thermal phenomena in an analytical expression of the Duffing equation. In chaos analysis, a modified Duffing equation chaos is studied by Eissa in 2000[7]. Other work has been completed for chaos of the Duffing equation driven by the velocity damping, e.g. by Novak S. in 1982[8] and by Luo, A. in 1999[9]. However, chaos induced by the thermal effect for the laminated structure such as PWB has not been studied in the known literature.

A case study is made for the simply supported PWB. Analytical solution is obtained for its dynamic response based on the nonlinear dynamics with thermal effect in multi-mode superposition. Under transient temperature variation, chaotic behavior in deflection is observed under harmonic excitation.

Equation of Motion

The equation of motion for the laminate can be derived based on the Von-Karman strain field definition for the nonlinear response. The equilibrium condition leads to the equation of motion for the out-of-plane deformation $w(x,y,t)$ in the form of:

$$D_{11}\nabla^4 w + I\ddot{w} - I_2\nabla^2\ddot{w} = q + \mathrm{N}(w) \quad (1a)$$

$$\text{where} \quad \mathrm{N}(w) = N_y w,_{yy} - 2N_{xy} w,_{xy} + N_x w,_{xx} \quad (1b)$$

and $\{N_x, N_y, N_{xy}\}$ is the in-plane force vector.

The compatibility equation of the strain field:

$$\varepsilon^0_{xx},_{yy} + \varepsilon^0_{yy},_{xx} - \gamma^0_{xy},_{xy} = w,_{xy}{}^2 - w,_{xx}\, w,_{yy} \quad (2a)$$

can be transformed into the following for the isotropic laminate:

$$\nabla^4 F = r^*\left(w,_{xy}{}^2 - w,_{xx}\, w,_{yy}\right) \qquad r^* = \frac{1}{A^*_{22}} \quad (2b)$$

and $F(x,y,t)$ is the Airy's stress function.

For the simply supported PWB, the solution is assumed to be:

$$w(x,y,t) = \sum_{n=1}^{\infty}\sum_{m=1}^{\infty} W_{mn}(t)\sin(\alpha_m x)\sin(\beta_n y)$$

$$\alpha_m = \frac{m\pi}{a} \quad \beta_n = \frac{n\pi}{b} \qquad x\in[0,\ a]\ \ y\in[0,\ b] \quad (3a)$$

which satisfies the boundary condition:

$$u_0(0,y,t)=0 \quad u_0(a,y,t)=0 \quad v_0(x,0,t)=0 \quad v_0(x,b,t)=0$$
$$w(x,0,t)=0 \quad w(x,b,t)=0 \quad w(0,y,t)=0 \quad w(a,y,t)=0 \quad (3b)$$

and $N_{xy}(x,y,t)=0$ on all the boundaries. (3c)

The spatial temperature variation includes function $T_0(x,y)$ for the in-plane variation and function $T_1(x,y)$ for the through thickness variation. The oscillation of the temperature field is assumed as:

$$T(x,y,z,t) = T_0(x,y) * f(t) + z * T_1(x,y) * g(t) \quad (4a)$$

where z is the through thickness coordinate of the lamina. It corresponds to the linear increment of the temperature variation through the thickness. Function $T_1(x,y)$ is chosen as a constant. The temporal oscillation $f(t)$ and $g(t)$ are assumed to be:

$$f(t) = \cos(vt) \qquad g(t) = 1 \quad (4b)$$

where v is the thermal frequency. Function $g(t)$ is set as a constant for the purpose of simplification due to its insignificance as analyzed below.

With above assumption, the equation of motion for the transient oscillation of the out-of-plane deformation $W(t)$ can be reduced to the following:

$$\ddot{W}_{mn} + \left(\omega_{mn}{}^2 + \chi_{mn}\cos(vt)\right)W_{mn} + r_{mn}W_{mn}{}^3 = q_{mn} \quad (5)$$

$$\text{where:} \quad \omega_{mn}{}^2 = \frac{D_{11}\left(\alpha_m^2 + \beta_n^2\right)^2}{\tilde{I}} \quad (5a)$$

$$\chi_{mn} = \frac{\eta}{4\tilde{I}} T^0_{mn}\left(\alpha_m^2 + \beta_n^2\right) \quad (5b)$$

$$r_{mn} = \frac{1}{16\tilde{I}}\left[4A_{12}\alpha_m^2\beta_n^2 + \frac{3A_{11}{}^2 - A_{12}{}^2}{A_{11}}\left(\alpha_m^4 + \beta_n^4\right)\right] \quad (5c)$$

$$q_{mn} = q_{0\_mn} + q_{T\_mn} \quad (5d)$$

$$q_{0\_mn} = \frac{Q_{mn}}{\tilde{I}} \qquad q_{T\_mn} = \frac{\left(\alpha_m^2 + \beta_n^2\right)T^1_{mn}}{\tilde{I}} \quad (5e)$$

Equation (5) is a modified Duffing equation with a hard spring, due to $r_{mn} > 0$, as determined by the stiffness coefficients of the laminate $A_{11} > A_{12} > 0$. It is noted that the in-plane temperature variation contributes to the frequency oscillation, while the Duffing system stiffness r_{mn} remains. Through-thickness temperature variation adds pressure load due to thermal moment.

The admissible function for the temperature field in the governing equation is:

$$T_0(x,y,t) = \sum_{n=1}^{\infty}\sum_{m=1}^{\infty}\left(T^c_{mn} + T_{mn}(t)\cos(2\alpha_m x)\cos(2\beta_n y)\right) \quad (6)$$

A thermal field $T_0(x,y)$ is shown in Fig.1 with the following coefficients:

$$T^0_{11} = 70 \quad T^c_{11} = 70 \quad T^0_{12} = 60 \quad T^c_{12} = 55$$
$$T^0_{21} = 12 \quad T^c_{21} = 70 \quad T^0_{22} = 0.3 \quad T^c_{22} = 3.5 \quad (7)$$

It is seen from T^0_{mn} that the first 2 modes are dominant for generating the thermal field, hence it dictates that the modal analysis should take the first two modes into account. The maximum in-plane temperature rise from the two-mode superposition is about 100 ^{0}C at the central portion. It is gradually smooth down at edges of PWB. Noted that a sharp temperature rise exists at the corner points. The maximum value at these points is about 125 ^{0}C, 25% higher than the maximum temperature assumed for $T_0(x,y)$. This is the

Gibbs phenomena associated with the admissible function for the temperature field. This would be an error for the case when the corner temperature subsides. The assumed temperature variation through the thickness of the PWB in 1 ^{0}C/mm introduces thermal load of q_{11}^T=0.09 N/cm^2, which is insignificant compared to the contribution of the in-plane temperature variation.

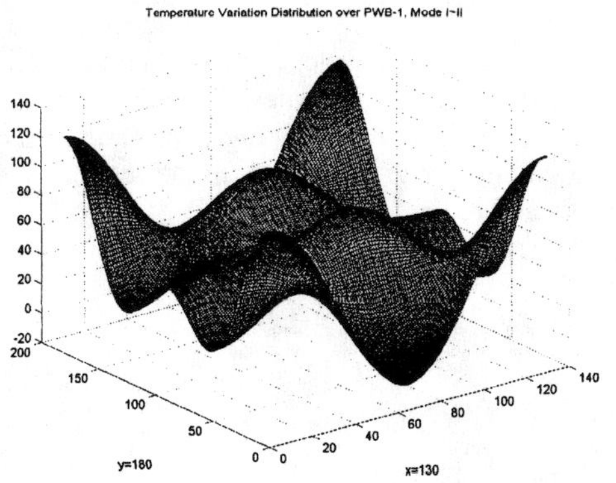

Fig.1 First 2 terms of $T_0(x, y)$

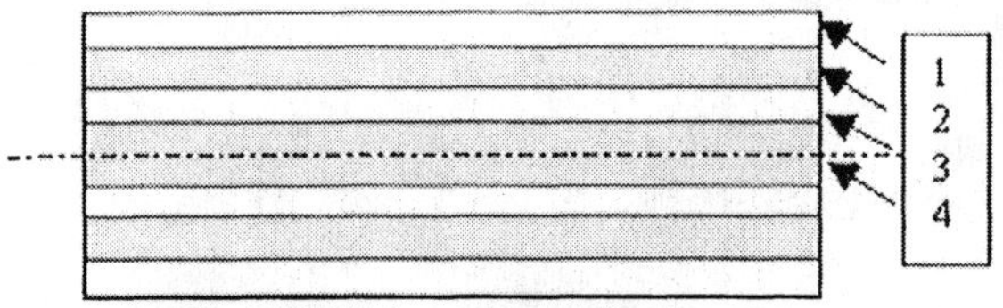

Fig.2 PWB in symmetric structure

Table 1 Material Properties of PWB

Layer	Material	Specific weight (g/cm^3)	Young's Module (MPa)	Poission ratio	Thick -ness (mm)
1	Cu/FR-4	7.599	$2.98*10^4$	0.205	0.030
2	FR-4	1.200	$2.0*10^4$	0.190	0.200
3	Cu	8.310	$1.18*10^5$	0.340	0.035
4	FR-4	1.200	$2.0*10^4$	0.190	1.000

Table 2 System Parameters for 2-mode Analysis

Mode	Mode I, m=n=1	Mode-II, m=1,n=2
ω_{mn} [Hz]	1,708.0	3,466.0
$r_{mn}\left[(mm.s)^{-2}\right]$	1,551,840	6,490,828.3
$\chi_{mn}\left[s^{-2}\right]$	5,167,903.5	8,983,273.3
$\left[\omega_{mn}^T\right]_{max}$ [Hz]	2,844.0	4,582.2
T_{mn}^1 [^0C]	16.2	8.10
q_{mn}^T [mm/s^2]	973,416.8	987,042.6

The PWB under analysis is in symmetry as shown in Fig.2. It is made of Cu and FR-4 material for the induction layer and the dielectrics. The material properties are listed in Table 1. The Duffing equation parameters for the two modes are listed in Table 2.

In the following, the effect of the temperature variation is studied. Since the through thickness temperature variation generates trivial constant load q^T as analyzed above, the thermal analysis is focused on the in-plane temperature variation only. For simplification, the response without in-plane temperature variation is referred to as "isothermal response", and the response with in-plane thermal effect is "anisothermal response". Numerical results are generated from the Duffing equation computation in 4th order Rung-Kutta method with time step in less than $10^{(-4)}$s.

Response under Constant Loading

The maximum deflection with respect to the increasing temperature frequency under the load Q=0 and Q=2.4 N/cm^2, respectively, is plotted in Fig.3 and Fig.4 for the first mode with thermal frequency range between j=[0 10] and j=[0 100], respectively. The thermal frequency is in a multiple of the modal frequency ω_{11}, i.e. $v = j*\omega_{11}$ for $T_0(x, y, t) = T_0(x, y)\cos(vt)$. The response without in-plane temperature variation is shown as an horizontal line, i.e. the isothermal line representing the maximum deflection without transient thermal load. Its periodic behavior has been studied by Suhir[1] and He[2]. The isothermal and the anisothermal responses come to convergence at certain frequency range of v, which means that the maximum deflection is not affected by these temperature oscillation frequencies. Further increasing the thermal frequency, as in Fig.4, sees the anisothermal response regains its maximum magnitude almost periodically.

For mode-II response with respect to thermal frequency $v = j*\omega_{12}$, j=[0 100] as shown in Fig.5, repetitive resonance of the anisothermal response appears to occur at a shorter frequency interval than that of the first mode. The interval for the index k between the two adjacent peak centers is about $\Delta k_1 = 36$ for the first mode in Fig.4 and $\Delta k_2 = 18$ for the second mode in Fig.5. From modal parameters listed in Table 2, it is known that higher mode oscillates at a higher frequency, i.e. $\omega_{12} \cong 2\omega_{11}$. Therefore, the frequency interval is in fact about the same given that $\Delta\omega_{11} \cong \Delta k_1\omega_{11}$, and $\Delta\omega_{12} \cong \Delta k_2\omega_{12}$. This means that the resonance of both modes occur at about the same excitation frequency range. The superposition of the two modes strengthens the resonance deflection in these frequency ranges.

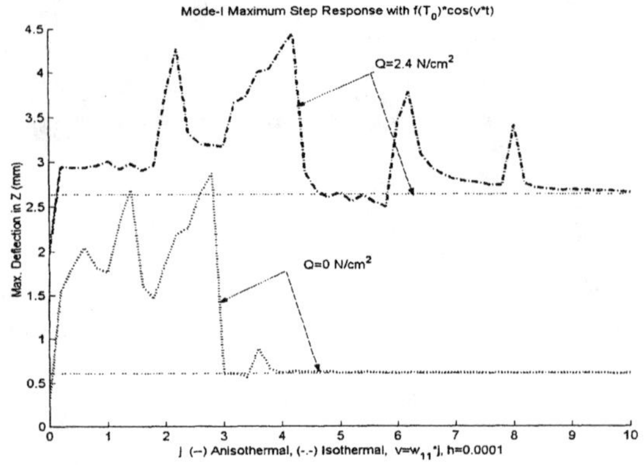

Fig.3 Maximum deflection in mode-1 at increasing $v = j * \omega_{11}$ at Q=0, 2.4 N/cm²

Fig.3 also shows the response induced by thermal effect only, i.e. Q=0. The maximum deflection is in the nonlinear range, or over half of the PWB thickness. This means that deflection under transient high temperature rise is not negligible. Large deflection can occur due to high rise temperature. This transient thermal effect can also be observed for mode-II response.

The temporal oscillation at resonance for the first mode under the load Q=2.4 N/cm², j=2.2 is plotted in Fig.6 for the deflection and in Fig.7 for the deflection-velocity diagram for the anisothermal response. The quasi-periodic behavior with transient thermal effect is seen from Fig.7. As time evolves, its deflection oscillation and the velocity-deflection diagram almost periodically repeats itself.

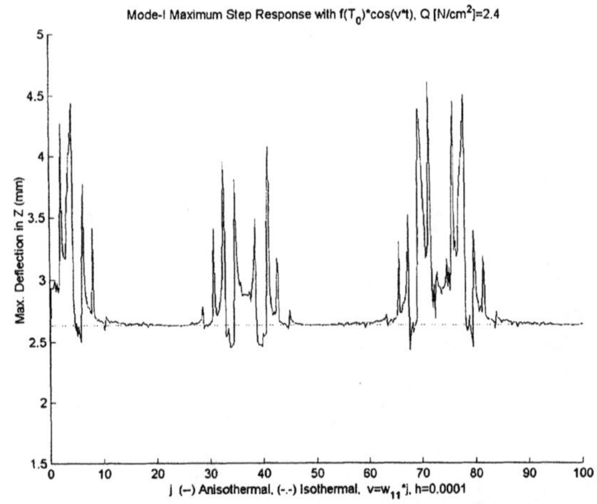

Fig.4 Maximum deflection in mode-1 at increasing $v = j * \omega_{11}$ at Q=2.4 N/cm²

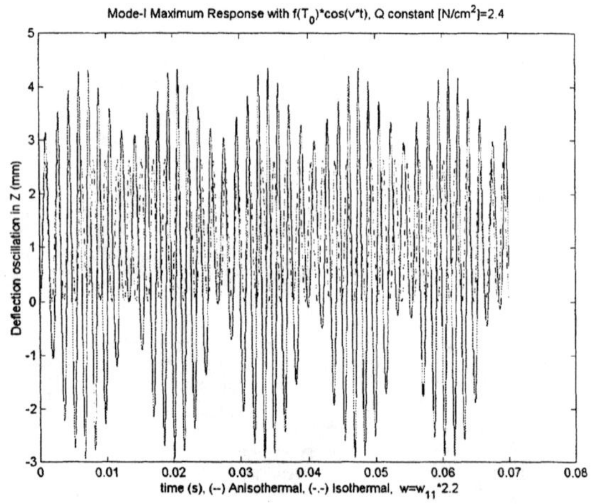

Fig.6 Peak response under constant load of 1st mode at Q=2.4 N/cm², $v = j * \omega_{11}$ j=2.2

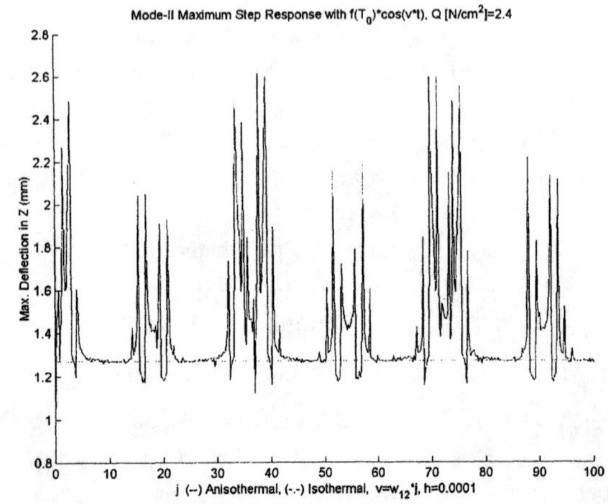

Fig.5 Maximum deflection in mode-2 at increasing $v = j * \omega_{11}$ at Q=2.4 N/cm²

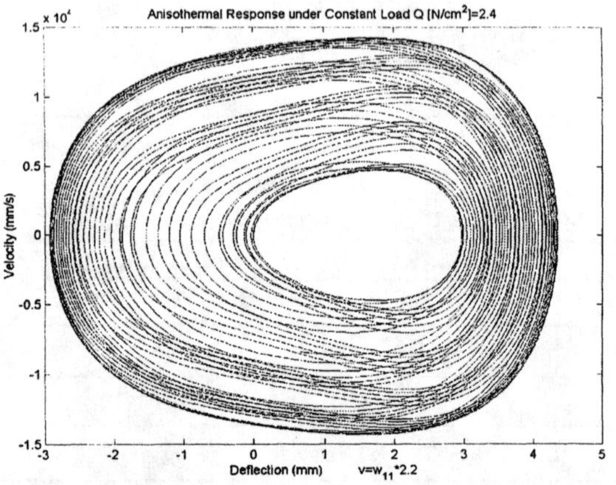

Fig.7 Velocity/Deflection in anisothermal response of 1st mode under constant load at Q=2.4 N/cm², $v = j * \omega_{11}$ j=2.2

This is different from the periodic response without thermal effect, which exhibits a single loop only in the velocity-deflection diagram. From Equ.(5), it is seen that the square frequency term with thermal effect, $\left(\omega_{mn}^{2}\right)^{T}$, varies between $[\omega_{mn}^{2} - \chi_{mn}, \omega_{mn}^{2} + \chi_{mn}]$ in each cycle of $T = \left(j\omega_{mn}\right)^{-1}$. Table 2 indicates that $\omega_{mn}^{2} < \chi_{mn}$, which means that system undergoes unstable state in each cycle. It oscillates between stable and unstable state. When the frequency term $\left(\omega_{mn}^{T}\right)^{2}$ is negative, the unstable system behaves as an isothermal system with a harmonic load applied, caused by thermal term. Note that the unstable condition is temporary and well contained. The frequency can not reach infinity as in a completely unstable system. As a consequence of the cyclic thermal frequency, the periodic response under constant load is transformed into the almost periodic oscillation.

Harmonic Response and Chaos

Maximum deflection of harmonic response with respect to the increased excitation frequency are studied in Fig.8 for k=[0 10]. Curves plotted has the thermal frequency j=0, 1, and 2 respectively. From thermal frequency $v = j * \omega_{11}$, response at j=0 represents the case of time-independent thermal effect. In each of the plotted curve, the initial point at k=0 is the response under constant load.

Fig.8 shows that the maximum harmonic deflection with certain values of j is higher than that of both isothermal, $T_0(x, y, t) = 0$, and the time-independent thermal effect, i.e. $T_0(x, y, t) \neq 0$, and j=0. Besides, Unlike the convergence occurred between the isothermal response and the response under time-independent thermal effect, where the maximum deflection diminishes in certain frequency range, the maximum deflection of the transient thermal response with j=1, 2 is constantly in a higher magnitude throughout the frequency domain.

The phase diagram with respect to the variation of the thermal frequency $v = j * \omega_{11}$ is shown in Fig.9 under harmonic load with k=1.8, Q=0 and 2.4 N/cm^2, respectively. The varying index j is within the range of j=[0 100]. It is noted that resonance occurs at certain range of j. In some other range, the thermal effect and that of the isothermal effect converges to a flat line, similar to the convergence in a diminished deflection in Fig.8. This reflects the fact that some thermal frequency contributes significantly to the increased deflection while others has trivial effect. Also shown in Fig.9 is the load effect. Higher load causes higher deflection, and resonance frequency range can be varied due to the load variation.

In the extended frequency range of j, resonance occurs in an almost repetitively pattern with respect to the increased thermal frequency. Similar repetitive resonance can be observed with respect to the increased harmonic excitation frequency with k=[0 100]. This quasi-repetitiveness in resonance behavior is observed under both constant load response and the harmonic response. It characterizes the resonance behavior of the system in a modified Duffing equation.

Fig.10 to Fig.12 showed the temporal oscillation for isothermal and anisothermal harmonic response with Q=0.24 N/cm^2, k=1.8 and j=2. It corresponds to the peak deflection in the phase diagram shown in Fig.8. Its maximum deflection is higher than that of the isothermal response, as evident from both Fig.8 and Fig. 10.

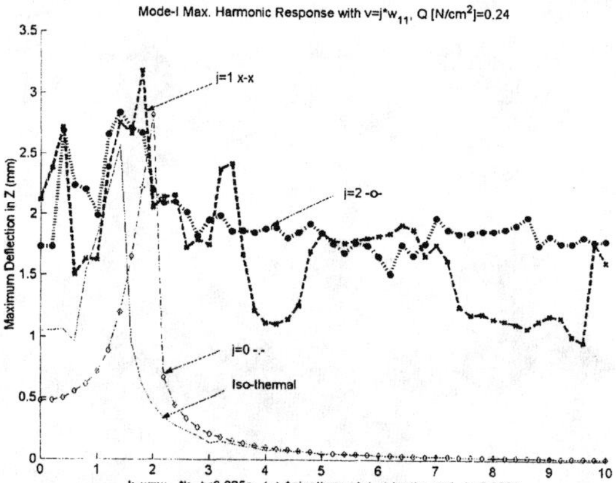

Fig.8 Mode-1 harmonic resonance at $\omega = k * \omega_{11}$ Q=0.24 N/cm^2, k=[0, 10]

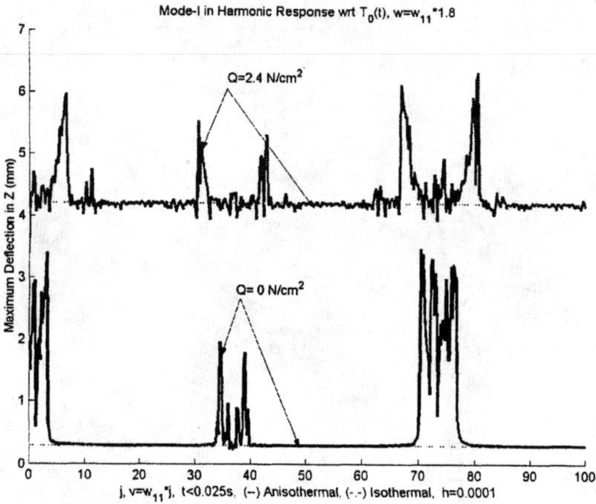

Fig.9 Mode-1 harmonic resonance at $v = j * \omega_{11}$ Q=0.24 N/cm^2, j=[0, 100], k=1.8

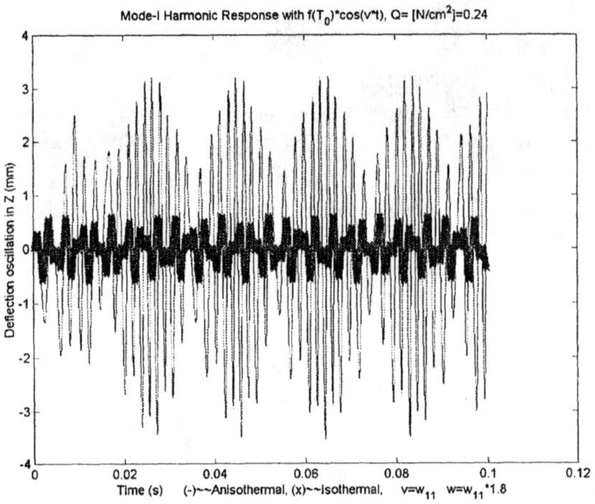

Fig.10 Mode-1 resonance at Q=0.24 N/cm^2 j=1, k=1.8

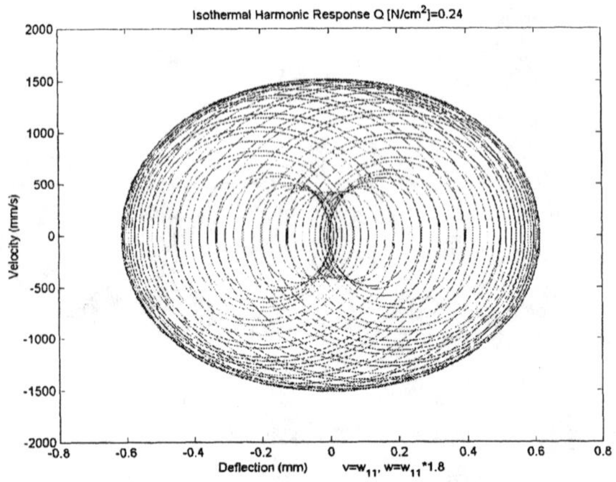

Fig.11 Mode-1 isothermal resonance at Q=0.24 N/cm^2, j=1, k=1.8

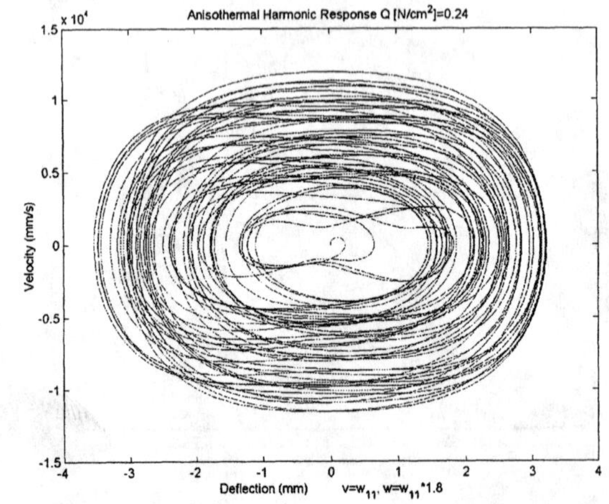

Fig.12 Mode-1 anisothermal resonance at Q=0.24 N/cm^2, j=1, k=1.8

In the velocity-deflection diagram for isothermal response, certain regularity with respect to time is observed in the accumulated data. However, for the anisothermal response, chaotic behavior is occurred, i.e. the exact value of the response at a specific time is unpredictable. Fig.12 shows the chaotic pattern for its velocity-deflection diagram.

Without the transient thermal effect, the Duffing system response is almost periodic. The interaction of the harmonic thermal frequency and the excitation frequency leads to chaos. From comparison between Fig.11 and Fig.12, it is noted that the anisothermal response velocity range is increased due to the transient thermal effect.

Discussion

The transient thermal effect drastically altered the system behavior in its deflection response. It makes the response under general loading conditions tends to be irregular. The almost periodic response under the constant load and the chaotic behavior under harmonic load all attributed to the transient thermal oscillation. It also makes the deflection and velocity range increased under general loading conditions. The chaos is stable and in a confined range for its deflection and velocity range.

The repetitive resonance behavior in the phase diagram with respect to both the thermal frequency and the excitation frequency means that the resonance can occur under various combinations of thermal effect and mechanical load.

The modal analysis demonstrates that arbitrary thermal field can be simulated by using Fourier series, i.e. $\sum_{k=1}^{N} \chi_k \cos(v_k t)$. Resonance frequency range can be identified for the given thermal effect and loading condition. The modal response also indicates that higher mode analysis is needed given its dominant role in representing arbitrary thermal field and its significant contribution to the increased deflection associated with the chaotic behavior.

Conclusion

Transient temperature variation has more detrimental effect on the reliability of the PWB than the case when the temperature variation is time-independent, due to the high deflection induced. It also makes the deflection chaotic under harmonic excitation. The quasi-repetitive resonance behavior means that the failure associated with the large deflection and stress interaction should be analyzed on a wide frequency domain. The chaotic behavior and the resonance character are essential in predicting response under general loading conditions as well as in an arbitrary temperature variation.

The modal approach demonstrated the flexibility in representing and analyzing the response under alternative thermal fields and load. The analytical expression of the governing equation of motion, i.e. the Duffing equation, and the numerical results for the deflection provide the basis for further stress induced failure analysis, such as fatigue, overstress and other physical failure mechanism. PWB reliability can be optimized through material and structure evaluation in favor of its thermal and mechanical response.

References

1. Suhir, E., 1992, "Nonlinear Dynamic Response of a Printed Circuit Board to Shock Loads Applied to Its Support Contour", Journal of Electronic Packaging, **114**, December, pp368-377.
2. He, X., 2000, "A Theoretical Framework for Dynamic Response of a Printed Wiring Board", Ph.D thesis, Georgia Institute of Technology, Atlanta.
3. Whitney, J. and Leissa, A., 1969, "Analysis of Heterogeneous Anisotropic Plates", Journal of Applied Mechanics, June, pp261-266.
4. Leissa, A., 1969, "Vibration of Plates", NASA Report, SP-160.
5. Chia, C., 1980, Nonlinear Analysis of Plates, McGraw-Hill, New York, NY.
6. Pal, M.,1970, "Large Amplitude Free Vibration of Rectangular Plates Subjected to Aerodynamic Heating", *Journal of Engineering Mathematics,* **4(1)**, *January,* pp39-49.
7. Eissa, M., 2000, "Dynamics and Resonance of a Nonlinear Mechanical Oscillator Subjected to Parametric and External Excitation", *Integral Methods in Science and Engineering*, Chapman & Hall/CRC.
8. Novak, S., 1982, "Transiention to Chaos in the Duffing Oscillator", Physics Review A, **26(6)**, pp3660-3663.
9. Luo, A., 1999, "Analytical Predictions of Chaos in a Nonlinear Rod", Journal of Sound and Vibration, **227(3)**, pp523-544.
10. Jones, R., 1999, Mechanics of Composites, 2nd edition,: Taylor & Francis. Philadelphia, PA.
11. R. Gilmore and J. McCallum, 1995, "Structure and the Bifurcation Diagram of the Duffing Oscillator", Physics Review E, **51(2)**, pp935-954.

Proceedings of
2001 ASME International Mechanical Engineering Congress and Exposition
November 11-16, 2001, New York, NY
EPP-Vol. 1

IMECE2001/EPP-24734

Analysis of Thermal Stresses in a Layered Circuit Board

James C. Gerdeen, Ph.D.
Professor

Department of Mechanical Engineering
University of Colorado at Denver

ABSTRACT

When choosing materials for a printed circuit board (PCB), it appears that a common misconception is that the coefficients of thermal expansion should be matched as closely as possible to minimize the stresses. In this paper it is shown that this is not necessarily true. A one-dimensional thermal stress analysis is considered for a printed circuit board (PCB) with N layers. Bending is neglected and average axial normal stresses and shear stresses are accounted for. It is shown that the solution for the shear stresses between layers can be written in the form of a difference equation (recurrence relation).

For the first example, a two-layer board is considered. The solution leads to a design guideline. If there is a shear stress design limit then based on this a plot can be made to choose the optimum ratio of CTEs for the layers depending upon whether the conduction heat flow q is zero or not and depending upon magnitude of temperature change.. If q = 0, then matching the coefficients of expansion α_1 and α_2 reduces the stress to zero. However if q is not zero at some times, then a choice of the CTE ratio α_1/α_2 should be made to keep the stress between the design limits.

For three or more layers with different properties a design optimization scheme can be used to choose values of the coefficients of thermal expansion.

Analysis of Thermal Stresses in a Layered Circuit Board

Consider a one-dimensional analysis in each layer. Bending will be neglected for now, and only average axial normal stresses and shear stresses will be accounted for.Consider N layers as shown in Figure 1. All layers have the same length L. The thickness of an individual layer *n* is t_n. Let us look at the stresses in layer *n*, as shown in Figure 2. For equilibrium the shear stresses on the top and bottom faces change sign and oppose each other for equilibrium. Assume the edges at x=0 and x=L are stress free. Assume the average normal stress has a maximum σ_{on} at x=L/2 as shown. In order to satisfy the boundary conditions at x=0 and x=L, and the symmetry condition, assume the distribution of stresses as shown inin Figure 3. Approximate the stresses as sine functions as follows:

$$\sigma_n = \sigma_{on} \sin^2 (\pi x/L), \qquad \text{(1 a,b)}$$

$$\tau_n = \tau_{on} \sin \frac{2\pi x}{L}$$

(This form of the stress equations satisfy the stress function equations of elasticity [1], when boundary conditions, at x=0 and x = L, are accounted for. It can be shown that a singularity exists in the elasticity solution between layers of dissimilar materials at the edges [2,3]. But this singularity occurs over an infinitesimal region and is removed by local yielding. The above equations are considered a good approximate statically admissible solution.)

For axial equilibrium of forces (Figure 2), we must have

$$\sigma_{on} t_n = \int_0^{L/2} \tau_n dx - \int_0^{L/2} \tau_{n-1} dx$$

From the above equation, there results

$$\sigma_{on} = \frac{L}{\pi t_n} (\tau_{on} - \tau_{on-1}) \qquad (2)$$

Displacement u

Assume u=u(x) only, i.e. assume L/t >> 1 such that $\sigma_{on} >> \tau_{on}$, so that u results from σ only. The total strain ε is a sum of strain due to stress and strain due to a change in temperature ΔT. Therefore

$$\frac{du}{dx} = \varepsilon = \frac{\sigma}{E} + \alpha \Delta T \qquad (3)$$

Assume ΔT is constant in each layer.
Thus,

$$u_n = \int_o^L \frac{\sigma}{E_n} dx + \alpha_n \Delta T_n L = \frac{\sigma_{on}}{E_n}\left(\frac{L}{2}\right) + \alpha_n \Delta T_n L$$

$$u_n = \frac{L}{2E_n} * \frac{L}{\pi t_n}\left(\tau_{on} - \tau_{on-1}\right) + \alpha_n \Delta T_n L \qquad (4)$$

Rewrite (4) as: $u_n = A_n\left(\tau_{on} - \tau_{on-1}\right) + B_n \Delta T_n$ (5)

This is in the form of a difference equation.
Assume $u_n = u_{n-1} = u_1 = u_2 = \ldots = u =$ Const = average expansion
Then

$$\tau_{on} = \tau_{on-1} - \frac{B_n}{A_n}\Delta T_n + \frac{u}{A_n} \qquad (6)$$

Assume, ΔT_n is the average temperature change in each layer, from a heat heat transfer analysis. The displacement $u_n = u$ is not known but can be solved.

From (5):

$$\begin{aligned} u_1 &= A_1\left(\tau_{01} - \tau_{00}\right) + B_1 \Delta T_1 \\ u_2 &= A_2\left(\tau_{02} - \tau_{01}\right) + B_2 \Delta T_2 \\ u_3 &= A_3\left(\tau_{03} - \tau_{02}\right) + B_3 \Delta T_3 \\ u_N &= A_N\left(\tau_{ON} - \tau_{0N-1}\right) + B_N \Delta T_N \end{aligned} \qquad (7)$$

Assume the top and bottom surfaces of the circuit board are stress free:

$$\tau_{00} = 0, \quad \tau_{ON} = 0. \qquad (8a,b)$$

Add the above equations (7):

$$\frac{u_1}{A_1} + \frac{u_2}{A_2} + \frac{u_3}{A_{31}} + \ldots = \frac{B_1}{A_1}\Delta T + \frac{B_2}{A_2}\Delta T_2 + \ldots + \frac{B_N \Delta T_N}{A_N}$$

$$= u\left(\frac{1}{A_1} + \frac{1}{A_2} + \ldots \frac{1}{A_N}\right) = \ldots$$

$$\therefore \; uA = B \; , \; u = \frac{B}{A} \qquad (9)$$

Rewrite (6) as:

$$\tau_{on} - \tau_{on-1} = C_n \; , \; where \; C_n = const \qquad (10)$$

Equation (10) is the form of a difference equation and can be solved as follows:

(

)

$$\begin{aligned} \tau_{on} - \tau_{on-1} &= C_n \\ \tau_{on-1} - \tau_{on-2} &= C_n - 1 \\ &\;\;\vdots \\ \tau_{02} - \tau_{01} &= C_2 \\ \tau_{01} - \tau_{00} &= C_1 \end{aligned} \qquad (11)$$

Adding the above equations gives:

$$\tau_{0n} - \tau_{00} = \tau_{0n} = \sum_{k=1}^{n} C_k \qquad (12)$$

where

$$C_k = \frac{u}{A_k} - \frac{B_k}{A_k}\Delta T_k \;\ldots\ldots\ldots\ldots \qquad (13)$$

Heat Transfer Analysis

Consider steady state heat conduction through a multi-layered circuit board as shown in Figure 4.

The heat flux is q applied at the top surface, that may come from a source like a resistor. Let the temperatures on the layers be designated as follows:

T_n = temperature at top surface of layer *n*, and
T_{n+1} = temperature at the bottom surface of layer *n*.

The steady state heat conduction equation for one dimensional heat flow is:

$$q_1 = \frac{k_1 A}{t_1}\left(T_1 - T_2\right) = q_2 = \frac{k_2 A}{t_2}\left(T_2 - T_3\right) = \ldots$$

$$= q_n = \frac{k_n A}{t_n}\left(T_n - T_{n+1}\right) \qquad (14)$$

The average temperature in a layer is:

$$T_{na} = \frac{T_n + T_{n+1}}{2} \qquad (15)$$

The change in temperature in a layer, from a common reference temperature is:

$$\Delta T_n = T_{na} - T_{ref} \qquad (16)$$

T_{ref} may be room temperature or the ambient temperature.)
The difference in ΔT between two adjacent layers is

$$\Delta T_n - \Delta T_{n+1} = T_{na} - T_{n+1a} = \frac{T_n + T_{n+1} - (T_{n+1} + T_{n+2})}{2} = \frac{T_n - T_{n+2}}{2} \quad (17)$$

Define thermal resistance of layer n as:

$$R_n = {t_n}/{(k_n A)} \quad (18)$$

Consider the heat flow through the first two layers:

$$q = \frac{T_1 - T_2}{R_1} = \frac{T_2 - T_3}{R_2} \quad (19)$$

In general it can be shown that

$$q = (T_1 - T_{n+1}) / \sum_{i=1}^{n} R_i \quad (20)$$

Equation (20) can be rewritten as

$$T_1 = T_{n+1} + q \sum_{i=1}^{n} R_i \quad (21)$$

Which shows that T_1 can be minimized by minimizing the thermal resistance $\sum R_i$. A common design criteria is the maximum surface temperature T_1 allowed on the board.

The next step is to combine the above heat transfer analysis with the thermal stress analysis above.

Example 1

Consider a two-layered board. From Equation (5) the displacements are:

$$u_1 = A_1 \tau_{o1} + B_1 \Delta T_1 \quad , \quad u_2 = A_2(-\tau_{o1}) + B_2 \Delta T_2 \quad (22\text{ a,b})$$

Since $u_1 = u_2$,

$$\tau_{o1}(A_1 + A_2) = B_2 \Delta T_2 - B_1 \Delta T_1 \quad (23)$$

where

$$A_1 + A_2 = \frac{L^2}{2\pi}\left[\frac{1}{E_1 t_1} + \frac{1}{E_2 t_2}\right] = \frac{L^2}{2\pi E} \quad (24)$$

and where

$$B_2 \Delta T_2 - B_1 \Delta T_1 = L(\alpha_2 \Delta T_2 - \alpha_1 \Delta T_1) \quad (25)$$

Substitution of (24) and (25) into (23) gives:

$$\tau_{o1} = \frac{2\pi}{L^2} EL(\alpha_2 \Delta T_2 - \alpha_1 \Delta T_1) \quad (26)$$

From (20):

$$\alpha_2 \Delta T_2 - \alpha_1 \Delta T_1 = \alpha_2\left[\Delta T_1 - q\frac{(R_1 + R_2)}{2}\right] - \alpha_1 \Delta T_1$$

(27)

From (27), if q=0, (no heat flow):

$$\tau_{01} \sim (\alpha_2 - \alpha_1)\Delta T_1 \quad (28)$$

If $q \neq 0$,

$$\tau_{01} \sim (\alpha_2 - \alpha_1)\Delta T_1 - \alpha_2 q(R_1 + R_2)/2 \quad (29)$$

From (20), $q(R_1 + R_2) = T_1 - T_3$, and (29) can be rewritten for $q \neq 0$, as

$$\tau_{01} \sim \alpha_2\left[\left(1 - {\alpha_1}/{\alpha_2}\right)\Delta T_1 - (T_1 - T_3)/2\right]$$

or

$$\tau_{01} \sim \alpha_2 (TF) \quad (30)$$

where TF is a thermal stress factor.

Numerical Example:

As an example take
$\Delta T_1 = 80°C$, $\Delta T_2 = 10°C$, and

$$\frac{T_1 - T_3}{2} = \Delta T_1 - \Delta T_2 = 70°C.$$

Construct the table shown in Table 1 for $\alpha_1 \le \alpha_2$, $\Delta T_1 = 80°C$, $\Delta T_2 = 10°C$, $(T_1 - T_3)/2 = 70°C$, What this table shows is that if $q \neq 0$ that the maximum stress occurs when the CTEs are not matched, $(\alpha_1 \neq \alpha_2)$. This is contrary to common expectations. On the other hand if $\Delta T_1 = \Delta T_2$, $(q = 0)$, then the minimum stress is zero with $\alpha_1 = \alpha_2$. Consider the opposite case with $\alpha_1 > \alpha_2$ in Table 2. Table 2 shows that if $q \neq 0$ and $\alpha_1 > \alpha_2$ then the minimum stress (absolute value) occurs when $\alpha_1 = \alpha_2$.

From Table 1 and Table 2 together for the case when q is such that $(T_1 - T_3)/2 = 70°C$, it is seen that the minimum stress (minimum absolute value of TF) is zero and occurs for $\alpha_1/\alpha_2 = 0.125$. The above also shows that for positive q, that $\alpha_1 < \alpha_2$ is preferable over $\alpha_1 > \alpha_2$.
This leads to a design guideline shown in Figure 5. If there is a shear stress design limit

$$\tau_{01} \le \tau_{des} \tag{31}$$

Then based on this, a plot can be made to choose α_1/α_2 as shown in Figure 5, depending upon whether q is zero or not and depending upon magnitude of ΔT_1.

Example 2

Consider the problem of a superconducting ceramic $Yba_2Cu_3O_{7-x}$ mounted on a Si substrate. This problem was solved in [4] using a finite element analysis with hundreds of elements. The data are:
$E_1 = 9.6$ Gpa, $t_1 = 1$ micron, $\alpha_1 = 11.8 \times 10^{-6}$/K, $\Delta T_1 = -823$K,

$E_2 = 131$ Gpa, $t_2 = 200$ microns, $\alpha_2 = 2.9 \times 10^{-6}$/K, $\Delta T_2 = -823$K.

The results using Equations (2) and (26) above are

$$\tau_{o1} = 440 MPa \text{ and } \sigma_{on} = 1401 MPa$$

The maximum stresses from the finite element analysis were

$\tau = 524$ MPa and $\sigma = 931$ Mpa.

The results show that the equations derived in this paper give a good approximation to the stresses in layered packages without the expense of elaborate finite element analysis.

References

[1] Timoshenko, S.P. and J.N. Goodier, J.N., Theory of Elasticity, CR McGraw Hill,1951,1970,1987, ISBN 0-07-064720-8, pp 53-54.
[2] Papadakis, P., "Computational Aspects of the Determination of Stress Intensity Factors for Two-Dimensional Elasticity", Ph.D. Dissertation Univ. of Maryland 1989
[3] Hui, C.Y., H.D. Conway and Y.Y. Lin, "A Reexamination of Residual Stresses in Thin Films and of the Validity of Stoney's Estimate",Jnl of Electronic Packaging, ASME, Sept. 2000, Vol. 122, pp 267-273.
[4] Lau,J., C.P. Wong, J.L. Prince and W. Nakayama, Electronic Packaging, Design, Materials, Process and Reliability, CR McGraw Hill, 1998, ISBN 0-07-037135-0, p 288-290.

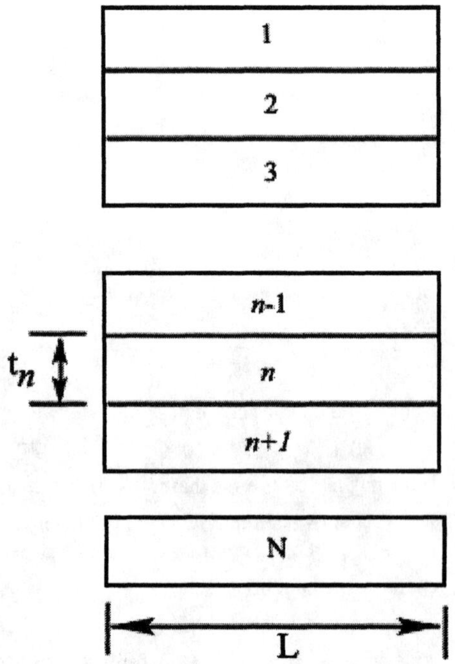

Figure 1. Circuit Board with N Layers

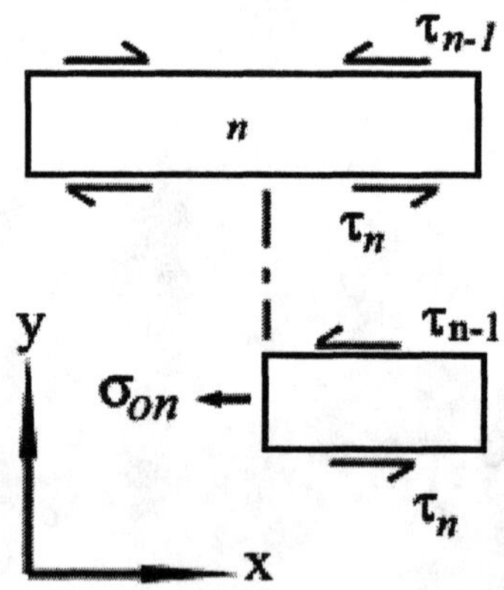

Figure 2. Normal and Shear Stresses in Layer *n*.

Table 1. Thermal Stress Factor TF for Two Layered Board for Steady State Heat Flow q

α_1/α_2	$(1-\alpha_1/\alpha_2)\Delta T_1$	$(T_1-T_3)/2$	TF
0	80	70	10
0.1	72	70	2
0.2	64	70	-6
0.3	56	70	-14
0.4	48	70	-22
0.5	40	70	-30
0.6	3L	70	-38
0.8	16	70	-54
1.0	0	70	-70

$\Delta T_1 = 80°C$, $\Delta T_2 = 10°C$, $(T_1 - T_3)/2 = 70°C$, *and* $\alpha_1 < \alpha_2$

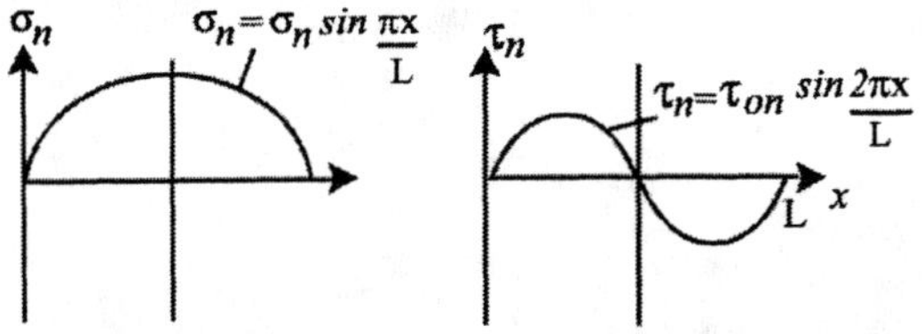

Figure 3. Distribution of Stresses

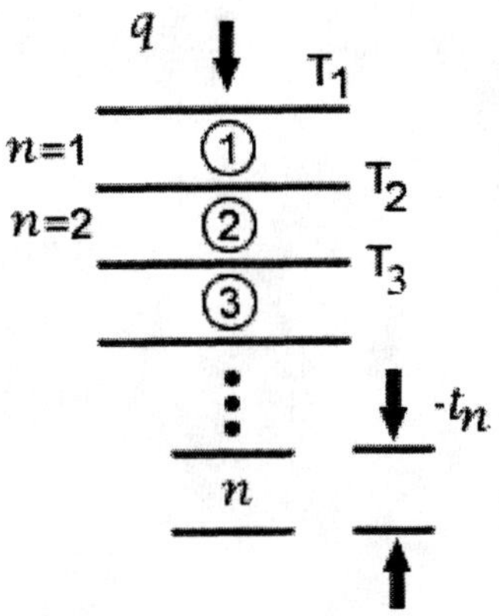

Figure 4. Heat Conduction through a Multi Layered Board

Table 2. Thermal Stress Factor TF for Two Layered Board.

α_1/α_2	$(1-\alpha_1/\alpha_2)\Delta T_1$	$(T_1-T_3)/2$	TF
1.0	0	70	-70
1.2	-16	70	-86
1.4	-32	70	-102
1.6	-48	70	-118
1.8	-64	70	-134
2.0	-80	70	-150

$\Delta T_1 = 80°C$, $\Delta T_2 = 10°C$, $(T_1 - T_3)/2 = 70°C$, *and* $\alpha_1 > \alpha_2$

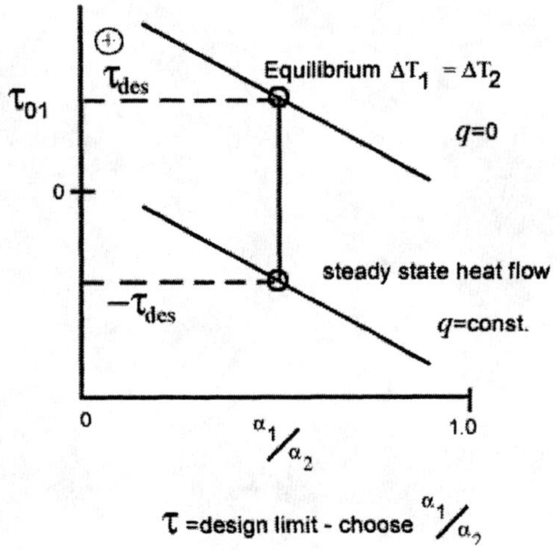

Figure 5 Design Limits for CTE Ratio Depending Upon Heat Flow

Proceedings of
2001 ASME International Mechanical Engineering Congress and Exposition
November 11–16, 2001, New York, NY
EPP-Vol. 1

IMECE2001/EPP-24735

HIGH DIGITAL SPEED LEVEL-2 INTERCONNECTIONS VIA A NEW MICROCONTACT DESIGN

Dariusz R. Pryputniewicz*, **Dimitry G. Grabbe**, and **Ryszard J. Pryputniewicz**
NEST – NanoEngineering, Science, and Technology
CHSLT – Center for Holographic Studies and Laser micro-mechaTronics
Mechanical Engineering Department
Worcester Polytechnic Institute
Worcester, MA 01609

* Current address: Draper Laboratory, Cambridge, MA, 02139

ABSTRACT

Requirements for high digital speed, high density, level-2 interconnections have led to development of a new microcontact. Design of this microcontact allows for separable and reusable interconnections. In this paper, we discuss methodology used to develop the microcontact, allowing 100% material utilization, present its design, including analysis and process optimization, and summarize its characteristics as they relate to electronic packaging.

Keywords: microcontact, separable, reusable, high digital speed, high density, level-2 interconnection, analysis, process optimization, electronic packaging.

1. INTRODUCTION

Recent competitive trends to miniaturize systems have introduced extraordinary challenges concerning design of today's connectors and have led designers to develop a number of different concepts for structures and components at virtually all levels of complexity [1]. For practical considerations, many connectors are put together from subassemblies which require some kind of foundation, joining, and/or safety mechanisms, Fig. 1. Many of these mechanisms are electromechanical in nature, e.g., proximity fuses, limiting switches, active foundations, microcontacts, etc., and for reasons of economy, are to a large degree mechanical in design, requiring ever tighter tolerances to satisfy ever increasing performance specifications.

By necessity, the materials and configurations employed for these mechanisms are selected on the basis of their electrical, mechanical, thermal, magnetic, and/or some other individual or coupled performance characteristics. Therefore, the task is to work within the specific performance constraints of available materials to design configurations assuring operationally functional and reliable mechanisms. This paper addresses development of one of such mechanisms, a microcontact for separable, reusable, level-2, high density, and high digital speed interconnections. More specifically, this microcontact, known as "micro-v", is a microspring, Fig. 2.

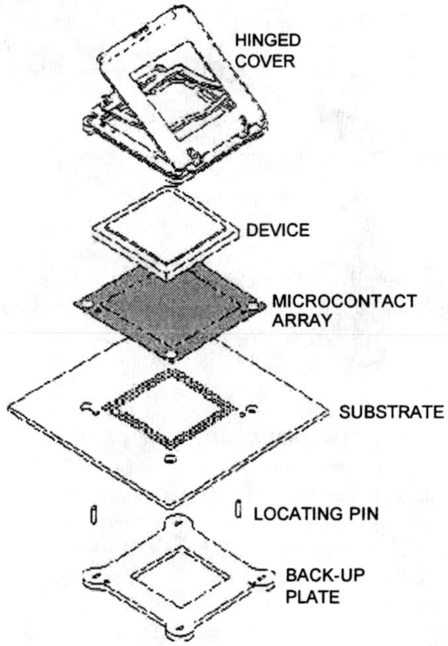

Fig. 1. A connector assembly.

In order to achieve their full operational characteristics, the initial micro-v contacts were manufactured out of a flat stock of special noble metal alloys, which are very expensive, but have very attractive combination of electrical and mechanical properties. In addition, because of the micro-v's geometric configuration, material utilization during manufacture was only 14%. Although the micro-v contacts were produced and tested to demonstrate that they have satisfied the required operational characteristics they were too expensive to manufacture. To overcome the cost, while approximating the geometry, which is very functional, the idea was to use the same material in the form of 100 Nm (0.004 in.) diameter wire (the microwire) to achieve 100% of material utilization during formation of the microcontacts.

Two manufacturing approaches were considered to produce the microcontacts from the microwire. The first approach, based on making loose pieces out of the microwire and then assembling them into plastic carriers (housings) of specific configurations, as required by a given application, proved to be too difficult. The second approach was to form in-place the desired shapes of these microcontacts using the method of wiping pins. This required inserting short segments of the microwire into the plastic carrier (i.e., housing) and wiping both sides, protruding on top and bottom of the housing, to form the desired shape, which would have spring-like characteristics while delivering required contact forces over very large deflections [1,2].

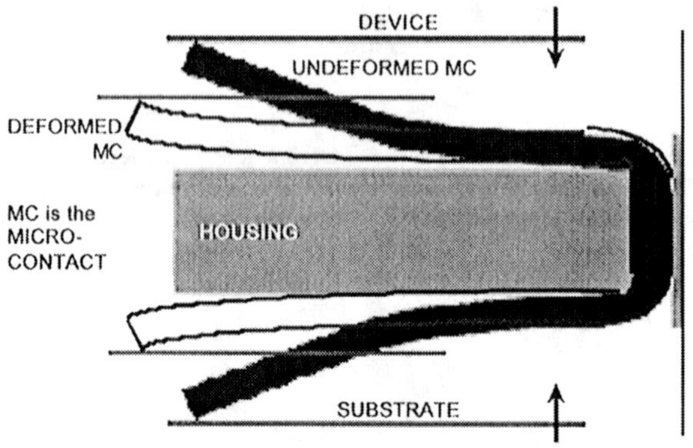

Fig. 2. The micro-v contact.

In the following sections, the methodology used to develop the microcontacts is outlined, results of modeling/simulation are summarized, and pertinent performance characteristics of the microcontacts are discussed.

2. DEVELOPMEMT OF THE MICROCONTACT

Development of the microcontact has been facilitated by an approach to design and analysis of microcomponents based on a hybrid methodology, which unifies, or hybridizes, analytical, computational, and experimental solutions (ACES) methodology [1-5], to obtain results where they would be difficult to obtain using any one of the methodologies alone. The ACES methodology uses one or more methods – as needed – from each of the three categories: (1) analytical, (2) computational, and (3) experimental to solve a specific problem. In this case, it is used to develop a new microcontact.

Today, many connectors are put together from subassemblies, Fig. 1, in which some components are made in-situ and require non-rigid, or elastic, supports facilitating modularity of the complex structures. In general, modular assemblies are intended to facilitate the production of complex and costly subassemblies or, as importantly, provide enabling mechanisms for numerous easy to engage operations between various components comprising the complex structure of interest, while controlling properties such as load-deflection characteristics in the microcontacts.

A non-rigid, elastic support must provide a number of features assuring a reliable, durable, and manufacturable interconnection. A good microcontact must have deflection compliance sufficiently large to accommodate the operational requirements of a given system. This implies a spring-like function and, therefore, places a requirement that a microcontact should exhibit a spring property. The physics of spring-like microcontacts stresses the importance of adequate load-deflection characteristics needed to achieve reliable and durable functionality. It also stresses the importance of selection of appropriate materials, which satisfy specific requirements for a good mechanism and produce a manufacturable product.

One way to manufacture a product that would deliver a controlled and pre-determined force and maintain this force during its life is to use a wiping pin method to produce a microcontact shape as described in the pending AMP patent application. It is stressed here that the shape produced will depend on the desired force. Before the microcontact could be manufactured it should first be analyzed after it has been designed. Because of the complexities due to elastic-plastic characteristics required in the microcontact, the most efficient way to perform such analyses is to use the computational method, which can be either finite element method, or boundary element method, or finite difference method. In this study, finite element method was used.

3. FINITE ELEMENT METHOD

Analyzing a process of a microcontact formation proved to be a complex, time consuming, and difficult task. Analytical, closed form solutions of this problem cannot be obtained practically. The success in the solution of complex 3D elastic-plastic problems is concurrent with the development of numerical methods, in particular, the finite element method (FEM). To aid in this task as well as to provide graphical representations, the FEM is used to model the forming process and to characterize the final shape of the microcontact in terms of its load-deflection characteristics.

Mathematically, FEM formulations are written as

$$\{\mathbf{F}\} \quad [\mathbf{K}]\{\mathbf{r}\}\ , \qquad (1)$$

where $\{\mathbf{F}\}$ is the vector of global nodal forces of the assembly, $\{\mathbf{r}\}$ is the vector of global nodal displacements of the assembly, and $[\mathbf{K}]$ is the assembled system global stiffness matrix [1, 6-9], which results in a number of matrix operations and calculations.

The analysis of plastic action in metal formation, through the use of the FEM, depends on three basic concepts: (1) yield criterion, (2) flow rule, and (3) hardening rule [1], which were incorporated in the modeling and simulation to obtained results presented in this paper.

In this paper, a half-model was used, because of the symmetry of the microcontact considered, Fig. 2, and the desire to speed the solution process. Figure 3 shows the 3D geometry of the FEM half-model and the corresponding mesh of a section of the microwire prior to the formation process of the microcontact.

Working on a half-model allows for faster solution time while retaining all-important and accurate information. In fact, because of the symmetry, the results obtained are identical to those that would be obtained if a complete 3D model were used.

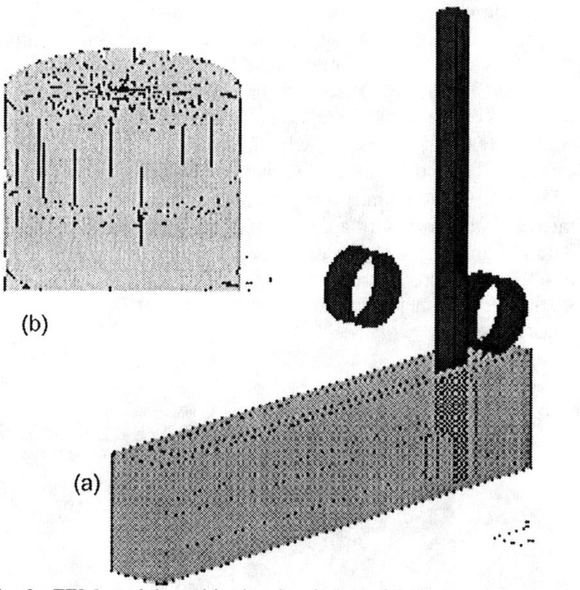

Fig. 3. FEM model used in the simulation of the microcontact formation process: (a) half-model, (b) mesh detail of the microwire.

4. MATERIAL PROPERTIES

FEM analyses depend on input parameters defining boundary, initial, and loading (BIL) conditions, as well as material properties, each of which must be well defined in order to obtain reliable results. Although the BIL conditions can usually be reasonably well defined, material properties for samples in small geometries, such as those found in the microcontact discussed herein, are not readily available, at this time.

More specifically, in order to model elastic-plastic deformations of the microwire as the microcontact is being formed, mechanical properties of its material - as a function of radial position across the microwire - should be known.

Unfortunately, the only properties that were available from the producer of the microwire were average values based on the entire cross section of the microwire. FEM simulation of the formation process based on average values of the material properties showed that the process would not produce desired characteristics in the finished microcontact.

To overcome this deficiency, nanoindentation and ACES methodologies [10-13] were used to measure material properties of the microwire [1].

Nanoindentation is a process in which a force is applied to a surface of a test sample (having dimensions of the microcontact) using an indenter of a known geometry and a resulting penetration of the indenter below the initial interface is measured as a function of the applied force. During these measurements, the forces and the corresponding penetration depths are measured with the NN and nm accuracies, respectively. Based on the force-penetration depth measurements, local material properties are determined.

The fundamental relationship between the measured force-penetration depth characteristics and the material properties of the test sample are given by the equation

$$\frac{dF}{dh} \quad E\sqrt{A}\,\frac{E}{1 \quad \nu^2}\ , \qquad (2)$$

where dF/dh represents slope of the force-penetration curve at the specific radial position on the test sample, E is the constant characterizing the indenter used, A is the area of the specific indentation site, E is the modulus of elasticity of the test sample, and ν is its Poisson's ratio.

Use of the material properties measured as a function of the radial position across the 100 Nm diameter microwire [1], based on Eq. 2, yielded results with attractive microcontact characteristics.

5. FORMATION PROCESS

The formation process starts with insertion of a microwire into a hole in the plastic housing; the tolerance on the hole diameter is such that it tightly holds the microwire.

The microwire is pushed through the hole until its one end extends a specific length beyond the housing, e.g., beyond its bottom surface. The final shape and the size of the microcontact, that is to be formed, specify the length of the microwire extending from the hole. Next, the other end of the microwire is cut off to appropriate length. Finally, two pairs of pins are brought into contact with the microwire, one pair on top, Fig. 4, and the other pair on the bottom of the housing, and are moved along its surfaces forming the microspring by the wiping action on the microwire.

Parameters shown in Fig. 4 specify configuration of the wiping pins and control the final shape of the microcontact formed during the process. These parameters are particularly suitable for optimization of the formation process of the microcontact [1].

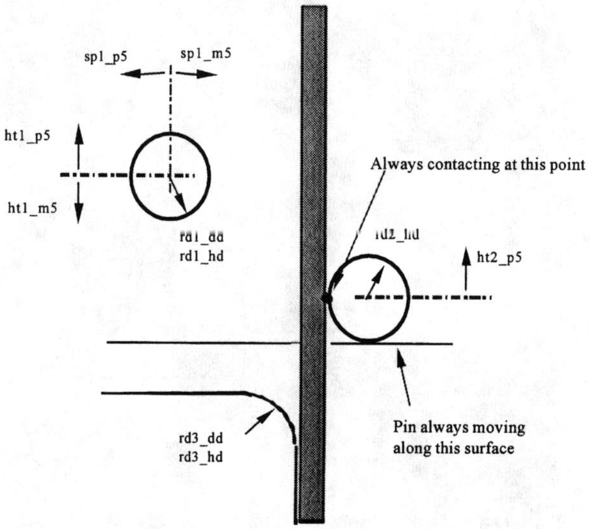

Fig. 4. Parameters affecting quality of formation of the microcontact, for the top pair of the wiping pins.

The forming process takes place on both sides, i.e., top and bottom, of the housing simultaneously. Center-to-center distance between the pins and their locations relative to the microwire are based on the final shape and size of the microcontact and have to be determined [1]. The pin on the right side of the wire extending above the top surface of the housing, Fig. 4, is used to form the cusp, while the pin on the left is used to form the tip of the microcontact.

6. RESULTS AND DISCUSSION

Using the models shown in Figs 3 and 4, simulations of the microcontact formation process were performed [1].

Representative results of the simulations, showing the microcontact in different stages of formation, are displayed in Fig. 5. As expected, the areas of highest stress concentrations occur at the bends of the component. The values shown correspond to plastic deformations when the microcontact is formed. Based on the 3D FEM results, shapes of the "activated" and "released" microcontacts are as shown in Fig. 6. These results correspond to the representation of the microcontact action displayed in Fig. 2.

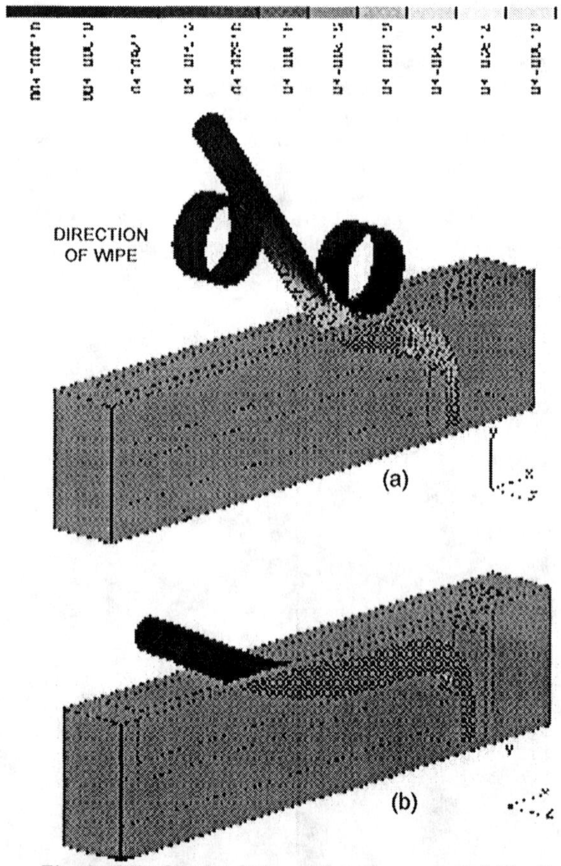

Fig. 5. FEM 3D simulation of the microcontact: (a) beginning of the formation process, (b) final shape of the formed microcontact.

Final centerline shape of the nominal geometry of the microcontact, based on the 3D FEM model, is shown in Fig. 7.

Following formation of the microcontact, it is necessary to determine its load-deflection characteristics, as they will influence the specific interconnection application. For the microcontact considered in this paper, the load-deflection characteristics, based on the 3D FEM analysis [1], are shown in Fig. 8. These characteristics indicate the contact load of approximately 25 grams for the maximum tip deflections of about 230 Nm (9 mils). This deflection, when compared with the 100 Nm (4 mil) diameter of the microwire out of which the microcontacts were formed, indicates large deformations. These large deformations are nonlinear functions of the contact load. According to Fig. 8, initially the microcontacts behave as soft springs exhibiting low stiffness, then they become hard springs with much higher spring constant.

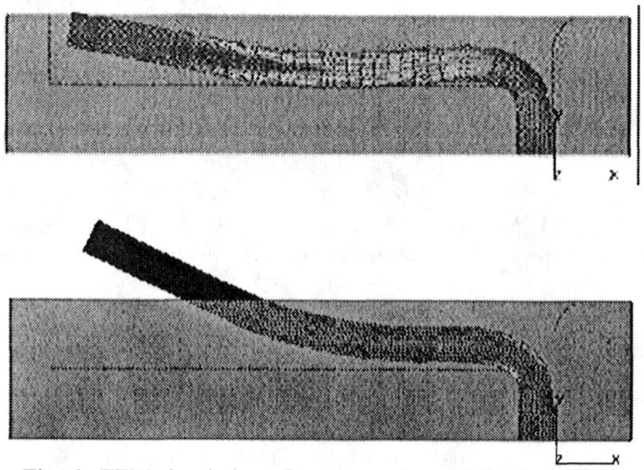

Fig. 6. FEM simulation of a microspring: (a) "activated" or loaded state, and (b) "released" or free state.

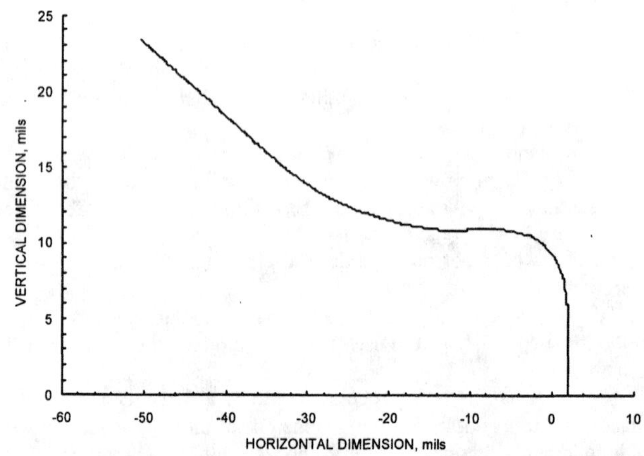

Fig. 7. Final centerline shape of the nominal geometry of the microcontact, corresponding to the shape shown in Fig. 6b, based on the 3D FEM model.

The horizontal movement of the tips of the microcontact, i.e., the microcontact wipe, was determined to be 75 Nm (3 mils) while the tip displacement ranged from zero to 230 Nm (9 mils) [1]. The magnitude of wipe directly influences quality of the interconnection.

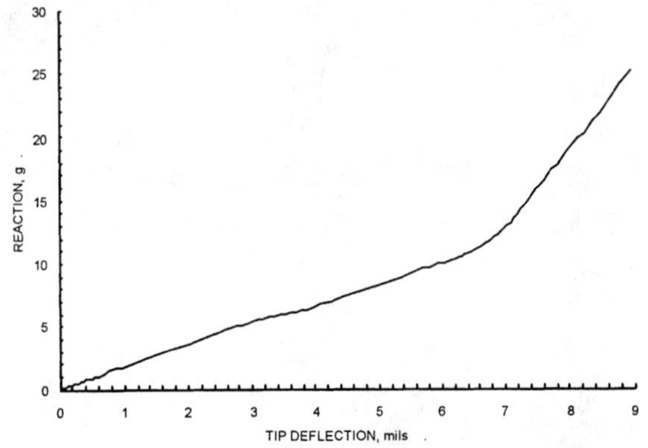

Fig. 8. Load-deflection characteristics of the microcontacts based on the 3D FEM model.

7. CONCLUSIONS

The results presented in this paper show the validity of using a hybrid approach in analyzing the formation process of microcontacts.

Determination of material properties was viewed as being of primary importance. Many times, analyses are performed by using standard "book" values. These book values, though, are based on samples much larger then the 100 Nm (4 mil) diameter ones used to form the microcontacts discussed in this paper. Therefore, material properties had to be measured in order to make the analyses possible. Furthermore, because of the nature of elasto-plastic formation of the microcontact, the material properties had to be known as the function of radial position across the sample. For this purpose, nanoindentation was used in this study.

The comparison between analytical and computational modeling showed both, the strengths and weaknesses of each one. The analytical modeling, though less time consuming and much simpler then the computational one, is only, and at best, an approximation of the component. However, it can give useful insight into the values to expect. Computational modeling is subject to the same problems as the analytical modeling, because it is also an approximation of the component behavior, though more sophisticated. It allows the analysis of much more complex and realistic structures then analytical modeling. However, it is very sensitive to input parameters such as material properties, boundary conditions, and shape discretization.

Results of 3D FEM simulation of the microcontact indicate that the maximum tip deflections of 230 Nm are achievable, while the corresponding wipe length is 75 Nm, for the contact load of 25 grams. Furthermore, the microcontact clearly exhibits nonlinear load-deflection behavior characterized by two slopes, or spring constants. Initially, for loads up to 12 grams, the microcontact behaves as a soft spring with tip deflections up to 173 Nm (6.8 mils). Then, at loads above 12 grams, the microcontact becomes a hard spring with the spring constant about 3.35 times higher than the initial one.

The microcontacts developed in this paper are particularly suitable for level-2, high digital speed interconnections requiring separable, reusable characteristics in electronic packaging.

8. REFERENCES

1. Pryputniewicz, D. R., 1997, ACES *approach to the development of microcomponents*, MS Thesis, Worcester Polytechnic Institute, Worcester, MA.
2. Pryputniewicz, D. R., Grabbe, D. G., and Pryputniewicz, R. J., 2001, "Development of a new microcontact design for separable, reusable, high digital speed level-2 interconnections," *Proc. Internat. Congress on Experimental and Applied Mechanics for Emerging Technologies*, SEM, Bethel, CT, pp. 832-835.
3. Pryputniewicz, D. R., and Pryputniewicz, R. J., 1996, "Hybrid approach to development of a separable interconnection for high density applications," *Proc. ASME Symp. on Electronic Packaging*, Atlanta.
4. Pryputniewicz, R. J., 1994, "A hybrid approach to deformation analysis," *SPIE Proc.*, 2342:282-296.
5. Pryputniewicz, R. J., and Grabbe, D. G., 1991, "Developments in micromechanics through analysis and experimentation," Ch 24 in *Proc. of the 10th Internat. Invitational UFEM Symp.*, Worcester Polytechnic Institute, Worcester, MA, pp. 505-532.
6. Grandin, H. J., 1986, *Fundamentals of the finite element method*, Waveland Press, Prospect Heights, IL.
7. Hughes, T. J. R., 1987, *The finite element method: linear statics and dynamic finite element analysis*, Prentice Hall, Englewood Cliffs, NJ.
8. Zienkiewicz, O. C., and Taylor, R. L., 1989, *The finite element method: basic formulation and linear problems*, Mc-Graw Hill, New York.
9. Zienkiewicz, O. C., and Taylor, R. L., 1989, *The finite element method: solid and fluid mechanics, dynamics, and nonlinearity*, Mc-Graw Hill, New York.
10. Pryputniewicz, D. R., and Pryputniewicz, R. J., 1995, "Nanoindentation approach to characterization of materials for electronic packaging," *Advances in Electronic Packaging*, EEP-Vol. 10.2, ASME, New York.
11. Furlong, C., and Pryputniewicz, R. J., 1995, "Study of the elasto-plastic behavior of the spherical nanoindentation process," *Proc. SEM Spring Conference*, SEM, Bethel, CT, pp. 727-734.
12. Furlong, C., and Pryputniewicz, R. J., 2001, "Measurements and modeling of a nanoindentation process for MEMS," *Proc. Internat. Symp on MEMS: Mechanics and Materials*, Portland, OR, pp. 28-33.
13. Pryputniewicz, D. R., Furlong, C., and Pryputniewicz, R. J., 2001, "ACES approach to the study of material properties of MEMS," *Proc. Internat. Symp on MEMS: Mechanics and Materials*, Portland, OR, pp. 80-83.

Proceedings of
2001 ASME International Mechanical Engineering Congress and Exposition
November 11–16, 2001, New York, NY
EPP-Vol. 1

IMECE2001/EPP-24736

MECHANICAL CONSIDERATIONS OF SHIN-ETSU ELASTOMER AS A Z-AXIS INTERCONNECT

C. Cox, W.F. Schmidt, and M.H. Gordon*
Department of Mechanical Engineering
University of Arkansas
Fayetteville, AR 72703
*501-575-4458, Email: mhg@engr.uark.edu

W. Marsh**, G. Bates, and M. Lucas
Northrop Grumman Corporation
Electronic Sensors and Systems Sector
Baltimore, Maryland 21203
**410-993-8681, Email: william_j_marsh@mail.northgrum.com

ABSTRACT

This research effort addresses the integration of a dense z-axis interconnect technology, Shin-Etsu, into a 3D thermo-mechanical multi-chip module with two to eight stack layers. Shin-Etsu is a matrix of vertical beryllium-copper wires embedded in silicone. Since it is a composite material, accurately determining Shin-Etsu's mechanical properties is extremely important for estimating long-term reliability of the electronic package.

Five percent compression of the Shin-Etsu elastomer was specified to maintain proper electrical contact between substrates. To determine the corresponding force, Tinius-Olsen compression tests up to 15% were conducted on both single and double layer samples of Shin-Etsu. In addition, these data were used to infer Shin-Etsu's Young's modulus. Both series of test were also conducted with and without alumina (Al_2O_3) to simulate the contact surface boundary condition between Shin-Etsu and the LTCC substrate (2 pieces were used for the single Shin-Etsu case, and 3 pieces were used for the 2 layer Shin-Etsu case). All compression tests follow the same trends, though the sample to sample scatter is relatively large (+/- 75%). Because Shin-Etsu's Poisson's ratio is near 0.5 (volume is conserved), the frictional forces between the contact surfaces is important. By carefully accounting for these frictional forces, we infer a Young's modulus for Shin-Etsu of 3.25 MPa. Using this value with the appropriate contact model (we found that a simpler model with infinite friction is suitable in our case), we are able to successfully design a two-board test vehicle which incorporates Shin-Etsu.

Introduction

Motivation remains high to continually pack more electronic capability into a smaller volume. Packaging engineers meet this challenge by miniaturizing electronic systems, while simultaneously reducing I/O requirements and parasitics. These changes improve the overall electrical performance of the electronic package, but the problem of linking the electronic devices becomes a serious issue. Current interconnect technology does not effectively utilize substrate real estate. Wire bonding and tape automated bonding (TAB), for example, are limited by solder ball and bonding pad pitch distance requirements. Consequently, maximum performance of today's electronic systems is primarily limited by the interconnections between components and subsystems. [3]

Some packaging techniques, such as COG (chip-on-glass) and COB (chip-on-board) require technology that makes electrical connection in the z-direction only [5]. Z-axis interconnect technology allows engineers to decrease the length of wiring needed between IC's (integrated circuited). Thus, more electronic components and subsystems can be placed in a given volume.

Elastomeric connector technology is relatively new compared to conventional methods, such as PIH (pin-in-hole), interposer dendrite, solder ball, and mechanical cantilever spring connectors. Elastomeric connectors are a dense array of wires embedded in a flexible polymer, usually silicone. These connectors provide redundant contacts per pad, and may have an x-y pitch of 0.05mm and contact resistances as low as 0.100Ω. [2,3] Consequently, z-axis interconnect technology, utilizing elastomeric connectors, plays a prime role in the advancement of these MCMs.

Each multichip module requires an appropriate chip-to-substrate interconnection for its specific application. Unfortunately, no universal solution exists that can be applied to every application. Therefore, the type of interconnect technology used will depend primarily on the application requirements of the system in which it will be integrated. Electrical and mechanical factors such as electrical conductivity, alignment, required contact force, contact resistance, and operating temperature must be considered.

Interconnect design objectives for this research work include,

- High-density interconnection matrices.
- Reliable electrical performance with short interconnection lengths.
- Minimal noise and crosstalk.

- Reliability in a typical military environment.
- Demateability

Many of the interconnection techniques mentioned previously require precision alignment of the interconnect media to the chip or substrate. Reflow equipment for detaching chips from the substrate is also needed. We chose Shin-Etsu connectors since exact alignment is unnecessary due to the high wire density and redundant contacts per pad and since mechanical bonding is accomplished by simple compression.

Silicone and beryllium-copper are used extensively in the electronics industry, and their material properties are widely known. However, elastomers are basically silicone "doped" with beryllium-copper wires. [1] Thus, this new material may exhibit viscoelastic and solid material property characteristics. [4] Conductor materials such as carbon fibers, or particles, silver, and nickel are also often used. Consequently, the lowest level of combined resistance is a function of the kind of metal used in the connector, the surface finish, and the contact pressure. Therefore, successful integration of Shin-Etsu, or any other interconnect material, into our final product, a two-board test vehicle (TBTV), requires knowledge of its mechanical properties. In this study, we focus on determining these properties, and using them in our design process.

Experimental Test Setup and Procedure

A Tinius-Olsen series 5000 Universal Testing Machine (UTM) (Figure 1) was used to conduct various compression tests on the Shin-Etsu elastomer samples. Tests were conducted at 0.5 mm/min with a 200 kg load cell set at 10 percent of full scale. We thus had a compression range of –20 kg to +20 kg. However, by pre-compressing the system to –20 kg and then zeroing the force, we were able to move the compression range to -40 kg to 0 kg, optimal settings for our compression-only test. With the 12-bit processor, our single bit resolution was then 9.8g.

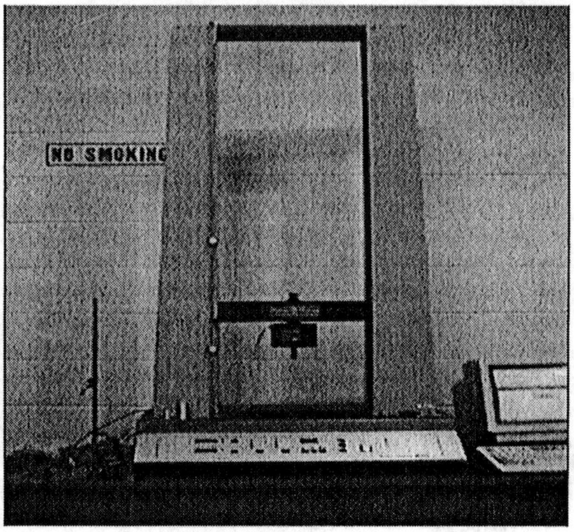

Figure 1 *Tinius-Olsen series 5000 Universal Testing Machine (UTM)*

To determine the onset of compression, we normalized all data to start when the compression force exceeded 5000 Pa (about 6 bits). This normalization allowed for better comparisons between samples and runs due to non-uniformity in each sample's surface.

Two aluminum cylinders, 25.4mm in diameter and 50.8mm long, were fabricated for use with the testing machine. One cylinder, attached to the upper crosshead unit, is used to apply pressure to the samples, while the second cylinder serves as a base. Shin-Etsu and alumina (Al_2O_3) dimensions are roughly 10mm x 10mm x 1.7mm and 17.8mm x 17.8mm x 0.52mm, respectively.

The following three test configurations were conducted:

1. Individually compress 9 single Shin-Etsu samples.
2. Individually compress 2 single Shin-Etsu samples between 2 alumina pieces.
3. Compress two Shin-Etsu samples interspersed with 3 alumina pieces.

The first test allowed us to check for sample to sample variation. The second test allowed us to check the compressibility of Shin-Etsu between alumina (closely matching the LTCC used in the TBTV). The third test best simulated the TBTV (which would be a stacked device) and allowed comparison to individual compression. Note that the polished alumina was chosen as the contact surface because it has mechanical properties that are similar to LTCC, the substrate material used in the actual model, yet costs significantly less.

Finite Element Analysis

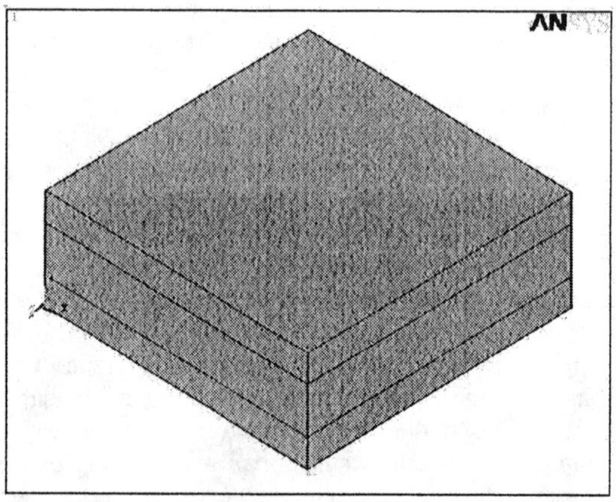

Figure 2 *3D 1/8th model of Tinius-Olsen compression test*

Finite element analysis (ANSYS 5.6) is used to simulate both the Tinuis-Olsen tests and the final TBTV. Model size is reduced by taking advantage of symmetry. Figure 2 shows a 3D 1/8th-model used to simulate the Tinius-Olsen compression tests. Note that we only simulate compression case 2, and that we model the alumina with the same cross-sectional area as the Shin-Etsu. Although the alumina is larger than the Shin-Etsu, there is very little effect on the compression, and the model is much simpler with the same dimensions. Although we only model case 2, we really are also modeling case 1 as well. Both the alumina and aluminum are significantly stiffer than the Shin-Etsu so that they are numerically equivalent (numerical simulations affirm this). As we will demonstrate, the coefficient of friction at the contact surface is the most dominant factor, and we investigate that parametrically.

Figure 3a shows a 3D half-model of the key mechanical aspects of the final TBTV – diamond, for example, was not modeled since its role was thermal and has no force applied to it. In Figure 3b, The thickest top and bottom blocks represent 6.25mm thick kovar which serves as a support frame. Just above the bottom Kovar piece is one of the two LTCC boards (each 1.74mm thick). Above the LTCC board is a Shin-Etsu layer (2 9.5mmX24mm pieces per layer, each 1.7mm thick). Above that is another LTCC board, and above that is the other Shin-Etsu layer.

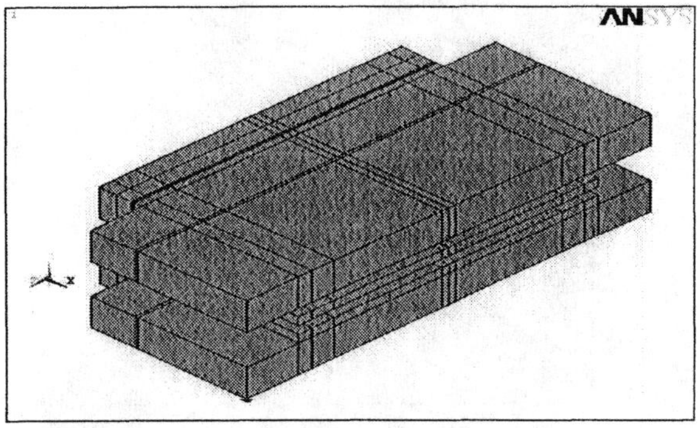

Figure 3a *Isometric view of 3D 1/2 model of TBTV*

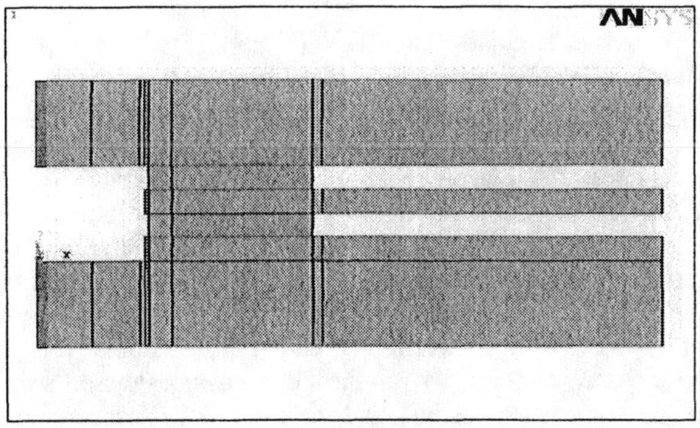

Figure 3b *Side view of 3D 1/2 model of TBTV*

For the Tinius-Olsen model, the bottom nodes (representing the aluminum-alumina contact surace) are displaced in the y-direction one-half of the desired compression, and the top nodes (Shin-Etsu) are constrained to represent the symmetry plane of the one-eighth model. The other symmetry planes are the two visible ones in Figure 2. Contact elements are used to mate the alumina and Shin-Etsu surfaces.

For the TBTV model, the right hand side of Figure 3b is the symmetry plane. Three nodes on the top and bottom Kovar pieces (near the left edge) are constrained one-half of the desired displacement (downward on the top piece, and upward on the bottom piece). These nodes represent the screws used to compress the assembly.

Results and Discussion

Figure 4 shows the compression test results for case 1. Note that one curve has symbols to signify the number of data acquired; the other curves have the same number of data, but symbols are not shown for clarity. Because it has been found that 5% compression gives desirable electrical performance for our device, we focus on this value. Figure 4 clearly shows the large scatter from sample to sample. At 5% compression (or 0.05 strain), the stress varies from 200 to 600 kPa. However, we suspect that the variation has more to do with how the data are assigned zero-strain. As stated, we zero the strain once the stress exceeds 5 kPa for consistency. Yet, were other values chose, the curves could better align with each other. In any case, we are most interested in the implied Young modulus. Again, because 5% compression is our reference, we evaluated the Young's modulus for the data between 3% and 7%. The nine values range between 13.4 and 20.2 MPa with an average value of 15.8 MPa.

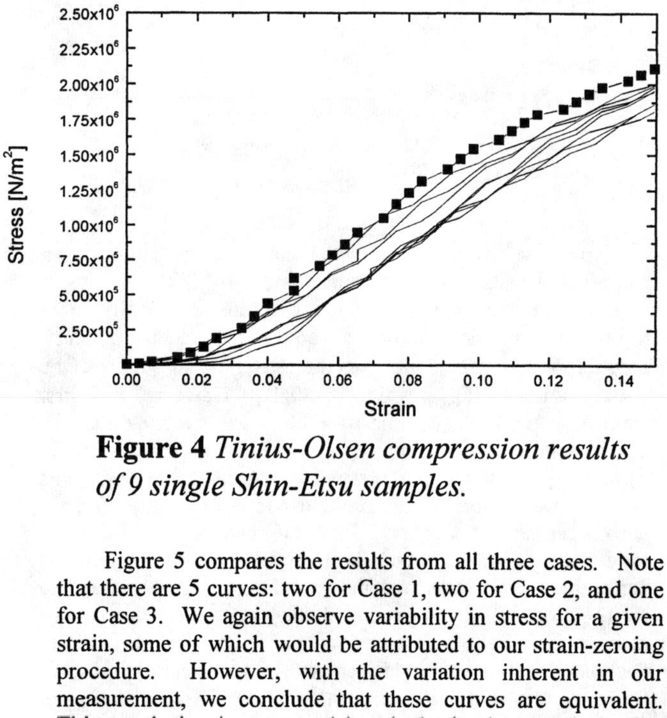

Figure 4 *Tinius-Olsen compression results of 9 single Shin-Etsu samples.*

Figure 5 compares the results from all three cases. Note that there are 5 curves: two for Case 1, two for Case 2, and one for Case 3. We again observe variability in stress for a given strain, some of which would be attributed to our strain-zeroing procedure. However, with the variation inherent in our measurement, we conclude that these curves are equivalent. This conclusion is not surprising; both aluminum and alumina are similar in hardness (compared with the Shin-Etsu), so that Cases 1 and 2 should vary little. There is also no theoretical reason to expect Case 3 to vary either. If each piece of Shin-Etsu between 2 pieces of alumina is thought of as a spring, then Case 3 represents compressing two identical springs twice the distance as either spring alone and no difference in required pressure (or stress) is expected.

For modeling purposes, we need to select a value for the Shin-Etsu's Young's modulus. For these 5 curves in Figure 5, the values range from 11.0 to 22.5 MPa with an average value of 16.4 MPa. Recalling that the previous 9 samples gave an average Young's modulus of 15.8 MPa, we choose 16.0 MPa. Note that numerically, this value will result in a perfectly straight stress-strain curves in contrast to Figures 4 and 5. This difference will result in numerically higher compression forces required than we measure (safer for design purposes). However, when the TBTV is assembled, the desired 5% compression is measured after the assembly screws are hand-tight. Thus, combined with the relatively heavy top assembly plate, we estimate at least 50 kPa of initial stress. 50 kPa of stress corresponds to about 3% strain, or where our curves are nearly linear. Thus, using perfectly straight stress-strain relations should reasonably simulate our actual assembly.

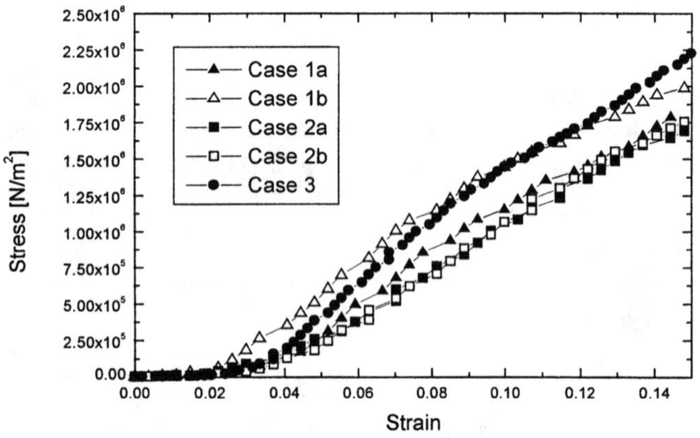

Figure 5 *Comparison of the three compression test cases*

Our first, and simplest, model simulated the Tinius-Olsen compression with a Young's modulus of 16 MPa. One might expect that the predicted force for this essentially one-dimensional problem would be obtained from the Young's modulus (E), the amount of compression strain (δ), and the surface area A ($F_{expected}=E*\delta*A=80N$). However, our first simulation produced a force of 428N, over five times larger. The reason for this large discrepancy is that our ideal calculation assumes no friction. Our first model had no contact elements and thus was equivalent to infinite friction. The reason friction is so important is because the Shin-Etsu material has a Poisson's ratio near 0.5; thus volume is conserved. If the material is constrained by friction and thus not free to expand, then the required force for compression will increase as we observe. Our actual case, of course, is somewhere in between zero and infinite friction. Ideally, we would measure the coefficient of friction. Since we were not able to measure this, we introduced contact elements into our model and simulated a range of values for the coefficient of friction.

Figure 6 shows the required force as a function of the coefficient of friction. This curve asymptotically approaches the two limits previously described (80N for zero friction and 428N for infinite friction). To more accurately predict the actual force required to compress the TBTV, we note that the reported coefficient of friction measured between rubber and concrete (the closest pair to our Shin-Etsu and Aluminum pair that we could find) is 0.8. [6] If the wires in the Shin-Etsu dominate the contact, then we note that the coefficient of friction between brass and steel is 0.5. Consulting Figure 6, we see that at 0.8, there is virtually no change from the maximum force. For 0.5, there is a 15% drop in force. These results impact our inference of the Young's modulus. We implicitly assume zero friction in our measurement. However, we now know that our measured force is nearly 5 times larger because of friction. Thus, to still get 80N of force at 5% compression with a coefficient of friction of 0.8, the Young's modulus would be 3 MPa. Using a coefficient of friction of 0.5, the Young's modulus would be 3.5MPa. Since the simplest ANSYS model (one which has no contact elements and thus assumes infinite friction) is very nearly equivalent to either friction case, we choose to simulate the TBTV with a Young's modulus of 3.25MPa.

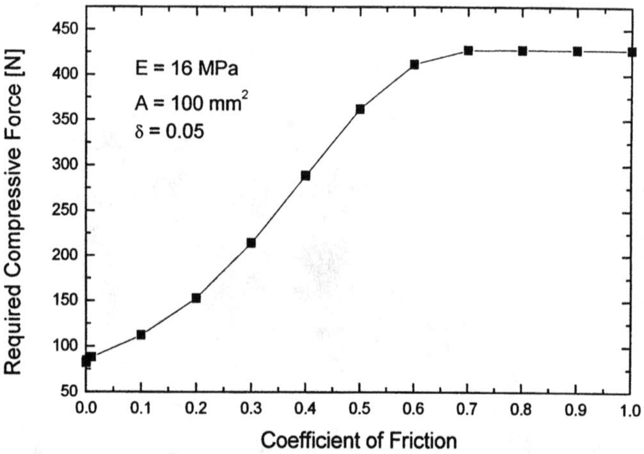

Figure 6 *5% Compression force as a function of the coefficient of friction.*

Now, we simulate the TBTV (Figure 3) with a Young's modulus of 3.25 MPa for Shin-Etsu. To simulate 5% compression for both 1.7 mm thick layers of Shin-Etsu which are in series, 0.17 mm is the imposed y-deflection. Again, half this distance is displaced on each of the top and bottom Kovar layers. The resulting required compression force is 1040 N. The minimum (zero-friction) force is 74 N. These numbers differ more than the equivalent two did in our compression test simulation. However, the Shin-Etsu used in the TBTV is larger (each piece is 9.5mmX24mm) and this accounts for the difference. As a check, we ran the TBTV with a Poisson's ratio of 0 (no volume expansion), and the force is 74N as expected. For safety, we designed the TBTV to handle 1320 N, and the actual device performed mechanically as desired. Uniform pressure, and thus good electrical contact, across the Shin-Etsu in addition to low stress in the LTCC was observed.

Conclusions

Experimental and numerical data have been combined to characterize the mechanical performance of Shin-Etsu, a z-axis interconnect material. Results indicate that the predicted compressive force can vary by more than a factor of five depending on the coefficient of friction between the Shin-Etsu and the mating material. This variation is due to the fact that Shin-Etsu is a polymer with a Poisson ration near 0.5 (volume is conserved limit). After inferring the appropriate value for

Young's modulus based on a coefficient of friction between 0.5-0.8, simulations lead to successful designs of a two-board test vehicle which incorporated the Shin-Etsu.

REFERENCES

1. W.D. Brown, Advanced Electronic Packaging, With Emphasis on Multichip Modules, IEEE Press Series on Microelectronic Systems, IEEE Press, Piscataway, New Jersey, 1999

2. Technical Documentation, Shin-Etsu Inter-Connector, GB-Matrix-4X Type, July 5, 2000, Shin-Etsu Polymer America, Inc, 5600 Mowry School Road, Newark, CA 94560, TEL: 510-623-8116.

3. G.L. Ginsberg, Multichip Modules And Related Technologies; MCM, TAB, and COB Designs, McGraw-Hill, Inc., New York, 1994.

4. J.H. Lupinski, Polymeric Materials For Electronics Packaging And Interconnection, American Chemical Society, Washington, DC, 1989, p 7.

5. P. Viswanadham and P. Singh, Failure Modes and Mechanisms in Electronic Packages, Chapman and Hall, New York, New York, 10003, p 43.

6. T. Baumeister, Marks' Standard Handbook for Mechanical Engineers, Eight Edition. McGraw-Hill, New York, New York, 1979

Proceedings of
2001 ASME International Mechanical Engineering Congress and Exposition
November 11-16, 2001, New York, NY
EPP-Vol. 1

IMECE2001/EPP-24737

RELIABILITY OF PLASTIC PACKAGE WITH PARTIALLY EXPOSED DIE

Tiao Zhou
STMicroelectronics, Inc.

ABSTRACT

Reliability issues are discussed for a plastic laminate-based package with partially exposed die active area. Thermal-mechanical stresses are calculated with finite element analysis to examine the effect of encapsulant in comparison with a similar package with die fully encapsulated. Different material and geometry configurations are studied to assess the optimization of the package design.

Keywords: Exposed die, stress, finite element analysis.

INTRODUCTION

Many solid state sensors are developed in the industry. Some require the Si die active area being exposed to the environment. Examples are fingerprint and pressure sensors. The reliability requirement due to exposed die active area include:

- Die resistance to solvent and moisture;
- Die resistance to mechanical contact such as scratch;
- Package robustness under temperature cycles and moisture conditions.

The first two requirements are addressed by choice of die passivation. This study focuses on package reliability due to temperature cycling. TouchChip finger print sensor package is considered.

TouchChip package is BT substrate based module. It consists of chip on board (COB) on top and surface mount components on the bottom side of the substrate (figure 1). Bond wires are on one side of the die. The sensor die is 13x20.8 mm. Sensor area of approximately 13x18.5 mm is exposed. 13x2.3 mm die area at wire bond side is encapsulated along with the die edges with mold compound (MC). This study evaluates die and MC stresses with finite element method and investigates approaches to reduce the risk of die and MC cracking.

The problem statement is given first. The stress difference between COB's with partially exposed die and fully encapsulated die is discussed next, followed by analysis of package material and geometry impacts. Conclusions are made at the end.

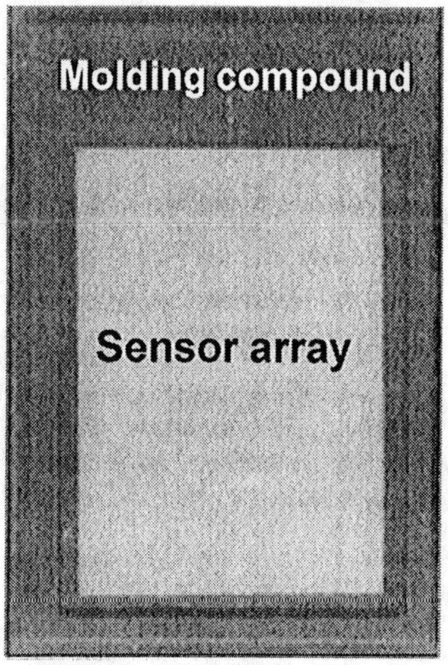

Figure 1.TouchChip finger print sensor COB.

NOMENCLATURE

COB – Chip on Board

TouchChip – Solid state finger print sensor trade marked by STMicroelectronics

CTE – Coefficient of thermal expansion

PROBLEM STATEMENT

This study focuses on the stress analysis of the overmolded COB with exposed silicon active area. As shown by figure 1, the key components of the COB are: silicon die, MC with overall dimension of 27x18.4 mm and the BT substrate. The die attach (DA) bondline thickness is assumed to be 50 μm. The material properties are listed in Table 1.

Table 1. Material properties of COB components.

	Elastic modulus (Gpa)	CTE (10^{-6} m/m/°C)	Linear shrinkage at cure
Si die	130	2.6	0
MC 1	17	17	0.4%
MC 2	21.3	11	0.1%
DA 1	5	60	0
DA 2	0.892	110	0
Substrate	19	15 in xy 57 in z	0

The height of the MC over the die is considered to be 0.4 mm. Molding temperature of 175 °C is considered as the reference (stress free) temperature. For comparison purpose, stresses at –40 °C are calculated for selected designs options.

Two problems are addressed in following discussion:

1. Effect of molding compound.
2. Optimization in material and package geometry

EFFECT OF THE ENCAPSULATION - COMPARISON BETWEEN COB'S WITH FULLY ENCAPSULATED DIE AND PARTIALLY ENCAPSULATED DIE

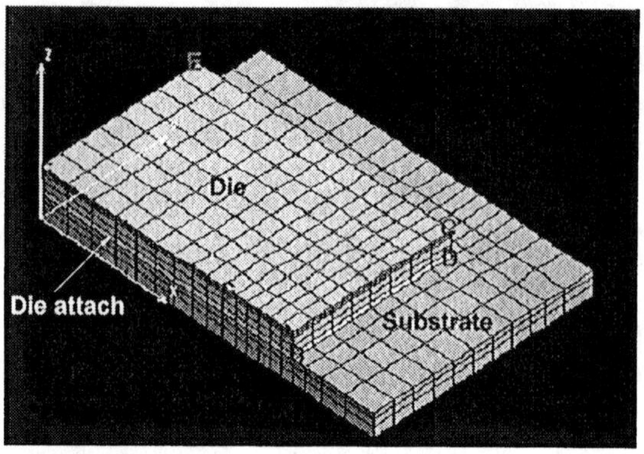

Figure 2. Key points.

To simplify the analysis, the top left quadrant of the COB (Figure 1) is modeled only. Symmetrical boundary conditions are assumed at the longitudinal and transversal centerlines. The key points are defined and depicted in figure 1. Points A and B are at center of die wire bond edge. C and D are at die the corner at wire bond side. E and F are at end and middle of die transversal centerline. G is at the die longitudinal centerline and next to the MC over the bond wires.

Four COB structures are considered to understand effect of MC. They are described by table 2. In the analysis, MC 1and DA glue 1 are considered. The die, DA and substrate thickness are 0.68, 0.05 and 0.8 mm, respectively. The outline of the mold compound is 18.4x27 mm. The thickness of mold compound over die is 0.4 mm.

Table 2. Description of selected models.

Case	Description
1	No MC.
2	MC around die only
3	MC around die and on top of bond wire area
4	Die fully encapsulated by MC

At this point die tensile stresses perpendicular to its crystal planes σ_x and σ_y are critical to die cracking. While for MC, Von Mises stress σ_{von} is considered to quantify the risk of MC cracking. Figures 3 - 6 show the finite element models and stress plots. The vertical (z) displacements of the die and molding compound along the package longitudinal centerline are plotted in Figure 7.

Table 3. Maximum stress value and locations for cases 1 - 4.

Case	1	2	3	4
Die max σ_x value (Mpa)	**110**	**141**	94.3	0
Die max σ_x location	F	A	G	N/a
Die max σ_y value (Mpa)	109	111	104	0
Die max σ_y location	F	E	E	N/a
Resin max σ_{von} value (Mpa)		248	225	235
Resin max σ_{von} location		C	C	C&D

As is seen that when no MC is applied, the maximum stress is at die center due to thermal mismatch with substrate. The COB has positive warpage. When MC is applied on substrate around the die (figure 4), shrinkage of the MC reduces the package warpage. The center of die edge at wire side shows maximum stress. As MC is applied to die area at wire side, package warpage is further changed with respect to previous case. Die under the MC shows convex curvature. When the die is fully encapsulated, the die is under compression due to the shrinkage of the MC and substrate; witch gives the die lowest chance of cracking.

Therefore, the MC on top of die has two effects. The first one is to reduce the die tensile stress by its shrinkage. The second effect is to change the package warpage.

When the die is partially exposed, the die loses some compression from MC and shows higher tensile stress, compared to a fully encapsulated die. Shrinkage balance among MC, die and substrate is different between fully and partially encapsulated dice.

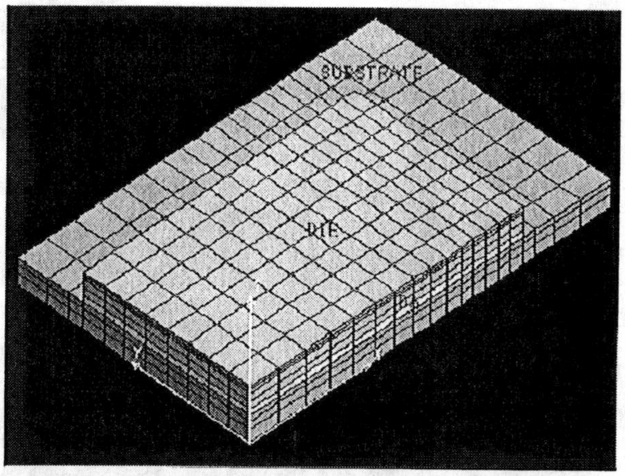

(a) Finite element model (no molding)

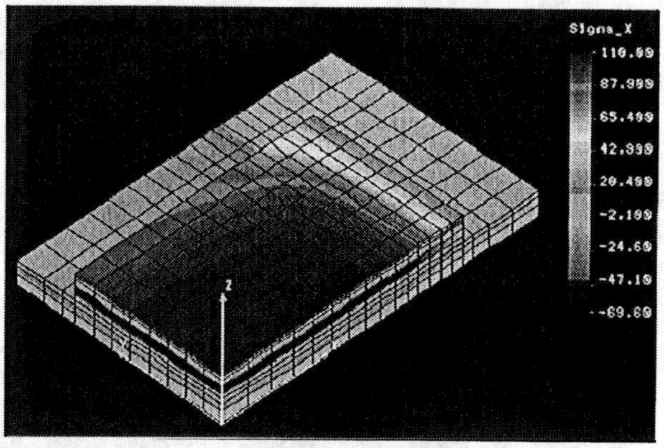

(b) Plot of tensile stress in x direction

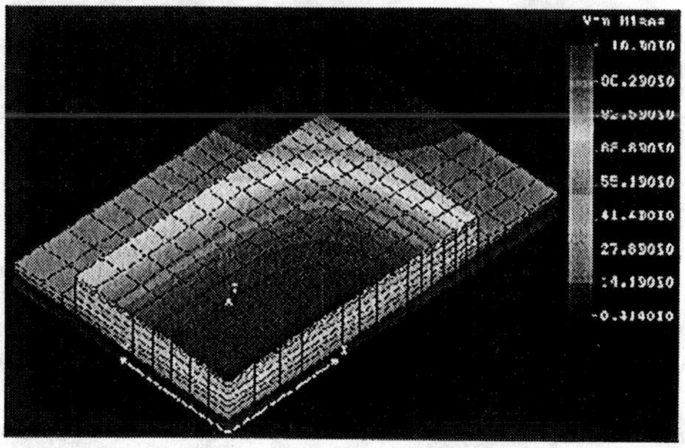

(c) Plot of Von Mises stress

Figure 3. Finite element model and stress plot for case with die attached to substrate only.

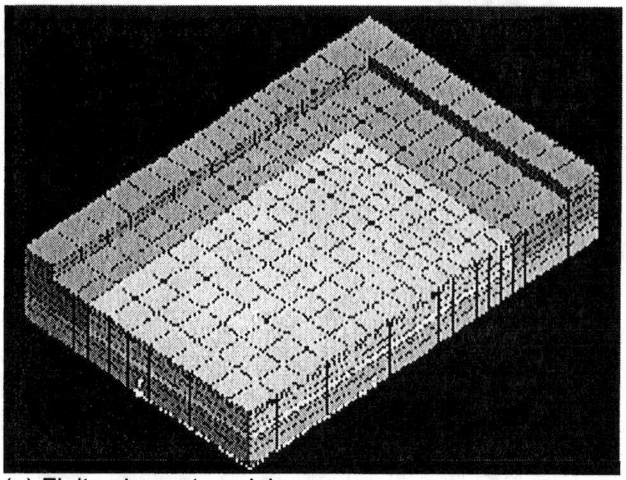

(a) Finite element model

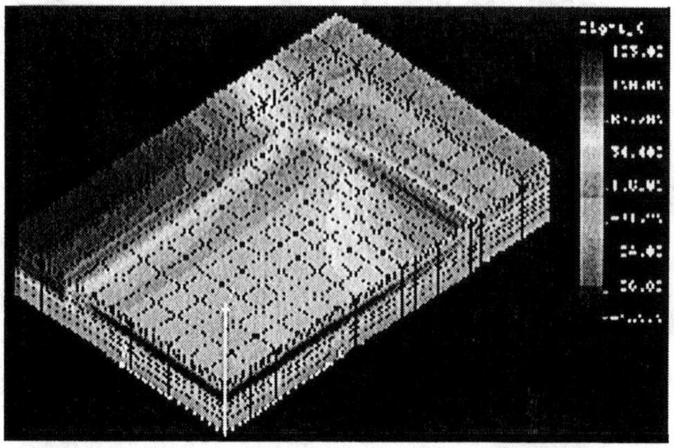

(b) σx plot

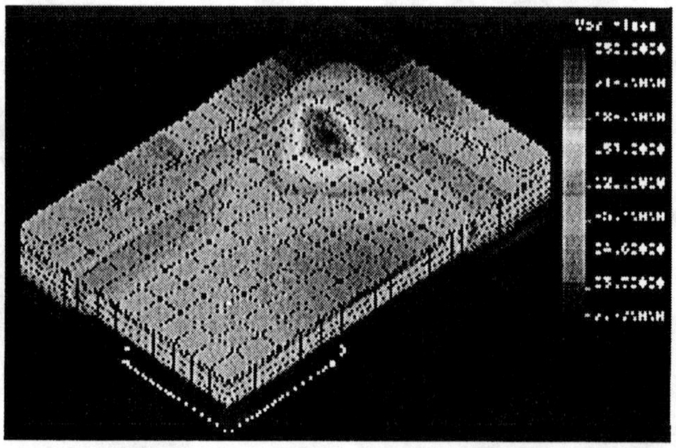

(c) Von Mises stress plot

Figure 4. Finite element model and stress plots for the case with molding compound around die only.

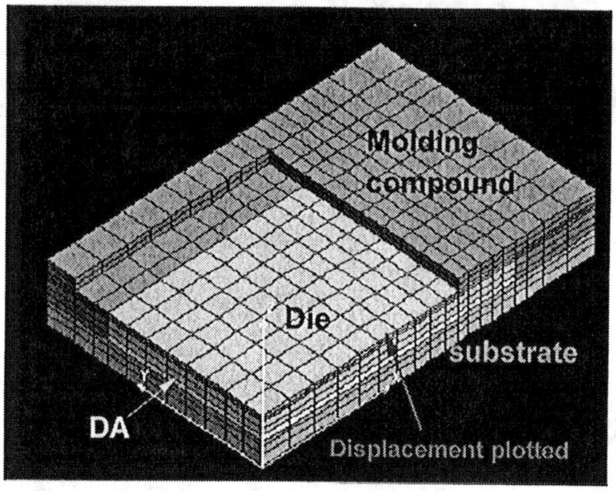

(a) Finite element model

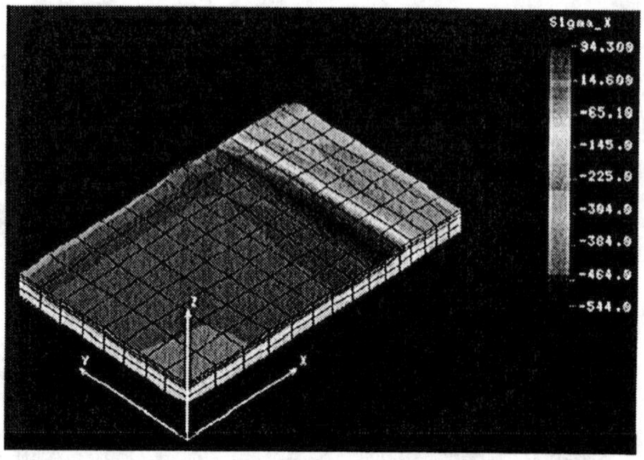

(b) σx plot

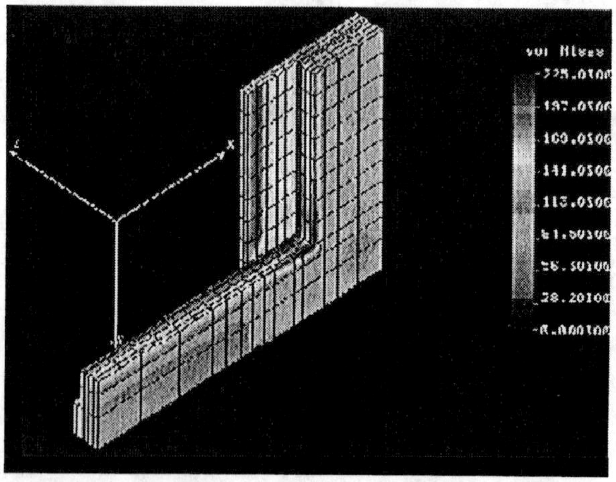

(c) Stress of MC

Figure 5. Finite element model and stresses for case with partially encapsulated die.

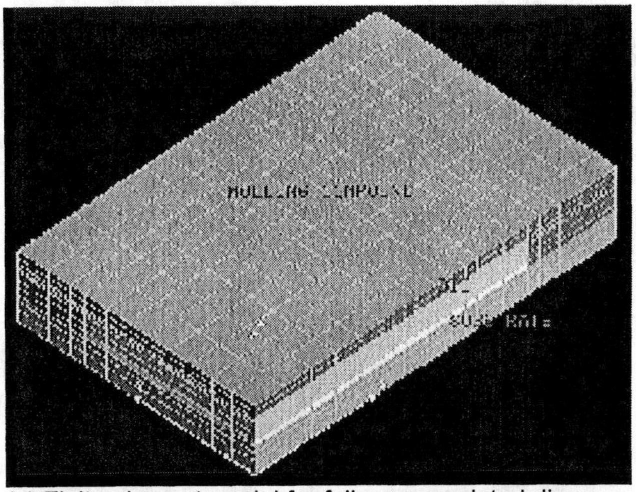

(a) Finite element model for fully encapsulated die

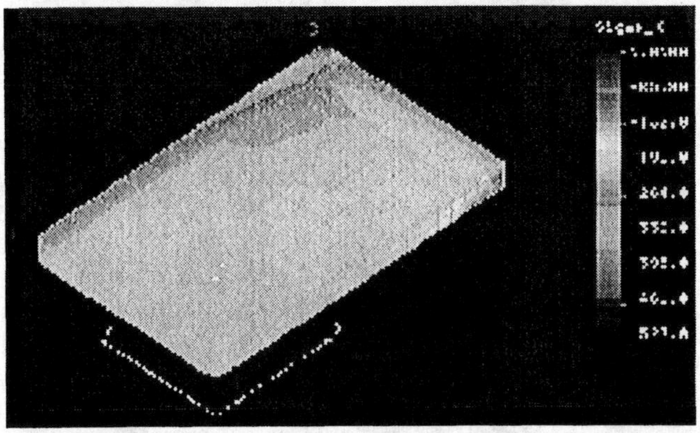

(b) Die σx plot

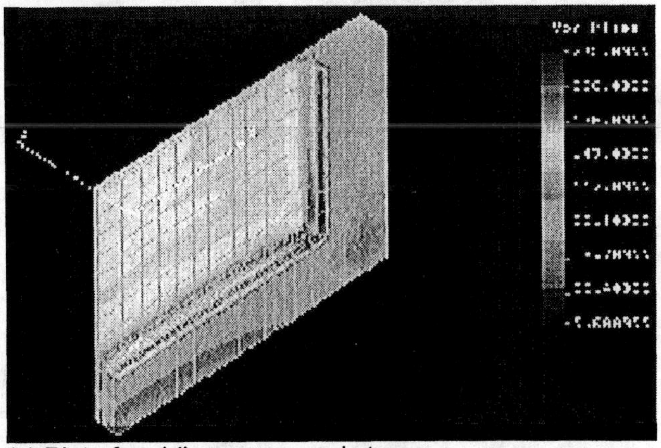

(c) Plot of molding compound stress

Figure 6. Finite element model and stress plot for fully encapsulated die.

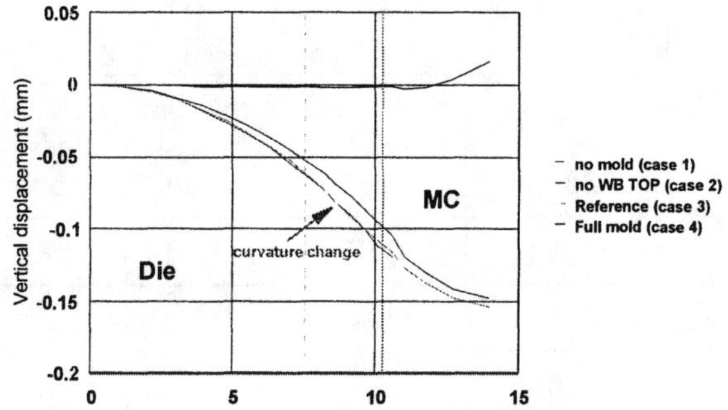

Figure 7. Displacement (warpage) along longitudinal centerline.

PACKAGE OPTIMIZATION

Four variables are considered:

1. MC material
2. Substrate thickness
3. Die thickness
4. DA glue material

Eight models are considered. The variables are listed in table 4. The objective here is to minimize die and MC stresses. The analysis results are listed in table 5. The marginal means[1] for die and MC stresses are plotted in figure 8.

Table 4. Variables considered.

Run	MC type	Substrate thickness (mm)	Si die thickness (mm)	DA type
1	1	0.8	0.68	1
2	1	0.8	0.68	2
3	1	0.36	0.375	1
4	1	0.36	0.375	2
5	2	0.8	0.375	1
6	2	0.8	0.375	2
7	2	0.36	0.68	1
8	2	0.36	0.68	2

Table 5. Maximum die and MC stresses and locations.

Run	Max die stress (Mpa)	Max MC stress (Mpa)	Max die stress location	Max MC stress location
1	104	225	E	C
2	103	261	E	D
3	43.7	252	E	D
4	55	237	E	D
5	59	136	G	D
6	73	114	G	D
7	127	137	E	D
8	71	145	E	D

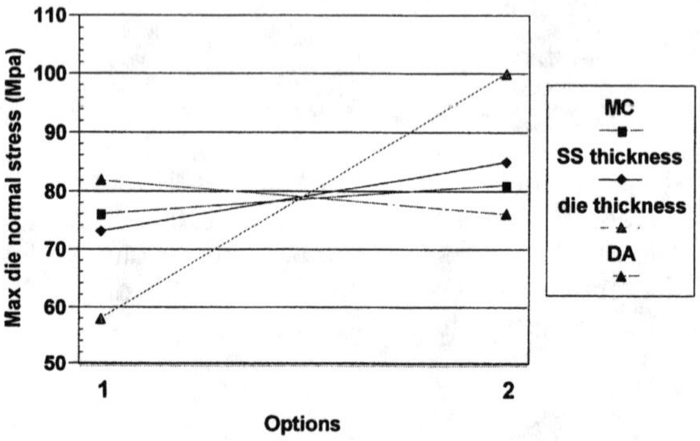

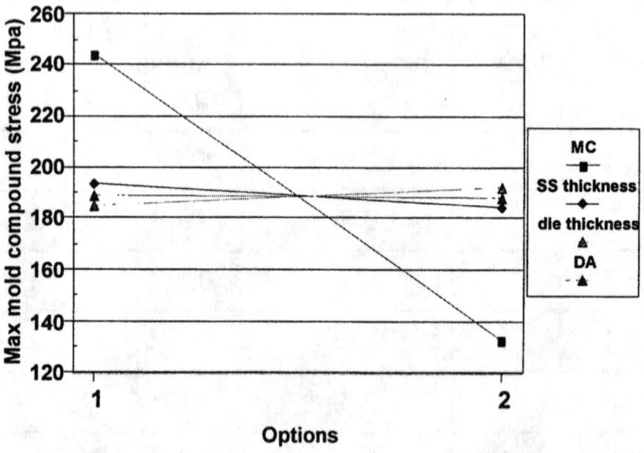

Figure 8. Sensitivity of die and MC stresses to selected variables.

It is seen that mold compound choice has dominant effect on MC stress level. MC 2 produces much lower stress on mold compound and the die. Other variables considered have minimum effect on MC stress. Die stress is most sensitive to die thickness. Thinner die gives lower die stress. DA 2 and thinner substrate produce slightly lower die stress. Therefore, the optimum variable combination is:

- MC 2
- DA 2
- 375 μm thick die
- 0.36 mm thick substrate

Simulation shows that in this case, the die is under compression and maximum MC stress is 125 Mpa.

CONCLUSIONS

Stress simulation is performed to study the thermal mechanical stresses of a plastic package with partially exposed die in comparison to a package with fully encapsulated die. Material and geometric variations for the COB materials are studied to minimize the die and mold compound stress. The following can be concluded:

1. Higher die tensile stress is expected when die is partially exposed, compared to fully encapsulated die.
2. The package warpage has curvature change next to the edge of mold compound window.
3. Reducing die thickness significantly lowers the die stress.
4. Mold compound with lower CTE significantly reduces the mold compound stress.
5. For plastic package with partially exposed die, material and package geometry optimization can minimize stress. With the right configuration, die can be put in compression.

REFERENCES

[1] S. R. Schmide and R. G. Launsby, Understanding industrial designed experiments, CQG Ltd Printing, 1988.

Proceedings of
2001 ASME International Mechanical Engineering Congress and Exposition
November 11–16, 2001, New York, NY
EPP-Vol. 1

IMECE2001/EPP-24738

Selection of Method for Fatigue Life Prediction under Random Loading

Amit Awasthi
Department of Mechanical Engineering
State University of New York at Binghamton

Stephen Mckeown
BAE Systems of North America
Johnson City, New York

Dr. C. Sahay, P.E.
Chair, Department of Mechanical Engineering
University of Hartford

Dr. Bahgat Sammakia
Department of Mechanical Engineering
State University of New York at Binghamton

Abstract: Solder fatigue continues to be one the primary reliability concerns in electronic packages. Three basic factors are necessary to cause fatigue failure. These are (i) a stress of sufficiently high value, (ii) varying amplitude stress, and (iii) a number of cycles of applied stress. Fatigue failure of a solder joint may also occur under cyclic thermal loading. The goal of this project is to compare the experimental results of fatigue testing with the various theoretical approaches (Rainflow counting method, Steinberg's 3 band approach, and Rayleigh's method) and come up with a formulation that would predict solder fatigue failure under random vibration loading.

Experiments have been performed to collect data. Initially various components having frequency range from 20-2000Hz were combined at random phase to produce a random profile. Rainflow cycle counting was done on 125 random profiles using an algorithm developed by downing (1982). The basic assumption made in this algorithm is that the strain-time history starts with either the maximum peak or the minimum valley. Further each cycle was segregated into one of the three bands depending on the value of the range of the cycle. It was found that the results matched very closely with Gaussian distribution curve.

The first phase of testing was to observe the behavior of aluminum and soldered copper (forming a lap joint) specimens under completely reversed sinusoidal loading in order to evaluate *S-N* curve. Three different stress levels were chosen and all experiments were carried out at a frequency of 20Hz. The second phase of testing consisted of observing the behavior of both the samples under semi-random loading. Random waveform was generated through a waveform function generator and was externally fed to the MTS machine.

The present paper presents the result of experiments along with the applicability to the three approaches to prediction.

Introduction: Repeated cyclic loading of structures can cause failure even though the stresses are well below the ultimate stress. This phenomenon, called fatigue, is caused by the formation and growth of microscopic cracks within the material during each load cycle. Fatigue is generally divided into low and high-cycle fatigue modes. Low-cycle fatigue is characterized by higher loads and quicker failure than high-cycle fatigue. The reason failure is quick is because the material is stressed beyond its elastic limit in each cycle, usually resulting in fatigue lives below 10,000 cycles. Stresses that cause high-cycle fatigue generally fall within the elastic range of the material. This paper focuses on high-cycle fatigue.

For many engineering structures and components the primary mode of failure can be attributed to fatigue damage resulting from the application of time varying loading i.e. random loading. The most loadings to which the components are exposed in their service life are random in their nature. The objective of this paper is to observe the behavior of solder under semi-random vibration loading and to come up with a simple yet reliable formulation that would predict the service life accurately.

In virtually every engineering application where fatigue is an important failure mode, the alternating stress amplitude may be expected to vary or change in some way during the service life. Such variations and changes in load amplitude, often referred to as *spectrum loading history*, make the direct use of standard S-N curves inapplicable because these curves are developed and presented for constant stress amplitude operation. Therefore, it becomes important to an engineer to have available a theory or hypothesis, verified by experimental observations, that will permit good design estimates to be made for operation under conditions of random loading using the standard constant-amplitude *S-N* curves that are more readily available.

Constant amplitude completely reversed fatigue experiments were performed at three stress levels, 10 samples each, in order to evaluate the *S-N* curve of aluminum and soldered copper specimens (both specimens described under *Experimental Setup*). For both the samples the maximum stress was kept well below the yield point so we can safely assume the plastic component of strain to be insignificant. The second phase of testing consisted of observing the behavior of both the specimens under semi-random loading. In semi-random loading a random spectrum of loading was repeated until the failure occurred. Random spectrum was generated by combining sinusoidal components, having different amplitude and frequency range from 20-2000Hz, at a random phase. Theoretical life of the specimens was calculated using three methods, Rainflow counting method, Gaussian method and Rayleigh method. When experimental life of the sample under semi-random loading was compared with theoretical life it was observed that Rayleigh method predicts the life most accurately.

Literature survey: The linear damage rule was first proposed by Palmgren (1924) and was further developed by Miner (1945). Today this method is commonly referred to as Miner's rule, which states

$$\frac{n}{N} = \text{Cycle ratio}$$

Where n is the number of cycles at stress level S and N is the fatigue life in cycles at stress level S.

The damage fraction, D, is defined as the fraction of life used up by an event or a series of events. Failure in any of the cumulative damage theories is assumed to occur when the summation of damage fractions equals 1.

The results of Miner's original tests (Sines et al., 1959) showed that the cycle ratio corresponding to failure ranged from 0.61 to 1.45. Other researchers have shown variations as large as 0.18 to 23.0. Most results tend to fall between 0.5 and 2.0. In most cases the average value is close to Miner's proposed value of 1. Miner's damage rule has two main shortcomings when it comes to describing observed material behavior. First, it does not consider sequence effects. Second, the linear damage rule is amplitude independent. It predicts that the rate of damage accumulation is independent of stress level.

Many nonlinear damage theories have been proposed which attempt to overcome the shortcomings of Miner's rule. Several of these methods (Henry, Gatts, Corten-Dolan, Marin, Manson-double linear) have been described by Collins (1981). There are some practical problems involved when trying to use these methods:

1. They require material and shaping constants, which must be determined from a series of step tests. In some cases this requires a considerable amount of testing.
2. Since some of the methods take into account sequence effects, the number of calculations and bookkeeping can become a problem in complicated histories.

For most situations where there is a pseudo-random load history, Miner's rule is adequate for predicting fatigue life. Most load histories are made up of a random distribution of loads of various magnitudes. Tests using random histories with several stress levels show very good correlation with Miner's rule. The other nonlinear methods do not give significantly more reliable life predictions. These theories also require material and shaping constants, which may not be available (Bannantine et al., 1990).

According to Dowling (1972) in order to predict the fatigue life for complicated stress-strain histories where large mean stresses are present for a significant number of cycles, it is necessary to use some method of accounting for mean stress in combination with a cycle counting method. Sometimes an average mean stress is defined for the entire history (Raser, 1969). The use of the average mean stress necessitates the assumption that equal numbers of cycles an equal amount above and below the average mean have a canceling effect. But Dowling (1972) showed experimentally that this is not so. Jacoby (1967) and Leve (1969) suggested that the average mean stress be defined for sections of the stress-strain history. Such a procedure could be applied to simpler sequences but for more irregular histories arbitrary divisions of the history into sections would have to be made and life predictions could vary considerably depending on how the history is divided into sections. When cycles are defined according to the rainflow counting method, then mean stress of each cycle can be computed individually.

Dowling (1972) has mentioned several methods to include the effect of mean stress on fatigue life as proposed by other authors. We would be using the method proposed by Smith (1942) which defines the effective stress for a cycle having σ_0 mean stress as:

$$\sigma_{eff} = \frac{\Delta\sigma / 2}{1 - \sigma_o / \sigma_u}$$

Gaussian or Normal distribution curve:

The distribution most often encountered, and the one that lends most readily to analysis, is the Gaussian distribution, which is defined by:

$$Y = \frac{e^{\left(-x^2/2\sigma^2\right)}}{\sigma\sqrt{2\pi}}$$

The right side of the above equation represents the probability density function, or the probability, per unit of X, for the ratio of the instantaneous acceleration or displacement (X) to the RMS acceleration (σ). The Gaussian distribution curve represents the probability for the value of the instantaneous acceleration levels at any time. The abscissa is the ratio of the instantaneous acceleration to RMS acceleration, and the ordinate is the probability density. The area under the curve between any two points directly represents the probability that the acceleration will be between these two points. Figure 1 below shows how the Gaussian distribution relates to the magnitude of the acceleration levels expected for random vibration. The instantaneous acceleration will be between the $+1\sigma$ and the -1σ values 68.3% of the time. It will be between the $+2\sigma$ and the -2σ values 95.4% of the time and between the $+3\sigma$ and the -3σ values 99.73% of the time.

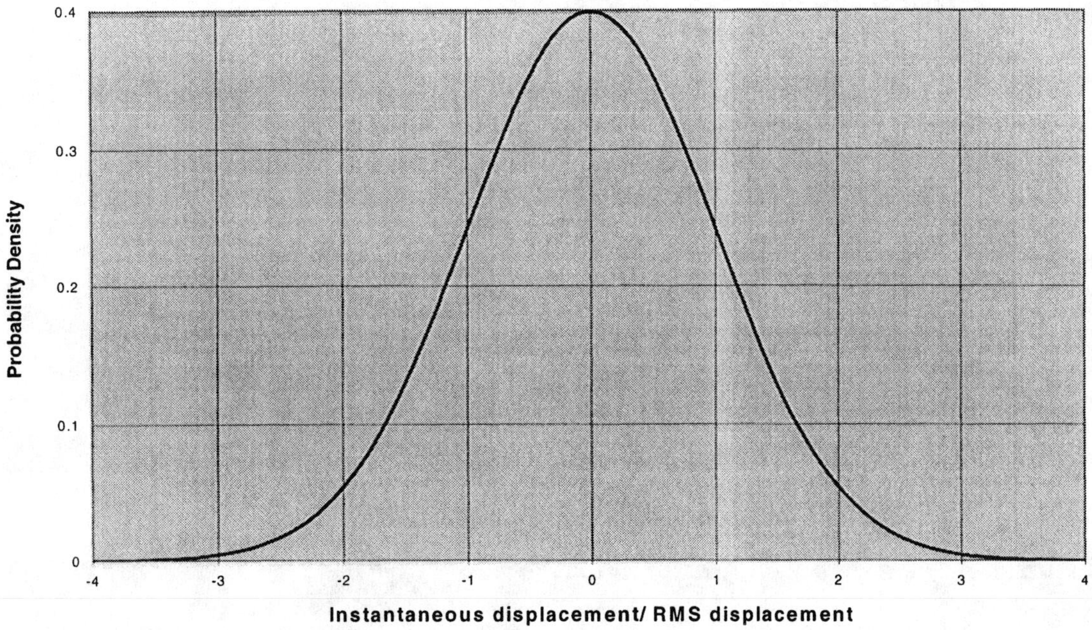

Figure 1: *Gaussian distribution curve. The probability of a range of X is given by the area under the curve in that range.*

Another way of interpreting the Gaussian distribution is that the instantaneous acceleration will exceed the 1σ value, which is the RMS value, 31.7% of the time. It will exceed the 2σ value, which is two times the RMS value, 4.6% of the time. It will exceed the 3σ value, which is three times the RMS value, 0.27% of the time.

Rayleigh Distribution Function:

The Rayleigh distribution shows the probability of the distribution of the peak accelerations in a random-vibration environment. The curve is shown in figure 2. The total area under the curve is unity. The area under the curve between the two points then represents the probability that peak amplitude will be between these two points. Rayleigh distribution is defined as:

$$Y = \frac{X}{\sigma^2}\exp\left(\frac{-X^2}{2\sigma^2}\right)$$

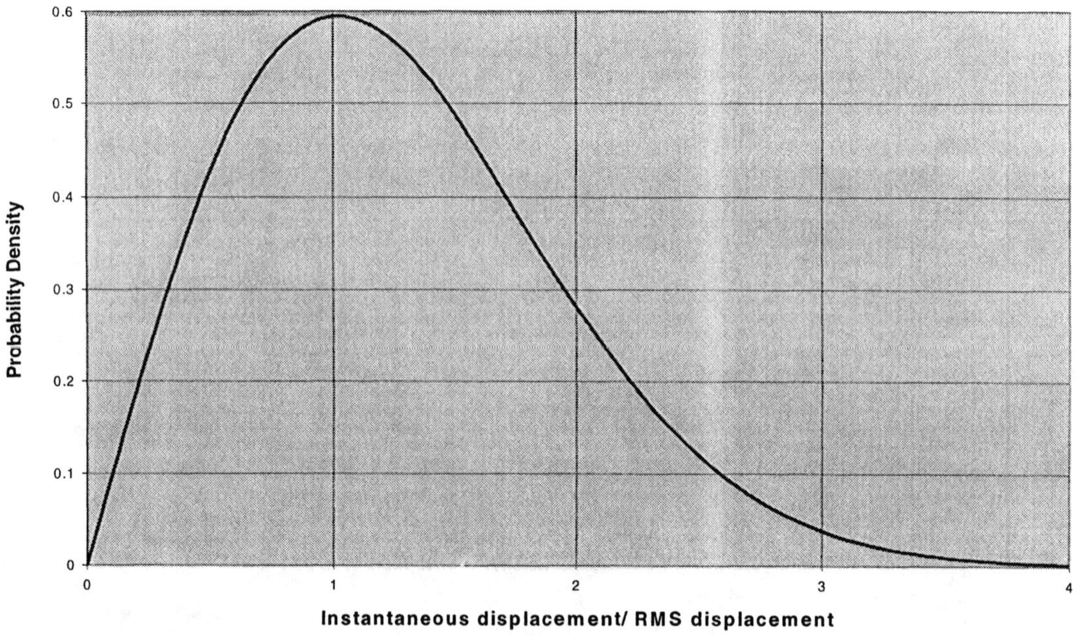

Figure 2: *Rayleigh probability distribution curve*

Three-band Technique:

The basis for the three-band technique is the Gaussian distribution. The instantaneous accelerations between $+1\sigma$ and -1σ are assumed to act at the 1σ level 68.3% of the time. The instantaneous accelerations between the $+2\sigma$ and the -2σ are assumed to act at the 2σ level 95.4 - 68.3 or 27.1% of the time. The instantaneous accelerations between $+3\sigma$ and -3σ are assumed to act at 3σ level 99.73 - 95.4, or the 4.33%, of the time. These values are shown below for three bands:

1σ values occur 68.3% of the time
2σ values occur 27.1% of the time
3σ values occur 4.33% of the time

Steinberg's 3-band Approach: This approach considers the probability of the occurrence of ranges and groups them into 3 bands and damage is calculated based on 1, 2 and 3σ stress levels. It uses gaussian probability distribution to determine the number of cycles required for each stress level. Stress levels greater than 3σ are not taken into account as they are expected to occur only 0.27% of the time. According to this approach:

$CDI_{total} = CDI_{1\sigma} + CDI_{2\sigma} + CDI_{3\sigma}$

Where $CDI_{i\sigma}$ is the cumulative damage accrued at i^{th} σ stress level

$$\text{Total life, } N = \frac{1}{\frac{0.6826}{N_1} + \frac{0.2718}{N_2} + \frac{0.0456}{N_3}}$$

Where N_1, N_2 and N_3 are the lives when the stress is assumed to be at 1, 2 and 3σ levels.

Experimental Set-up: Fatigue tests were carried out on two materials in order to develop comparative data at first hand. Materials tested were Al-6061 T6 aluminum, and two pieces of copper, in the shape of dog bone, soldered together by eutectic solder forming a lap joint. Sinusoidal and random vibration testing of both the materials was carried on an MTS closed loop electro-hydraulic materials testing system. In the MTS machine, tests ranging from strength tests up to fatigue tests under determinate or randomized loading can be performed either by feeding the signal through the in-built microprofiler or by externally feeding the signal through random waveform generator. All tests were conducted at room temperature and under uniaxial stress control. Before carrying out the elaborate testing, static tension test was conducted on aluminum sample to make sure stresses during the subsequent testing do not cross the elastic range.

The geometry of aluminum and soldered copper specimen is described below.

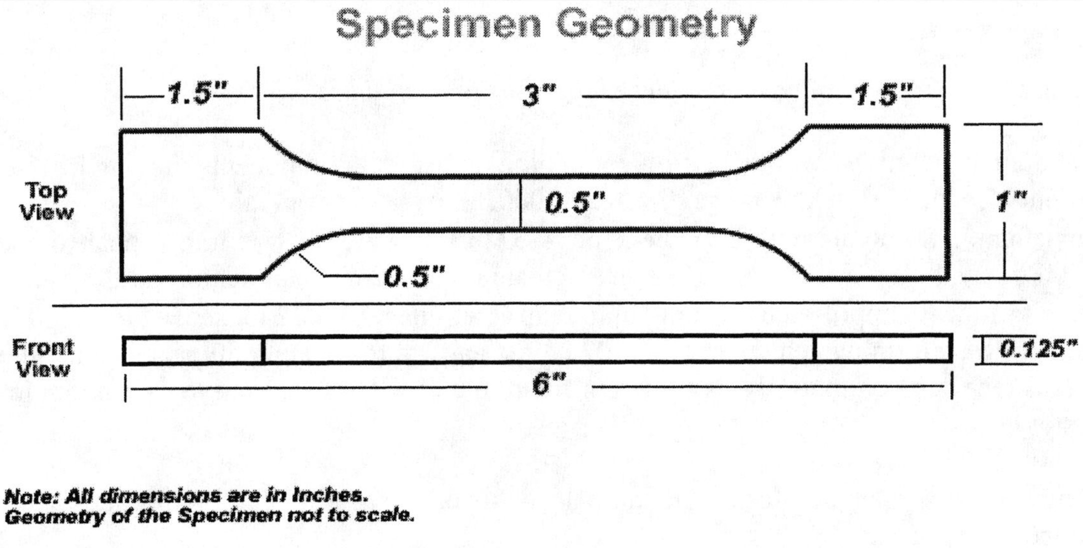

Figure 3: *Geometry of Aluminum Specimen*

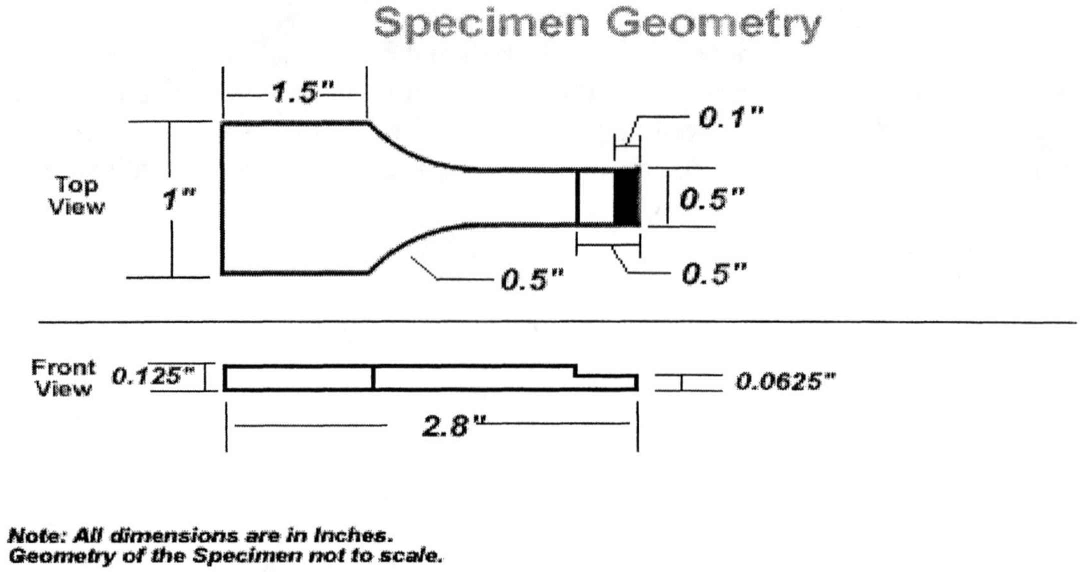

Figure 4: *Geometry of half-Copper Specimen.*

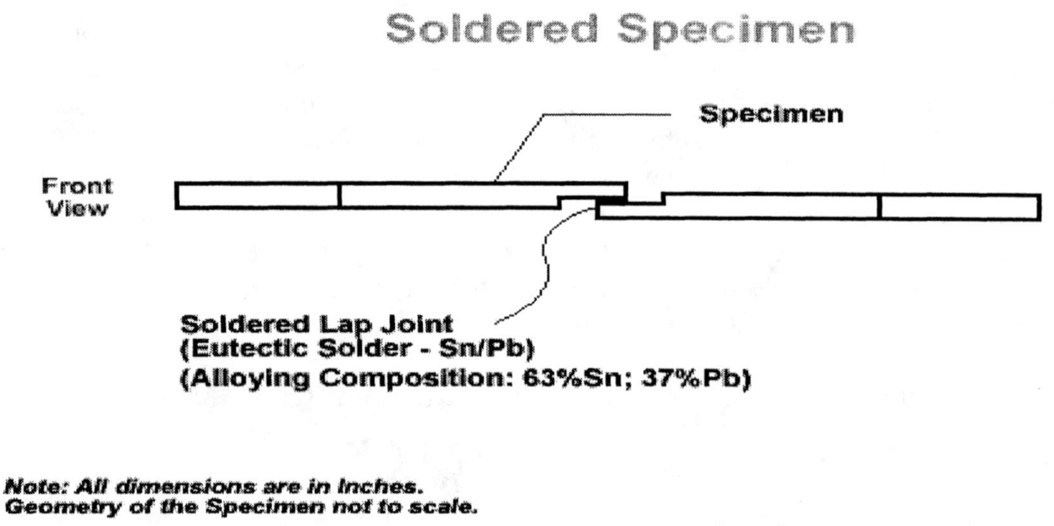

Figure 5: *Two pieces of copper soldered together to form a lap joint.*

Random spectrum was generated by combining sinusoidal components, having different amplitude and frequency range from 20-2000Hz, at a random phase. This spectrum contained 2400 points with an increment of 62.5μsec. Since it was impractical to enter 2400 points in the waveform generator to create one random load history so the random waveform was simplified by picking up points at an interval of 1250μsec. This simplified random waveform, which consists of 122 points and approximately 50 peaks and valleys that is repeated continuously, is shown in figure 6 and will be referred to as random load history.

Below are the statistics for the random load history used for aluminum and soldered copper specimens.

	Aluminum Specimen	Soldered Copper Specimen
Maximum Stress (Tensile)	198Mpa - 211Mpa	28Mpa - 38Mpa
Minimum Stress (Compressive)	174Mpa - 161Mpa	28Mpa - 38Mpa
RMS Stress	62Mpa - 86Mpa	10Mpa - 18Mpa

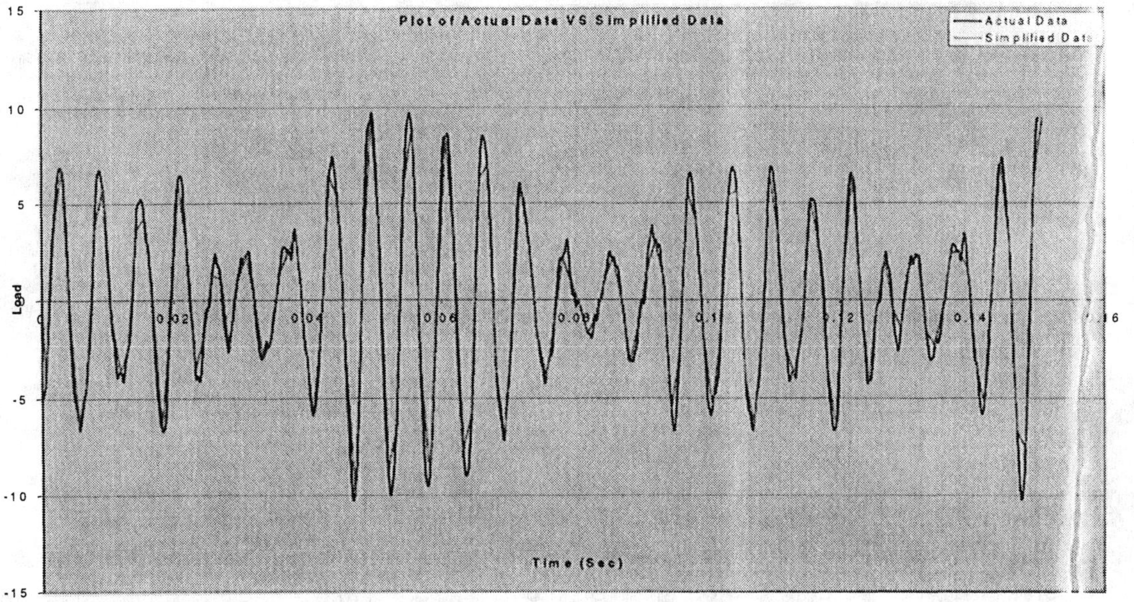

Figure 6: *A Typical Random Load History*

Results for Aluminum:

Axially loaded unnotched specimens of 6061-T6 aluminum with rectangular test section having an area of $0.0625in^2$ (width = 1/2 in and thickness=1/8 in) and length of 6 inches were tested. Initially conventional stress-controlled fatigue tests were carried out under completely reversed sinusoidal loading at three different stress levels of 155, 136.4 and 124 Mpa with 10 samples at each stress level in order to obtain *S-N* curve.

Random vibration fatigue testing with aluminum specimens was carried out under different random loading histories with a minimum of 2 samples tested under each random loading history. A total number of 12 random loading histories were applied on 30 aluminum specimens in order to gather data for aluminum fatigue failure under random loading. A plot of *S-N* for sinusoidal and random loading is shown in Figure 7.

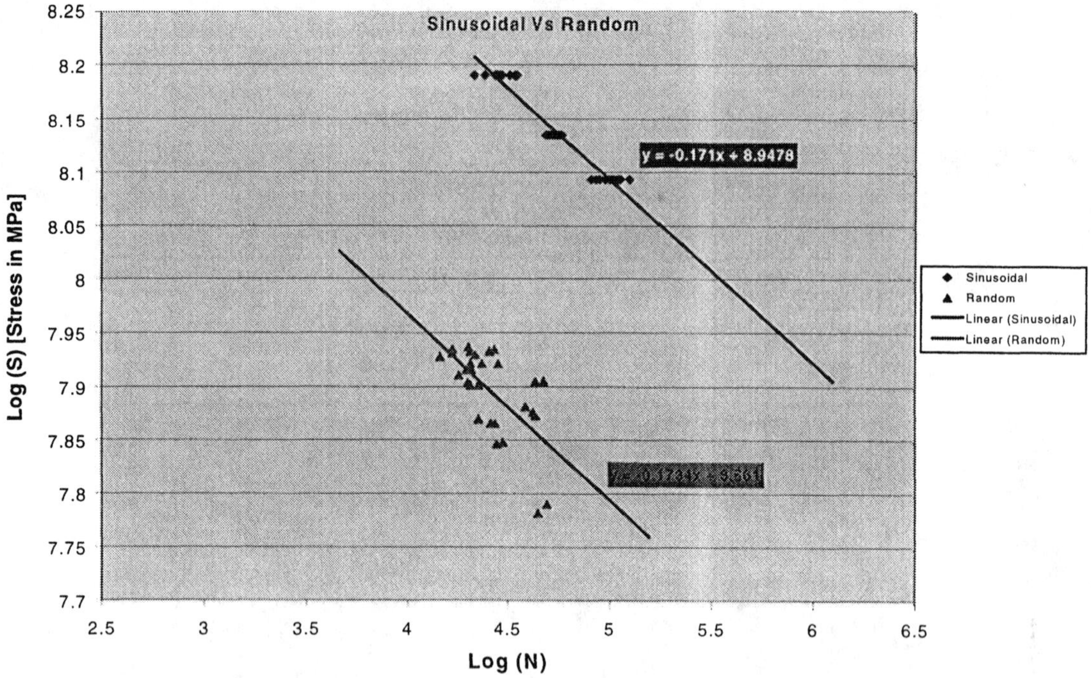

Figure 7: *log S - log N Plot for Sinusoidal and Random Loading*

Theoretical life of the aluminum specimens, tested under random fatigue loading, was calculated using three approaches, Rainflow counting method, Gaussian method and Rayleigh method. A comparison plot of theoretical life and experimental life under fatigue loading is made in figure 8.

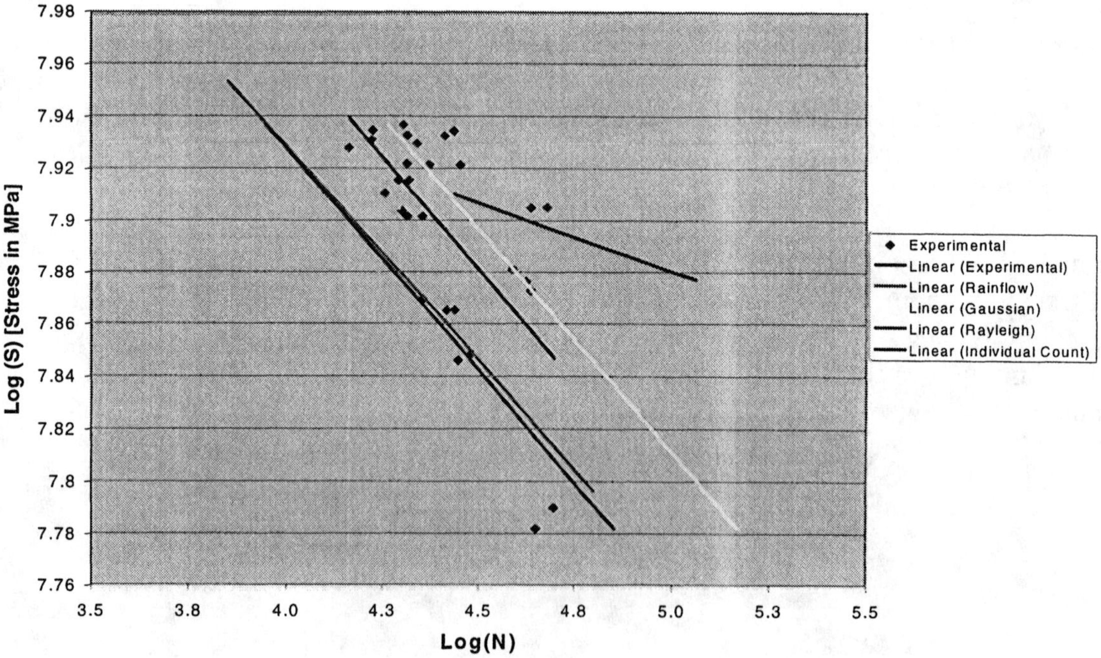

Figure 8: *Plot of log S - log N for three different approaches.*

Hazard analysis, McKeown (1999), was performed for sinusoidal loading as well as random loading of aluminum. Figure 9 shows the hazard curve/ failure distribution for aluminum under sinusoidal loading and figure 10 shows the same for aluminum under random loading.

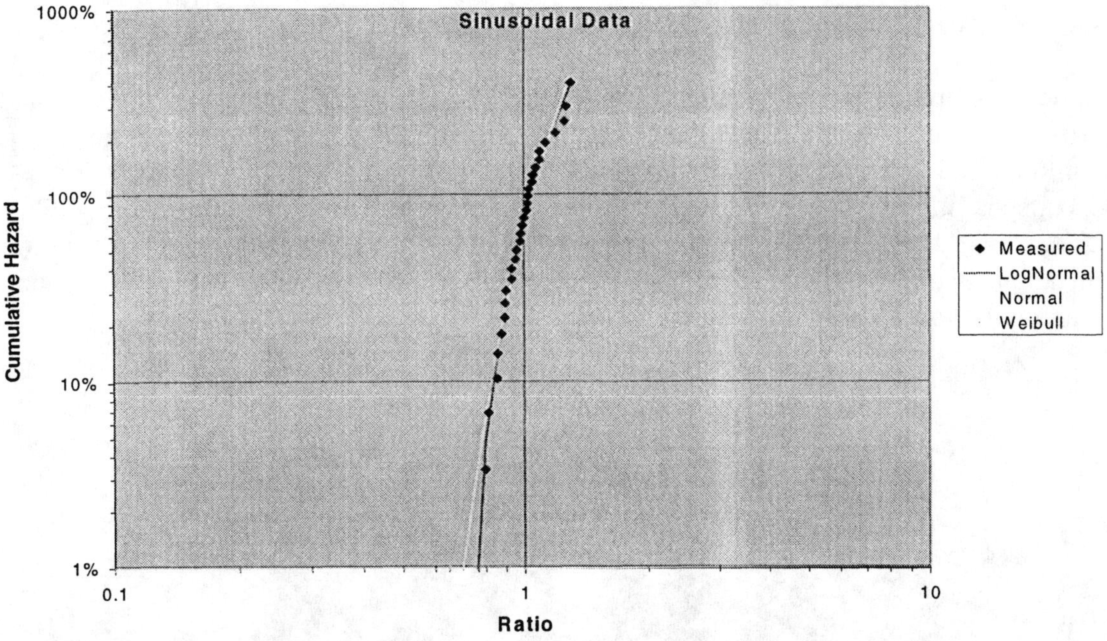

Figure 9: *Failure distribution for aluminum under sinusoidal loading with normal, lognormal and Weibull curves.*

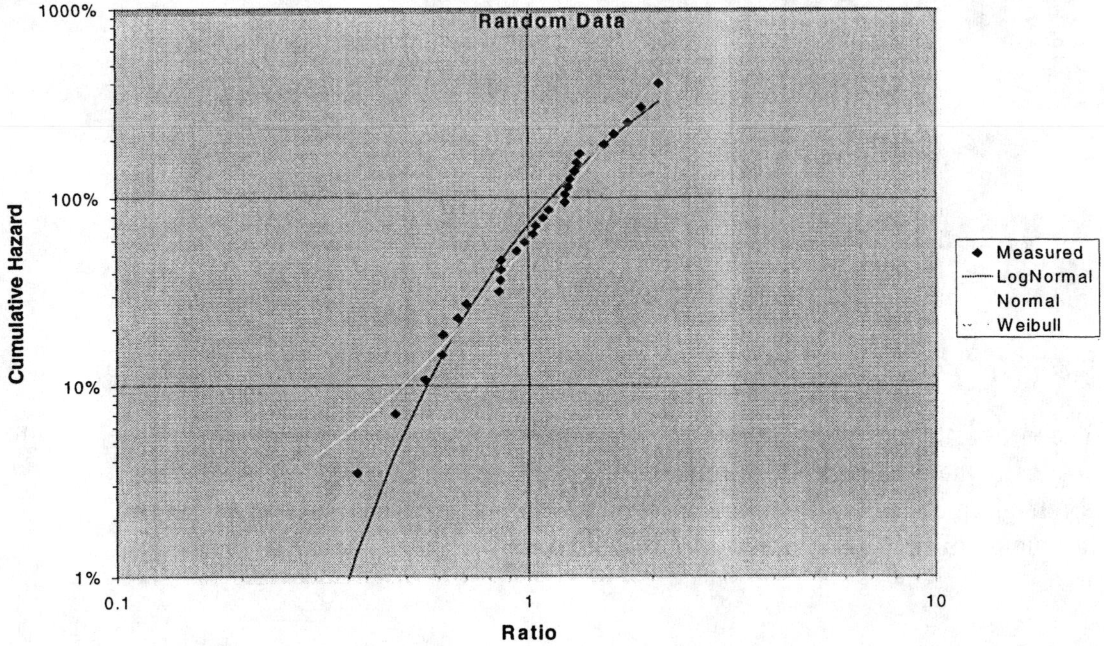

Figure 10: *Failure distribution for aluminum under random loading with normal, lognormal and Weibull curves*

Results for Solder:

Axially loaded unnotched specimens of soldered copper with rectangular test section having an area of $0.03125 in^2$ (width = 1/2 in and thickness=1/16 in) and length of 6 inches with an average overlapping soldered area of $0.045\ in^2$ were tested. Initially conventional stress-controlled fatigue tests were carried out under completely reversed sinusoidal loading at three different stress levels of 19.5, 17.0 and 16.5 Mpa with 10 samples at each stress level in order to obtain *S-N* curve.

Random vibration fatigue testing with soldered copper specimens was carried out under different random loading histories with a minimum of 3 samples tested under each random loading history. A total number of 9 random loading histories were applied on 27 specimens in order to gather data for solder fatigue failure under random loading. Out of 27 specimens the data for four specimens was not taken into consideration, as they were misaligned so they failed earlier than expected. A plot of *S-N* for sinusoidal and random loading is shown in Figure 11.

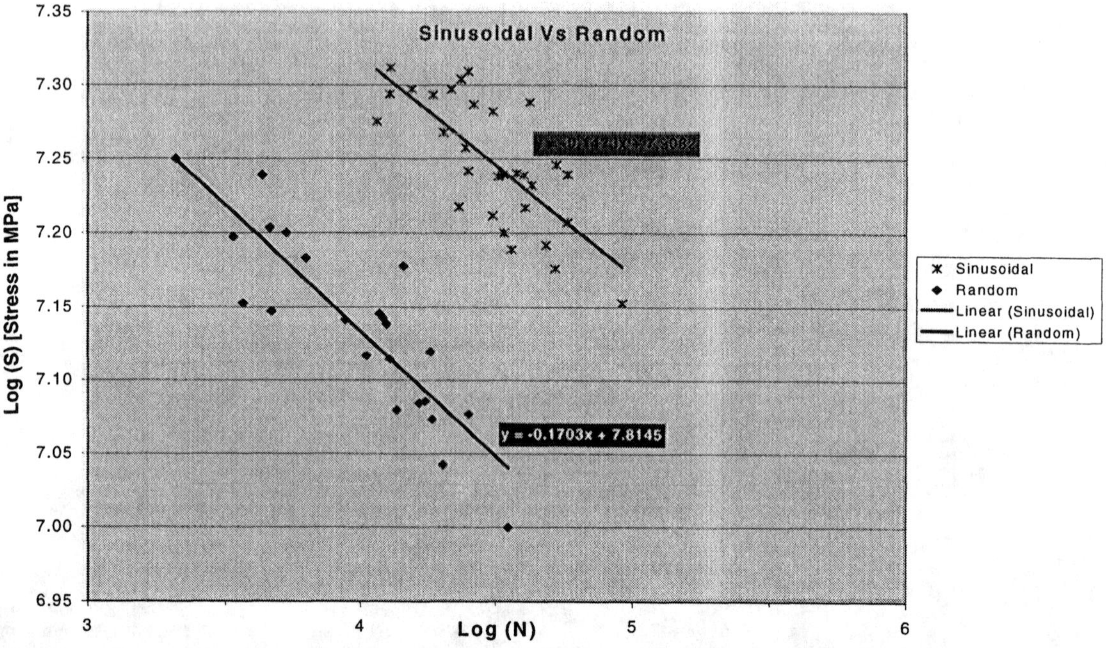

Figure 11: *log S - log N Plot for Sinusoidal and Random Loading*

Theoretical life of the soldered copper specimens, tested under random fatigue loading, was calculated using three approaches, Rainflow counting method, Gaussian method and Rayleigh method. A comparison plot of theoretical life and experimental life under fatigue loading is made in figure 12.

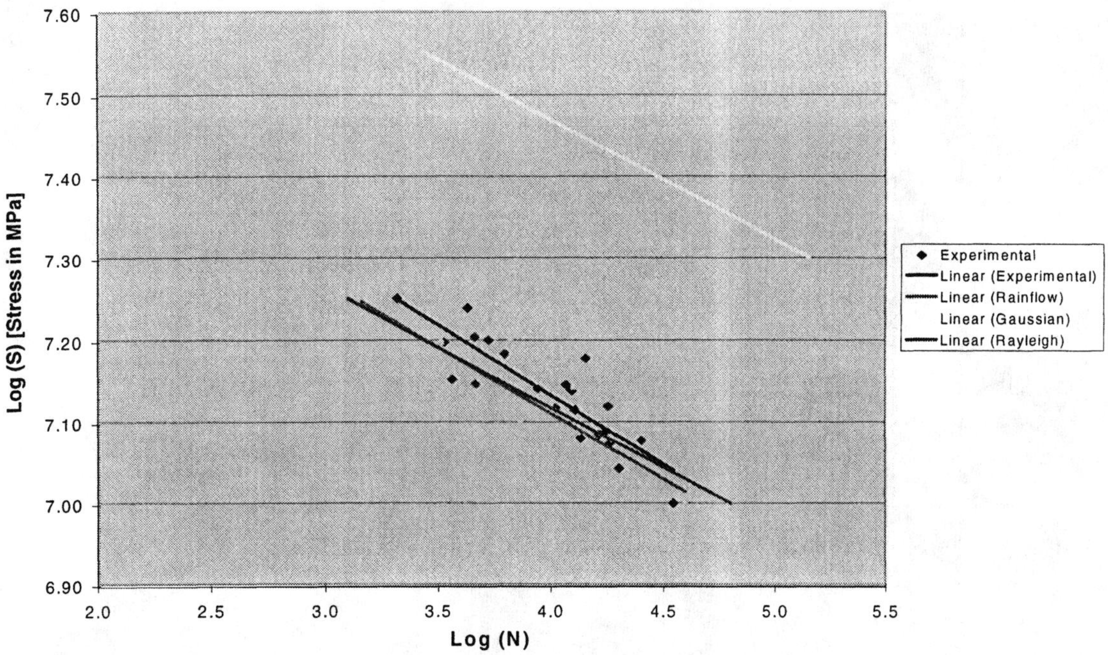

Figure 12: *Plot of log S- log N for solder using three different approaches*

Hazard analysis was performed for sinusoidal loading as well as random loading of soldered copper specimen. Figure 13 shows the hazard curve/ failure distribution for solder under sinusoidal loading and figure 14 shows the same under random loading.

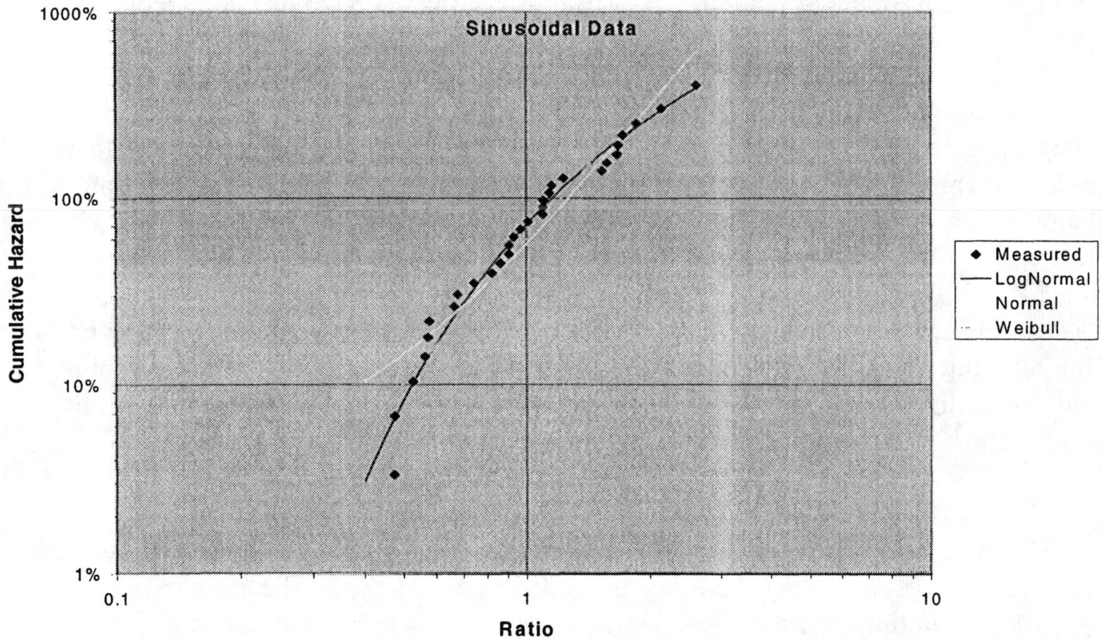

Figure 13: *Failure distribution for solder under sinusoidal loading with normal, lognormal and Weibull curves*

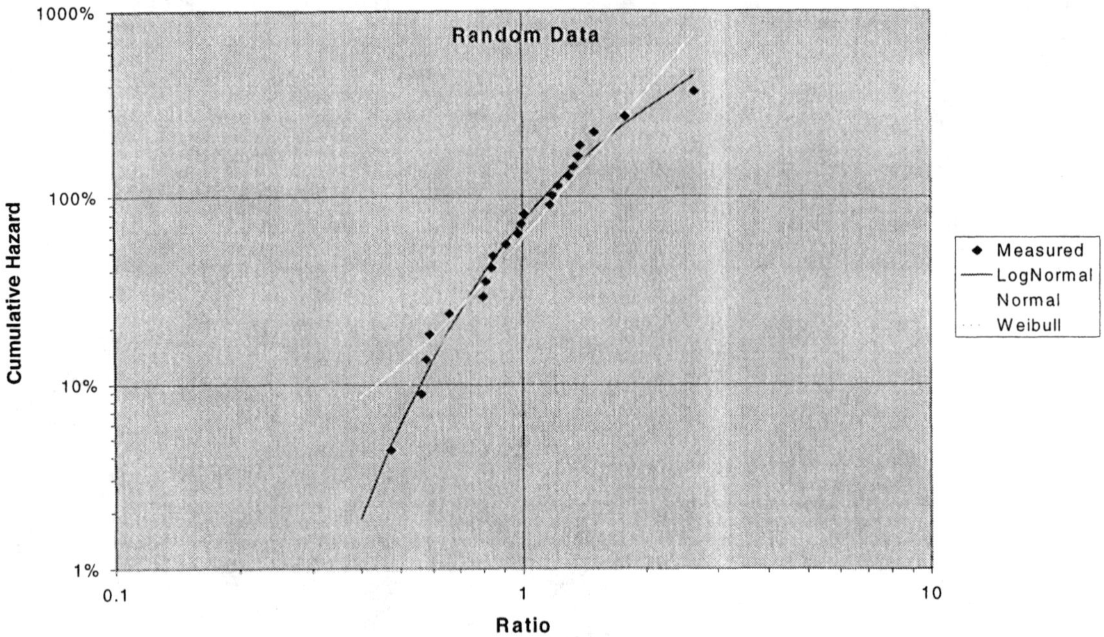

Figure 14: *Failure distribution for solder under random loading with normal, lognormal and Weibull curves*

Conclusions:

A. For Aluminum:

Both the sinusoidal and random stress-life fatigue curves have the same slope parameter values for aluminum, as suggested by Lambert (1993).

The predicted lives, using Rainflow counting method, are within a factor of three of the actual lives for all of the tests. The predicted lives using Rayleigh method are within a factor of three for 93% of the tests. The same is true when the life prediction is made using Gaussian method but Gaussian method overestimates the life for almost 65% of the tests.

As evident from figure 8, Rainflow counting method and Rayleigh method predict fatigue life of aluminum on the conservative side. Failure distribution for aluminum under sinusoidal loading (figure 9) fits very well with lognormal distribution where as failure distribution under random loading (figure 10) fits well with Weibull curve.

B. For Solder:

It is evident from figure 12 that Rainflow counting and Rayleigh method make better life prediction for solder than made by using Gaussian method. The predicted lives, using Rainflow counting method and Rayleigh method, are within a factor of three of the actual lives for 95% of the tests.

Hazard analysis for solder produced very good correlation with the lognormal distribution under sinusoidal loading (figure 13) as well as random loading (figure 14).

References:

Bannantine, Julie A., Comer, Jess J., and Handrock, James L., 1990, Fundamentals of Metal Fatigue Analysis, Prentice-Hall, New Jersey.

Collins, J.A., 1981, Failure of Materials in Mechanical Design, Wiley-Interscience, New York.

Dowling, N.E., 1972, "Fatigue Failure Predictions for Complicated Stress-Strain Histories," Journal of Materials, JMLSA, Vol. 7, No. 1, pp. 71-87.

Downing, S.D., and Socie, D.F., 1982, "Simplified Rainflow Counting Algorithms," International Journal of Fatigue, Vol. 4, No. 1, pp. 31-40.

IPC, July 1992, "Guidelines for Accelerated Reliability Testing of Surface Mount Solder Attachments", IPC-SM-785, Institute for Interconnecting and Packaging Electronic Circuits, Lincolnwood, IL, July 1992.

Jacoby, G.H., Aug. 1967, "Fatigue Life Estimation Processes Under conditions of Irregularly Varying Loads," AFML TR 67-215, Air Force Materials Lab., Wright-Patterson AFB, Ohio.

Lambert, Ronald G., May/June 1993, "Fatigue Damage Prediction for Combined Random and Static Mean Stresses", Journal of IES, pp.26.

Leve, H.L., 1969, in Metal Fatigue: Theory and Design, A.F. Madayag, Ed., John Wiley & Sons, pp. 170-208.

McKeown, Stephen A., 1999, Mechanical Analysis of Electronic Packaging Systems, Marcel Dekker Inc.

Miner, M.A., 1945, "Cumulative Damage in Fatigue," Journal of Applied Mechanics, Vol. 12, Trans. ASME, Vol. 67, pp. A159-A164

Palmgren, A., 1924, "Durability of Ball Bearings," ZVDI, Vol. 68, No. 14, pp. 339-341.

Raser, W.H., Jr., 1969, in Metal Fatigue: Theory and Design, A.F. Madayag, Ed., John Wiley & Sons, pp. 247-288.

Sines, G., and Waisman, J.L., (eds.), 1959, Metal Fatigue, McGraw-Hill, New York.

Smith, J.O., Feb. 1942"The Effect of Range of Stress on the Fatigue Strength of Metals," Bulletin No. 334, University of Illinois, Engineering Experiment Station, Urbana.

Steinberg, D., August 23, 1989 "Preventing Thermal Cycling and Vibration Failures in Electronic Equipment," presented at McDonnell Douglas, St. Louis, MO

Steinberg, D., 2000, Vibration Analysis for Electronic Equipment, John Wiley & Sons, Third Edition, Figure 3.2, page 44.

AUTHOR INDEX

EPP-Vol. 1
Packaging, Reliability and Manufacturing Issues Associated With Electronic and Photonic Products

Book Number: I00533